SI Prefixes

Multiple	Exponential Form	Prefix	SI Symbol
1 000 000 000	10^9	giga	G
1 000 000	10^6	mega	M
1 000	10^3	kilo	k

Submultiple			
0.001	10^{-3}	milli	m
0.000 001	10^{-6}	micro	μ
0.000 000 001	10^{-9}	nano	n

ENGINEERING MECHANICS
Statics

SI EDITION

R. C. HIBBELER

S. C. FAN

PRENTICE HALL
Singapore   New York   London   Toronto   Sydney   Tokyo

First published in 1997 by
Prentice Hall
Simon & Schuster (Asia) Pte Ltd
317 Alexandra Road
#04-01 IKEA Building
Singapore 159965

Original edition published by:
Prentice-Hall, Inc.
A Simon & Schuster Company
Copyright © 1995, 1992, 1989, 1986, 1983, 1978, and 1974 by R. C. Hibbeler

PRINTED IN SINGAPORE

1 2 3 4 5   01 00 99 98 97

ISBN 0-13-599598-1

Prentice-Hall International (UK) Limited, London
Prentice-Hall of Australia Pty. Limited, Sydney
Prentice-Hall Canada Inc., Toronto
Prentice-Hall Hispanoamericana, S.A., Mexico
Prentice-Hall of India Private Limited, New Delhi
Prentice-Hall of Japan, Inc., Tokyo
Editora Prentice-Hall do Brasil, Ltda., Rio de Janeiro
Prentice-Hall, Upper Saddle River, New Jersey

TO THE STUDENT

With the hope that this work
will stimulate an interest in Engineering Mechanics
and provide an acceptable guide to its understanding.

Preface

The main purpose of this book is to provide the student with a clear and thorough presentation of the theory and applications of engineering mechanics. To achieve this objective the author has by no means worked alone, for to a large extent this book has been shaped by the comments and suggestions of more than a hundred reviewers in the teaching profession as well as many of the author's students.

Continued improvements have been made to this the seventh edition. Previous users of the book may first notice that the art work has been enhanced in a multi-color presentation in order to provide the reader with a more realistic and understandable sense of the material. Also, the problem sets have been greatly expanded. Often, several problem statements refer to the same drawing, so that the instructor can reinforce concepts discussed in class. The problem sets also provide a wider variation in the degree of difficulty in problem solutions, and instructors can now select problems that focus on design rather than on analysis.

Although the contents of the book have remained in the same order, the details of some topics have been expanded, some examples have been changed and others have been replaced with new ones. Also, the explanation of many topics has been improved by a careful rewording of selected sentences. The hallmarks of the book, however, remain the same: where necessary, a strong emphasis is placed on drawing a free-body diagram, and the importance of selecting an appropriate coordinate system and associated sign convention for vector components is stressed when the equations of mechanics are applied.

Organization and Approach. The contents of each chapter are organized into well-defined sections. Selected groups of sections contain an explanation of specific topics, illustrative example problems, and a set of home-work problems. The topics within each section are placed into subgroups defined by boldface titles. The purpose of this is to present a structured method for introducing each new definition or concept, and to make the book convenient for later reference and review.

A "procedure for analysis" is given at the end of many sections of the book in order to provide the student with a review or summary of the material and a logical and orderly method to follow when applying the theory. As in the previous editions, the example problems are solved using this outlined method in order to clarify its numerical application. It is to be understood, however, that once the relevant principles have been mastered and enough confidence and judgment have been obtained, the student can then develop his or her own procedures for solving problems. In most cases, it is felt that the first step in any procedure should require drawing a diagram. By doing so, the student forms the habit of tabulating the necessary data while focusing on the physical aspects of the problem and its associated geometry. If this step is correctly performed, applying the relevant equations of mechanics becomes somewhat methodical, since the data can be taken directly from the diagram. This step is particularly important when solving problems involving equilib-rium, and for this reason, drawing a free-body diagram is strongly emphasized throughout the book.

Since mathematics provides a systematic means of applying the principles of mechanics, the student is expected to have prior knowledge of algebra, geometry, trigonometry, and, for complete coverage, some calculus. Vector analysis is introduced at points where it is most applicable. Its use often provides a convenient means for presenting concise derivations of the theory, and it makes possible a simple and systematic solution of many complicated three-dimensional problems. Occasionally, the example problems are solved using more than one method of analysis so that the student develops the ability to use mathematics as a tool whereby the solution of any problem may be carried out in the most direct and effective manner.

Problems. The majority of problems in the book depict realistic situations encountered in engineering practice. It is hoped that this realism will both stimulate the student's interest in engineering mechanics and provide a means for developing the skill to reduce any such problem from its physical descrip-tion to a model or symbolic representation to which the principles of mechan-ics may be applied. As in the previous edition, an effort has been made to include some problems which may be solved using a numerical procedure executed on either a desktop computer or a programmable pocket calculator. Suitable numerical techniques along with associated computer programs are given in Appendix B. The intent here is to broaden the student's capacity for using other forms of mathematical analysis *without* sacrificing the time needed to focus on the application of the principles of mechanics. Problems of this type which either can or must be solved using numerical procedures are identified by a "square" symbol (■) preceding the problem number.

Throughout the text all problems use SI units. Furthermore, in any set, an attempt has been made to arrange the problems in order of increasing difficulty.* The answers to all but every fourth problem are listed in the back of the book. To alert the user to a problem without a reported answer, an asterisk (*) is placed before the problem number.

Contents. The book is divided into 11 chapters, in which the principles introduced are first applied to simple situations. Most often, each principle is applied first to a particle, then to a rigid body subjected to a coplanar system of forces, and finally to the general case of three-dimensional force systems acting on a rigid body.

The text begins in Chapter 1 with an introduction to mechanics and a discussion of units. The notion of a vector and the properties of a concurrent force system are introduced in Chapter 2. This theory is then applied to the equilibrium of particles in Chapter 3. Chapter 4 contains a general discussion of both concentrated and distributed force systems and the methods used to simplify them. The principles of rigid-body equilibrium are developed in Chapter 5 and then applied to specific problems involving the equilibrium of trusses, frames, and machines in Chapter 6, and to the analysis of internal forces in beams and cables in Chapter 7. Applications to problems involving frictional forces are discussed in Chapter 8, and topics related to the center of gravity and centroid are treated in Chapter 9. If time permits, sections concerning more advanced topics, indicated by stars (★), may be covered. Most of these topics are included in Chapter 10 (area and mass moments of inertia) and Chapter 11 (virtual work and potential energy). Note that this material also provides a suitable reference for basic principles when it is discussed in more advanced courses.

At the discretion of the instructor, some of the material may be presented in a different sequence with no loss in continuity. For example, it is possible to introduce the concept of a force and all the necessary methods of vector analysis by first covering Chapter 2 and Sec. 4.1. Then, after covering the rest of Chapter 4 (force and moment systems), the equilibrium methods in Chapters 3 and 5 can be discussed.

Acknowledgments. I have endeavored to write this book so that it will appeal to both the student and instructor. Through the years many people have helped in its development and I should like to acknowledge their valued suggestions and comments. Specifically, I wish to personally thank the following individuals who have contributed to this edition, namely, William Palm, University of Rhode Island, James R. Matthews, University of New Mexico, J. K.

*Review problems, at the end of each chapter, are presented in random order.

Al-Abdulla, University of Wisconsin-Madison, Nicholas P. Dadario, GMI Engineering and Management Institute, Larry A. Stauffer, University of Idaho, Derle Thorpe, Utah State University, and Gerald W. May, University of New Mexico. A particular note of thanks is also to be given to Professor Will Lidell, Jr., Auburn University at Montgomery, for his help and support.

Many thanks are also extended to all my students, a colleague, Bob Wang, and to members of the teaching profession who have freely taken the time to send me their suggestions and comments. Since the list is too long to mention, it is hoped that those who have given help in this manner will accept this anonymous recognition. Lastly, I should like to acknowledge the assistance of my wife, Conny, during the time it has taken to prepare the manuscript for publication.

Russell Charles Hibbeler

Contents

3

Equilibrium of a Particle   77

4

Force System Resultants   107

5

Equilibrium of a Rigid Body   181

6

Structural Analysis   241

7

Internal Forces   303

8

Friction   355

9

Center of Gravity and Centroid 411

10

Moments of Inertia 469

11

Virtual Work 519

APPENDIXES

A

B

Statics

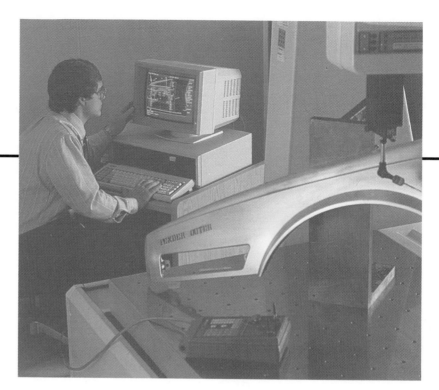

Although computers are often used in engineering, the design and analysis of any structural or mechanical part requires a fundamental understanding of the principles of engineering mechanics.

1

General Principles

This chapter provides an introduction to many of the fundamental concepts in mechanics. It includes a discussion of models or idealizations that are used to apply the theory, a statement of Newton's laws of motion, upon which this subject is based, and a general review of the principles for applying the SI system of units. Standard procedures for performing numerical calculations are then discussed. At the end of the chapter we will present a general guide that should be followed for solving problems.

1.1 Mechanics

Mechanics can be defined as that branch of the physical sciences concerned with the state of rest or motion of bodies that are subjected to the action of forces. In general, this subject is subdivided into three branches: *rigid-body mechanics, deformable-body mechanics,* and *fluid mechanics.* This book treats only rigid-body mechanics, since it forms a suitable basis for the design and analysis of many types of structural, mechanical, or electrical devices encountered in engineering. Also, rigid-body mechanics provides part of the necessary background for the study of the mechanics of deformable bodies and the mechanics of fluids.

Rigid-body mechanics is divided into two areas: statics and dynamics. *Statics* deals with the equilibrium of bodies, that is, those that are either at rest or move with a constant velocity; whereas *dynamics* is concerned with the accelerated motion of bodies. Although statics can be considered as a special case of dynamics, in which the acceleration is zero, statics deserves separate treatment in engineering education, since many objects are designed with the intention that they remain in equilibrium.

Historical Development. The subject of statics developed very early in history, because the principles involved could be formulated simply from measurements of geometry and force. For example, the writings of Archimedes (287–212 B.C.) deal with the principle of the lever. Studies of the pulley, inclined plane, and wrench are also recorded in ancient writings—at times when the requirements of engineering were limited primarily to building construction.

Since the principles of dynamics depend on an accurate measurement of time, this subject developed much later. Galileo Galilei (1564–1642) was one of the first major contributors to this field. His work consisted of experiments using pendulums and falling bodies. The most significant contributions in dynamics, however, were made by Isaac Newton (1642–1727), who is noted for his formulation of the three fundamental laws of motion and the law of universal gravitational attraction. Shortly after these laws were postulated, important techniques for their application were developed by Euler, D'Alembert, Lagrange, and others.

1.2 Fundamental Concepts

Before beginning our study of rigid-body mechanics, it is important to understand the meaning of certain fundamental concepts and principles.

Basic Quantities. The following four quantities are used throughout rigid-body mechanics.

Length. *Length* is needed to locate the position of a point in space and thereby describe the size of a physical system. Once a standard unit of length is defined, one can then quantitatively define distances and geometric properties of a body as multiples of the unit length.

Time. *Time* is conceived as a succession of events. Although the principles of statics are time independent, this quantity does play an important role in the study of dynamics.

Mass. *Mass* is a property of matter by which we can compare the action of one body with that of another. This property manifests itself as a gravitational attraction between two bodies and provides a quantitative measure of the resistance of matter to a change in velocity.

Force. In general, *force* is considered as a "push" or "pull" exerted by one body on another. This interaction can occur when there is direct contact between the bodies, such as a person pushing on a wall, or it can occur through a distance when the bodies are physically separated. Examples of the latter type include gravitational, electrical, and magnetic forces. In any case, a force is completely characterized by its magnitude, direction, and point of application.

Idealizations. Models or idealizations are used in mechanics in order to simplify application of the theory. A few of the more important idealizations will now be defined. Others that are noteworthy will be discussed at points where they are needed.

Particle. A *particle* has a mass but a size that can be neglected. For example, the size of the earth is insignificant compared to the size of its orbit, and therefore the earth can be modeled as a particle when studying its orbital motion. When a body is idealized as a particle, the principles of mechanics reduce to a rather simplified form, since the geometry of the body will not be involved in the analysis of the problem.

Rigid Body. A *rigid body* can be considered as a combination of a large number of particles in which all the particles remain at a fixed distance from one another both before and after applying a load. As a result, the material properties of any body that is assumed to be rigid will not have to be considered when analyzing the forces acting on the body. In most cases the actual deformations occurring in structures, machines, mechanisms, and the like are relatively small, and the rigid-body assumption is suitable for analysis.

Concentrated Force. A *concentrated force* represents the effect of a loading which is assumed to act at a point on a body. We can represent this effect by a concentrated force, provided the area over which the load is applied is very small compared to the overall size of the body.

Newton's Three Laws of Motion. The entire subject of rigid-body mechanics is formulated on the basis of Newton's three laws of motion, the validity of which is based on experimental observation. They apply to the motion of a particle as measured from a nonaccelerating reference frame and may be briefly stated as follows.

First Law. A particle originally at rest, or moving in a straight line with constant velocity, will remain in this state provided the particle is *not* subjected to an unbalanced force.

Second Law. A particle acted upon by an *unbalanced force* $\mathbf{F}$ experiences acceleration $\mathbf{a}$ that has the same direction as the force and a magnitude that is directly proportional to the force.* If $\mathbf{F}$ is applied to a particle of mass m, this law may be expressed mathematically as

$$\mathbf{F} = m\mathbf{a} \tag{1-1}$$

Third Law. The mutual forces of action and reaction between two particles are equal, opposite, and collinear.

---

*Stated another way, the unbalanced force acting on the particle is proportional to the time rate of change of the particle's linear momentum.

Newton's Law of Gravitational Attraction. Shortly after formulating his three laws of motion, Newton postulated a law governing the gravitational attraction between any two particles. Stated mathematically,

$$F = G\frac{m_1 m_2}{r^2} \tag{1-2}$$

where F = force of gravitation between the two particles
G = universal constant of gravitation; according to experimental evidence, $G = 66.73(10^{-12})$ m^3/(kg · s^2)
m_1, m_2 = mass of each of the two particles
r = distance between the two particles

Weight. According to Eq. 1–2, any two particles or bodies have a mutual attractive (gravitational) force acting between them. In the case of a particle located at or near the surface of the earth, however, the only gravitational force having any sizable magnitude is that between the earth and the particle. Consequently, this force, termed the *weight*, will be the only gravitational force considered in our study of mechanics.

From Eq. 1–2, we can develop an approximate expression for finding the weight W of a particle having a mass $m_1 = m$. If we assume the earth to be a nonrotating sphere of constant density and having a mass m_2, then if r is the distance between the earth's center and the particle, we have

$$W = G\frac{mm_2}{r^2}$$

Letting $g = Gm_2/r^2$ yields

$$W = mg \tag{1-3}$$

By comparison with Eq. 1–1, we term g the acceleration due to gravity. Since it depends on r, it can be seen that the weight of a body is *not* an absolute quantity. Instead, its magnitude is determined from where the measurement was made. For most engineering calculations, however, g is determined at sea level and at a latitude of 45°, which is considered the "standard location."

1.3 Units of Measurement

The four basic quantities—length, time, mass, and force—are not all independent from one another; in fact, they are *related* by Newton's second law of motion, **F** = m**a.** Hence, the *units* used to define force, mass, length, and time cannot *all* be selected arbitrarily. The equality **F** = m**a** is maintained only if three of the four units, called *base units,* are *arbitrarily defined* and the fourth unit is *derived* from the equation.

SI Units. The International System of units, abbreviated SI after the French "Système International d'Unités," is a modern version of the metric system which has received worldwide recognition. The SI system specifies length in meters (m), time in seconds (s), and mass in kilograms (kg). The unit of force, called a newton (N), is *derived* from $\mathbf{F} = m\mathbf{a}$. Thus, 1 newton is equal to a force required to give 1 kilogram of mass an acceleration of 1 m/s^2 ($N = kg \cdot m/s^2$).

If the weight of a body located at the "standard location" is to be determined in newtons, then Eq. 1–3 must be applied. Here $g = 9.806\ 65 \text{ m/s}^2$; however, for calculations, the value $g = 9.81 \text{ m/s}^2$ will be used. Thus,

$$W = mg \qquad (g = 9.81 \text{ m/s}^2) \qquad\qquad (1\text{–}4)$$

Therefore, a body of mass 1 kg has a weight of 9.81 N, a 2-kg body weighs 19.62 N, and so on.

1.4 The International System of Units

The SI system of units is used extensively in this book since it is intended to become the worldwide standard for measurement. Consequently, the rules for its use and some of its terminology relevant to mechanics will now be presented.

Prefixes. When a numerical quantity is either very large or very small, units used to define its size may be modified by using a prefix. Some of the prefixes used in the SI system are shown in Table 1–1. Each represents a multiple or submultiple of a unit which, if applied successively, moves the decimal point of a numerical quantity to every third place.* For example, $4\ 000\ 000 \text{ N} = 4\ 000 \text{ kN}$ (kilo-newton) $= 4 \text{ MN}$ (mega-newton), or $0.005 \text{ m} = 5 \text{ mm}$ (milli-meter). Notice that the SI system does not include the multiple deca (10) or the submultiple centi (0.01), which form part of the metric system. Except for some volume and area measurements, the use of these prefixes is to be avoided in science and engineering.

---

*The kilogram is the only base unit that is defined with a prefix.

Table 1–1   Prefixes

	Exponential Form	*Prefix*	*SI Symbol*
Multiple			
1 000 000 000	10^9	giga	G
1 000 000	10^6	mega	M
1 000	10^3	kilo	k
Submultiple			
0.001	10^{-3}	milli	m
0.000 001	10^{-6}	micro	μ
0.000 000 001	10^{-9}	nano	n

Rules for Use. The following rules are given for the proper use of the various SI symbols:

1. A symbol is *never* written with a plural "s," since it may be confused with the unit for second (s).

2. Symbols are always written in lowercase letters, with the following exceptions: symbols for the two largest prefixes shown in Table 1–3, giga and mega, are capitalized as G and M, respectively; and symbols named after an individual are also capitalized, e.g., N.

3. Quantities defined by several units which are multiples of one another are separated by a *dot* to avoid confusion with prefix notation, as indicated by $N = kg \cdot m/s^2 = kg \cdot m \cdot s^{-2}$. Also, $m \cdot s$ (meter-second), whereas ms (milli-second).

4. The exponential power represented for a unit having a prefix refers to both the unit *and* its prefix. For example, $\mu N^2 = (\mu N)^2 = \mu N \cdot \mu N$. Likewise, mm^2 represents $(mm)^2 = mm \cdot mm$.

5. Physical constants or numbers having several digits on either side of the decimal point should be reported with a *space* between every three digits rather than with a comma; e.g., 73 569.213 427. In the case of four digits on either side of the decimal, the spacing is optional; e.g., 8537 or 8 537. Furthermore, always try to use decimals and avoid fractions; that is, write 15.25, *not* $15\frac{1}{4}$.

6. When performing calculations, represent the numbers in terms of their *base or derived units* by converting all prefixes to powers of 10. The final result should then be expressed using a *single prefix*. Also, after calculation, it is best to keep numerical values between 0.1 and 1000; otherwise, a suitable prefix should be chosen. For example,

$$(50 \text{ kN})(60 \text{ nm}) = [50(10^3) \text{ N}][60(10^{-9}) \text{ m}]$$
$$= 3000(10^{-6}) \text{ N} \cdot \text{m} = 3(10^{-3}) \text{ N} \cdot \text{m} = 3 \text{ mN} \cdot \text{m}$$

7. Compound prefixes should not be used; e.g., $k\mu s$ (kilo-micro-second) should be expressed as ms (milli-second) since $1 \text{ k}\mu\text{s} = 1(10^3)(10^{-6}) \text{ s} = 1(10^{-3}) \text{ s} = 1 \text{ ms}$.

8. With the exception of the base unit the kilogram, in general avoid the use of a prefix in the denominator of composite units. For example, do not write N/mm, but rather kN/m; also, m/mg should be written as Mm/kg.

9. Although not expressed in multiples of 10, the minute, hour, etc., are retained for practical purposes as multiples of the second. Furthermore, plane angular measurement is made using radians (rad). In this book, however, degrees will often be used, where $180° = \pi \text{ rad}$.

1.5 Numerical Calculations

Numerical work in engineering practice is most often performed by using hand-held calculators and computers. It is important, however, that the answers to any problem be reported with both justifiable accuracy and appropriate significant figures. In this section we will discuss these topics together with some other important aspects involved in all engineering calculations.

Dimensional Homogeneity. The terms of any equation used to describe a physical process must be *dimensionally homogeneous;* that is, each term must be expressed in the same units. Provided this is the case, all the terms of an equation can then be combined if numerical values are substituted for the variables. Consider, for example, the equation $s = vt + \frac{1}{2}at^2$, where, in SI units, s is the position in meters, m, t is time in seconds, s, v is velocity in m/s, and a is acceleration in m/s^2. Regardless of how this equation is evaluated, it maintains its dimensional homogeneity. In the form stated each of the three terms is expressed in meters [m, (m/s̸)s̸, (m/s̸²)s̸²], or solving for a, $a = 2s/t^2 - 2v/t$, the terms are each expressed in units of m/s^2 [m/s^2, m/s^2, (m/s)/s].

Since problems in mechanics involve the solution of dimensionally homogeneous equations, the fact that all terms of an equation are represented by a consistent set of units can be used as a partial check for algebraic manipulations of an equation.

Significant Figures. The accuracy of a number is specified by the number of significant figures it contains. A *significant figure* is any digit, including a zero, provided it is not used to specify the location of the decimal point for the number. For example, the numbers 5604 and 34.52 each have four significant figures. When numbers begin or end with zeros, however, it is difficult to tell how many significant figures are in the number. Consider the number 40. Does it have one (4), or perhaps two (40) significant figures? In order to clarify this situation, the number should be reported using powers of 10. There are two ways of doing this. The format for *scientific notation* specifies one digit to the left of the decimal point, with the remaining digits to the right; for example, 40 expressed to one significant figure would be $4(10^1)$. Using *engineering notation,* which is preferred here, the exponent is displayed in multiples of three in order to facilitate conversion of SI units to those having an appropriate prefix. Thus, 40 expressed to one significant figure would be $0.04(10^3)$. Likewise, 2500 and 0.00546 expressed to three significant figures would be $2.50(10^3)$ and $5.46(10^{-3})$.

Rounding Off Numbers. For numerical calculations, the accuracy obtained from the solution of a problem generally can never be better than the accuracy of the problem data. This is what is to be expected, but often hand-held calculators or computers involve more figures in the answer than the number of significant figures used for the data. For this reason, a calculated result should always be "rounded off" to an appropriate number of significant figures.

To ensure accuracy, the following rules for rounding off a number to n significant figures apply:

1. If the $n + 1$ digit is *less than 5,* the $n + 1$ digit and others following it are dropped. For example, 2.326 and 0.451 rounded off to $n = 2$ significant figures would be 2.3 and 0.45.

2. If the $n + 1$ digit is equal to 5 with zeros following it, then round off the nth digit to an *even number.* For example, 1245 and 0.8655 rounded off to $n = 3$ significant figures become 1240 and 0.866.

3. If the $n + 1$ digit is *greater than 5* or equal to 5 with any nonzero digits following it, then increase the nth digit by 1 and drop the $n + 1$ digit and others following it. For example, 0.723 87 and 565.500 3 rounded off to $n = 3$ significant figures become 0.724 and 566.

Calculations. As a general rule, to ensure accuracy of a final result when performing calculations with numbers of unequal accuracy, always retain one extra significant figure in the more accurate numbers than in the least accurate number *before* beginning the computations. Then round off the final result so that it has the same number of significant figures as the least accurate number. If possible, try to work out the computations so that numbers which are approximately equal are not subtracted, since accuracy is often lost from this calculation.

In engineering we generally round off final answers to *three* significant figures since the data for geometry, loads, and other measurements are often reported with this accuracy.* Consequently, in this book the intermediate calculations for the examples are often worked out to four significant figures and the answers are generally reported to *three* significant figures.

The following example illustrates application of the principles just discussed as related to the proper use of units.

---

*Of course, some numbers, such as π, e, or numbers used in derived formulas, are exact and are therefore accurate to an infinite number of significant figures.

Example 1–1

Evaluate each of the following and express with SI units having an appropriate prefix: (a) (50 mN)(6 GN), (b) (400 mm)(0.6 MN)^2, (c) 45 MN^3/900 Gg.

SOLUTION

First convert each number to base units, perform the indicated operations, then choose an appropriate prefix (see Rule 6 on p. 8).

Part (a)

$$(50 \text{ mN})(6 \text{ GN}) = [50(10^{-3}) \text{ N}][6(10^9) \text{ N}]$$
$$= 300(10^6) \text{ N}^2$$
$$= 300(10^6) \text{ N}^2\left(\frac{1 \text{ kN}}{10^3 \text{ N}}\right)\left(\frac{1 \text{ kN}}{10^3 \text{ N}}\right)$$
$$= 300 \text{ kN}^2 \qquad\qquad Ans.$$

Note carefully the convention kN^2 = (kN)^2 = 10^6 N^2 (Rule 4 on p. 8).

Part (b)

$$(400 \text{ mm})(0.6 \text{ MN})^2 = [400(10^{-3}) \text{ m}][0.6(10^6) \text{ N}]^2$$
$$= [400(10^{-3}) \text{ m}][0.36(10^{12}) \text{ N}^2]$$
$$= 144(10^9) \text{ m} \cdot \text{N}^2$$
$$= 144 \text{ Gm} \cdot \text{N}^2 \qquad\qquad Ans.$$

We can also write

$$144(10^9) \text{ m} \cdot \text{N}^2 = 144(10^9) \text{ m} \cdot \text{N}^2\left(\frac{1 \text{ MN}}{10^6 \text{ N}}\right)\left(\frac{1 \text{ MN}}{10^6 \text{ N}}\right)$$
$$= 0.144 \text{ m} \cdot \text{MN}^2$$

Part (c)

$$45 \text{ MN}^3/900 \text{ Gg} = \frac{45(10^6 \text{ N})^3}{900(10^6) \text{ kg}}$$
$$= 0.05(10^{12}) \text{ N}^3/\text{kg}$$
$$= 0.05(10^{12}) \text{ N}^3\left(\frac{1 \text{ kN}}{10^3 \text{ N}}\right)^3 \frac{1}{\text{kg}}$$
$$= 0.05(10^3) \text{ kN}^3/\text{kg}$$
$$= 50 \text{ kN}^3/\text{kg} \qquad\qquad Ans.$$

Here we have used Rules 4 and 8 on pp. 8 and 9.

1.6 General Procedure for Analysis

The most effective way of learning the principles of engineering mechanics is to *solve problems.* To be successful at this, it is important always to present the work in a *logical* and *orderly manner,* as suggested by the following sequence of steps:

1. Read the problem carefully and try to correlate the actual physical situation with the theory studied.

2. Draw any necessary diagrams and tabulate the problem data.

3. Apply the relevant principles, generally in mathematical form.

4. Solve the necessary equations algebraically as far as practical, then, making sure they are dimensionally homogeneous, use a consistent set of units and complete the solution numerically. Report the answer with no more significant figures than the accuracy of the given data.

5. Study the answer with technical judgment and common sense to determine whether or not it seems reasonable.

6. Once the solution has been completed, review the problem. Try to think of other ways of obtaining the same solution.

In applying this general procedure, do the work as neatly as possible. Being neat generally stimulates clear and orderly thinking, and vice versa.

PROBLEMS

1–1. What is the weight in newtons of an object that has a mass of (a) 8 kg, (b) 0.04 g, and (c) 760 Mg?

1–2. Using Table 1–1, determine your own mass in kilograms, your weight in newtons, and your height in meters.

*1–3. Represent each of the following combinations of units in the correct SI form using an appropriate prefix: (a) m/ms, (b) μkm, (c) ks/mg, and (d) km · μN.

1–4. Represent each of the following as a number between 0.1 and 1000 using an appropriate prefix: (a) 45 320 kN, (b) 568(10^5) mm, and (c) 0.00563 mg.

1–5. Evaluate each of the following and express with an appropriate prefix: (a) $(430 \text{ kg})^2$, (b) $(0.002 \text{ mg})^2$, and (c) $(230 \text{ m})^3$.

1–6. Represent each of the following combinations of units in the correct SI form: (a) GN · μm, (b) kg/μm, (c) N/ks², and (d) kN/μs.

*1–7. Represent each of the following combinations of units in the correct SI form: (a) kN/μs, (b) Mg/mN, and (c) MN/(kg · ms).

1–8. Using the base units of the SI system, show that Eq. 1–2 is a dimensionally homogeneous equation which gives F in newtons. Compute the gravitational force acting between two identical spheres that are touching each other. The mass of each sphere is 150 kg and the radius is 275 mm.

*1–9. Two particles have a mass of 8 kg and 12 kg, respectively. If they are 800 mm apart, determine the force of gravity acting between them. Compare this result with the weight of each particle.

1–10. Evaluate each of the following to three significant figures and express each answer in SI units using an appropriate prefix: (a) $(212 \text{ mN})^2$, (b) $(52\ 800 \text{ ms})^2$, and (c) $[548(10^6)]^{1/2}$ ms.

1–11. Evaluate each of the following and express each answer in SI units using an appropriate prefix: (a) $(684\ \mu\text{m})/43$ ms, (b) (28 ms)(0.0458 Mm)/(348 mg), and (c) (2.68 mm)(426 Mg).

*1–12. Determine the mass in kilograms of an object that has a weight of (a) 20 mN, (b) 150 kN, and (c) 60 MN. Express each answer using an appropriate prefix.

This communications tower is stabilized by cables that exert resultant forces at the three points of connection. In this chapter we will show how to determine the magnitudes and directions of these resultant forces.

2

Force Vectors

In this chapter we will introduce the concept of a concentrated force and give the procedures for adding forces, resolving them into components, and projecting them along an axis. Since force is a vector quantity, we must use the rules of vector algebra whenever forces are considered. We will begin our study by defining scalar and vector quantities and then develop some of the basic rules of vector algebra.

2.1 Scalars and Vectors

Most of the physical quantities in mechanics can be expressed mathematically by means of scalars and vectors.

Scalar. A quantity characterized by a positive or negative number is called a *scalar*. Mass, volume, and length are scalar quantities often used in statics. In this book, scalars are indicated by letters in italic type, such as the scalar A. The mathematical operations involving scalars follow the same rules as those of elementary algebra.

Vector. A *vector* is a quantity that has both a magnitude and a direction. In statics the vector quantities frequently encountered are position, force, and moment. For handwritten work, a vector is generally represented by a letter with an arrow written over it, such as $\vec{A}$. The magnitude is designated $|\vec{A}|$ or simply A. In this book vectors will be symbolized in boldface type; for example, **A** is used to designate the vector ''A''. Its magnitude, which is always a positive quantity, is symbolized in italic type, written as $|A|$, or simply A when it is understood that A is a positive scalar.

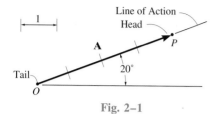

Fig. 2–1

A vector is represented graphically by an arrow, which is used to define its magnitude, direction, and sense. The *magnitude* of the vector is indicated by the length of the arrow, the *direction* is defined by the angle between a reference axis and the arrow's line of action, and the *sense* is indicated by the arrowhead. For example, the vector **A** shown in Fig. 2–1 has a magnitude of 4 units, a direction which is 20° measured counterclockwise from the horizontal axis, and a sense which is upward and to the right. The point O is called the *tail* of the vector, the point P is the *tip* or *head.*

2.2 Vector Operations

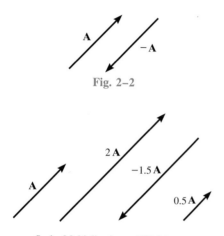

Fig. 2–2

Scalar Multiplication and Division

Fig. 2–3

Multiplication and Division of a Vector by a Scalar. The product of vector **A** and scalar a, yielding a**A**, is defined as a vector having a magnitude $|aA|$. The *sense* of a**A** is the *same* as **A** provided a is *positive;* it is *opposite* to **A** if a is *negative.* Consequently, the negative of a vector is formed by multiplying the vector by the scalar (-1), Fig. 2–2. Division of a vector by a scalar can be defined using the laws of multiplication, since $A/a = (1/a)A$, $a \neq 0$. Graphic examples of these operations are shown in Fig. 2–3.

Vector Addition. Two vectors **A** and **B** of the same type, Fig. 2–4a, may be added to form a "resultant" vector **R** = **A** + **B** by using the *parallelogram law.* To do this, **A** and **B** are joined at their tails, Fig. 2–4b. Parallel lines drawn from the head of each vector intersect at a common point, thereby forming the adjacent sides of a parallelogram. As shown, the resultant **R** is the diagonal of the parallelogram, which extends from the tails of **A** and **B** to the intersection of the lines.

We can also add **B** to **A** using a *triangle construction,* which is a special case of the parallelogram law, whereby vector **B** is added to vector **A** in a "head-to-tail" fashion, i.e., by connecting the head of **A** to the tail of **B,** Fig. 2–4c. The resultant **R** extends from the tail of **A** to the head of **B.** In a similar manner, **R** can also be obtained by adding **A** to **B,** Fig. 2–4d. By

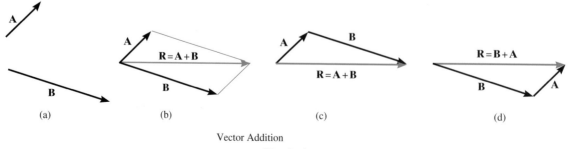

(a)   (b)   (c)   (d)

Vector Addition

Fig. 2–4

comparison, it is seen that vector addition is commutative; in other words, the vectors can be added in either order, i.e., $\mathbf{R} = \mathbf{A} + \mathbf{B} = \mathbf{B} + \mathbf{A}$.

As a special case, if the two vectors $\mathbf{A}$ and $\mathbf{B}$ are *collinear,* i.e., both have the same line of action, the parallelogram law reduces to an *algebraic* or *scalar addition* $R = A + B$, as shown in Fig. 2–5.

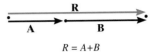

$$R = A+B$$

Addition of Collinear Vectors

Fig. 2–5

Vector Subtraction. The resultant *difference* between two vectors $\mathbf{A}$ and $\mathbf{B}$ of the same type may be expressed as

$$\mathbf{R'} = \mathbf{A} - \mathbf{B} = \mathbf{A} + (-\mathbf{B})$$

This vector sum is shown graphically in Fig. 2–6. Subtraction is therefore defined as a special case of addition, so the rules of vector addition also apply to vector subtraction.

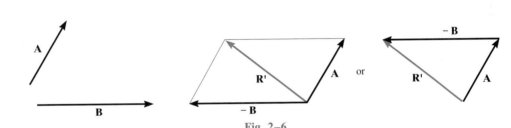

Fig. 2–6

Resolution of a Vector. A vector may be resolved into two "components" having known lines of action by using the parallelogram law. For example, if $\mathbf{R}$ in Fig. 2–7a is to be resolved into components acting along the lines a and b, one starts at the *head* of $\mathbf{R}$ and extends a line *parallel* to a until it intersects b. Likewise, a line parallel to b is drawn from the *head* of $\mathbf{R}$ to the point of intersection with a, Fig. 2–7a. The two components $\mathbf{A}$ and $\mathbf{B}$ are then drawn such that they extend from the tail of $\mathbf{R}$ to the points of intersection, as shown in Fig. 2–7b.

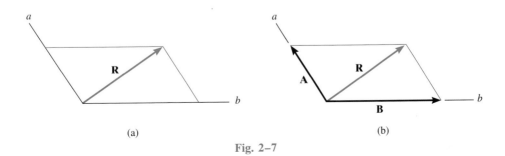

(a)

(b)

Fig. 2–7

2.3 Vector Addition of Forces

Experimental evidence has shown that a force is a vector quantity since it has a specified magnitude, direction, and sense and it adds according to the parallelogram law. Two common problems in statics involve either finding the resultant force, knowing its components, or resolving a known force into two components. As described in Sec. 2–2, both of these problems require application of the parallelogram law.

If more than two forces are to be added, successive applications of the parallelogram law can be carried out in order to obtain the resultant force. For example, if three forces $\mathbf{F}_1$, $\mathbf{F}_2$, $\mathbf{F}_3$ act at point O, Fig. 2–8, the resultant of any two of the forces is found—say, $\mathbf{F}_1 + \mathbf{F}_2$—and then this resultant is added to the third force, yielding the resultant of all three forces; i.e., $\mathbf{F}_R = (\mathbf{F}_1 + \mathbf{F}_2) + \mathbf{F}_3$. Using the parallelogram law to add more than two forces, as shown here, often requires extensive geometric and trigonometric calculation to determine the numerical values for the magnitude and direction of the resultant. Instead, problems of this type are easily solved by using the "rectangular-component method," which is explained in Sec. 2.4.

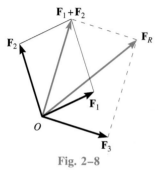

Fig. 2–8

PROCEDURE FOR ANALYSIS

Problems that involve the addition of two forces and contain at most *two unknowns* can be solved by using the following procedure:

Parallelogram Law. Make a sketch showing the vector addition using the parallelogram law. If possible, determine the interior angles of the parallelogram from the geometry of the problem. Recall that the sum total of these angles is 360°. Unknown angles, along with known and unknown force magnitudes, should be clearly labeled on this sketch. Redraw a half portion of the constructed parallelogram to illustrate the triangular head-to-tail addition of the components.

Trigonometry. By using trigonometry, the two unknowns can be determined from the data listed on the triangle. If the triangle does *not* contain a 90° angle, the law of sines and/or the law of cosines may be used for the solution. These formulas are given in Fig. 2–9 for the triangle shown.

The following examples illustrate this method numerically.

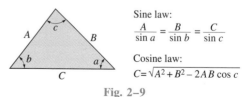

Sine law:
$$\frac{A}{\sin a} = \frac{B}{\sin b} = \frac{C}{\sin c}$$

Cosine law:
$$C = \sqrt{A^2 + B^2 - 2AB \cos c}$$

Fig. 2–9

Example 2–1

The screw eye in Fig. 2–10a is subjected to two forces, $\mathbf{F}_1$ and $\mathbf{F}_2$. Determine the magnitude and direction of the resultant force.

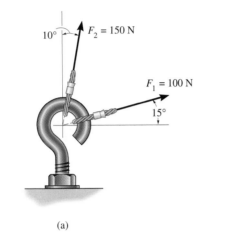

(a)

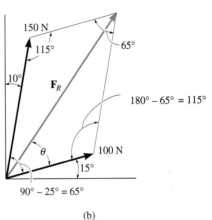

(b)

SOLUTION

The resultant force is formed from the parallelogram law.

Parallelogram Law. The addition is shown in Fig. 2–10b. The two unknowns are the magnitude of $\mathbf{F}_R$ and the angle θ (theta). From Fig. 2–10b, the vector triangle, Fig. 2–10c, is constructed.

Trigonometry. F_R is determined by using the law of cosines:

$$F_R = \sqrt{(100 \text{ N})^2 + (150 \text{ N})^2 - 2(100 \text{ N})(150 \text{ N}) \cos 115°}$$
$$= \sqrt{10\,000 + 22\,500 - 30\,000(-0.4226)} = 212.6 \text{ N}$$
$$= 213 \text{ N} \qquad\qquad\qquad Ans.$$

The angle θ is determined by applying the law of sines, using the computed value of F_R.

$$\frac{150 \text{ N}}{\sin \theta} = \frac{212.6 \text{ N}}{\sin 115°}$$

$$\sin \theta = \frac{150 \text{ N}}{212.6 \text{ N}} (0.9063)$$

$$\theta = 39.8°$$

Thus, the direction ϕ (phi) of $\mathbf{F}_R$, measured from the horizontal, is

$$\phi = 39.8° + 15.0° = 54.8° \quad \angle^{\phi} \qquad\qquad Ans.$$

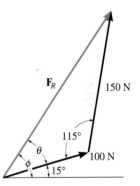

(c)

Fig. 2–10

Example 2–2

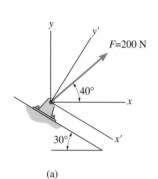

(a)

Resolve the 200-N force shown acting on the pin, Fig. 2–11a, into components in the (a) x and y directions, and (b) x' and y directions.

SOLUTION

In each case the parallelogram law is used to resolve **F** into its two components, and then the vector triangle is constructed to determine the numerical results by trigonometry.

Part (a).   The vector addition $\mathbf{F} = \mathbf{F}_x + \mathbf{F}_y$ is shown in Fig. 2–11b. In particular, note that the length of the components is scaled along the x and y axes by first constructing lines parallel to the axes in accordance with the parallelogram law. From the vector triangle, Fig. 2–11c,

$$F_x = 200 \text{ N} \cos 40° = 153 \text{ N} \qquad Ans.$$
$$F_y = 200 \text{ N} \sin 40° = 129 \text{ N} \qquad Ans.$$

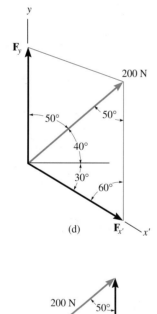

(d)

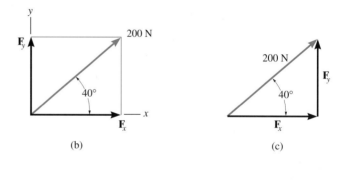

(b)   (c)

Fig. 2–11

Part (b).   The vector addition $\mathbf{F} = \mathbf{F}_{x'} + \mathbf{F}_y$ is shown in Fig. 2–11d. Note carefully how the parallelogram is constructed. Applying the law of sines and using the data listed on the vector triangle, Fig. 2–11e, yields

$$\frac{F_{x'}}{\sin 50°} = \frac{200 \text{ N}}{\sin 60°}$$

$$F_{x'} = 200 \text{ N}\left(\frac{\sin 50°}{\sin 60°}\right) = 177 \text{ N} \qquad Ans.$$

$$\frac{F_y}{\sin 70°} = \frac{200 \text{ N}}{\sin 60°}$$

$$F_y = 200 \text{ N}\left(\frac{\sin 70°}{\sin 60°}\right) = 217 \text{ N} \qquad Ans.$$

(e)

Example 2–3

The force **F** acting on the frame shown in Fig. 2–12*a* has a magnitude of 500 N and is to be resolved into two components acting along struts *AB* and *AC*. Determine the angle θ, measured *below* the horizontal, so that the component $\mathbf{F}_{AC}$ is directed from *A* toward *C* and has a magnitude of 400 N.

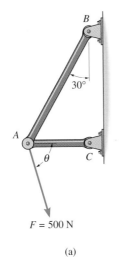

(a)

Fig. 2–12

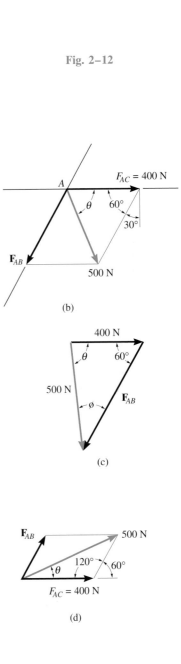

(b)

(c)

(d)

SOLUTION

By using the parallelogram law, the vector addition of the two components yielding the resultant is shown in Fig. 2–12*b*. Note carefully how the resultant force is resolved into the two components $\mathbf{F}_{AB}$ and $\mathbf{F}_{AC}$, which have specified lines of action. The corresponding vector triangle is shown in Fig. 2–12*c*. The angle ϕ can be determined by using the law of sines:

$$\frac{400 \text{ N}}{\sin \phi} = \frac{500 \text{ N}}{\sin 60°}$$

$$\sin \phi = \left(\frac{400 \text{ N}}{500 \text{ N}}\right) \sin 60° = 0.6928$$

$$\phi = 43.9°$$

Hence,

$$\theta = 180° - 60° - 43.9° = 76.1° \qquad Ans.$$

Using this value for θ, apply the law of cosines and show that $\mathbf{F}_{AB}$ has a magnitude of 561 N.

Notice that **F** can also be directed at an angle θ *above* the horizontal, as shown in Fig. 2–12*d*, and still produce the required component $\mathbf{F}_{AC}$. Show that in this case $\theta = 16.1°$ and $F_{AB} = 161$ N.

Example 2–4

The ring shown in Fig. 2–13a is subjected to two forces, $\mathbf{F}_1$ and $\mathbf{F}_2$. If it is required that the resultant force have a magnitude of 1 kN and be directed vertically downward, determine (a) the magnitudes of $\mathbf{F}_1$ and $\mathbf{F}_2$ provided $\theta = 30°$, and (b) the magnitudes of $\mathbf{F}_1$ and $\mathbf{F}_2$ if F_2 is to be a minimum.

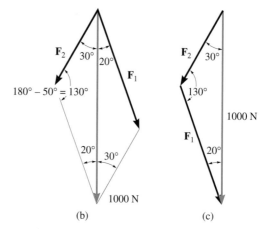

(a)   (b)   (c)

SOLUTION

Part (a).   A sketch of the vector addition according to the parallelogram law is shown in Fig. 2–13b. From the vector triangle constructed in Fig. 2–13c, the unknown magnitudes F_1 and F_2 can be determined by using the law of sines.

$$\frac{F_1}{\sin 30°} = \frac{1000 \text{ N}}{\sin 130°}$$

$$F_1 = 653 \text{ N} \qquad\qquad Ans.$$

$$\frac{F_2}{\sin 20°} = \frac{1000 \text{ N}}{\sin 130°}$$

$$F_2 = 446 \text{ N} \qquad\qquad Ans.$$

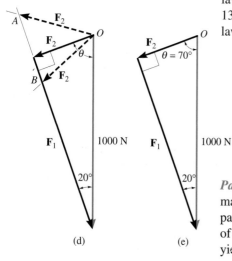

(d)   (e)

Part (b).   If θ is not specified, then by the vector triangle, Fig. 2–13d, $\mathbf{F}_2$ may be added to $\mathbf{F}_1$ in various ways to yield the resultant 1000-N force. In particular, the *minimum* length or magnitude of $\mathbf{F}_2$ will occur when its line of action is *perpendicular to* $\mathbf{F}_1$. Any other direction, such as OA or OB, yields a larger value for F_2. Hence, when $\theta = 90° - 20° = 70°$, F_2 is minimum. From the triangle shown in Fig. 2–13e, it is seen that

$$F_1 = 1000 \sin 70° \text{ N} = 940 \text{ N} \qquad Ans.$$

$$F_2 = 1000 \sin 20° \text{ N} = 342 \text{ N} \qquad Ans.$$

Fig. 2–13

PROBLEMS

2–1. Determine the magnitude of the resultant force $\mathbf{F}_R = \mathbf{F}_1 + \mathbf{F}_2$ and its direction, measured counterclockwise from the positive x axis.

2–2. Determine the magnitude of the resultant force $\mathbf{F}_R = \mathbf{F}_1 + \mathbf{F}_3$ and its direction, measured counterclockwise from the positive x axis.

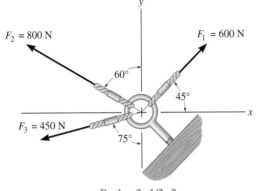

$F_2 = 800$ N $F_1 = 600$ N

$60°$

$45°$

$F_3 = 450$ N

$75°$

Probs. 2–1/2–2

2–3. Determine the magnitude of the resultant force $\mathbf{F}_R = \mathbf{F}_1 + \mathbf{F}_2$ and its direction, measured counterclockwise from the positive x axis.

*2–4. Determine the magnitude of the resultant force $\mathbf{F}_R = \mathbf{F}_1 - \mathbf{F}_2$ and its direction, measured counterclockwise from the positive x axis.

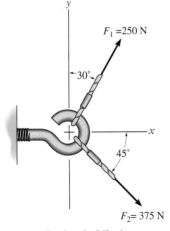

$F_1 = 250$ N

$30°$

x

$45°$

$F_2 = 375$ N

Probs. 2–3/2–4

2–5. Determine the magnitude of the resultant force and its direction, measured counterclockwise from the positive x axis.

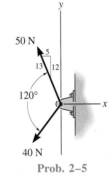

50 N

5

13 12

$120°$

x

40 N

Prob. 2–5

2–6. Determine the magnitude of the resultant force $\mathbf{F}_R = \mathbf{F}_1 + \mathbf{F}_2$ and its direction, measured clockwise from the positive u axis.

2–7. Resolve the force $\mathbf{F}_1$ into components acting along the u and v axes and determine the magnitudes of the components.

*2–8. Resolve the force $\mathbf{F}_2$ into components acting along the u and v axes and determine the magnitudes of the components.

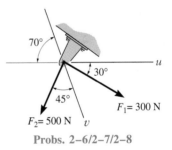

$70°$

$30°$

u

$45°$

$F_1 = 300$ N

$F_2 = 500$ N v

Probs. 2–6/2–7/2–8

2–9. The V-grooved wheel is used to run along the track. If the track exerts a vertical force of 200 N on the wheel, determine the components of this force acting along the a and b axes, which are perpendicular to the sides of the groove.

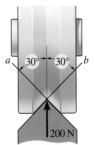

a $30°$ $30°$ b

200 N

Prob. 2–9

2–10.   Resolve the 60-N force into components acting along the *u* and *v* axes and determine the magnitudes of the components.

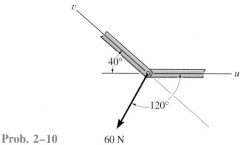

Prob. 2–10   60 N

2–11.   The wind is deflected by the sail of a boat such that it exerts a resultant force of $F = 110$ N perpendicular to the sail. Resolve this force into two components, one parallel and one perpendicular to the keel *aa* of the boat. *Note:* The ability to sail into the wind is known as tacking, made possible by the force parallel to the boat's keel. The perpendicular component tends to tip the boat or push it over.

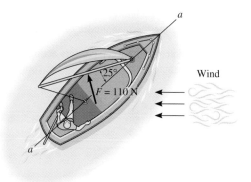

Prob. 2–11

*2–12.   The hook supports the two cable forces $F_1 = 500$ N and $F_2 = 300$ N. If the resultant of these forces acts vertically downward and has a magnitude of $F_R = 750$ N, determine the angles θ and ϕ of the cables.

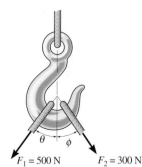

Prob. 2–12 $F_1 = 500$ N $F_2 = 300$ N

2–13.   The vertical force of $F = 60$ N acts downward at *A* on the two-member frame. Determine the magnitudes of the two components of **F** directed along the axes of members *AB* and *AC*. Set $\theta = 45°$.

2–14.   The vertical force of $F = 60$ N acts downward at *A* on the two-member frame. Determine the angle θ ($0° \leq \theta \leq 90°$) of member *AB* so that the component of **F** acting along the axis of *AB* is 80 N. What is the magnitude of the force component acting along the axis of member *AC*?

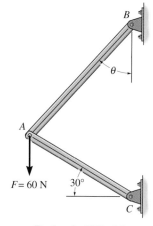

Probs. 2–13/2–14

2–15.   The plate is subjected to the two forces at *A* and *B* as shown. If $\theta = 60°$, determine the magnitude of the resultant of these two forces and its direction measured clockwise from the positive *x* axis.

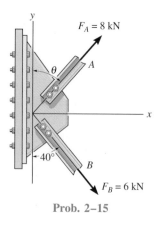

Prob. 2–15

*2–16. Resolve the 50-N force into components acting along (a) the x and y axes, and (b) the x and y' axes.

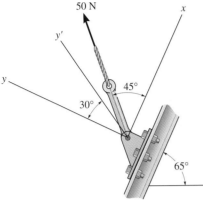

Prob. 2–16

2–17. The force acting on the gear tooth is $F = 20$ N. Resolve this force into two components acting along the lines aa and bb.

2–18. The component of force $\mathbf{F}$ acting along line aa is required to be 30 N. Determine the magnitude of $\mathbf{F}$ and its component along line bb.

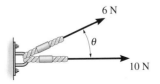

Probs. 2–17/2–18

2–19. Two forces having a magnitude of 10 N and 6 N act on the ring. If the largest magnitude of the resultant force the ring can support is 14 N, determine the angle θ between the forces.

*2–20. Determine the angle θ ($0° \le \theta \le 90°$) between the two forces so that the magnitude of the resultant force acting on the ring is a minimum. What is the magnitude of the resultant force?

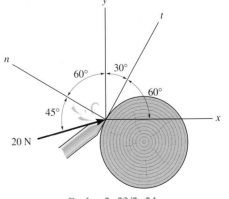

Probs. 2–19/2–20

2–21. The post is to be pulled out of the ground using two ropes A and B. Rope A is subjected to a force of 600 N and is directed at 60° from the horizontal. Determine the force T in rope B if the post starts to lift out when $\theta = 20°$. For this to occur, the resultant force on the post is to be directed vertically upward. Also calculate the magnitude of the resultant force.

2–22. The post is to be pulled out of the ground using two ropes A and B. Rope A is subjected to a force of 600 N and is directed at 60° from the horizontal. If the resultant force acting on the post is to be 1200 N, vertically upward, determine the force T in rope B and the corresponding angle θ.

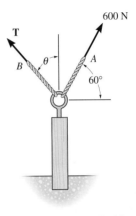

Probs. 2–21/2–22

2–23. The chisel exerts a force of 20 N on the wood dowel rod which is turning in a lathe. Resolve this force into components acting (a) along the n and t axes and (b) along the x and y axes.

*2–24. The chisel exerts a force of 20 N on the wood dowel rod which is turning in a lathe. Resolve this force into components acting (a) along the n and y axes and (b) along the x and t axes.

Probs. 2–23/2–24

2–25. If $\theta = 20°$ and $\phi = 35°$, determine the magnitudes of $\mathbf{F}_1$ and $\mathbf{F}_2$ so that the resultant force has a magnitude of 20 N and is directed along the positive x axis.

2–26. If $F_1 = F_2 = 30$ N, determine the angles θ and ϕ so that the resultant force is directed along the positive x axis and has a magnitude of $F_R = 20$ N.

2–29. Determine the design angle θ $(0° \le \theta \le 90°)$ for strut AB so that the 400-N horizontal force has a component of 500 N directed from A towards C. What is the component of force acting along member AB? Take $\phi = 40°$.

2–30. Determine the design angle ϕ $(0° \le \phi \le 90°)$ between struts AB and AC so that the 400-N horizontal force has a component of 600 N which acts up to the left, in the same direction as from B towards A. Take $\theta = 30°$.

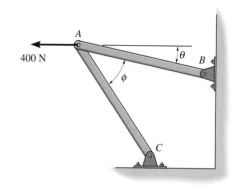

Probs. 2–29/2–30

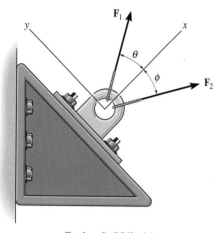

Probs. 2–25/2–26

2–27. Determine the magnitude and direction of the resultant $\mathbf{F}_R = \mathbf{F}_1 + \mathbf{F}_2 + \mathbf{F}_3$ of the three forces by first finding the resultant $\mathbf{F}' = \mathbf{F}_1 + \mathbf{F}_2$ and then forming $\mathbf{F}_R = \mathbf{F}' + \mathbf{F}_3$.

***2–28.** Determine the magnitude and direction of the resultant $\mathbf{F}_R = \mathbf{F}_1 + \mathbf{F}_2 + \mathbf{F}_3$ of the three forces by first finding the resultant $\mathbf{F}' = \mathbf{F}_2 + \mathbf{F}_3$ and then forming $\mathbf{F}_R = \mathbf{F}' + \mathbf{F}_1$.

2–31. The log is being towed by two tractors A and B. Determine the magnitudes of the two towing forces $\mathbf{F}_A$ and $\mathbf{F}_B$ if it is required that the resultant force have a magnitude $F_R = 10$ kN and be directed along the x axis. Set $\theta = 15°$.

***2–32.** If the resultant $\mathbf{F}_R$ of the two forces acting on the log is to be directed along the positive x axis and have a magnitude of 10 kN, determine the angle θ of the cable attached to B such that the force $\mathbf{F}_B$ in this cable is a *minimum*. What is the magnitude of force in each cable for this situation?

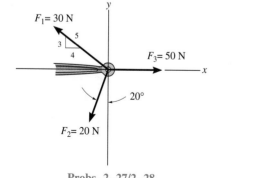

Probs. 2–27/2–28

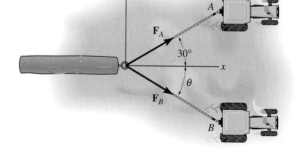

Probs. 2–31/2–32

2.4 Addition of a System of Coplanar Forces

When the resultant of more than two forces has to be obtained, it is easier to find the components of each force along specified axes, add these components algebraically, and then form the resultant, rather than form the resultant of the forces by successive application of the parallelogram law as discussed in Sec. 2.3. In this section we will resolve each force into its rectangular components $\mathbf{F}_x$ and $\mathbf{F}_y$, which lie along the x and y axes, respectively, Fig. 2–14a. Although the axes are shown here to be horizontal and vertical, they may in general be directed at any inclination, as long as they remain perpendicular to one another, Fig. 2–14b. In either case, by the parallelogram law, we require:

and
$$\mathbf{F} = \mathbf{F}_x + \mathbf{F}_y$$
$$\mathbf{F}' = \mathbf{F}'_x + \mathbf{F}'_y$$

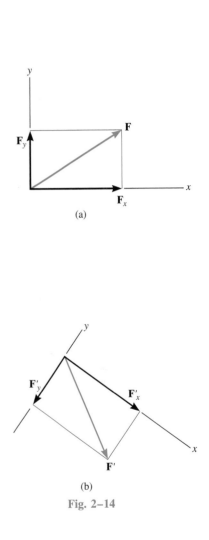

(a)

Fig. 2–14

As shown in Fig. 2–14, the sense of direction of each component is represented *graphically* by the *arrowhead*. For *analytical* work, however, we must establish a notation for representing the directional sense of the rectangular components for each coplanar vector. This can be done in one of two ways.

Scalar Notation. Since the x and y axes have designated positive and negative directions, the magnitude and directional sense of the rectangular components of a force can be expressed in terms of *algebraic scalars*. For example, the components of $\mathbf{F}$ in Fig. 2–14a can be represented by positive scalars F_x and F_y since their sense of direction is along the *positive x* and *y* axes, respectively. In a similar manner, the components of $\mathbf{F}'$ in Fig. 2–14b are F'_x and $-F'_y$. Here the y component is negative, since $\mathbf{F}'_y$ is directed along the negative y axis. It is important to keep in mind that this scalar notation is to be used only for computational purposes, not for graphical representations in figures. Throughout the text, the *head of a vector arrow* in any figure indicates the sense of the vector *graphically;* algebraic signs are not used for this purpose. Thus, the vectors in Figs. 2–14a and 2–14b are designated by using boldface (vector) notation.* Whenever italic symbols are written near vector arrows in figures, they indicate the *magnitude* of the vector, which is *always* a *positive* quantity.

*Negative signs are used only in figures with boldface notation when showing equal but opposite pairs of vectors as in Fig. 2–2.

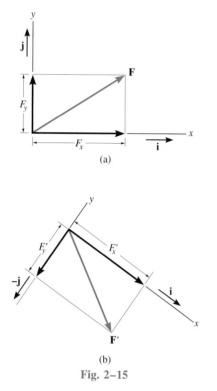

(a)

(b)

Fig. 2–15

Cartesian Vector Notation. It is also possible to represent the components of a force in terms of Cartesian unit vectors. By doing this the methods of vector algebra are easier to apply, and we will see that this becomes particularly advantageous for solving problems in three dimensions. In two dimensions the *Cartesian unit vectors* $\mathbf{i}$ and $\mathbf{j}$ are used to designate the *directions* of the x and y axes, respectively, Fig. 2–15a.* These vectors have a dimensionless magnitude of unity, and their sense (or arrowhead) will be described analytically by a plus or minus sign, depending on whether they are pointing along the positive or negative x or y axis.

As shown in Fig. 2–15a, the *magnitude* of each component of $\mathbf{F}$ is *always a positive quantity*, which is represented by the (positive) scalars F_x and F_y. Therefore, having established notation to represent the magnitude and the direction of each component, we can express $\mathbf{F}$ in Fig. 2–15a as a *Cartesian vector*, i.e.,

$$\mathbf{F} = F_x\mathbf{i} + F_y\mathbf{j}$$

And in the same way, $\mathbf{F}'$ in Fig. 2–15b can be expressed as

$$\mathbf{F}' = F_x'\mathbf{i} + F_y'(-\mathbf{j})$$

or simply

$$\mathbf{F}' = F_x'\mathbf{i} - F_y'\mathbf{j}$$

Coplanar Force Resultants. Either of the two methods just described for representing the rectangular components of a force can be used to determine the resultant of several *coplanar forces.* To do this, each force is first resolved into its x and y components and then the respective components are added using *scalar algebra* since they are collinear. The resultant force is then formed by adding the resultants of the x and y components using the parallelogram law. For example, consider the three forces in Fig. 2–16a, which have x and y components as shown in Fig. 2–16b. To solve this problem using *Cartesian vector notation,* each force is first represented as a Cartesian vector, i.e.,

$$\mathbf{F}_1 = F_{1x}\mathbf{i} + F_{1y}\mathbf{j}$$
$$\mathbf{F}_2 = -F_{2x}\mathbf{i} + F_{2y}\mathbf{j}$$
$$\mathbf{F}_3 = F_{3x}\mathbf{i} - F_{3y}\mathbf{j}$$

---

*For handwritten work, unit vectors are usually indicated using a circumflex, e.g., $\hat{\mathbf{i}}$ and $\hat{\mathbf{j}}$.

The vector resultant is therefore

$$\mathbf{F}_R = \mathbf{F}_1 + \mathbf{F}_2 + \mathbf{F}_3$$
$$= F_{1x}\mathbf{i} + F_{1y}\mathbf{j} - F_{2x}\mathbf{i} + F_{2y}\mathbf{j} + F_{3x}\mathbf{i} - F_{3y}\mathbf{j}$$
$$= (F_{1x} - F_{2x} + F_{3x})\mathbf{i} + (F_{1y} + F_{2y} - F_{3y})\mathbf{j}$$
$$= (F_{Rx})\mathbf{i} + (F_{Ry})\mathbf{j}$$

If *scalar notation* is used, then, from Fig. 2–16*b*, since x is positive to the right and y is positive upward, we have

$(\overset{+}{\rightarrow})$ $\qquad\qquad\qquad F_{Rx} = F_{1x} - F_{2x} + F_{3x}$

$(+\uparrow)$ $\qquad\qquad\qquad F_{Ry} = F_{1y} + F_{2y} - F_{3y}$

These results are the *same* as the $\mathbf{i}$ and $\mathbf{j}$ components of $\mathbf{F}_R$ determined above.

In the general case, the x and y components of the resultant of any number of coplanar forces can be represented symbolically by the algebraic sum of the x and y components of all the forces, i.e.,

$$F_{Rx} = \Sigma F_x$$
$$F_{Ry} = \Sigma F_y \qquad\qquad (2\text{–}1)$$

When applying these equations it is important to use the *sign convention* established for the components; and that is, components having a directional sense along the positive coordinate axes are considered positive scalars, whereas those having a directional sense along the negative coordinate axes are considered negative scalars. If this convention is followed, then the signs of the resultant components will specify the sense of these components. For example, a positive result indicates that the component has a directional sense which is in the positive coordinate direction.

Once the resultant components are determined, they may be sketched along the x and y axes in their proper directions, and the resultant force can be determined from vector addition, as shown in Fig. 2–16*c*. From this sketch, the magnitude of $\mathbf{F}_R$ is then found from the Pythagorean theorem; that is,

$$F_R = \sqrt{F_{Rx}^2 + F_{Ry}^2}$$

Also, the direction angle θ, which specifies the orientation of the force, is determined from trigonometry.

$$\theta = \tan^{-1}\left|\frac{F_{Ry}}{F_{Rx}}\right|$$

The above concepts are illustrated numerically in the following examples.

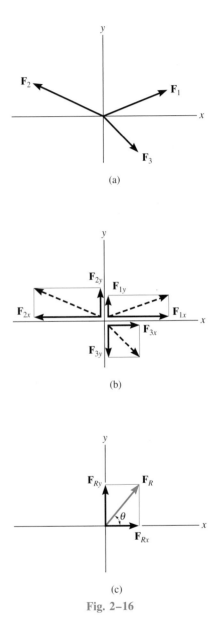

Fig. 2–16

Example 2–5

Determine the x and y components of $\mathbf{F}_1$ and $\mathbf{F}_2$ shown in Fig. 2–17a. Express each force as a Cartesian vector.

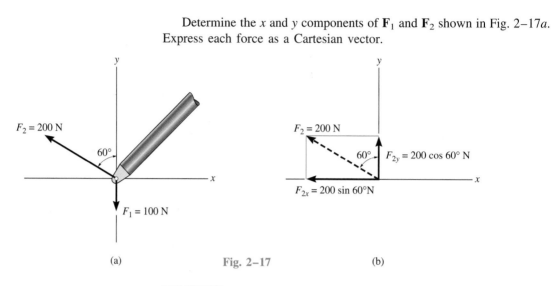

Fig. 2–17

(a)   (b)

SOLUTION

Scalar Notation. Since $\mathbf{F}_1$ acts along the negative y axis, and the magnitude of $\mathbf{F}_1$ is 100 N, the components written in scalar form are

$$F_{1x} = 0, \qquad F_{1y} = -100 \text{ N} \qquad\qquad Ans.$$

or, alternatively,

$$F_{1x} = 0, \qquad F_{1y} = 100 \text{ N} \downarrow \qquad\qquad Ans.$$

By the parallelogram law, $\mathbf{F}_2$ is resolved into x and y components, Fig. 2–17b. The magnitude of each component is determined by trigonometry. Since $\mathbf{F}_{2x}$ acts in the $-x$ direction, and $\mathbf{F}_{2y}$ acts in the $+y$ direction, we have

$$F_{2x} = -200 \sin 60° \text{ N} = -173 \text{ N} = 173 \text{ N} \leftarrow \qquad Ans.$$
$$F_{2y} = 200 \cos 60° \text{ N} = 100 \text{ N} = 100 \text{ N} \uparrow \qquad Ans.$$

Cartesian Vector Notation. Having computed the magnitudes of the components of $\mathbf{F}_2$, Fig. 2–17b, we can express each force as a Cartesian vector.

$$\mathbf{F}_1 = 0\mathbf{i} + 100 \text{ N}(-\mathbf{j})$$
$$= \{-100\mathbf{j}\} \text{ N} \qquad\qquad Ans.$$

and

$$\mathbf{F}_2 = 200 \sin 60° \text{ N}(-\mathbf{i}) + 200 \cos 60° \text{ N}(\mathbf{j})$$
$$= \{-173\mathbf{i} + 100\mathbf{j}\} \text{ N} \qquad\qquad Ans.$$

Example 2–6

Determine the x and y components of the force $\mathbf{F}$ shown in Fig. 2–18a.

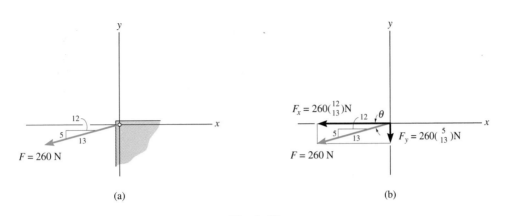

(a)   (b)

Fig. 2–18

SOLUTION

The force is resolved into its x and y components as shown in Fig. 2–18b. Here the *slope* of the line of action for the force is indicated. From this "slope triangle" we could obtain the direction angle θ, e.g., $\theta = \tan^{-1}(\frac{5}{12})$, and then proceed to determine the magnitudes of the components in the same manner as for $\mathbf{F}_2$ in Example 2–5. An easier method, however, consists of using proportional parts of similar triangles, i.e.,

$$\frac{F_x}{260\text{ N}} = \frac{12}{13} \qquad F_x = 260\text{ N}\left(\frac{12}{13}\right) = 240\text{ N}$$

Similarly,

$$F_y = 260\text{ N}\left(\frac{5}{13}\right) = 100\text{ N}$$

Notice that the magnitude of the *horizontal component, F_x,* was obtained by multiplying the force magnitude by the ratio of the *horizontal leg* of the slope triangle divided by the hypotenuse; whereas the magnitude of the *vertical component, F_y,* was obtained by multiplying the force magnitude by the ratio of the *vertical leg* divided by the hypotenuse. Hence, using scalar notation,

$$F_x = -240\text{ N} = 240\text{ N} \leftarrow \qquad\qquad Ans.$$
$$F_y = -100\text{ N} = 100\text{ N} \downarrow \qquad\qquad Ans.$$

If $\mathbf{F}$ is expressed as a Cartesian vector, we have

$$\mathbf{F} = \{-240\mathbf{i} - 100\mathbf{j}\}\text{ N} \qquad\qquad Ans.$$

Example 2–7

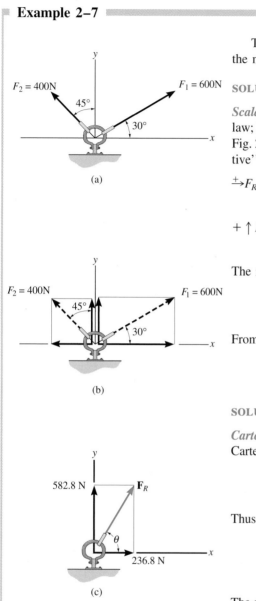

(a)

(b)

(c)

Fig. 2–19

The link in Fig. 2–19a is subjected to two forces $\mathbf{F}_1$ and $\mathbf{F}_2$. Determine the magnitude and orientation of the resultant force.

SOLUTION I

Scalar Notation.   This problem can be solved by using the parallelogram law; however, here we will resolve each force into its x and y components, Fig. 2–19b, and sum these components algebraically. Indicating the "positive" sense of the x and y force components alongside Eqs. 2–1, we have

$$\xrightarrow{+}F_{Rx} = \Sigma F_x; \qquad F_{Rx} = 600 \cos 30° \text{ N} - 400 \sin 45° \text{ N}$$
$$= 236.8 \text{ N} \rightarrow$$

$$+ \uparrow F_{Ry} = \Sigma F_y; \quad F_{Ry} = 600 \sin 30° \text{ N} + 400 \cos 45° \text{ N}$$
$$= 582.8 \text{ N} \uparrow$$

The resultant force, shown in Fig. 2–19c, has a *magnitude* of

$$F_R = \sqrt{(236.8 \text{ N})^2 + (582.8 \text{ N})^2}$$
$$= 629 \text{ N} \qquad\qquad Ans.$$

From the vector addition, Fig. 2–19c, the direction angle θ is

$$\theta = \tan^{-1}\left(\frac{582.8 \text{ N}}{236.8 \text{ N}}\right) = 67.9° \qquad Ans.$$

SOLUTION II

Cartesian Vector Notation.   From Fig. 2–19b, each force expressed as a Cartesian vector is

$$\mathbf{F}_1 = \{600 \cos 30°\mathbf{i} + 600 \sin 30°\mathbf{j}\} \text{ N}$$
$$\mathbf{F}_2 = \{-400 \sin 45°\mathbf{i} + 400 \cos 45°\mathbf{j}\} \text{ N}$$

Thus

$$\mathbf{F}_R = \mathbf{F}_1 + \mathbf{F}_2 = (600 \cos 30° \text{ N} - 400 \sin 45° \text{ N})\mathbf{i}$$
$$+ (600 \sin 30° \text{ N} + 400 \cos 45° \text{ N})\mathbf{j}$$
$$= \{236.8\mathbf{i} + 582.8\mathbf{j}\} \text{ N}$$

The magnitude and direction of $\mathbf{F}_R$ are determined in the same manner as shown above.

Comparing the two methods of solution, it is seen that use of scalar notation is more efficient, since the scalar components can be found *directly,* without first having to express each force as a Cartesian vector before adding the components. Cartesian vector analysis, however, will later be shown to be more advantageous for solving three-dimensional problems.

Example 2–8

The end of the boom O in Fig. 2–20a is subjected to three concurrent and coplanar forces. Determine the magnitude and orientation of the resultant force.

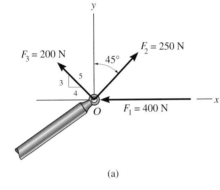

(a)

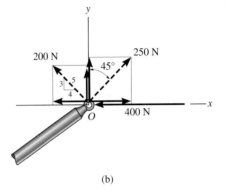

(b)

SOLUTION

Each force is resolved into its x and y components as shown in Fig. 2–20b. Summing the x components, we have

$$\overset{+}{\rightarrow} F_{Rx} = \Sigma F_x; \quad F_{Rx} = -400 \text{ N} + 250 \sin 45° \text{ N} - 200(\tfrac{4}{5}) \text{ N}$$
$$= -383.2 \text{ N} = 383.2 \text{ N} \leftarrow$$

The negative sign indicates that F_{Rx} acts to the left, i.e., in the negative x direction as noted by the small arrow. Summing the y components yields

$$+ \uparrow F_{Ry} = \Sigma F_y; \quad F_{Ry} = 250 \cos 45° \text{ N} + 200(\tfrac{3}{5}) \text{ N}$$
$$= 296.8 \text{ N} \uparrow$$

The resultant force, shown in Fig. 2–20c, has a *magnitude* of

$$F_R = \sqrt{(-383.2)^2 + (296.8)^2}$$
$$= 485 \text{ N} \qquad \qquad \textit{Ans.}$$

From the vector addition in Fig. 2–20c, the direction angle θ is

$$\theta = \tan^{-1}\left(\frac{296.8}{383.2}\right) = 37.8° \qquad \textit{Ans.}$$

Realize that the single force $\mathbf{F}_R$ shown in Fig. 2–20c creates the *same effect* on the boom as the three forces in Fig. 2–20a.

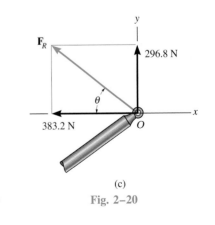

(c)

Fig. 2–20

PROBLEMS

2–33. Determine the x and y components of the 800-N force.

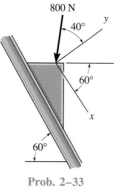

Prob. 2–33

2–34. Determine the magnitude of the resultant force and its direction, measured counterclockwise from the positive x axis.

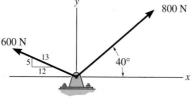

Prob. 2–34

2–35. Determine the magnitude of the resultant force and its direction, measured clockwise from the positive x axis.

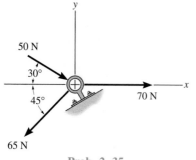

Prob. 2–35

*2–36. Express $\mathbf{F}_1$, $\mathbf{F}_2$, and $\mathbf{F}_3$ as Cartesian vectors.

2–37. Determine the magnitude of the resultant force and its direction, measured counterclockwise from the positive x axis.

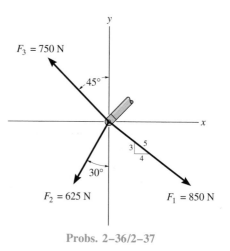

Probs. 2–36/2–37

2–38. Express $\mathbf{F}_1$ and $\mathbf{F}_2$ as Cartesian vectors.

2–39. Determine the magnitude of the resultant force and its direction, measured counterclockwise from the positive x axis.

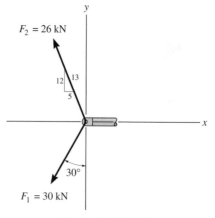

Probs. 2–38/2–39

*2–40. Determine the x and y components of $\mathbf{F}_1$ and $\mathbf{F}_2$.

2–41. Determine the magnitude of the resultant force and its direction, measured counterclockwise from the positive x axis.

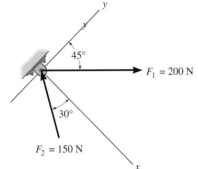

Probs. 2–40/2–41

2–42. Solve Prob. 2–1 by summing the rectangular or x, y components of the forces to obtain the resultant force.

2–43. Solve Prob. 2–2 by summing the rectangular or x, y components of the forces to obtain the resultant force.

*2–44. Solve Prob. 2–3 by summing the rectangular or x, y components of the forces to obtain the resultant force.

2–45. Solve Prob. 2–15 by summing the rectangular or x, y components of the forces to obtain the resultant force.

2–46. Solve Prob. 2–27 by summing the rectangular or x, y components of the forces to obtain the resultant force.

2–47. Determine the x and y components of each force acting on the *gusset plate* of the bridge truss. Show that the resultant force is zero.

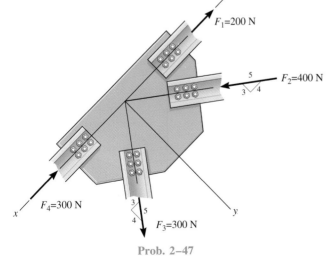

Prob. 2–47

*2–48. If $\theta = 60°$ and $F = 20$ kN, determine the magnitude of the resultant force and its direction measured clockwise from the positive x axis.

2–49. Determine the magnitude F and direction θ of force $\mathbf{F}$ so that the resultant of the three forces acting on the hook is zero.

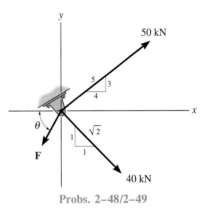

Probs. 2–48/2–49

2–50. Express each of the three forces acting on the column in Cartesian vector form and compute the magnitude of the resultant force.

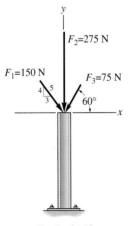

Prob. 2–50

2–51. Three forces act on the bracket. Determine the magnitude and direction θ of $\mathbf{F}_1$ so that the resultant force is directed along the positive x' axis and has a magnitude of 1 kN.

*2–52. If $F_1 = 300$ N and $\theta = 20°$, determine the magnitude and direction, measured counterclockwise from the x' axis, of the resultant force of the three forces acting on the bracket.

2–54. Determine the magnitude and direction θ of $\mathbf{F}_A$ so that the resultant force is directed along the positive x axis and has a magnitude of 1250 N.

2–55. If $F_A = 750$ N and $\theta = 45°$, determine the magnitude and direction, measured counterclockwise from the positive x axis, of the resultant force acting on the ring at O.

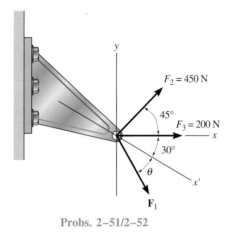

Probs. 2–51/2–52

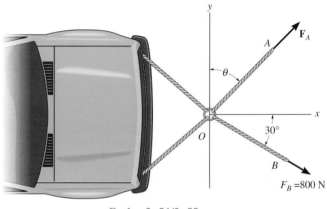

Probs. 2–54/2–55

2–53. Three forces act on the ring. Determine the range of values for the magnitude of $\mathbf{P}$ so that the magnitude of the resultant force does not exceed 2500 N. Force $\mathbf{P}$ is always directed to the right.

*2–56. Three forces act on the bracket. Determine the magnitude and direction θ of $\mathbf{F}_1$ so that the resultant force is directed along the positive x' axis and has a magnitude of 800 N.

2–57. If $F_1 = 300$ N and $\theta = 10°$, determine the magnitude and direction, measured counterclockwise from the positive x' axis, of the resultant force acting on the bracket.

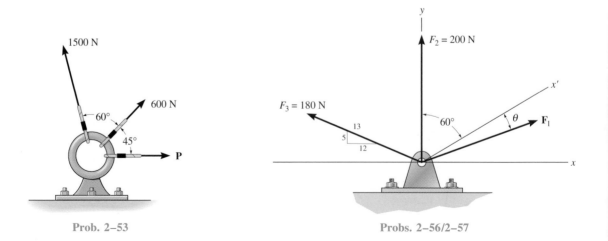

Prob. 2–53   Probs. 2–56/2–57

2–58. Express each of the three forces acting on the bracket in Cartesian vector form with respect to the x and y axes. Determine the magnitude and direction θ of $\mathbf{F}_1$ so that the resultant force is directed along the positive x' axis and has a magnitude of $F_R = 600$ N.

*2–60. Determine the direction θ of the cable and the required tension F_1 so that the resultant force is directed vertically upward and has a magnitude of 800 N.

2–61. Determine the magnitude and direction of the resultant force of the three forces acting on the ring A. Take $F_1 = 500$ N and θ = 20°.

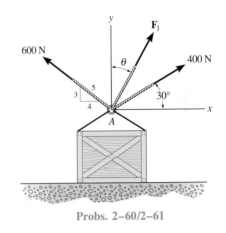

Prob. 2–58

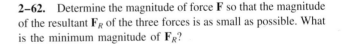

Probs. 2–60/2–61

2–59. The three concurrent forces acting on the post produce a resultant force $\mathbf{F}_R = 0$. If $F_2 = \frac{1}{2}F_1$, and $\mathbf{F}_1$ is to be 90° from $\mathbf{F}_2$ as shown, determine the required magnitude F_3 expressed in terms of F_1 and the angle θ.

2–62. Determine the magnitude of force $\mathbf{F}$ so that the magnitude of the resultant $\mathbf{F}_R$ of the three forces is as small as possible. What is the minimum magnitude of $\mathbf{F}_R$?

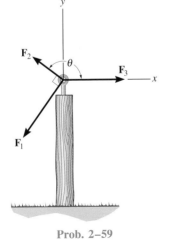

Prob. 2–59

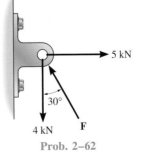

Prob. 2–62

2.5 Cartesian Vectors

The operations of vector algebra, when applied to solving problems in *three dimensions,* are greatly simplified if the vectors are first represented in Cartesian vector form. In this section we will present a general method for doing this, then in Sec. 2.6 we will apply this method to solving problems involving the addition of forces. Similar applications will be illustrated for the position and moment vectors given in later sections of the text.

Right-Handed Coordinate System. A right-handed coordinate system will be used for developing the theory of vector algebra that follows. A rectangular or Cartesian coordinate system is said to be *right-handed* provided the thumb of the right hand points in the direction of the positive z axis when the right-hand fingers are curled about this axis and directed from the positive x toward the positive y axis, Fig. 2–21. Furthermore, according to this rule, the z axis for a two-dimensional problem as in Fig. 2–20 would be directed outward, perpendicular to the page.

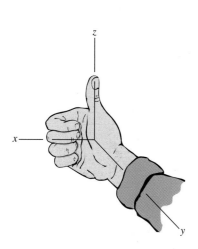

Right-handed coordinate system

Fig. 2–21

Rectangular Components of a Vector. A vector $\mathbf{A}$ may have one, two, or three rectangular components along the x, y, z coordinate axes, depending on how the vector is oriented relative to the axes. In general, though, when $\mathbf{A}$ is directed within an octant of the x, y, z frame, Fig. 2–22, then by two successive applications of the parallelogram law, we may resolve the vector into components as $\mathbf{A} = \mathbf{A}' + \mathbf{A}_z$ and then $\mathbf{A}' = \mathbf{A}_x + \mathbf{A}_y$. Combining these equations, $\mathbf{A}$ is represented by the vector sum of its *three* rectangular components,

$$\mathbf{A} = \mathbf{A}_x + \mathbf{A}_y + \mathbf{A}_z \qquad (2\text{–}2)$$

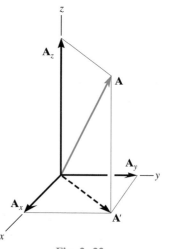

Fig. 2–22

Unit Vector. In general, a *unit vector* is a vector having a magnitude of 1. If **A** is a vector having a magnitude $A \neq 0$, then a unit vector having the *same direction* as **A** is represented by

$$\mathbf{u}_A = \frac{\mathbf{A}}{A} \qquad (2\text{–}3)$$

Rewriting this equation gives

$$\mathbf{A} = A\mathbf{u}_A \qquad (2\text{–}4)$$

Since vector **A** is of a certain type, e.g., a force vector, it is customary to use the proper set of units for its description. The magnitude A also has this same set of units; hence, from Eq. 2–3, the *unit vector will be dimensionless* since the units will cancel out. Equation 2–4 therefore indicates that vector **A** may be expressed in terms of both its magnitude and direction *separately;* i.e., A (a positive scalar) defines the *magnitude* of **A**, and $\mathbf{u}_A$ (a dimensionless vector) defines the *direction* and sense of **A**, Fig. 2–23.

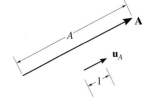

Fig. 2–23

Cartesian Unit Vectors. In three dimensions, the set of Cartesian unit vectors, **i, j, k,** is used to designate the directions of the *x, y, z* axes respectively. As stated in Sec. 2–4, the *sense* (or arrowhead) of these vectors will be described analytically by a plus or minus sign, depending on whether they are pointing along the positive or negative *x, y,* or *z* axis. Thus the positive unit vectors are shown in Fig. 2–24.

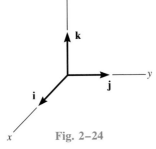

Fig. 2–24

Cartesian Vector Representation. Using Cartesian unit vectors, the three vector components of Eq. 2–2 may be written in "Cartesian vector form." Since the components act in the positive **i, j,** and **k** directions, Fig. 2–25, we have

$$\mathbf{A} = A_x\mathbf{i} + A_y\mathbf{j} + A_z\mathbf{k} \qquad (2\text{–}5)$$

There is a distinct advantage to writing vectors in terms of their Cartesian components. Since each of these components has the same form as Eq. 2–4, the *magnitude* and *direction* of each *component vector* are *separated,* and it will be shown that this will simplify the operations of vector algebra, particularly in three dimensions.

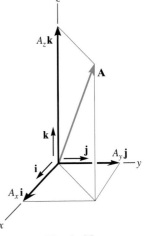

Fig. 2–25

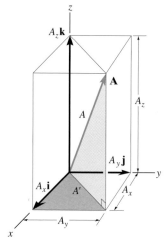

Fig. 2–26

Magnitude of a Cartesian Vector.

Magnitude of a Cartesian Vector. It is always possible to obtain the magnitude of vector **A** provided the vector is expressed in Cartesian vector form. As shown in Fig. 2–26, from the colored right triangle, $A = \sqrt{A'^2 + A_z^2}$, and from the shaded right triangle, $A' = \sqrt{A_x^2 + A_y^2}$. Combining these equations yields

$$A = \sqrt{A_x^2 + A_y^2 + A_z^2} \tag{2–6}$$

*Hence, the magnitude of **A** is equal to the positive square root of the sum of the squares of its components.*

Direction of a Cartesian Vector. The *orientation* of vector **A** is defined by the *coordinate direction angles* α (alpha), β (beta), and γ (gamma), measured between the *tail* of **A** and the *positive x, y, z* axes located at the tail of **A,** Fig. 2–27. Note that regardless of where **A** is directed, each of these angles will be between 0° and 180°. To determine α, β, and γ, consider the projection of **A** onto the *x, y, z* axes, Fig. 2–28. Referring to the colored right triangles shown in each figure, we have

$$\cos\alpha = \frac{A_x}{A} \qquad \cos\beta = \frac{A_y}{A} \qquad \cos\gamma = \frac{A_z}{A} \tag{2–7}$$

These numbers are known as the *direction cosines* of **A.** Once they have been obtained, the coordinate direction angles α, β, γ can then be determined from the inverse cosines.

An easy way of obtaining the direction cosines of **A** is to form a unit vector in the direction of **A,** Eq. 2–3. Provided **A** is expressed in Cartesian

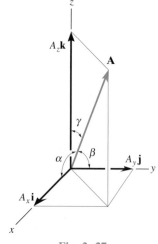

Fig. 2–27

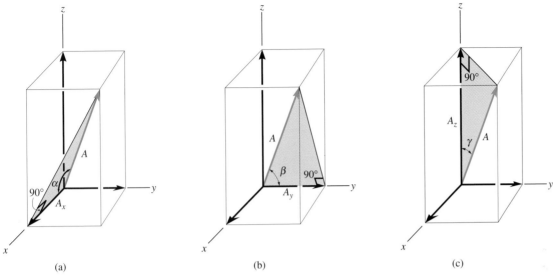

(a)   (b)   (c)

Fig. 2-28

vector form as $\mathbf{A} = A_x\mathbf{i} + A_y\mathbf{j} + A_z\mathbf{k}$ (Eq. 2–5), we have

$$\mathbf{u}_A = \frac{\mathbf{A}}{A} = \frac{A_x}{A}\mathbf{i} + \frac{A_y}{A}\mathbf{j} + \frac{A_z}{A}\mathbf{k} \qquad (2\text{–}8)$$

where $A = \sqrt{(A_x)^2 + (A_y)^2 + (A_z)^2}$ (Eq. 2–6). By comparison with Eqs. 2–7, it is seen that *the* **i, j,** *and* **k** *components of* $\mathbf{u}_A$ *represent the direction cosines of* **A,** i.e.,

$$\mathbf{u}_A = \cos\alpha\mathbf{i} + \cos\beta\mathbf{j} + \cos\gamma\mathbf{k} \qquad (2\text{–}9)$$

Since the magnitude of a vector is equal to the positive square root of the sum of the squares of the magnitudes of its components, and $\mathbf{u}_A$ has a magnitude of 1, then from Eq. 2–9 an important relation between the direction cosines can be formulated as

$$\cos^2\alpha + \cos^2\beta + \cos^2\gamma = 1 \qquad (2\text{–}10)$$

Provided vector **A** lies in a known octant, this equation can be used to determine one of the coordinate direction angles if the other two are known. (See Example 2–10.)

Finally, if the magnitude and coordinate direction angles of **A** are given, **A** may be expressed in Cartesian vector form as

$$\begin{aligned} \mathbf{A} &= A\mathbf{u}_A \\ &= A\cos\alpha\mathbf{i} + A\cos\beta\mathbf{j} + A\cos\gamma\mathbf{k} \qquad (2\text{–}11) \\ &= A_x\mathbf{i} + A_y\mathbf{j} + A_z\mathbf{k} \end{aligned}$$

2.6 Addition and Subtraction of Cartesian Vectors

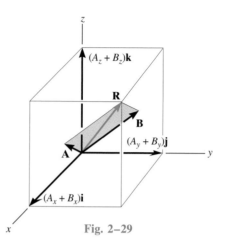

Fig. 2–29

The vector operations of addition and subtraction of two or more vectors are greatly simplified if the vectors are expressed in terms of their Cartesian components. For example, consider the two vectors $\mathbf{A}$ and $\mathbf{B}$, both of which are directed within the positive octant of the x, y, z frame, Fig. 2–29. If $\mathbf{A} = A_x\mathbf{i} + A_y\mathbf{j} + A_z\mathbf{k}$ and $\mathbf{B} = B_x\mathbf{i} + B_y\mathbf{j} + B_z\mathbf{k}$, then the resultant vector, $\mathbf{R}$, has components which represent the scalar sums of the $\mathbf{i}$, $\mathbf{j}$, and $\mathbf{k}$ components of $\mathbf{A}$ and $\mathbf{B}$, i.e.,

$$\mathbf{R} = \mathbf{A} + \mathbf{B} = (A_x + B_x)\mathbf{i} + (A_y + B_y)\mathbf{j} + (A_z + B_z)\mathbf{k}$$

Vector subtraction, being a special case of vector addition, simply requires a scalar subtraction of the respective $\mathbf{i}$, $\mathbf{j}$, and $\mathbf{k}$ components of either $\mathbf{A}$ or $\mathbf{B}$. For example,

$$\mathbf{R}' = \mathbf{A} - \mathbf{B} = (A_x - B_x)\mathbf{i} + (A_y - B_y)\mathbf{j} + (A_z - B_z)\mathbf{k}$$

Concurrent Force Systems. In particular, the above concept of vector addition may be generalized and applied to a system of several concurrent forces. In this case, the force resultant is the vector sum of all the forces in the system and can be written as

$$\mathbf{F}_R = \Sigma\mathbf{F} = \Sigma F_x\mathbf{i} + \Sigma F_y\mathbf{j} + \Sigma F_z\mathbf{k} \qquad (2\text{–}12)$$

Here ΣF_x, ΣF_y, and ΣF_z represent the algebraic sums of the respective x, y, z, or $\mathbf{i}, \mathbf{j}, \mathbf{k}$ components of each force in the system.

The following examples illustrate numerically the methods used to apply the above theory to the solution of problems involving force as a vector quantity.

Example 2–9

Determine the magnitude and the coordinate direction angles of the resultant force acting on the ring in Fig. 2–30a.

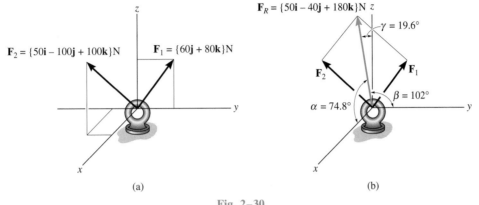

(a)

(b)

Fig. 2–30

SOLUTION

Since each force is represented in Cartesian vector form, the resultant force, shown in Fig. 2–30b, is

$$\mathbf{F}_R = \Sigma \mathbf{F} = \mathbf{F}_1 + \mathbf{F}_2 = \{60\mathbf{j} + 80\mathbf{k}\}\,\text{N} + \{50\mathbf{i} - 100\mathbf{j} + 100\mathbf{k}\}\,\text{N}$$
$$= \{50\mathbf{i} - 40\mathbf{j} + 180\mathbf{k}\}\,\text{N}$$

The magnitude of $\mathbf{F}_R$ is found from Eq. 2–6, i.e.,

$$F_R = \sqrt{(50)^2 + (-40)^2 + (180)^2}$$
$$= 191.0\,\text{N} \qquad\qquad Ans.$$

The coordinate direction angles α, β, γ are determined from the components of the unit vector acting in the direction of $\mathbf{F}_R$.

$$\mathbf{u}_{F_R} = \frac{\mathbf{F}_R}{F_R} = \frac{50}{191.0}\mathbf{i} - \frac{40}{191.0}\mathbf{j} + \frac{180}{191.0}\mathbf{k}$$
$$= 0.2617\mathbf{i} - 0.2094\mathbf{j} + 0.9422\mathbf{k}$$

so that

$$\cos\alpha = 0.2617 \qquad \alpha = 74.8° \qquad\qquad Ans.$$
$$\cos\beta = -0.2094 \qquad \beta = 102° \qquad\qquad Ans.$$
$$\cos\gamma = 0.9422 \qquad \gamma = 19.6° \qquad\qquad Ans.$$

These angles are shown in Fig. 2–30b. In particular, note that $\beta > 90°$ since the $\mathbf{j}$ component of $\mathbf{u}_{F_R}$ is negative.

Example 2–10

Express the force **F** shown in Fig. 2–31 as a Cartesian vector.

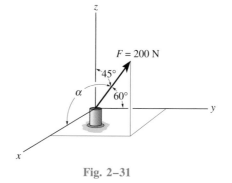

Fig. 2–31

SOLUTION

Since only two coordinate direction angles are specified, the third angle α is determined from Eq. 2–10; i.e.,

$$\cos^2 \alpha + \cos^2 \beta + \cos^2 \gamma = 1$$
$$\cos^2 \alpha + \cos^2 60° + \cos^2 45° = 1$$
$$\cos \alpha = \sqrt{1 - (0.707)^2 - (0.5)^2} = \pm 0.5$$

Hence,

$$\alpha = \cos^{-1}(0.5) = 60° \qquad \text{or} \qquad \alpha = \cos^{-1}(-0.5) = 120°$$

By inspection of Fig. 2–31, however, it is necessary that $\alpha = 60°$, since $\mathbf{F}_x$ is in the $+x$ direction.

Using Eq. 2–11, with $F = 200$ N, we have

$$\mathbf{F} = F \cos \alpha \mathbf{i} + F \cos \beta \mathbf{j} + F \cos \gamma \mathbf{k}$$
$$= 200 \cos 60° \text{ N}\mathbf{i} + 200 \cos 60° \text{ N}\mathbf{j} + 200 \cos 45° \text{ N}\mathbf{k}$$
$$= \{100.0\mathbf{i} + 100.0\mathbf{j} + 141.4\mathbf{k}\} \text{ N} \qquad\qquad Ans.$$

By applying Eq. 2–6, note that indeed the magnitude of $F = 200$ N.

$$F = \sqrt{F_x^2 + F_y^2 + F_z^2}$$
$$= \sqrt{(100.0)^2 + (100.0)^2 + (141.4)^2} = 200 \text{ N}$$

Example 2–11

Express the force **F** shown acting on the hook in Fig. 2–32*a* as a Cartesian vector.

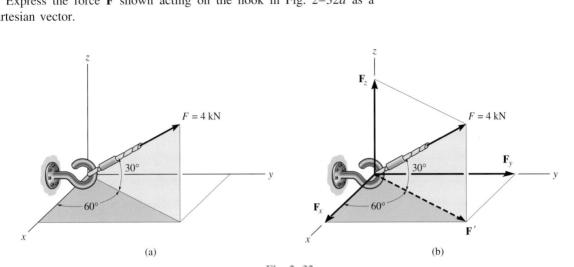

Fig. 2–32

SOLUTION

In this case the angles 60° and 30° defining the direction of **F** are *not* coordinate direction angles. Why? By two successive applications of the parallelogram law, however, **F** can be resolved into its *x, y, z* components as shown in Fig. 2–32*b*. First, from the colored triangle,

$$F' = 4 \cos 30° \text{ kN} = 3.46 \text{ kN}$$
$$F_z = 4 \sin 30° \text{ kN} = 2.00 \text{ kN}$$

Next, using **F'** and the shaded triangle,

$$F_x = 3.46 \cos 60° \text{ kN} = 1.73 \text{ kN}$$
$$F_y = 3.46 \sin 60° \text{ kN} = 3.00 \text{ kN}$$

Thus,

$$\mathbf{F} = \{1.73\mathbf{i} + 3.00\mathbf{j} + 2.00\mathbf{k}\} \text{ kN} \qquad \textit{Ans.}$$

As an exercise, show that the magnitude of **F** is indeed 4 kN, and that the coordinate direction angle $\alpha = 64.3°$.

Example 2–12

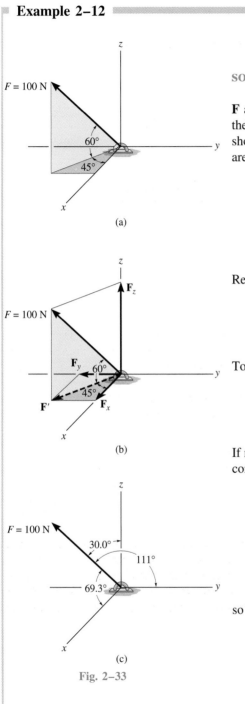

(a)

(b)

(c)

Fig. 2–33

Express force $\mathbf{F}$ shown in Fig. 2–33a as a Cartesian vector.

SOLUTION

As in Example 2–11, the angles of 60° and 45° defining the direction of $\mathbf{F}$ are *not* coordinate direction angles. The two successive applications of the parallelogram law needed to resolve $\mathbf{F}$ into its x, y, z components are shown in Fig. 2–33b. By trigonometry, the magnitudes of the components are

$$F_z = 100 \sin 60° \text{ N} = 86.6 \text{ N}$$
$$F' = 100 \cos 60° \text{ N} = 50 \text{ N}$$
$$F_x = 50 \cos 45° \text{ N} = 35.4 \text{ N}$$
$$F_y = 50 \sin 45° \text{ N} = 35.4 \text{ N}$$

Realizing that $\mathbf{F}_y$ has a direction defined by $-\mathbf{j}$, we have

$$\mathbf{F} = F_x \mathbf{i} + F_y \mathbf{j} + F_z \mathbf{k}$$
$$\mathbf{F} = \{35.4\mathbf{i} - 35.4\mathbf{j} + 86.6\mathbf{k}\} \text{ N} \qquad \textit{Ans.}$$

To show that the magnitude of this vector is indeed 100 N, apply Eq. 2–6,

$$F = \sqrt{F_x^2 + F_y^2 + F_z^2}$$
$$= \sqrt{(35.4)^2 + (-35.4)^2 + (86.6)^2} = 100 \text{ N}$$

If needed, the coordinate direction angles of $\mathbf{F}$ can be determined from the components of the unit vector acting in the direction of $\mathbf{F}$. Hence,

$$\mathbf{u} = \frac{\mathbf{F}}{F} = \frac{F_x}{F}\mathbf{i} + \frac{F_y}{F}\mathbf{j} + \frac{F_z}{F}\mathbf{k}$$

$$= \frac{35.4}{100}\mathbf{i} - \frac{35.4}{100}\mathbf{j} + \frac{86.6}{100}\mathbf{k}$$

$$= 0.354\mathbf{i} - 0.354\mathbf{j} + 0.866\mathbf{k}$$

so that

$$\alpha = \cos^{-1}(0.354) = 69.3°$$
$$\beta = \cos^{-1}(-0.354) = 111°$$
$$\gamma = \cos^{-1}(0.866) = 30.0°$$

These results are shown in Fig. 2–33c.

Example 2–13

Two forces act on the hook shown in Fig. 2–34a. Specify the coordinate direction angles of $\mathbf{F}_2$ so that the resultant force $\mathbf{F}_R$ acts along the positive y axis and has a magnitude of 800 N.

SOLUTION

To solve this problem, the resultant force and its two components, $\mathbf{F}_1$ and $\mathbf{F}_2$, will each be expressed in Cartesian vector form. Then, as shown in Fig. 2–34b, it is necessary that $\mathbf{F}_R = \mathbf{F}_1 + \mathbf{F}_2$.

Applying Eq. 2–11,

$$\mathbf{F}_1 = F_1\mathbf{u}_{F_1} = F_1 \cos \alpha_1 \mathbf{i} + F_1 \cos \beta_1 \mathbf{j} + F_1 \cos \gamma_1 \mathbf{k}$$
$$= 300 \cos 45° \,\mathrm{N}\mathbf{i} + 300 \cos 60° \,\mathrm{N}\mathbf{j} + 300 \cos 120°300 \,\mathrm{N}\mathbf{k}$$
$$= \{212.1\mathbf{i} + 150\mathbf{j} - 150\mathbf{k}\} \,\mathrm{N}$$
$$\mathbf{F}_2 = F_2\mathbf{u}_{F_2} = F_{2x}\mathbf{i} + F_{2y}\mathbf{j} + F_{2z}\mathbf{k}$$

According to the problem statement, the resultant force $\mathbf{F}_R$ has a magnitude of 800 N and acts in the $+\mathbf{j}$ direction. Hence,

$$\mathbf{F}_R = (800 \,\mathrm{N})(+\mathbf{j}) = \{800\mathbf{j}\} \,\mathrm{N}$$

We require

$$\mathbf{F}_R = \mathbf{F}_1 + \mathbf{F}_2$$
$$800\mathbf{j} = 212.1\mathbf{i} + 150\mathbf{j} - 150\mathbf{k} + F_{2x}\mathbf{i} + F_{2y}\mathbf{j} + F_{2z}\mathbf{k}$$
$$800\mathbf{j} = (212.1 + F_{2x})\mathbf{i} + (150 + F_{2y})\mathbf{j} + (-150 + F_{2z})\mathbf{k}$$

To satisfy this equation, the corresponding $\mathbf{i}$, $\mathbf{j}$, and $\mathbf{k}$ components on the left and right sides must be equal. This is equivalent to stating that the x, y, z components of $\mathbf{F}_R$ be equal to the corresponding x, y, z components of $(\mathbf{F}_1 + \mathbf{F}_2)$. Hence,

$$0 = 212.1 + F_{2x} \qquad F_{2x} = -212.1 \,\mathrm{N}$$
$$800 = 150 + F_{2y} \qquad F_{2y} = 650 \,\mathrm{N}$$
$$0 = -150 + F_{2z} \qquad F_{2z} = 150 \,\mathrm{N}$$

Since the magnitudes of $\mathbf{F}_2$ and its components are known, we can use Eq. 2–11 to determine α, β, γ.

$$-212.1 = 700 \cos \alpha_2; \qquad \alpha_2 = \cos^{-1}\left(\frac{-212.1}{700}\right) = 108° \quad Ans.$$

$$650 = 700 \cos \beta_2; \qquad \beta_2 = \cos^{-1}\left(\frac{650}{700}\right) = 21.8° \quad Ans.$$

$$150 = 700 \cos \gamma_2; \qquad \gamma_2 = \cos^{-1}\left(\frac{150}{700}\right) = 77.6° \quad Ans.$$

These results are shown in Fig. 2–34b.

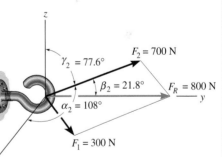

(a)

(b)

Fig. 2–34

PROBLEMS

2–63. The cable at the end of the crane boom exerts a force of $F = 250$ N on the boom as shown. Express **F** as a Cartesian vector.

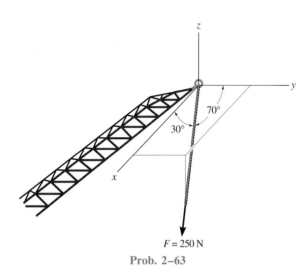

$F = 250$ N

Prob. 2–63

*2–64. The force **F** acting on the stake has a component of 40 N acting in the *x-y* plane as shown. Express **F** as a Cartesian vector.

2–65. Determine the magnitude and coordinate direction angles of the force **F** acting on the stake.

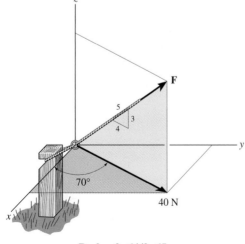

40 N

Probs. 2–64/2–65

2–66. The stock *S* mounted on the lathe is subjected to a force of 60 N, which is caused by a die. Determine the coordinate direction angle β and express the force as a Cartesian vector.

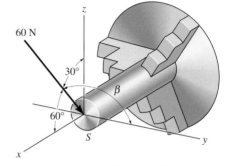

Prob. 2–66

2–67. Express each force as a Cartesian vector and then determine the resultant force $\mathbf{F}_R$. Find the magnitude and coordinate direction angles and sketch this vector on the coordinate system.

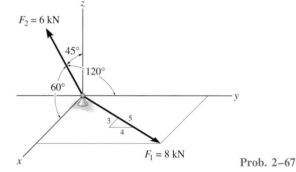

$F_1 = 8$ kN

Prob. 2–67

*2–68. Determine the magnitude and coordinate direction angles of the resultant force and sketch this vector on the coordinate system.

2–69. Specify the coordinate direction angles of $\mathbf{F}_1$ and $\mathbf{F}_2$ and express each force as a Cartesian vector.

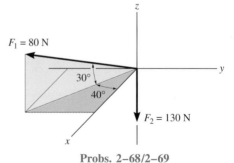

$F_2 = 130$ N

Probs. 2–68/2–69

2–70. Express each force as a Cartesian vector.

2–71. Determine the magnitude and coordinate direction angles of the resultant force and sketch this vector on the coordinate system.

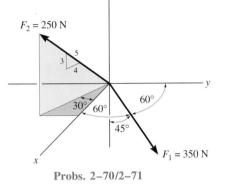

Probs. 2–70/2–71

*2–72. Determine the magnitude and coordinate direction angles of the resultant force.

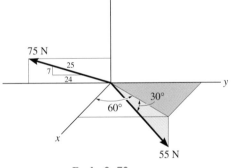

Prob. 2–72

2–73. The beam is subjected to the two forces shown. Express each force in Cartesian vector form and determine the magnitude and coordinate direction angles of the resultant force.

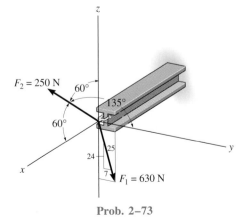

Prob. 2–73

2–74. The mast is subjected to the three forces shown. Determine the coordinate direction angles α_1, β_1, γ_1 of $\mathbf{F}_1$ so that the resultant force acting on the mast is $\mathbf{F}_R = \{350\mathbf{i}\}$ N.

2–75. The mast is subjected to the three forces shown. Determine the coordinate direction angles α_1, β_1, γ_1 of $\mathbf{F}_1$ so that the resultant force acting on the mast is zero.

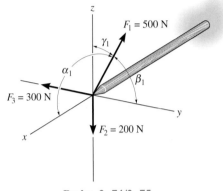

Probs. 2–74/2–75

*2–76. The two forces $\mathbf{F}_1$ and $\mathbf{F}_2$ acting at A have a resultant force of $\mathbf{F}_R = \{-100\mathbf{k}\}$ N. Determine the magnitude and coordinate direction angles of $\mathbf{F}_2$.

2–77. Determine the coordinate direction angles of the force $\mathbf{F}_1$ and indicate them on the figure.

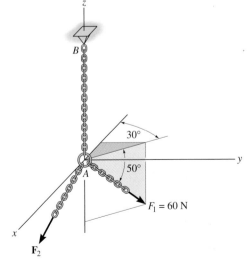

Probs. 2–76/2–77

2–78. The pole is subjected to the force **F**, which has components acting along the *x, y, z* axes as shown. If the magnitude of **F** is 3 kN, and $\beta = 30°$ and $\gamma = 75°$, determine the magnitudes of its three components.

2–79. The pole is subjected to the force **F** which has components $F_x = 1.5$ kN and $F_z = 1.25$ kN. If $\beta = 75°$, determine the magnitudes of **F** and **F**$_y$.

2–81. The bolt is subjected to the force **F**, which has components acting along the *x, y, z* axes as shown. If the magnitude of **F** is 80 N, and $\alpha = 60°$ and $\gamma = 45°$, determine the magnitudes of its components.

2–82. The bolt is subjected to the force **F** which has components $F_x = 20$ N, $F_z = 20$ N. If $\beta = 120°$, determine the magnitudes of **F** and **F**$_y$.

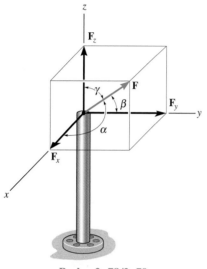

Probs. 2–78/2–79

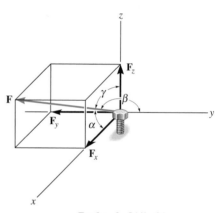

Probs. 2–81/2–82

*2–80. A force **F** is applied at the top of the tower at *A*. If it acts in the direction shown, such that one of its components lying in the shaded *y-z* plane has a magnitude of 80 N, determine the magnitude of **F** and its coordinate direction angles α, β, γ.

2–83. Two forces **F**$_1$ and **F**$_2$ act on the bolt. If the resultant force **F**$_R$ has a magnitude of 50 N and coordinate direction angles $\alpha = 110°$ and $\beta = 80°$, as shown, determine the magnitude of **F**$_2$ and its coordinate direction angles.

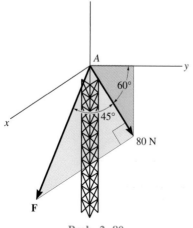

Prob. 2–80

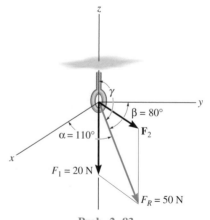

Prob. 2–83

2.7   Position Vectors

In this section we will introduce the concept of a position vector. It will be shown in Sec. 2.8 that this vector is of importance in formulating a Cartesian force vector directed between any two points in space, and later, in Chapter 4, we will use it for finding the moment of a force.

x, y, z **Coordinates.**   Throughout this text we will use a *right-handed* coordinate system to reference the location of points in space. Furthermore, we will use the convention followed in many technical books, and that is to require the positive *z* axis to be directed *upward* (the zenith direction) so that it measures the height of an object or the altitude of a point. The *x, y* axes then lie in the horizontal plane, Fig. 2–35. Points in space are located relative to the origin of coordinates, *O*, by successive measurements along the *x, y, z* axes. For example, in Fig. 2–35 the coordinates of point *A* are obtained by starting at *O* and measuring $x_A = +4$ m along the *x* axis, $y_A = +2$ m along the *y* axis, and $z_A = -6$ m along the *z* axis. Thus, $A(4, 2, -6)$. In a similar manner, measurements along the *x, y, z* axes from *O* to *B* yield the coordinates of *B*, i.e., $B(0, 2, 0)$. Also notice that $C(6, -1, 4)$.

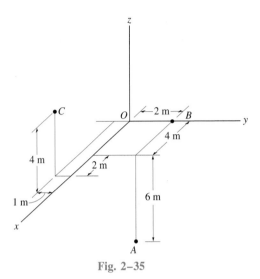

Fig. 2–35

Position Vector.   The *position vector* **r** is defined as a fixed vector which locates a point in space relative to another point. For example, if **r** extends from the origin of coordinates, *O*, to point $P(x, y, z)$, Fig. 2–36a, then **r** can be expressed in Cartesian vector form as

$$\mathbf{r} = x\mathbf{i} + y\mathbf{j} + z\mathbf{k}$$

In particular, note how the head-to-tail vector addition of the three components yields vector **r**, Fig. 2–36b. Starting at the origin *O*, one travels *x* in the $+\mathbf{i}$ direction, then *y* in the $+\mathbf{j}$ direction, and finally *z* in the $+\mathbf{k}$ direction to arrive at point $P(x, y, z)$.

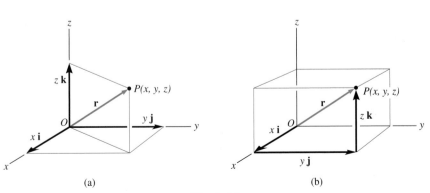

Fig. 2–36

In the more general case, the position vector may be directed from point A to point B in space, Fig. 2–37a. As noted, this vector is also designated by the symbol **r**. As a matter of convention, however, we will *sometimes* refer to this vector with *two subscripts* to indicate from and to the point where it is directed, thus, **r** can also be designated as $\mathbf{r}_{AB}$. Also, note that $\mathbf{r}_A$ and $\mathbf{r}_B$ in Fig. 2–37a are referenced with only one subscript since they extend from the origin of coordinates.

From Fig. 2–37a, by the head-to-tail vector addition, we require

$$\mathbf{r}_A + \mathbf{r} = \mathbf{r}_B$$

Solving for **r** and expressing $\mathbf{r}_A$ and $\mathbf{r}_B$ in Cartesian vector form yields

$$\mathbf{r} = \mathbf{r}_B - \mathbf{r}_A = (x_B\mathbf{i} + y_B\mathbf{j} + z_B\mathbf{k}) - (x_A\mathbf{i} + y_A\mathbf{j} + z_A\mathbf{k})$$

or

$$\mathbf{r} = (x_B - x_A)\mathbf{i} + (y_B - y_A)\mathbf{j} + (z_B - z_A)\mathbf{k} \qquad (2\text{–}13)$$

Thus, the **i, j, k** *components of the position vector* **r** *may be formed by taking the coordinates of the tail of the vector, $A(x_A, y_A, z_A)$, and subtracting them from the corresponding coordinates of the head, $B(x_B, y_B, z_B)$. Again note how the head-to-tail addition of these three components yields* **r,** i.e., going from A to B, Fig. 2–37b, one first travels $(x_B - x_A)$ in the $+\mathbf{i}$ direction, then $(y_B - y_A)$ in the $+\mathbf{j}$ direction, and finally $(z_B - z_A)$ in the $+\mathbf{k}$ direction.

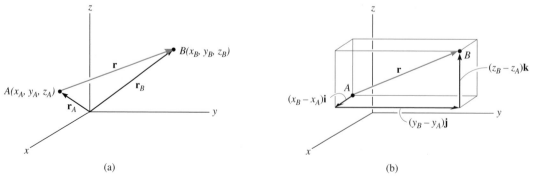

Fig. 2–37

Example 2–14

An elastic rubber band is attached to points A and B as shown in Fig. 2–38a. Determine its length and its direction measured from A toward B.

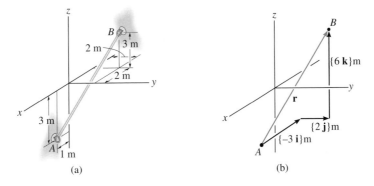

(a)   (b)

SOLUTION

We first establish a position vector from A to B, Fig. 2–38b. In accordance with Eq. 2–13, the coordinates of the tail $A(1 \text{ m}, 0, -3 \text{ m})$ are subtracted from the coordinates of the head $B(-2 \text{ m}, 2 \text{ m}, 3 \text{ m})$, which yields

$$\mathbf{r} = (-2 \text{ m} - 1 \text{ m})\mathbf{i} + (2 \text{ m} - 0)\mathbf{j} + [3 \text{ m} - (-3 \text{ m})]\mathbf{k}$$
$$= \{-3\mathbf{i} + 2\mathbf{j} + 6\mathbf{k}\} \text{ m}$$

As shown, the three components of $\mathbf{r}$ represent the direction and distance one must go along each axis in order to move from A to B, i.e., along the x axis $\{-3\mathbf{i}\}$ m, along the y axis $\{2\mathbf{j}\}$ m, and finally along the z axis $\{6\mathbf{k}\}$ m.

The magnitude of $\mathbf{r}$ represents the length of the rubber band.

$$r = \sqrt{(-3)^2 + (2)^2 + (6)^2} = 7 \text{ m} \qquad \textit{Ans.}$$

Formulating a unit vector in the direction of $\mathbf{r}$, we have

$$\mathbf{u} = \frac{\mathbf{r}}{r} = \frac{-3}{7}\mathbf{i} + \frac{2}{7}\mathbf{j} + \frac{6}{7}\mathbf{k}$$

The components of this unit vector yield the coordinate direction angles

$$\alpha = \cos^{-1}\left(\frac{-3}{7}\right) = 115° \qquad \textit{Ans.}$$

$$\beta = \cos^{-1}\left(\frac{2}{7}\right) = 73.4° \qquad \textit{Ans.}$$

$$\gamma = \cos^{-1}\left(\frac{6}{7}\right) = 31.0° \qquad \textit{Ans.}$$

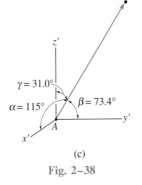

(c)

Fig. 2–38

These angles are measured from the *positive axes* of a localized coordinate system placed at the tail of $\mathbf{r}$, point A, as shown in Fig. 2–38c.

2.8 Force Vector Directed Along a Line

Quite often in three-dimensional statics problems, the direction of a force is specified by two points through which its line of action passes. Such a situation is shown in Fig. 2–39, where the force **F** is directed along the cord *AB*. We can formulate **F** as a Cartesian vector by realizing that it has the *same direction* and *sense* as the position vector **r** directed from point *A* to point *B* on the cord. This common direction is specified by the *unit vector* **u** = **r**/*r*. Hence,

$$\mathbf{F} = F\mathbf{u} = F\left(\frac{\mathbf{r}}{r}\right)$$

Although we have represented **F** symbolically in Fig. 2–39, note that it has units of force, and unlike **r**, or coordinates *x, y, z*, which have units of length, **F** cannot be scaled along the coordinate axes.

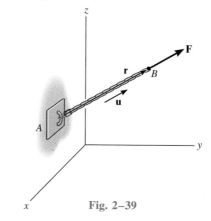

Fig. 2–39

PROCEDURE FOR ANALYSIS

When **F** is directed along a line which extends from point *A* to point *B*, then **F** can be expressed in Cartesian vector form as follows:

Position Vector.   Determine the position vector **r** directed from *A* to *B*, and compute its magnitude *r*.

Unit Vector.   Determine the unit vector **u** = **r**/*r* which defines the *direction* and *sense* of *both* **r** and **F**.

Force Vector.   Determine **F** by combining its magnitude *F* and direction **u**, i.e., **F** = *F***u**.

This procedure is illustrated numerically in the following example problems.

Example 2–15

The man shown in Fig. 2–40a pulls on the cord with a force of 70 N. Represent this force, acting on the support A, as a Cartesian vector and determine its direction.

SOLUTION

Force **F** is shown in Fig. 2–40b. The *direction* of this vector, **u**, is determined from the position vector **r**, which extends from A to B, Fig. 2–40b. To formulate **F** as a Cartesian vector we use the following procedure.

Position Vector.   The coordinates of the end points of the cord are $A(0, 0, 7.5$ m$)$ and $B(3$ m, -2 m, 1.5 m$)$. Forming the position vector by subtracting the corresponding x, y, and z coordinates of A from those of B, we have

$$\mathbf{r} = (3 \text{ m} - 0)\mathbf{i} + (-2 \text{ m} - 0)\mathbf{j} + (1.5 \text{ m} - 7.5 \text{ m})\mathbf{k}$$
$$= \{3\mathbf{i} - 2\mathbf{j} - 6\mathbf{k}\} \text{ m}$$

Show on Fig. 2–40a how one can write **r** *directly* by going from A $\{3\mathbf{i}\}$ m, then $\{-2\mathbf{j}\}$ m, and finally $\{-6\mathbf{k}\}$ m to get to B.

The magnitude of **r**, which represents the *length* of cord AB, is

$$r = \sqrt{(3)^2 + (-2)^2 + (-6)^2} = 7 \text{ m}$$

Unit Vector.   Forming the unit vector that defines the direction and sense of both **r** and **F** yields

$$\mathbf{u} = \frac{\mathbf{r}}{r} = \frac{3}{7}\mathbf{i} - \frac{2}{7}\mathbf{j} - \frac{6}{7}\mathbf{k}$$

Force Vector.   Since **F** has a *magnitude* of 70 N and a *direction* specified by **u**, then

$$\mathbf{F} = F\mathbf{u} = 70 \text{ N} \left(\frac{3}{7}\mathbf{i} - \frac{2}{7}\mathbf{j} - \frac{6}{7}\mathbf{k}\right)$$
$$= \{30\mathbf{i} - 20\mathbf{j} - 60\mathbf{k}\} \text{ N} \qquad\qquad Ans.$$

As shown in Fig. 2–40b, the coordinate direction angles are measured between **r** (or **F**) and the *positive axes* of a localized coordinate system with origin placed at A. From the components of the unit vector:

$$\alpha = \cos^{-1}\left(\frac{3}{7}\right) = 64.6° \qquad\qquad Ans.$$

$$\beta = \cos^{-1}\left(-\frac{2}{7}\right) = 107° \qquad\qquad Ans.$$

$$\gamma = \cos^{-1}\left(-\frac{6}{7}\right) = 149° \qquad\qquad Ans.$$

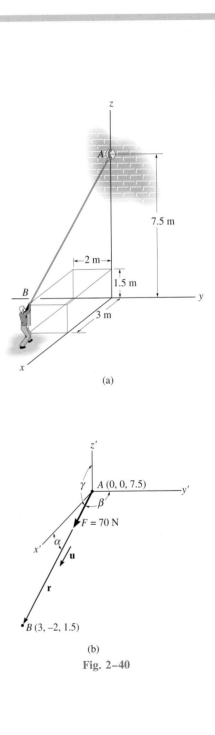

(a)

(b)

Fig. 2–40

Example 2–16

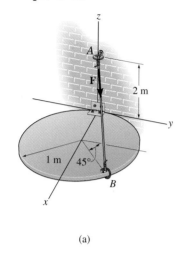

(a)

The circular plate shown in Fig. 2–41a is partially supported by the cable AB. If the force of the cable on the hook at A is $F = 500$ N, express $\mathbf{F}$ as a Cartesian vector.

SOLUTION

As shown in Fig. 2–41b, $\mathbf{F}$ has the same direction and sense as the position vector $\mathbf{r}$, which extends from A to B.

Position Vector.   The coordinates of the end points of the cable are $A(0, 0, 2$ m$)$ and $B(1.707$ m, 0.707 m, $0)$, as indicated in the figure. Thus,

$$\mathbf{r} = (1.707 \text{ m} - 0)\mathbf{i} + (0.707 \text{ m} - 0)\mathbf{j} + (0 - 2 \text{ m})\mathbf{k}$$
$$= \{1.707\mathbf{i} + 0.707\mathbf{j} - 2\mathbf{k}\} \text{ m}$$

Note how one can calculate these components *directly* by going from A, $\{-2\mathbf{k}\}$ m along the z axis, then $\{1.707\mathbf{i}\}$ m along the x axis, and finally $\{0.707\mathbf{j}\}$ m along the y axis to get to B.

The magnitude of $\mathbf{r}$ is

$$r = \sqrt{(1.707)^2 + (0.707)^2 + (-2)^2} = 2.72 \text{ m}$$

Unit Vector.   Thus,

$$\mathbf{u} = \frac{\mathbf{r}}{r} = \frac{1.707}{2.72}\mathbf{i} + \frac{0.707}{2.72}\mathbf{j} - \frac{2}{2.72}\mathbf{k}$$
$$= 0.627\mathbf{i} + 0.260\mathbf{j} - 0.735\mathbf{k}$$

(b)

Fig. 2–41

Force Vector.   Since $F = 500$ N and $\mathbf{F}$ has the direction $\mathbf{u}$, we have

$$\mathbf{F} = F\mathbf{u} = 500 \text{ N}(0.627\mathbf{i} + 0.260\mathbf{j} - 0.735\mathbf{k})$$
$$= \{314\mathbf{i} + 130\mathbf{j} - 368\mathbf{k}\} \text{ N} \qquad \qquad Ans.$$

Using these components, notice that indeed the magnitude of $\mathbf{F}$ is 500 N; i.e.,

$$F = \sqrt{(314)^2 + (130)^2 + (-368)^2} = 500 \text{ N}$$

Show that the coordinate direction angle $\gamma = 137°$, and indicate this angle on the figure.

Example 2–17

The cables exert forces $F_{AB} = 100$ N and $F_{AC} = 120$ N on the ring at A as shown in Fig. 2–42a. Determine the magnitude of the resultant force acting at A.

SOLUTION

The resultant force $\mathbf{F}_R$ is shown graphically in Fig. 2–42b. We can express this force as a Cartesian vector by first formulating $\mathbf{F}_{AB}$ and $\mathbf{F}_{AC}$ as Cartesian vectors and then adding their components. The directions of $\mathbf{F}_{AB}$ and $\mathbf{F}_{AC}$ are specified by forming unit vectors $\mathbf{u}_{AB}$ and $\mathbf{u}_{AC}$ along the cables. These unit vectors are obtained from the associated position vectors $\mathbf{r}_{AB}$ and $\mathbf{r}_{AC}$. With reference to Fig. 2–42b for $\mathbf{F}_{AB}$ we have

$$\mathbf{r}_{AB} = (4\text{ m} - 0)\mathbf{i} + (0 - 0)\mathbf{j} + (0 - 4\text{ m})\mathbf{k}$$
$$= \{4\mathbf{i} - 4\mathbf{k}\}\text{ m}$$
$$r_{AB} = \sqrt{(4)^2 + (-4)^2} = 5.66\text{ m}$$
$$\mathbf{F}_{AB} = 100\text{ N}\left(\frac{\mathbf{r}_{AB}}{r_{AB}}\right) = 100\text{ N}\left(\frac{4}{5.66}\mathbf{i} - \frac{4}{5.66}\mathbf{k}\right)$$
$$\mathbf{F}_{AB} = \{70.7\mathbf{i} - 70.7\mathbf{k}\}\text{ N}$$

For $\mathbf{F}_{AC}$ we have

$$\mathbf{r}_{AC} = (4\text{ m} - 0)\mathbf{i} + (2\text{ m} - 0)\mathbf{j} + (0 - 4\text{ m})\mathbf{k}$$
$$= \{4\mathbf{i} + 2\mathbf{j} - 4\mathbf{k}\}\text{ m}$$
$$r_{AC} = \sqrt{(4)^2 + (2)^2 + (-4)^2} = 6\text{ m}$$
$$\mathbf{F}_{AC} = 120\text{ N}\left(\frac{\mathbf{r}_{AC}}{r_{AC}}\right) = 120\text{ N}\left(\frac{4}{6}\mathbf{i} + \frac{2}{6}\mathbf{j} - \frac{4}{6}\mathbf{k}\right)$$
$$= \{80\mathbf{i} + 40\mathbf{j} - 80\mathbf{k}\}\text{ N}$$

The resultant force is therefore

$$\mathbf{F}_R = \mathbf{F}_{AB} + \mathbf{F}_{AC} = \{70.7\mathbf{i} - 70.7\mathbf{k}\}\text{ N} + \{80\mathbf{i} + 40\mathbf{j} - 80\mathbf{k}\}\text{ N}$$
$$= \{150.7\mathbf{i} + 40\mathbf{j} - 150.7\mathbf{k}\}\text{ N}$$

The magnitude of $\mathbf{F}_R$ is thus

$$F_R = \sqrt{(150.7)^2 + (40)^2 + (-150.7)^2}$$
$$= 217\text{ N} \qquad\qquad Ans.$$

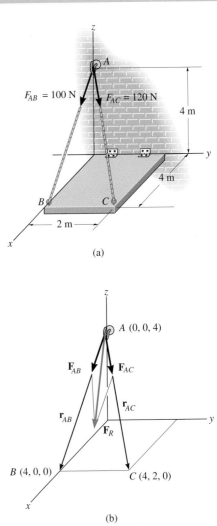

(a)

(b)

Fig. 2–42

PROBLEMS

*2–84.   Express the position vector **r** in Cartesian vector form; then determine its magnitude and coordinate direction angles.

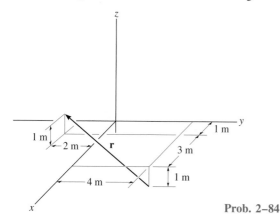

Prob. 2–84

2–85.   Express the position vector **r** in Cartesian vector form; then determine its magnitude and coordinate direction angles.

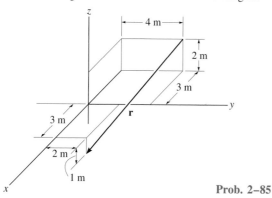

Prob. 2–85

2–86.   Express the position vector **r** in Cartesian vector form; then determine its magnitude and coordinate direction angles.

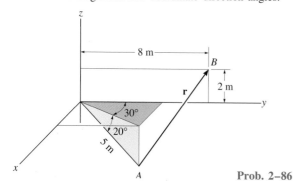

Prob. 2–86

2–87.   Determine the length of member AB of the truss by first establishing a Cartesian position vector from A to B and then determining its magnitude.

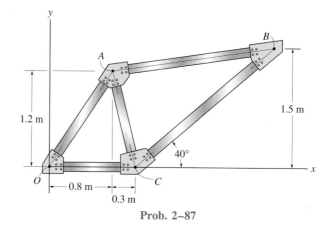

Prob. 2–87

*2–88.   The 8-m-long cable is anchored to the ground at A. If $x = 4$ m and $y = 2$ m, determine the coordinate z to the highest point of attachment along the column.

2–89.   The 8-m-long cable is anchored to the ground at A. If $z = 5$ m, determine the location $+x$, $+y$ of point A. Choose a value such that $x = y$.

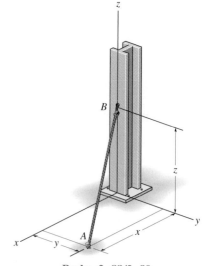

Probs. 2–88/2–89

2–90. Determine the length of the crankshaft *AB* by first formulating a Cartesian position vector from *A* to *B* and then determining its magnitude.

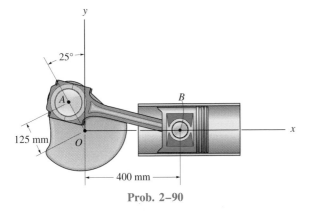

Prob. 2–90

2–91. At the instant shown, position vectors along the robotic arm from *O* to *B* and *B* to *A* are $\mathbf{r}_{OB} = \{100\mathbf{i} + 300\mathbf{j} + 400\mathbf{k}\}$ mm and $\mathbf{r}_{BA} = \{350\mathbf{i} + 225\mathbf{j} - 640\mathbf{k}\}$ mm, respectively. Determine the distance from *O* to the grip at *A*.

*2–92. If $\mathbf{r}_{OA} = \{0.5\mathbf{i} + 4\mathbf{j} + 0.25\mathbf{k}\}$ m and $\mathbf{r}_{OB} = \{0.3\mathbf{i} + 2\mathbf{j} + 2\mathbf{k}\}$ m, express $\mathbf{r}_{BA}$ as a Cartesian vector.

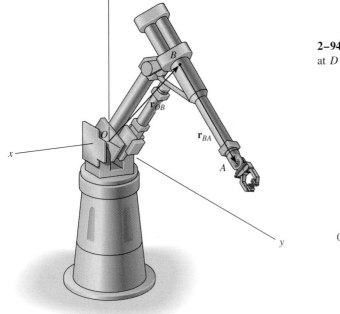

Probs. 2–91/2–92

2–93. At a given instant, the positions of a plane at *A* and a train at *B* are measured relative to a radar antenna at *O*. Determine the distance *d* between *A* and *B* at this instant. To solve the problem, formulate a position vector, directed from *A* to *B*, and then determine its magnitude.

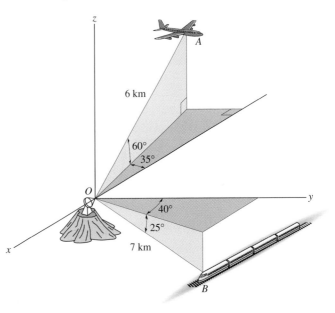

Prob. 2–93

2–94. Determine the lengths of wires *AD*, *BD*, and *CD*. The ring at *D* is midway between *A* and *B*.

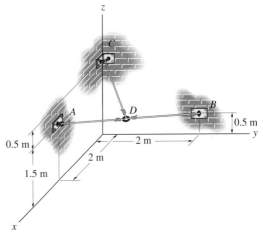

Prob. 2–94

2–95. Express force **F** as a Cartesian vector, then determine its coordinate direction angles.

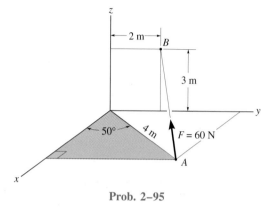

Prob. 2–95

2–97. Express each of the two forces in Cartesian vector form.

2–98. Determine the magnitude and coordinate direction angles of the resultant force acting at point A.

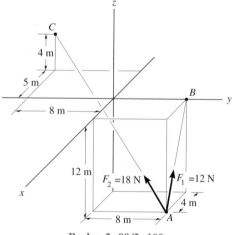

Probs. 2–97/2–98

*2–96. Express force **F** as a Cartesian vector; then determine its coordinate direction angles.

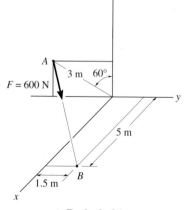

Prob. 2–96

2–99. Express each of the two forces in Cartesian vector form.

*2–100. Determine the magnitude and coordinate direction angles of the resultant force acting at point A.

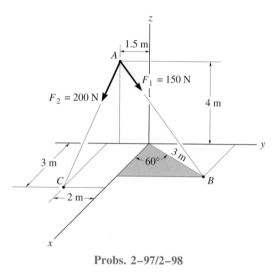

Probs. 2–99/2–100

2–101. The cord exerts a force of $\mathbf{F} = \{12\mathbf{i} + 9\mathbf{j} - 8\mathbf{k}\}$ N on the hook. If the cord is 8 m long, determine the location x, y of the point of attachment B, and the height z of the hook.

2–102. The cord exerts a force of $F = 30$ N on the hook. If the cord is 8 m long, $z = 4$ m, and the x component of the force is $F_x = 25$ N, determine the location x, y of the point of attachment B of the cord to the ground.

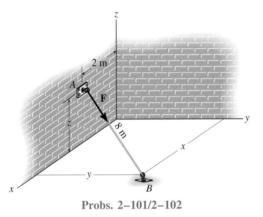

Probs. 2–101/2–102

2–103. The force $\mathbf{F}$ has a magnitude of 80 N and acts at the midpoint C of the thin rod. Express the force as a Cartesian vector.

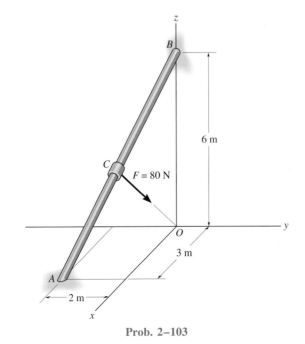

Prob. 2–103

*2–104. The window is held open by chain AB. Determine the length of the chain, and express the 50-N force acting at A along the chain as a Cartesian vector. Determine its coordinate direction angles.

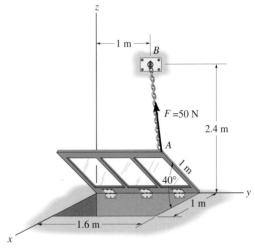

Prob. 2–104

2–105. The cable attached to the tractor at B exerts a force of 350 N on the framework. Express this force as a Cartesian vector.

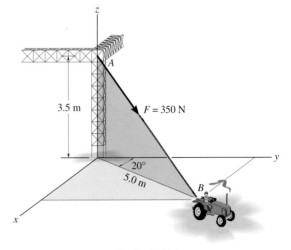

Prob. 2–105

2–106. Express force **F** as a Cartesian vector; then determine its coordinate direction angles.

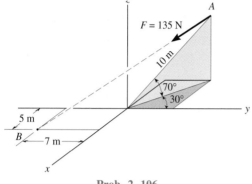

Prob. 2–106

2–107. Express each of the forces in Cartesian vector form and determine the magnitude and coordinate direction angles of the resultant force.

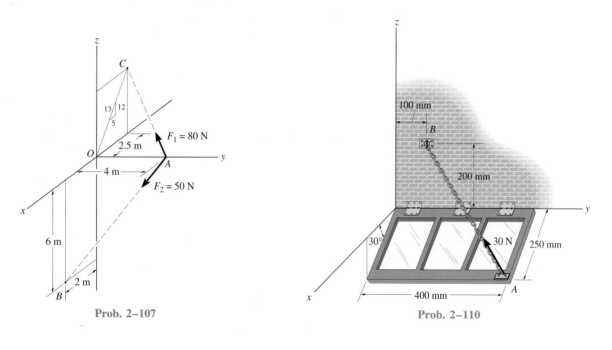

Prob. 2–107

*****2–108.** The three supporting cables exert the forces shown on the sign. Represent each force as a Cartesian vector.

2–109. Determine the magnitude and coordinate direction angles of the resultant force of the two forces acting on the sign at point A.

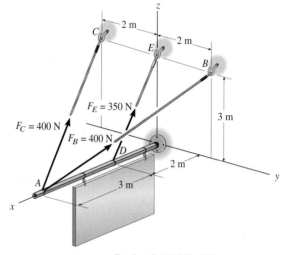

Probs. 2–108/2–109

2–110. The window is held open by chain AB. Determine the length of the chain and express the 30-N force acting at A along the chain as a Cartesian vector.

Prob. 2–110

2–111. Each of the four forces acting at E has a magnitude of 28 kN. Express each force as a Cartesian vector and determine the resultant force.

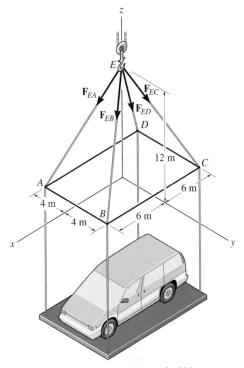

Prob. 2–111

*2–112. Express force F in Cartesian vector form if its acts at the midpoint B of the rod.

2–113. Express force F in Cartesian vector form if point B is located 3 m along the rod from end C.

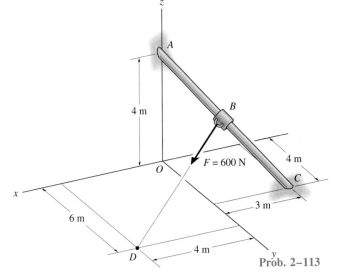

Prob. 2–113

2–114. The tower is held in place by three cables. If the force of each cable acting on the tower is shown, determine the position (x, y) for fixing cable DA so that the resultant force exerted on the tower is directed along its axis, from D toward O.

2–115. The tower is held in place by three cables. If the force of each cable acting on the tower is shown, determine the magnitude and coordinate direction angles α, β, γ of the resultant force. Take $x = 20$ m, $y = 15$ m.

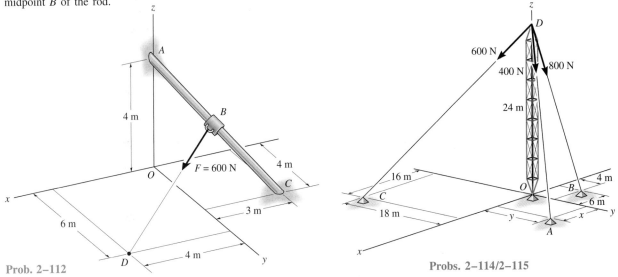

Prob. 2–112

Probs. 2–114/2–115

2.9 Dot Product

Occasionally in statics one has to find the angle between two lines or the components of a force parallel and perpendicular to a line. In two dimensions, these problems can readily be solved by trigonometry since the geometry is easy to visualize. In three dimensions, however, this is often difficult, and consequently vector methods should be employed for the solution. The dot product defines a particular method for "multiplying" two vectors and is used to solve the above-mentioned problems.

The *dot product* of vectors **A** and **B**, written **A** · **B**, and read "**A** dot **B**," is defined as the product of the magnitudes of **A** and **B** and the cosine of the angle θ between their tails, Fig. 2–43. Expressed in equation form,

$$\mathbf{A} \cdot \mathbf{B} = AB \cos \theta \qquad\qquad (2\text{–}14)$$

where $0° \le \theta \le 180°$. The dot product is often referred to as the *scalar product* of vectors, since the result is a *scalar* and not a vector.

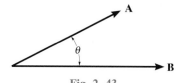

Fig. 2–43

Laws of Operation

1. Commutative law:

$$\mathbf{A} \cdot \mathbf{B} = \mathbf{B} \cdot \mathbf{A}$$

2. Multiplication by a scalar:

$$a(\mathbf{A} \cdot \mathbf{B}) = (a\mathbf{A}) \cdot \mathbf{B} = \mathbf{A} \cdot (a\mathbf{B}) = (\mathbf{A} \cdot \mathbf{B})a$$

3. Distributive law:

$$\mathbf{A} \cdot (\mathbf{B} + \mathbf{D}) = (\mathbf{A} \cdot \mathbf{B}) + (\mathbf{A} \cdot \mathbf{D})$$

It is easy to prove the first and second laws by using Eq. 2–14. The proof of the distributive law is left as an exercise (see Prob. 2–116).

Cartesian Vector Formulation. Equation 2–14 may be used to find the dot product for each of the Cartesian unit vectors. For example, $\mathbf{i} \cdot \mathbf{i} = (1)(1) \cos 0° = 1$ and $\mathbf{i} \cdot \mathbf{j} = (1)(1) \cos 90° = 0$. In a similar manner,

$$\mathbf{i} \cdot \mathbf{i} = 1 \qquad \mathbf{j} \cdot \mathbf{j} = 1 \qquad \mathbf{k} \cdot \mathbf{k} = 1$$
$$\mathbf{i} \cdot \mathbf{j} = 0 \qquad \mathbf{i} \cdot \mathbf{k} = 0 \qquad \mathbf{k} \cdot \mathbf{j} = 0$$

These results should not be memorized; rather, it should be clearly understood how each is obtained.

Consider now the dot product of two general vectors $\mathbf{A}$ and $\mathbf{B}$ which are expressed in Cartesian vector form. We have

$$\mathbf{A} \cdot \mathbf{B} = (A_x\mathbf{i} + A_y\mathbf{j} + A_z\mathbf{k}) \cdot (B_x\mathbf{i} + B_y\mathbf{j} + B_z\mathbf{k})$$
$$= A_xB_x(\mathbf{i} \cdot \mathbf{i}) + A_xB_y(\mathbf{i} \cdot \mathbf{j}) + A_xB_z(\mathbf{i} \cdot \mathbf{k})$$
$$+ A_yB_x(\mathbf{j} \cdot \mathbf{i}) + A_yB_y(\mathbf{j} \cdot \mathbf{j}) + A_yB_z(\mathbf{j} \cdot \mathbf{k})$$
$$+ A_zB_x(\mathbf{k} \cdot \mathbf{i}) + A_zB_y(\mathbf{k} \cdot \mathbf{j}) + A_zB_z(\mathbf{k} \cdot \mathbf{k})$$

Carrying out the dot-product operations, the final result becomes

$$\mathbf{A} \cdot \mathbf{B} = A_xB_x + A_yB_y + A_zB_z \qquad (2\text{–}15)$$

Thus, to determine the dot product of two Cartesian vectors, multiply their corresponding x, y, z components and sum their products algebraically. Since the result is a scalar, be careful *not* to include any unit vectors in the final result.

Applications. The dot product has two important applications in mechanics.

1. *The angle formed between two vectors or intersecting lines.* The angle θ between the tails of vectors $\mathbf{A}$ and $\mathbf{B}$ in Fig. 2–43 can be determined from Eq. 2–14 and written as

$$\theta = \cos^{-1} \left(\frac{\mathbf{A} \cdot \mathbf{B}}{AB} \right) \qquad 0° \le \theta \le 180°$$

Here $\mathbf{A} \cdot \mathbf{B}$ is computed from Eq. 2–15. In particular, notice that if $\mathbf{A} \cdot \mathbf{B} = 0$, $\theta = \cos^{-1} 0 = 90°$, so that $\mathbf{A}$ will be *perpendicular* to $\mathbf{B}$.

2. *The components of a vector parallel and perpendicular to a line.* The component of vector **A** parallel to or collinear with the line *aa'* in Fig. 2–44 is defined by **A**$_\parallel$, where $A_\parallel = A \cos \theta$. This component is sometimes referred to as the *projection* of **A** onto the line, since a right angle is formed in the construction. If the *direction* of the line is specified by the unit vector **u,** then, since $u = 1$, we can determine $A_\parallel$ directly from the dot product (Eq. 2–14); i.e.,

$$A_\parallel = A \cos \theta = \mathbf{A} \cdot \mathbf{u}$$

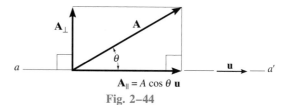

$$\mathbf{A}_\parallel = A \cos \theta \, \mathbf{u}$$

Fig. 2–44

Hence, the scalar projection of **A** *along a line is determined from the dot product of* **A** *and the unit vector* **u** *which defines the direction of the line.* Notice that if this result is positive, then **A**$_\parallel$ has a directional sense which is the same as **u,** whereas if $A_\parallel$ is a negative scalar, then **A**$_\parallel$ has the opposite sense of direction to **u.** The component **A**$_\parallel$ represented as a *vector* is therefore

$$\mathbf{A}_\parallel = A \cos \theta \, \mathbf{u} = (\mathbf{A} \cdot \mathbf{u})\mathbf{u}$$

Note that the component of **A** which is *perpendicular* to line *aa'* can also be obtained, Fig. 2–44. Since $\mathbf{A} = \mathbf{A}_\parallel + \mathbf{A}_\perp$, then $\mathbf{A}_\perp = \mathbf{A} - \mathbf{A}_\parallel$. There are two possible ways of obtaining $A_\perp$. The first would be to determine θ from the dot product, $\theta = \cos^{-1}(\mathbf{A} \cdot \mathbf{u}/A)$, then $A_\perp = A \sin \theta$. Alternatively, if $A_\parallel$ is known, then by the Pythagorean theorem we can also write $A_\perp = \sqrt{A^2 - A_\parallel^2}$.

The above two applications are illustrated numerically in the following example problems.

Example 2–18

The frame shown in Fig. 2–45a is subjected to a horizontal force **F** = {300**j**} N acting at its corner. Determine the magnitude of the components of this force parallel and perpendicular to member *AB*.

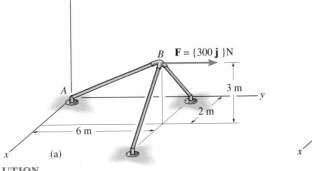

(a)

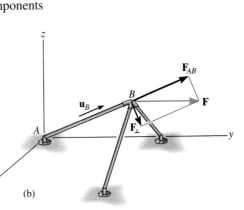

(b)

Fig. 2–45

SOLUTION

The magnitude of the component of **F** along *AB* is equal to the dot product of **F** and the unit vector **u**$_B$ which defines the direction of *AB*, Fig. 2–45b. Since

$$\mathbf{u}_B = \frac{\mathbf{r}_B}{r_B} = \frac{2\mathbf{i} + 6\mathbf{j} + 3\mathbf{k}}{\sqrt{(2)^2 + (6)^2 + (3)^2}} = 0.286\mathbf{i} + 0.857\mathbf{j} + 0.429\mathbf{k}$$

Then

$$F_{AB} = F \cos \theta = \mathbf{F} \cdot \mathbf{u}_B = (300\mathbf{j}) \cdot (0.286\mathbf{i} + 0.857\mathbf{j} + 0.429\mathbf{k})$$
$$= (0)(0.286) + (300)(0.857) + (0)(0.429)$$
$$= 257.1 \text{ N} \qquad\qquad\qquad Ans.$$

Since the result is a positive scalar, **F**$_{AB}$ has the same sense of direction as **u**$_B$, Fig. 2–45b.

Expressing **F**$_{AB}$ in Cartesian vector form, we have

$$\mathbf{F}_{AB} = F_{AB}\mathbf{u}_B = 257.1 \text{ N}(0.286\mathbf{i} + 0.857\mathbf{j} + 0.429\mathbf{k})$$
$$= \{73.5\mathbf{i} + 220\mathbf{j} + 110\mathbf{k}\} \text{ N} \qquad Ans.$$

The perpendicular component, Fig. 2–45b, is therefore

$$\mathbf{F}_\perp = \mathbf{F} - \mathbf{F}_{AB} = 300\mathbf{j} - (73.5\mathbf{i} + 220\mathbf{j} + 110\mathbf{k})$$
$$= \{-73.5\mathbf{i} + 80\mathbf{j} - 110\mathbf{k}\} \text{ N}$$

Its magnitude can be determined either from this vector or from the Pythagorean theorem, Fig. 2–45b:

$$F_\perp = \sqrt{F^2 - F_{AB}^2}$$
$$= \sqrt{(300)^2 - (257.1)^2}$$
$$= 155 \text{ N} \qquad\qquad\qquad Ans.$$

Example 2–19

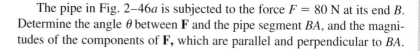

The pipe in Fig. 2–46a is subjected to the force $F = 80$ N at its end B. Determine the angle θ between $\mathbf{F}$ and the pipe segment BA, and the magnitudes of the components of $\mathbf{F}$, which are parallel and perpendicular to BA.

SOLUTION

Angle θ.   First we will establish position vectors from B to A and B to C, then we will determine the angle θ between the tails of these two vectors.

$$\mathbf{r}_{BA} = \{-2\mathbf{i} - 2\mathbf{j} + 1\mathbf{k}\} \text{ m}$$
$$\mathbf{r}_{BC} = \{-3\mathbf{j} + 1\mathbf{k}\} \text{ m}$$

Thus,

$$\cos \theta = \frac{\mathbf{r}_{BA} \cdot \mathbf{r}_{BC}}{r_{BA} r_{BC}} = \frac{(-2)(0) + (-2)(-3) + (1)(1)}{3\sqrt{10}}$$
$$= 0.7379$$
$$\theta = 42.5° \qquad \qquad Ans.$$

Components of $\mathbf{F}$.   The force $\mathbf{F}$ is resolved into components as shown in Fig. 2–46b. Since $F_{BA} = \mathbf{F} \cdot \mathbf{u}_{BA}$, we must first formulate the unit vector along BA and force $\mathbf{F}$ as Cartesian vectors.

$$\mathbf{u}_{BA} = \frac{\mathbf{r}_{BA}}{r_{BA}} = \frac{(-2\mathbf{i} - 2\mathbf{j} + 1\mathbf{k})}{3} = -\frac{2}{3}\mathbf{i} - \frac{2}{3}\mathbf{j} + \frac{1}{3}\mathbf{k}$$

$$\mathbf{F} = 80 \text{ N}\left(\frac{\mathbf{r}_{BC}}{r_{BC}}\right) = 80\left(\frac{-3\mathbf{j} + 1\mathbf{k}}{\sqrt{10}}\right) = -75.89\mathbf{j} + 25.30\mathbf{k}$$

Thus,

$$F_{BA} = \mathbf{F} \cdot \mathbf{u}_{BA} = (-75.89\mathbf{j} + 25.30\mathbf{k}) \cdot \left(-\frac{2}{3}\mathbf{i} - \frac{2}{3}\mathbf{j} + \frac{1}{3}\mathbf{k}\right)$$
$$= 0 + 50.60 + 8.43$$
$$= 59.0 \text{ N} \qquad \qquad Ans.$$

Since θ was calculated in Fig. 2–46b, this same result can also be obtained directly from trigonometry.

$$F_{BA} = 80 \cos 42.5° \text{ N} = 59.0 \text{ N} \qquad \qquad Ans.$$

The perpendicular component can be obtained by trigonometry,

$$F_{\perp} = F \sin \theta$$
$$= 80 \sin 42.5° \text{ N}$$
$$= 54.0 \text{ N} \qquad \qquad Ans.$$

Or, by the Pythagorean theorem,

$$F_{\perp} = \sqrt{F^2 - F_{BA}^2} = \sqrt{(80)^2 - (59.0)^2}$$
$$= 54.0 \text{ N} \qquad \qquad Ans.$$

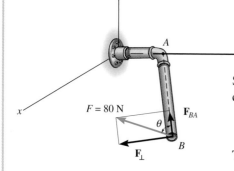

(a)

(b)

Fig. 2–46

PROBLEMS

*2–116. Given the three vectors **A**, **B**, and **D**, show that $\mathbf{A} \cdot (\mathbf{B} + \mathbf{D}) = (\mathbf{A} \cdot \mathbf{B}) + (\mathbf{A} \cdot \mathbf{D})$.

2–117. Determine the angle θ between the tails of the two vectors.

2–118. Determine the magnitude of the projection of $\mathbf{r}_1$ along $\mathbf{r}_2$, and the projected component of $\mathbf{r}_2$ along $\mathbf{r}_1$.

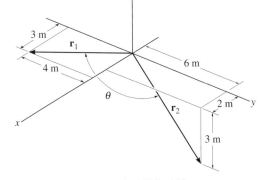

Probs. 2–117/2–118

2–119. Determine the angle θ between the tails of the two vectors.

*2–120. Determine the magnitude of the projected component of $\mathbf{r}_1$ along $\mathbf{r}_2$, and the projection of $\mathbf{r}_2$ along $\mathbf{r}_1$.

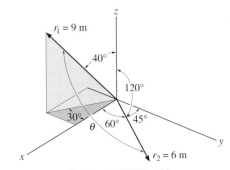

Probs. 2–119/2–120

2–121. Determine the two components of the force **F** along the lines Oa and Ob such that $\mathbf{F} = \mathbf{F}_A + \mathbf{F}_B$. Also find the projected component of **F** along Oa and Ob. Show graphically how the components and projections are constructed.

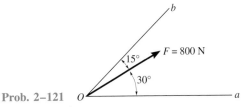

Prob. 2–121

2–122. Determine the angle θ between the edges of the sheet-metal bracket.

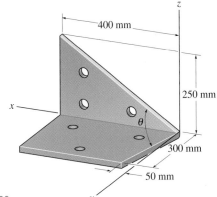

Prob. 2–122

2–123. Determine the magnitude of the projected component of the position vector **r** along the Oa axis.

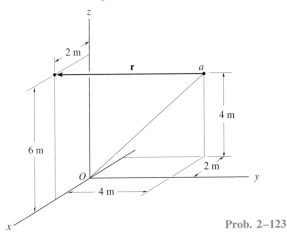

Prob. 2–123

*2–124.   Determine the projected component of the 80-N force acting along the axis AB of the pipe.

2–125.   Determine the angle θ between pipe segments BA and BC.

2–127.   The clamp is used on a jig. If the vertical force acting on the bolt is $\mathbf{F} = \{-500\mathbf{k}\}$ N, determine the magnitudes of the components $\mathbf{F}_1$ and $\mathbf{F}_2$ which act along the OA axis and perpendicular to it.

*2–128.   The clamp is used on a jig. Determine the angle θ between the line of action of $\mathbf{F}$ and the clamp axis OA.

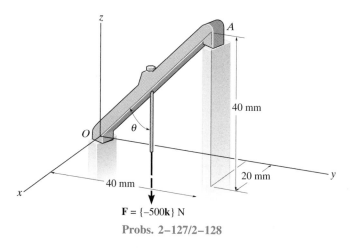

Probs. 2–127/2–128

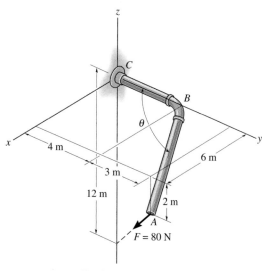

Probs. 2–124/2–125

2–129.   The cables each exert a force of 400 N on the post. Determine the magnitude of the projected component of $\mathbf{F}_1$ along the line of action of $\mathbf{F}_2$.

2–130.   Determine the angle θ between the two cables.

2–126.   The force $\mathbf{F}$ acts at the end A of the pipe assembly. Determine the magnitudes of the components $\mathbf{F}_1$ and $\mathbf{F}_2$ which act along the axis of AB and perpendicular to it.

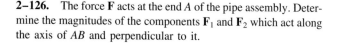

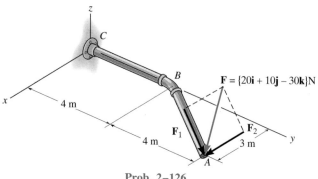

Prob. 2–126

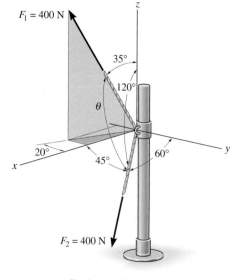

Probs. 2–129/2–130

2–131. Determine the components of **F** that act along rod AC and perpendicular to it. Point B is located at the midpoint of the rod.

*2–132. Determine the components of **F** that act along rod AC and perpendicular to it. Point B is located 3 m along the rod from end C.

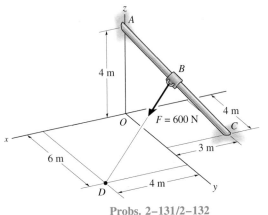

Probs. 2–131/2–132

2–133. Determine the angles θ and ϕ made between the axes OA of the flag pole and AB and AC, respectively, of each cable.

2–134. The two supporting cables exert the forces shown on the flag pole. Determine the projected component of each force acting along the axis OA of the pole.

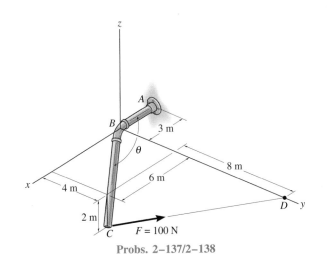

Probs. 2–133/2–134

2–135. Determine the angle θ cable OA makes with beam OC.

*2–136. Determine the angle ϕ cable OA makes with beam OD.

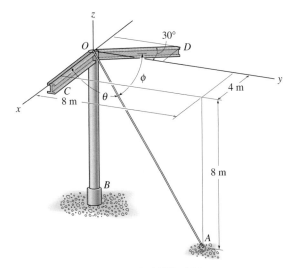

Probs. 2–135/2–136

2–137. Determine the magnitude of the projected component of the 100-N force acting along the axis BC of the pipe.

2–138. Determine the angle θ between pipe segments BA and BC.

Probs. 2–137/2–138

REVIEW PROBLEMS

2–139. The boat is to be pulled onto the shore using two ropes. Determine the magnitudes of forces **T** and **P** acting in each rope in order to develop a resultant force of 80 N, directed along the *aa* keel as shown. Take $\theta = 40°$.

*2–140. The boat is to be pulled onto the shore using two ropes. If the resultant force is to be 80 N, directed along the keel *aa*, as shown, determine the magnitudes of forces **T** and **P** acting in each rope and the angle θ of **P** so that the magnitude of **P** is a *minimum*. **T** acts at 30° from the keel as shown.

2–142. Determine the magnitude and coordinate direction angles of **F**₃ so that the resultant of the three forces acts along the positive *y* axis and has a magnitude of 600 N.

2–143. Determine the magnitude and coordinate direction angles of **F**₃ so that the resultant of the three forces is zero.

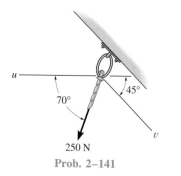

Probs. 2–139/2–140

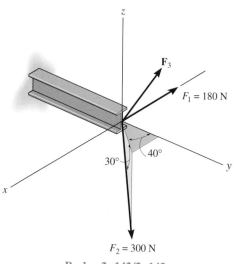

Probs. 2–142/2–143

2–141. Determine the components of the 250-N force acting along the *u* and *v* axes.

*2–144. Two forces **F**₁ and **F**₂ act on the hook. If their lines of action are at an angle θ apart and the magnitude of each force is $F_1 = F_2 = F$, determine the magnitude of the resultant force **F**_R and the angle between **F**_R and **F**₁.

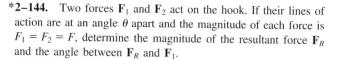

Prob. 2–141

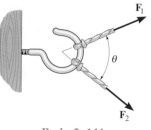

Prob. 2–144

2–145. Express $\mathbf{F}_1$ and $\mathbf{F}_2$ as Cartesian vectors.

2–146. Determine the magnitude of the resultant force and its direction, measured counterclockwise from the positive x axis.

*2–148. Determine the magnitudes of the projected components of the force $\mathbf{F} = \{60\mathbf{i} + 12\mathbf{j} - 40\mathbf{k}\}$ N in the direction of the cables AB and AC.

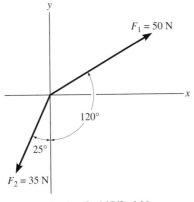

Probs. 2–145/2–146

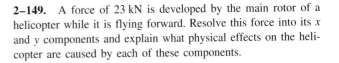

Prob. 2–148

2–147. Determine the angles θ and ϕ between the wire segments.

2–149. A force of 23 kN is developed by the main rotor of a helicopter while it is flying forward. Resolve this force into its x and y components and explain what physical effects on the helicopter are caused by each of these components.

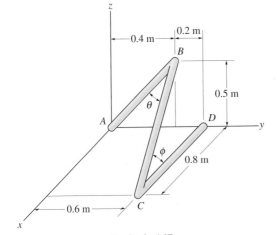

Prob. 2–147

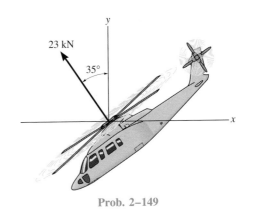

Prob. 2–149

Whenever cables are used for hoisting structural members they must be selected so that they do not fail when they are placed at their points of attachment. In this chapter we will show how to calculate cable loadings for such cases.

3

Equilibrium of a Particle

In this chapter the methods for resolving a force into components and expressing a force as a Cartesian vector will be used to solve problems involving the equilibrium of a particle. To simplify the discussion, particle equilibrium for a concurrent coplanar force system will be considered first. Then, in the last part of the chapter, equilibrium problems involving concurrent three-dimensional force systems will be considered.

3.1 Condition for the Equilibrium of a Particle

A particle is in *equilibrium* provided it is at rest if originally at rest, or has a constant velocity if originally in motion. Most often, however, the term "equilibrium" or, more specifically, "static equilibrium" is used to describe an object at rest. To maintain a state of equilibrium, it is *necessary* to satisfy Newton's first law of motion, which states that if the *resultant force* acting on a particle is *zero,* then the particle is in equilibrium. This condition may be stated mathematically as

$$\Sigma \mathbf{F} = \mathbf{0} \qquad (3\text{--}1)$$

where $\Sigma \mathbf{F}$ is the vector *sum of all the forces* acting on the particle.

Not only is Eq. 3–1 a necessary condition for equilibrium, it is also a *sufficient* condition. This follows from Newton's second law of motion, which can be written as $\Sigma \mathbf{F} = m\mathbf{a}.$ Since the force system satisfies Eq. 3–1, then $m\mathbf{a} = \mathbf{0},$ and therefore the particle's acceleration $\mathbf{a} = \mathbf{0},$ and consequently the particle indeed moves with constant velocity or remains at rest.

3.2 The Free-Body Diagram

To apply the equation of equilibrium correctly, we must account for *all* the known and unknown forces (ΣF) which act *on* the particle. The best way to do this is to draw the particle's *free-body diagram.* This diagram is a sketch of the particle which represents it as being isolated or "free" from its surroundings. On this sketch it is necessary to show *all* the forces that act *on* the particle. Once this diagram is drawn, it will then be easy to apply Eq. 3–1.

Before presenting a formal procedure as to how to draw a free-body diagram, we will first discuss two types of connections often encountered in particle equilibrium problems.

Springs. If a *linear elastic spring* is used for support, the length of the spring will change in direct proportion to the force acting on it. A characteristic that defines the "elasticity" of a spring is the *spring constant* or *stiffness k.* Specifically, the magnitude of force developed by a linear elastic spring which has a stiffness k, and is deformed (elongated or compressed) a distance s measured from its *unloaded* position, is

$$F = ks \qquad\qquad (3\text{–}2)$$

Note that s is determined from the difference in the spring's deformed length l and its undeformed length l_o, i.e., $s = l - l_o$. Thus, if s is positive, F "pulls" on the spring; whereas if s is negative, F must "push" on it. For example, the spring shown in Fig. 3–1 has an undeformed length $l_o = 0.4$ m and stiffness $k = 500$ N/m. To stretch it so that $l = 0.6$ m, a force $F = ks = (500 \text{ N/m})(0.6 \text{ m} - 0.4 \text{ m}) = 100$ N is needed. Likewise, to compress it to a length $l = 0.2$ m, a force $F = ks = (500 \text{ N/m})(0.2 \text{ m} - 0.4 \text{ m}) = -100$ N is required, Fig. 3–1.

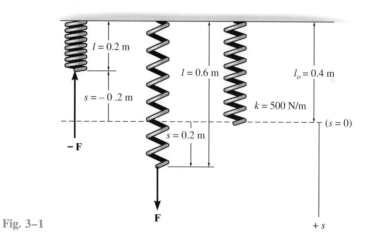

Fig. 3–1

Cables and Pulleys.   Throughout this book, except in Sec. 7.4, all cables (or cords) are assumed to have negligible weight and they cannot be stretched. A cable can support *only* a tension or "pulling" force, and this force always acts in the direction of the cable. In Chapter 5 it will be shown that the tension force developed in a *continuous cable* which passes over a frictionless pulley must have a *constant* magnitude to keep the cable in equilibrium. Hence, for any angle θ, shown in Fig. 3–2, the cable is subjected to a constant tension T throughout its length.

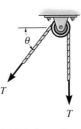

Cable is in tension

Fig. 3–2

PROCEDURE FOR DRAWING A FREE-BODY DIAGRAM

Since we must account for all the forces acting on the particle, the importance of drawing a free-body diagram before applying the equation of equilibrium to the solution of a problem cannot be overemphasized. To construct a free-body diagram, the following three steps are necessary.

Step 1.   Imagine the particle to be *isolated* or cut "free" from its surroundings. Hence the name "free-body" diagram. Draw or sketch its outlined shape.

Step 2.   Indicate on this sketch *all* the forces that act *on the particle.* These forces can be *active forces,* which tend to set the particle in motion, such as those caused by attached cables, weight, or magnetic and electrostatic interaction. Also, *reactive forces* will occur, such as those caused by the constraints or supports that tend to prevent motion. To account for all these forces, it may help to trace around the particle's boundary, carefully noting each force acting on it.

Step 3.   The forces that are *known* should be labeled with their proper magnitudes and directions. Letters are used to represent the magnitudes and directions of forces that are unknown. In particular, if a force has a known line of action but unknown magnitude, the "arrowhead," which defines the sense of the force, can be *assumed.* The correct sense will become apparent after solving for the unknown magnitude. By definition, the *magnitude* of a force is *always positive* so that, if the solution yields a "negative" scalar, the *minus sign* indicates that the arrowhead or sense of the force is opposite to that which was originally assumed.

Application of the above steps is illustrated in the following two examples.

Example 3–1

The crate in Fig. 3–3a has a weight of 20 N. Draw a free-body diagram of the crate, the cord *BD*, and the ring at *B*.

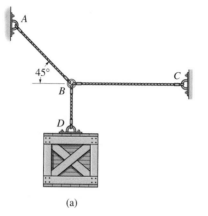

(a)

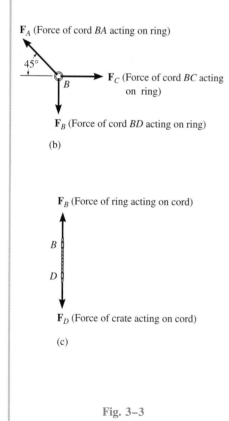

$\mathbf{F}_A$ (Force of cord *BA* acting on ring)

45°

$\mathbf{F}_C$ (Force of cord *BC* acting on ring)

B

$\mathbf{F}_B$ (Force of cord *BD* acting on ring)

(b)

$\mathbf{F}_B$ (Force of ring acting on cord)

B

D

$\mathbf{F}_D$ (Force of crate acting on cord)

(c)

SOLUTION

If we imagine the crate to be *isolated from its surroundings,* then by inspection there are only two forces acting on it, namely, the gravitational force or weight of 20 N, and the force of the cord *BD*. Thus the free-body diagram is shown in Fig. 3–3d.

If the cord *BD* is isolated from its surroundings, then there are only two forces acting on it, Fig. 3–3c, namely, the force of the crate, $\mathbf{F}_D$, and the force $\mathbf{F}_B$ caused by the ring. Since these forces tend to pull on the cord, we can state that the cord is in *tension.* (This must be the case, since compressive, or pushing, forces would cause the cord to collapse.)

When the ring at *B* is isolated from its surroundings, it should be noted that three forces act on it. All these forces are caused by the attached cords, Fig. 3–3b. Notice that $\mathbf{F}_B$ shown here is equal but opposite to that shown in Fig. 3–3c, a consequence of Newton's third law.

$\mathbf{F}_D$ (Force of cord acting on crate)

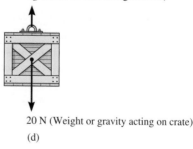

20 N (Weight or gravity acting on crate)

(d)

Fig. 3–3

Example 3–2

The sphere in Fig. 3–4*a* has a mass of 6 kg and is supported as shown. Draw a free-body diagram of the sphere and the knot at *C*.

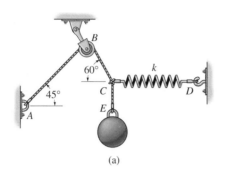

(a)

SOLUTION

There are two forces acting on the sphere, namely, its weight and the force **P** of the cord *CE*. The sphere has a weight of $(6 \text{ kg})(9.81 \text{ m/s}^2) = 58.9 \text{ N}$. Its free-body diagram is shown in Fig. 3–4*b*.

By inspection, three forces act on the knot at *C* when it is isolated. They are caused by cords *CBA* and *CE*, and the spring *CD*. Thus, the free-body diagram of the knot is shown in Fig. 3–4*c*.

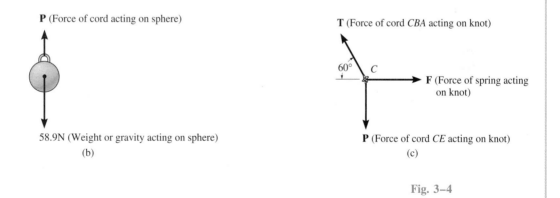

P (Force of cord acting on sphere)

58.9N (Weight or gravity acting on sphere)

(b)

T (Force of cord *CBA* acting on knot)

F (Force of spring acting on knot)

P (Force of cord *CE* acting on knot)

(c)

Fig. 3–4

3.3 Coplanar Force Systems

Many particle equilibrium problems involve a coplanar force system, Fig. 3–5. If the forces lie in the x–y plane, they can each be resolved into their

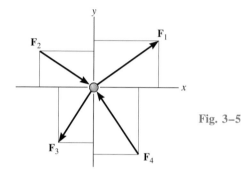

Fig. 3–5

respective $\mathbf{i}$ and $\mathbf{j}$ components and Eq. 3–1 can be written as

$$\Sigma \mathbf{F} = \mathbf{0}$$
$$\Sigma F_x \mathbf{i} + \Sigma F_y \mathbf{j} = \mathbf{0}$$

For this vector equation to be satisfied, both the x and y components must equal zero, otherwise $\Sigma \mathbf{F} \neq \mathbf{0}$. Hence, we require

$$\Sigma F_x = 0$$
$$\Sigma F_y = 0$$

(3–3)

These scalar equilibrium equations state that the algebraic sum of the x and y components of all the forces acting on the particle be equal to zero. As a result, Eqs. 3–3 can be solved for at most two unknowns, generally represented as angles and magnitudes of forces shown on the particle's free-body diagram.

Scalar Notation. Since each of the two equilibrium equations requires the resolution of vector components along a specified axis (x or y), we will use scalar notation to represent the components when applying these equations. By doing this, the sense of direction for each component is accounted for by an *algebraic sign* which corresponds to the arrowhead direction of the component as indicated graphically on the free-body diagram. In particular, if a force component has an unknown magnitude, then the arrowhead sense of the force on the free-body diagram can be *assumed*. Since the magnitude of a force is *always positive,* then if the *solution* yields a *negative scalar,* it indicates that the sense of the force as shown on the free-body diagram is opposite to that which was assumed.

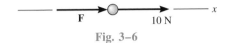

Fig. 3–6

For example, consider the free-body diagram of the particle subjected to the two forces as shown in Fig. 3–6. For the sake of discussion, we have *assumed* that the *unknown force* **F** acts to the right to maintain equilibrium. Application of the equation of equilibrium along the x axis yields

$$\overset{+}{\rightarrow}\Sigma F_x = 0; \qquad\qquad +F + 10\,\text{N} = 0$$

Both terms are "positive" since both forces act in the positive x direction, as indicated by the arrow placed alongside the equation. When this equation is solved, $F = -10$ N. Here the *negative sign* refers to the fact that **F** in Fig. 3–6 is shown in the opposite sense of the actual direction. In other words, **F** must act to the left to hold the particle in equilibrium. Notice that if the $+x$ axis in Fig. 3–6 was directed to the left both terms in the above equation would be negative, but again $F = -10$ N, indicating F would be directed to the left.

PROCEDURE FOR ANALYSIS

The following procedure provides a method for solving coplanar force problems involving particle equilibrium:

Free-Body Diagram.   Draw a free-body diagram of the particle. As outlined in Sec. 3.2, this requires that all the known and unknown force magnitudes and angles be labeled on the diagram. The sense of a force having an unknown magnitude can be assumed.

Equations of Equilibrium.   Establish the x, y axes in *any* suitable direction and apply the two equations of equilibrium, $\Sigma F_x = 0$, and $\Sigma F_y = 0$. For application, components are positive if they are directed along the positive axes, and negative if they are directed along the negative axes. If more than two unknowns exist and the problem involves a spring, apply $F = ks$ (Eq. 3–2) to relate the spring force to the deformation s of the spring.

The following example problems illustrate this solution procedure numerically.

Example 3–3

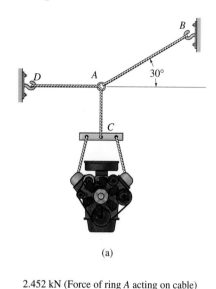

(a)

2.452 kN (Force of ring *A* acting on cable)

A

Equilibrium

C

2.452 kN (Force of engine support acting on cable)

(b)

2.452 kN (Force of cable acting on engine spreader bar)

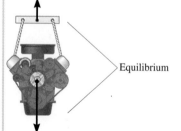

Equilibrium

2.452 kN (Weight or gravity acting on engine)

(c)

Fig. 3–7

Determine the tension in cables *AB* and *AD* for equilibrium of the 250-kg engine shown in Fig. 3–7*a*.

SOLUTION

Free-Body Diagram. To solve this problem we will investigate the equilibrium of the ring at *A*, because this "particle" is subjected to the forces of both cables *AB* and *AD*. First, however, note from the free-body diagram of the engine, Fig. 3–7*c*, that its weight $(250 \text{ kg})(9.81 \text{ m/s}^2) = 2.452 \text{ kN}$ is balanced by the 2.452-kN force of cable *CA* on the spreader bar, a consequence of equilibrium. Furthermore, by Newton's third law, the spreader bar exerts an equal but opposite force of 2.452 kN on the cable at *C*, Fig. 3–7*b*. And again, for equilibrium, the force of the ring at *A* on the cable must be 2.452 kN. Finally, as shown in Fig. 3–7*d*, there are three concurrent forces *acting on the ring*. The forces $\mathbf{T}_B$ and $\mathbf{T}_D$ have unknown magnitudes but known directions, and cable *AC* exerts a downward force on *A* equal to 2.452 kN. Why?

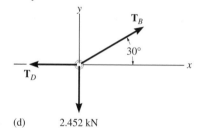

(d)   2.452 kN

Equations of Equilibrium. The two unknown magnitudes T_B and T_D can be obtained from the two scalar equations of equilibrium, $\Sigma F_x = 0$ and $\Sigma F_y = 0$. To apply these equations, the *x*, *y* axes are established on the free-body diagram and $\mathbf{T}_B$ is resolved into its dashed *x* and *y* components. Thus,

$$\xrightarrow{+} \Sigma F_x = 0; \qquad \qquad T_B \cos 30° - T_D = 0 \qquad (1)$$
$$+ \uparrow \Sigma F_y = 0; \qquad \qquad T_B \sin 30° - 2.452 \text{ kN} = 0 \qquad (2)$$

Solving Eq. (2) for T_B and substituting into Eq. (1) to obtain T_D yields

$$T_B = 4.91 \text{ kN} \qquad \qquad Ans.$$
$$T_D = 4.25 \text{ kN} \qquad \qquad Ans.$$

The accuracy of these results, of course, depends on the accuracy of the data, i.e., measurements of geometry and loads. For most engineering work involving a problem such as this, the data as measured to three significant figures would be sufficient. Also, note that here we have neglected the weights of the cables, a reasonable assumption since they would be small in comparison with the weight of the engine.

Example 3–4

If the sack at A in Fig. 3–8a has a weight of 20 N, determine the weight of the sack at B and the force in each cord needed to hold the system in the equilibrium position shown.

SOLUTION

Since the weight of A is known, the unknown tension in the two cords EG and EC can be determined by investigating the equilibrium of the ring at E. Why?

Free-Body Diagram.   There are three forces acting on E, as shown in Fig. 3–8b.

Equations of Equilibrium.   Establishing the x, y axes and resolving each force into its x and y components using trigonometry, we have

$$\xrightarrow{+}\Sigma F_x = 0; \qquad T_{EG}\sin 30° - T_{EC}\cos 45° = 0 \qquad (1)$$

$$+\uparrow\Sigma F_y = 0; \quad T_{EG}\cos 30° - T_{EC}\sin 45° - 20\text{ N} = 0 \qquad (2)$$

Solving Eq. 1 for T_{EG} in terms of T_{EC} and substituting the result into Eq. 2 allows a solution for T_{EC}. One then obtains T_{EG} from Eq. 1. The results are

$$T_{EC} = 38.6\text{ N} \qquad\qquad Ans.$$

$$T_{EG} = 54.6\text{ N} \qquad\qquad Ans.$$

Using the calculated result for T_{EC}, the equilibrium of the ring at C can now be investigated to determine the tension in CD and the weight of B.

Free-Body Diagram.   As shown in Fig. 3–8c, $T_{EC} = 38.6$ N "pulls" on C. The reason for this becomes clear when one draws the free-body diagram of cord CE and applies both equilibrium and the principle of action, equal but opposite force reaction (Newton's third law), Fig. 3–8d.

Equations of Equilibrium.   Establishing the x, y axes and noting the components of $\mathbf{T}_{CD}$ are proportional to the slope of the cord as defined by the 3–4–5 triangle, we have

$$\xrightarrow{+}\Sigma F_x = 0; \qquad 38.6\cos 45°\text{ N} - (\tfrac{4}{5})T_{CD} = 0 \qquad (3)$$

$$+\uparrow\Sigma F_y = 0; \qquad (\tfrac{3}{5})T_{CD} + 38.6\sin 45°\text{ N} - W_B = 0 \qquad (4)$$

Solving Eq. 3 and substituting the result into Eq. 4 yields

$$T_{CD} = 34.2\text{ N} \qquad\qquad Ans.$$

$$W_B = 47.8\text{ N} \qquad\qquad Ans.$$

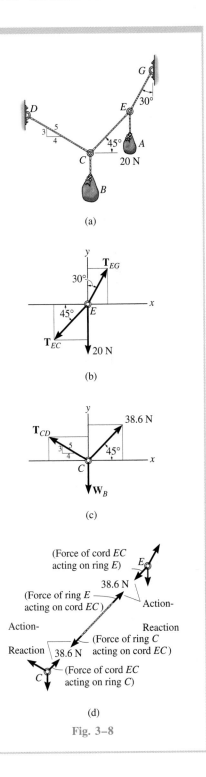

(a)

(b)

(c)

(Force of cord EC acting on ring E)

(Force of ring E acting on cord EC)

Action–

Reaction

Action–

Reaction

38.6 N

(Force of ring C acting on cord EC)

38.6 N

(Force of cord EC acting on ring C)

(d)

Fig. 3–8

Example 3–5

Determine the required length of cord AC in Fig. 3–9a so that the 8-kg lamp is suspended in the position shown. The *undeformed* length of the spring AB is $l'_{AB} = 0.4$ m, and the spring has a stiffness of $k_{AB} = 300$ N/m.

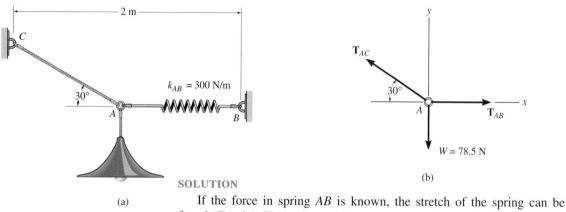

Fig. 3–9

(a)

(b)

SOLUTION

If the force in spring AB is known, the stretch of the spring can be found ($F = ks$). Using the problem geometry, it is then possible to calculate the required length of AC.

Free-Body Diagram. The lamp has a weight $W = 8(9.81) = 78.5$ N. The free-body diagram of the ring at A is shown in Fig. 3–9b.

Equations of Equilibrium Using the x, y axes,

$$\xrightarrow{+}\Sigma F_x = 0; \qquad\qquad T_{AB} - T_{AC} \cos 30° = 0$$
$$+\uparrow\Sigma F_y = 0; \qquad\qquad T_{AC} \sin 30° - 78.5 \text{ N} = 0$$

Solving, we obtain

$$T_{AC} = 157.0 \text{ N}$$
$$T_{AB} = 136.0 \text{ N}$$

The stretch of spring AB is therefore

$$T_{AB} = k_{AB} s_{AB}; \qquad 136.0 \text{ N} = 300 \text{ N/m}(s_{AB})$$
$$s_{AB} = 0.453 \text{ m}$$

so the stretched length is

$$l_{AB} = l'_{AB} + s_{AB}$$
$$l_{AB} = 0.4 \text{ m} + 0.453 \text{ m} = 0.853 \text{ m}$$

The horizontal distance from C to B, Fig. 3–9a, requires

$$2 \text{ m} = l_{AC} \cos 30° + 0.853 \text{ m}$$
$$l_{AC} = 1.32 \text{ m} \qquad\qquad\qquad Ans.$$

PROBLEMS

3–1. Determine the magnitudes of F_1 and F_2 so that the particle is in equilibrium.

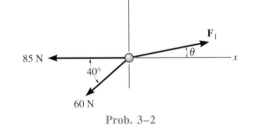

Prob. 3–1

3–2. Determine the magnitude and direction θ of F_1 so that the particle is in equilibrium.

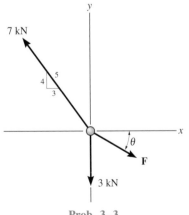

Prob. 3–2

3–3. Determine the magnitude and direction θ of F so that the particle is in equilibrium.

*3–4. Determine the magnitude and angle θ of F so that the particle is in equilibrium.

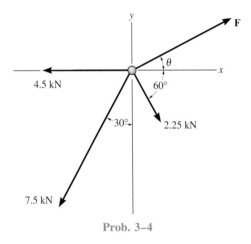

Prob. 3–4

3–5. The members of a truss are pin-connected at joint O. Determine the magnitudes of F_1 and F_2 for equilibrium. Set $\theta = 60°$.

3–6. The members of a truss are pin-connected at joint O. Determine the magnitude of F_1 and its angle θ for equilibrium. Set $F_2 = 6$ kN.

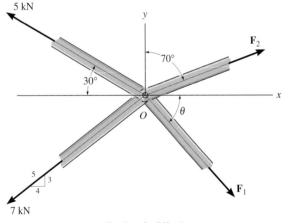

Prob. 3–3

Probs. 3–5/3–6

3–7. The sling is used to support a drum having a weight of 900 N. Determine the force in cords AB and AC for equilibrium. Take $\theta = 20°$.

*3–8. Cords AB and AC can each sustain a maximum tension of 800 N. If the drum has a weight of 900 N, determine the smallest angle θ at which they can be attached to the drum.

3–10. The spring ABC has a stiffness of 500 N/m and an unstretched length of 6 m. Determine the horizontal force **F** applied to the cord which is attached to the *small* pulley B so that the displacement of the pulley from the wall is $d = 1.5$ m.

■3–11. The spring ABC has a stiffness of 500 N/m and an unstretched length of 6 m. Determine the displacement d of the cord from the wall when a force $F = 175$ N is applied to the cord.

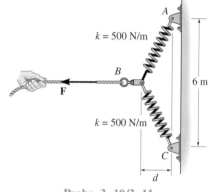

Probs. 3–10/3–11

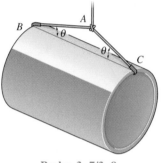

Probs. 3–7/3–8

*3–12. Determine the stretch in each spring for equilibrium of the 2-kg block. The springs are shown in the equilibrium position.

3–13. The unstretched length of spring AB is 2 m. If the block is held in the equilibrium position shown, determine the mass of the block at D.

3–9. Two electrically charged pith balls, each having a mass of 0.2 g, are suspended from light threads of equal length. Determine the resultant horizontal force of repulsion, F, acting on each ball if the measured distance between them is $r = 200$ mm.

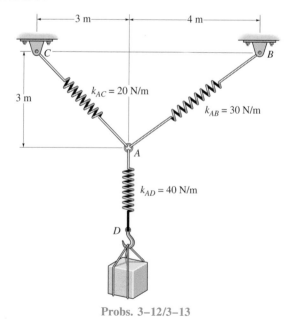

Probs. 3–12/3–13

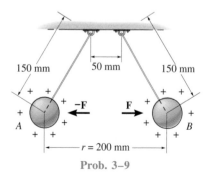

Prob. 3–9

3–14. Determine the stiffness k_T of the single spring such that the force **F** will stretch it by the same amount s as the force **F** stretches the two springs. Express k_T in terms of the stiffness k_1 and k_2 of the two springs.

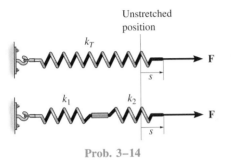

Prob. 3–14

3–15. The motor at B winds up the cord attached to the 65-N crate with a constant speed. Determine the force in cord CD supporting the pulley and the angle θ for equilibrium. Neglect the size of the pulley at C.

*3–16. The cords BCA and CD can each support a maximum load of 100 N. Determine the maximum weight of the crate that can be hoisted at constant velocity, and the angle θ for equilibrium.

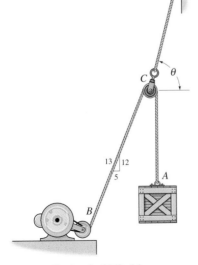

Probs. 3–15/3–16

3–17. Determine the mass that must be supported at A and the angle θ of the connecting cord in order to hold the system in equilibrium.

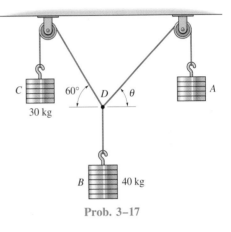

Prob. 3–17

3–18. The 500-N crate is hoisted using the ropes AB and AC. Each rope can withstand a maximum tension of 2500 N before it breaks. If AB always remains horizontal, determine the smallest angle θ to which the crate can be hoisted.

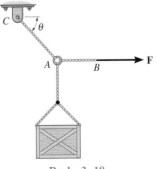

Prob. 3–18

3–19. The street-lights at A and B are suspended from the two poles as shown. If each light has a weight of 50 N, determine the tension in each of the three supporting cables and the required height h of the pole DE so that cable AB is horizontal.

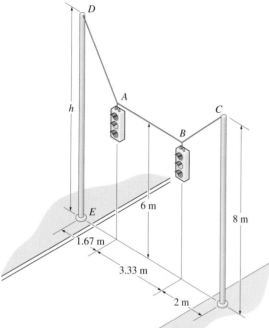

Prob. 3–19

*3–20. A nuclear-reactor vessel has a weight of $500(10^3)$ N. Determine the horizontal compressive force that the spreader bar AB exerts on point A and the force that each cable segment CA and AD exert on this point while the vessel is hoisted upward at constant velocity.

3–21. The block has a weight of 20 N and is being hoisted at uniform velocity. Determine the angle θ for equilibrium and the required force in each cord.

3–22. Determine the maximum weight W of the block that can be suspended in the position shown if each cord can support a maximum tension of 80 N. Also, what is the angle θ for equilibrium?

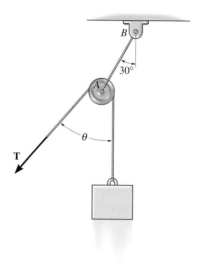

Probs. 3–21/3–22

3–23. The pipe is held in place by the vice. If the bolt exerts a force of 50 N on the pipe in the direction shown, determine the forces F_A and F_B that the smooth contacts at A and B exert on the pipe.

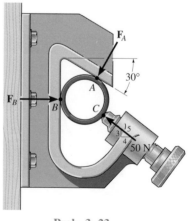

Prob. 3–20

Prob. 3–23

*3–24.   Blocks *D* and *F* weigh 5 N each and block *E* weighs 8 N.
Determine the sag *s* for equilibrium. Neglect the size of the pulleys.

3–25.   If blocks *D* and *F* weigh 5 N each, determine the weight of block *E* if the sag *s* = 3 m. Neglect the size of the pulleys.

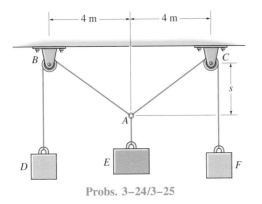

Probs. 3–24/3–25

■3–26.   A vertical force *P* = 10 N is applied to the ends of the 2-m cord *AB* and spring *AC*. If the spring has an unstretched length of 2 m, determine the angle *θ* for equilibrium. Take *k* = 15 N/m.

3–27.   Determine the unstretched length of spring *AC* if a force *P* = 80 N causes the angle *θ* = 60° for equilibrium. Cord *AB* is 2 m long. Take *k* = 50 N/m.

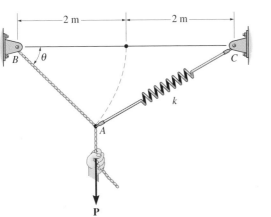

Probs. 3–26/3–27

*■3–28.   A car is to be towed using the rope arrangement shown. The towing force required is 600 N. Determine the minimum length *l* of rope *AB* so that the tension in either rope *AB* or *AC* does not exceed 750 N. *Hint:* Use the equilibrium condition at point *A* to determine the required angle *θ* for attachment, then determine *l* using trigonometry applied to triangle *ABC*.

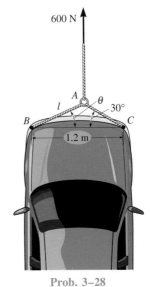

Prob. 3–28

3–29.   The sling *BAC* is used to lift the 100-N load with constant velocity. Determine the force in the sling and plot its value *T* (ordinate) as a function of its orientation *θ*, where 0 ≤ *θ* ≤ 90°.

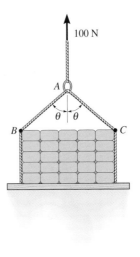

Prob. 3–29

3–30. When y is zero, the springs sustain a force of 60 N. Determine the magnitude of the applied vertical forces **F** and −**F** required to pull point A away from point B a distance of y = 2 m. The ends of cords CAD and CBD are attached to rings at C and D.

■3–31. When y is zero, the springs are each stretched 1.5 m. Determine the distance y if a force of F = 60 N is applied to points A and B as shown. The ends of cords CAD and CBD are attached to rings at C and D.

■3–33. The 10-N lamp fixture is suspended from two springs, each having an unstretched length of 4 m and stiffness of k = 5 N/m. Determine the angle θ for equilibrium.

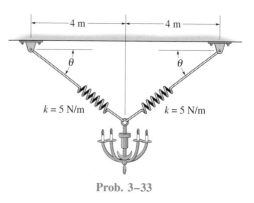

Prob. 3–33

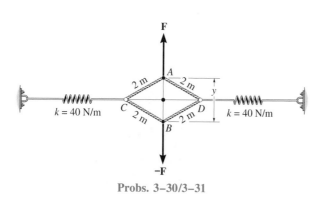

Probs. 3–30/3–31

*__3–32.__ Determine the maximum weight W that can be supported in the position shown if each cable AC and AB can support a maximum tension of 600 N before it fails.

3–34. If the cords suspend the two buckets in the equilibrium position shown, determine the weight of bucket B. Bucket A has a weight of 60 N.

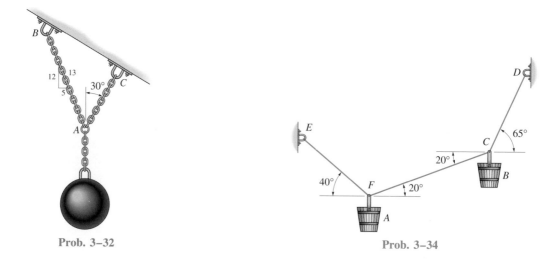

Prob. 3–32   Prob. 3–34

3–35. The 30-kg pipe is supported at A by a system of five cords. Determine the force in each cord for equilibrium.

■3–37. The cord AB of length 5 m is attached to the end B of a spring having a stiffness $k = 10$ N/m and an unstretched length of 5 m. The other end of the spring is attached to a roller C so that the spring remains horizontal as it stretches. If a 10-N weight is suspended from B, determine the angle θ of cord AB for equilibrium.

3–38. The cord AB has a length of 5 m and is attached to the end B of the spring having a stiffness $k = 10$ N/m. The other end of the spring is attached to a roller C so that the spring remains horizontal as it stretches. If a 10-N weight is suspended from B, determine the necessary unstretched length of the spring, so that $\theta = 40°$ for equilibrium.

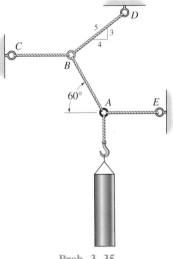

Prob. 3–35

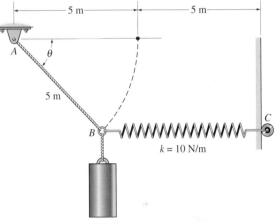

Probs. 3–37/3–38

*■3–36. The ring of negligible size is subjected to a vertical force of 200 N. Determine the required length l of cord AC such that the tension acting in AC is 160 N. Also, what is the force in cord AB? *Hint:* Use the equilibrium condition to determine the required angle θ for attachment, then determine l using trigonometry applied to triangle ABC.

3–39. The pail and its contents have a mass of 60 kg. If the cable is 15 m long, determine the elevation y of the pulley for equilibrium. Neglect the size of the pulley at A.

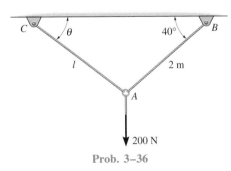

Prob. 3–36

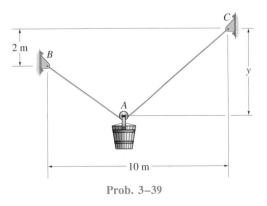

Prob. 3–39

3.4 Three-Dimensional Force Systems

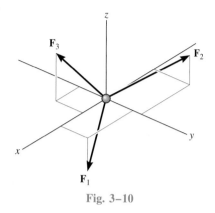

Fig. 3–10

It was shown in Sec. 3.1 that particle equilibrium requires

$$\Sigma \mathbf{F} = \mathbf{0} \qquad (3\text{–}4)$$

If the forces acting on the particle are resolved into their respective $\mathbf{i}, \mathbf{j}, \mathbf{k}$ components, Fig. 3–10, we can then write

$$\Sigma F_x \mathbf{i} + \Sigma F_y \mathbf{j} + \Sigma F_z \mathbf{k} = \mathbf{0}$$

To ensure that Eq. 3–4 is satisfied, we must therefore require that the following three scalar component equations be satisfied:

$$
\begin{aligned}
\Sigma F_x &= 0 \\
\Sigma F_y &= 0 \\
\Sigma F_z &= 0
\end{aligned}
\qquad (3\text{–}5)
$$

These equations represent the *algebraic sums* of the x, y, z force components acting on the particle. Using them we can solve for at most three unknowns, generally represented as angles or magnitudes of forces shown on the particle's free-body diagram.

PROCEDURE FOR ANALYSIS

The following procedure provides a method for solving three-dimensional force equilibrium problems.

Free-Body Diagram.   Draw a free-body diagram of the particle and label all the known and unknown forces on this diagram.

Equations of Equilibrium.   Establish the x, y, z coordinate axes with origin located at the particle and apply the equations of equilibrium. Use the three scalar Eqs. 3–5 in cases where it is easy to resolve each force acting on the particle into its x, y, z components. If this appears difficult, first express each force acting on the particle in Cartesian vector form, and then substitute these vectors into Eq. 3–4. By setting the respective $\mathbf{i}, \mathbf{j}, \mathbf{k}$ components equal to zero, the three scalar Eqs. 3–5 can be generated. If more than three unknowns exist and the problem involves a spring, consider using $F = ks$ to relate the spring force to the deformation s of the spring.

The following example problems numerically illustrate this solution procedure.

Example 3–6

A 90-N load is suspended from the hook shown in Fig. 3–11a. The load is supported by two cables and a spring having a stiffness $k = 500$ N/m. Determine the force in the cables and the stretch of the spring for equilibrium. Cable AD lies in the x–y plane and cable AC lies in the x–z plane.

SOLUTION

The stretch of the spring can be determined once the force in the spring is determined.

Free-Body Diagram.   The connection at A is chosen for the equilibrium analysis since the cable forces are concurrent at this point. The free-body diagram is shown in Fig. 3–11b.

Equations of Equilibrium.   By inspection, each force can easily be resolved into its x, y, z components, and therefore the three scalar equations of equilibrium can be directly applied. Considering components directed along the positive axes as "positive," we have

$$\Sigma F_x = 0; \qquad\qquad F_D \sin 30° - \tfrac{4}{5}F_C = 0 \qquad\qquad (1)$$

$$\Sigma F_y = 0; \qquad\qquad -F_D \cos 30° + F_B = 0 \qquad\qquad (2)$$

$$\Sigma F_z = 0; \qquad\qquad \tfrac{3}{5}F_C - 90 \text{ N} = 0 \qquad\qquad (3)$$

Solving Eq. 3 for F_C, then Eq. 1 for F_D, and finally Eq. 2 for F_B, we get

$$F_C = 150 \text{ N} \qquad\qquad Ans.$$
$$F_D = 240 \text{ N} \qquad\qquad Ans.$$
$$F_B = 208 \text{ N} \qquad\qquad Ans.$$

The stretch of the spring is therefore

$$F_B = ks_{AB}$$
$$208 \text{ N} = 500 \text{ N/m} \ (s_{AB})$$
$$s_{AB} = 0.416 \text{ m} \qquad\qquad Ans.$$

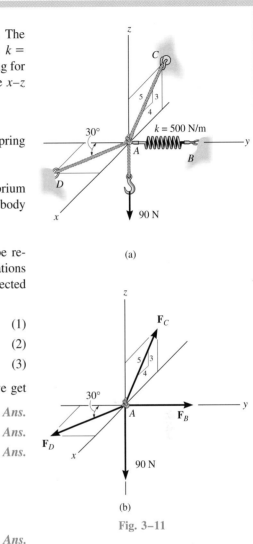

(a)

(b)

Fig. 3–11

Example 3–7

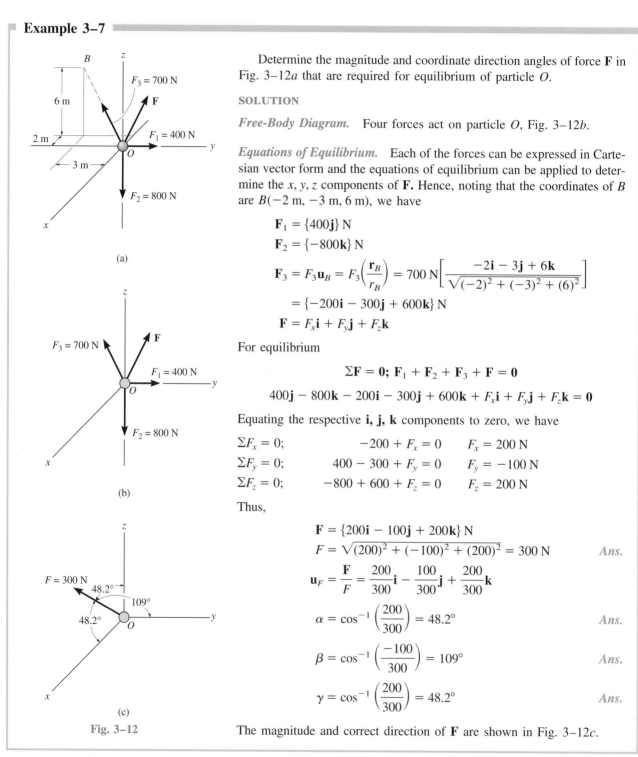

(a)

(b)

(c)

Fig. 3–12

Determine the magnitude and coordinate direction angles of force **F** in Fig. 3–12a that are required for equilibrium of particle O.

SOLUTION

Free-Body Diagram.   Four forces act on particle O, Fig. 3–12b.

Equations of Equilibrium.   Each of the forces can be expressed in Cartesian vector form and the equations of equilibrium can be applied to determine the x, y, z components of **F**. Hence, noting that the coordinates of B are B(−2 m, −3 m, 6 m), we have

$$\mathbf{F}_1 = \{400\mathbf{j}\}\ N$$

$$\mathbf{F}_2 = \{-800\mathbf{k}\}\ N$$

$$\mathbf{F}_3 = F_3\mathbf{u}_B = F_3\!\left(\frac{\mathbf{r}_B}{r_B}\right) = 700\ N\!\left[\frac{-2\mathbf{i} - 3\mathbf{j} + 6\mathbf{k}}{\sqrt{(-2)^2 + (-3)^2 + (6)^2}}\right]$$

$$= \{-200\mathbf{i} - 300\mathbf{j} + 600\mathbf{k}\}\ N$$

$$\mathbf{F} = F_x\mathbf{i} + F_y\mathbf{j} + F_z\mathbf{k}$$

For equilibrium

$$\Sigma\mathbf{F} = 0;\ \ \mathbf{F}_1 + \mathbf{F}_2 + \mathbf{F}_3 + \mathbf{F} = 0$$

$$400\mathbf{j} - 800\mathbf{k} - 200\mathbf{i} - 300\mathbf{j} + 600\mathbf{k} + F_x\mathbf{i} + F_y\mathbf{j} + F_z\mathbf{k} = 0$$

Equating the respective **i**, **j**, **k** components to zero, we have

$$\Sigma F_x = 0;\qquad -200 + F_x = 0\qquad F_x = 200\ N$$

$$\Sigma F_y = 0;\qquad 400 - 300 + F_y = 0\qquad F_y = -100\ N$$

$$\Sigma F_z = 0;\qquad -800 + 600 + F_z = 0\qquad F_z = 200\ N$$

Thus,

$$\mathbf{F} = \{200\mathbf{i} - 100\mathbf{j} + 200\mathbf{k}\}\ N$$

$$F = \sqrt{(200)^2 + (-100)^2 + (200)^2} = 300\ N \qquad\qquad Ans.$$

$$\mathbf{u}_F = \frac{\mathbf{F}}{F} = \frac{200}{300}\mathbf{i} - \frac{100}{300}\mathbf{j} + \frac{200}{300}\mathbf{k}$$

$$\alpha = \cos^{-1}\!\left(\frac{200}{300}\right) = 48.2° \qquad\qquad Ans.$$

$$\beta = \cos^{-1}\!\left(\frac{-100}{300}\right) = 109° \qquad\qquad Ans.$$

$$\gamma = \cos^{-1}\!\left(\frac{200}{300}\right) = 48.2° \qquad\qquad Ans.$$

The magnitude and correct direction of **F** are shown in Fig. 3–12c.

Example 3–8

Determine the force developed in each cable used to support the 40-N crate shown in Fig. 3–13a.

SOLUTION

Free-Body Diagram.   As shown in Fig. 3–13b, the free-body diagram of point *A* is considered in order to "expose" the three unknown forces in the cables, and by applying the condition for equilibrium we can obtain their magnitudes.

Equations of Equilibrium.   First we will express each force in Cartesian vector form. Since the coordinates of points *B* and *C* are $B(-0.75$ m, -1 m, 2 m) and $C(-0.75$ m, 1 m, 2 m), we have

$$\mathbf{F}_B = F_B\left[\frac{-0.75\mathbf{i} - 1\mathbf{j} + 2\mathbf{k}}{\sqrt{(-0.75)^2 + (-1)^2 + (2)^2}}\right]$$

$$= -0.318F_B\mathbf{i} - 0.424F_B\mathbf{j} + 0.848F_B\mathbf{k}$$

$$\mathbf{F}_C = F_C\left[\frac{-0.75\mathbf{i} + 1\mathbf{j} + 2\mathbf{k}}{\sqrt{(-0.75)^2 + (1)^2 + (2)^2}}\right]$$

$$= -0.318F_C\mathbf{i} + 0.424F_C\mathbf{j} + 0.848F_C\mathbf{k}$$

$$\mathbf{F}_D = F_D\mathbf{i}$$

$$\mathbf{W} = \{-40\mathbf{k}\} \text{ N}$$

Equilibrium requires

$$\Sigma\mathbf{F} = \mathbf{0}; \qquad \mathbf{F}_B + \mathbf{F}_C + \mathbf{F}_D + \mathbf{W} = \mathbf{0}$$

$$-0.318F_B\mathbf{i} - 0.424F_B\mathbf{j} + 0.848F_B\mathbf{k} - 0.318F_C\mathbf{i} + 0.424F_C\mathbf{j}$$
$$+ 0.848F_C\mathbf{k} + F_D\mathbf{i} - 40\mathbf{k} = \mathbf{0}$$

Equating the respective **i**, **j**, **k** components to zero yields

$$\Sigma F_x = 0; \qquad -0.318F_B - 0.318F_C + F_D = 0 \qquad (1)$$

$$\Sigma F_y = 0; \qquad -0.424F_B + 0.424F_C = 0 \qquad (2)$$

$$\Sigma F_z = 0; \qquad 0.848F_B + 0.848F_C - 40 = 0 \qquad (3)$$

Equation 2 states that $F_B = F_C$. Thus, solving Eq. 3 for F_B and F_C and substituting the result into Eq. 1 to obtain F_D, we have

$$F_B = F_C = 23.6 \text{ N} \qquad \qquad Ans.$$

$$F_D = 15.0 \text{ N} \qquad \qquad Ans.$$

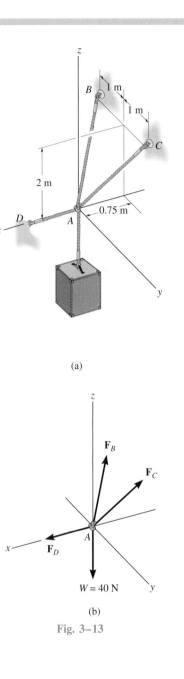

(a)

(b)

Fig. 3–13

Example 3–9

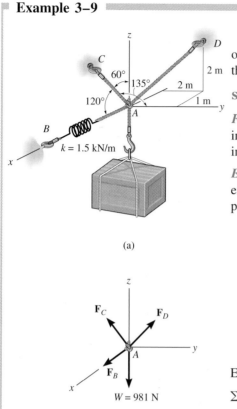

(a)

Fig. 3–14

(b)

The 100-kg box shown in Fig. 3–14a is supported by three cords, one of which is connected to a spring. Determine the tension in each cord and the stretch of the spring.

SOLUTION

Free-Body Diagram. The force in each of the cords can be determined by investigating the equilibrium of point A. The free-body diagram is shown in Fig. 3–14b. The weight of the cylinder is $W = 100(9.81) = 981$ N.

Equations of Equilibrium. Each vector on the free-body diagram is first expressed in Cartesian vector form. Using Eq. 2–11 for $\mathbf{F}_C$, and noting point $D(-1$ m, 2 m, 2 m) for $\mathbf{F}_D$, we have

$$\mathbf{F}_B = F_B\mathbf{i}$$

$$\mathbf{F}_C = F_C \cos 120°\mathbf{i} + F_C \cos 135°\mathbf{j} + F_C \cos 60°\mathbf{k}$$

$$= -0.5F_C\mathbf{i} - 0.707F_C\mathbf{j} + 0.5F_C\mathbf{k}$$

$$\mathbf{F}_D = F_D\left[\frac{-1\mathbf{i} + 2\mathbf{j} + 2\mathbf{k}}{\sqrt{(-1)^2 + (2)^2 + (2)^2}}\right]$$

$$= -0.333F_D\mathbf{i} + 0.667F_D\mathbf{j} + 0.667F_D\mathbf{k}$$

$$\mathbf{W} = \{-981\mathbf{k}\}\,\text{N}$$

Equilibrium requires

$$\Sigma\mathbf{F} = \mathbf{0}; \qquad \mathbf{F}_B + \mathbf{F}_C + \mathbf{F}_D + \mathbf{W} = \mathbf{0}$$

$$F_B\mathbf{i} - 0.5F_C\mathbf{i} - 0.707F_C\mathbf{j} + 0.5F_C\mathbf{k} - 0.333F_D\mathbf{i} + 0.667F_D\mathbf{j}$$

$$+ 0.667F_D\mathbf{k} - 981\mathbf{k} = \mathbf{0}$$

Equating the respective $\mathbf{i}, \mathbf{j}, \mathbf{k}$ components to zero,

$$\Sigma F_x = 0; \qquad F_B - 0.5F_C - 0.333F_D = 0 \qquad (1)$$

$$\Sigma F_y = 0; \qquad -0.707F_C + 0.667F_D = 0 \qquad (2)$$

$$\Sigma F_z = 0; \qquad 0.5F_C + 0.667F_D - 981 = 0 \qquad (3)$$

Solving Eq. 2 for F_D in terms of F_C and substituting into Eq. 3 yields F_C. F_D is determined from Eq. 2. Finally, substituting the results into Eq. 1 yields F_B. Hence,

$$F_C = 813\,\text{N} \qquad\qquad Ans.$$

$$F_D = 862\,\text{N} \qquad\qquad Ans.$$

$$F_B = 693.7\,\text{N} \qquad\qquad Ans.$$

The stretch of the spring is therefore

$$F = ks; \qquad\qquad 693.7 = 1500s$$

$$s = 0.462\,\text{m} \qquad\qquad Ans.$$

PROBLEMS

*3–40. Determine the magnitudes of $\mathbf{F}_1$, $\mathbf{F}_2$, and $\mathbf{F}_3$ for equilibrium of the particle.

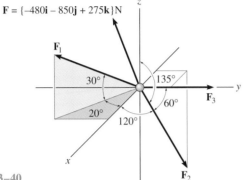

$\mathbf{F} = \{-480\mathbf{i} - 850\mathbf{j} + 275\mathbf{k}\}\text{N}$

Prob. 3–40

3–41. Determine the magnitudes of $\mathbf{F}_1$, $\mathbf{F}_2$, and $\mathbf{F}_3$ for equilibrium of the particle.

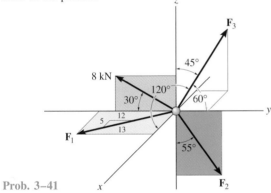

Prob. 3–41

3–42. Determine the magnitudes of $\mathbf{F}_1$, $\mathbf{F}_2$, and $\mathbf{F}_3$ for equilibrium of the particle.

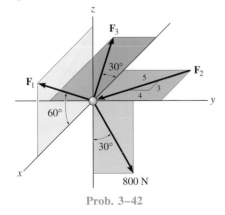

Prob. 3–42

3–43. Determine the magnitudes of $\mathbf{F}_1$, $\mathbf{F}_2$, and $\mathbf{F}_3$ for equilibrium of the particle.

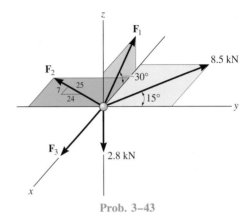

Prob. 3–43

*3–44. The 25-kg flowerpot is supported at A by the three cords. Determine the force acting in each cord for equilibrium.

3–45. If each cord can sustain a maximum tension of 50 N before it fails, determine the greatest weight of the flowerpot the cords can support.

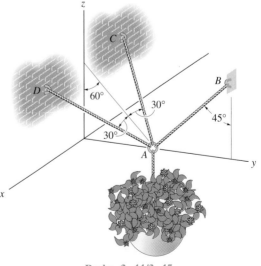

Probs. 3–44/3–45

3–46. Determine the tension developed in the three cables required to support the traffic light, which has a mass of 15 kg. Take $h = 4$ m.

3–47. Determine the tension developed in the three cables required to support the traffic light, which has a mass of 20 kg. Take $h = 3.5$ m.

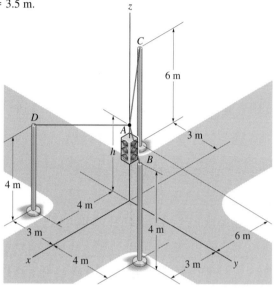

Probs. 3–46/3–47

*3–48. If the bucket and its contents have a total weight of 20 N, determine the force in the supporting cables *DA*, *DB*, and *DC*.

3–49. If each cable can sustain a maximum tension of 600 N, determine the greatest weight of the bucket and its contents that can be supported.

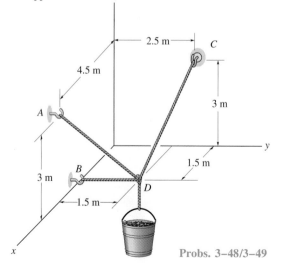

Probs. 3–48/3–49

3–50. The three cables are used to support the 800-N lamp. Determine the force developed in each cable for equilibrium.

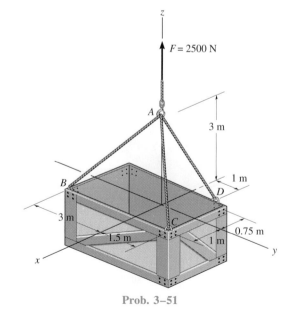

Prob. 3–50

■3–51. The 2500-N crate is to be hoisted with constant velocity from the hold of a ship using the cable arrangement shown. Determine the tension in each of the three cables for equilibrium.

Prob. 3–51

*■3–52. The lamp has a mass of 15 kg and is supported by a pole *AO* and cables *AB* and *AC*. If the force in the pole acts along its axis, determine the forces in *AO*, *AB*, and *AC* for equilibrium.

3–53. Cables *AB* and *AC* can sustain a maximum tension of 500 N, and the pole can support a maximum compression of 300 N. Determine the maximum weight of the lamp that can be supported in the position shown. The force in the pole acts along the axis of the pole.

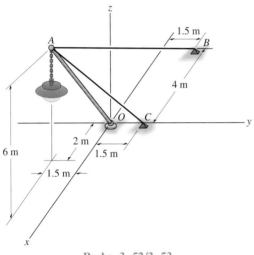

Probs. 3–52/3–53

3–54. The mast *OA* is supported by three cables. If cable *AB* is subjected to a tension of 500 N, determine the tension in cables *AC* and *AD* and the vertical force F_{AO} which the mast exerts along its axis at *A*.

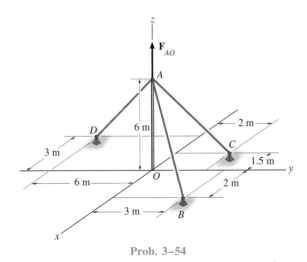

Prob. 3–54

3–55. The 500-N crate is suspended from the cable system shown. Determine the force in each segment of the cable, i.e., *AB*, *AC*, *CD*, *CE*, and *CF*. *Hint:* First analyze the equilibrium of point *A*, then using the result for *AC*, analyze the equilibrium of point *C*.

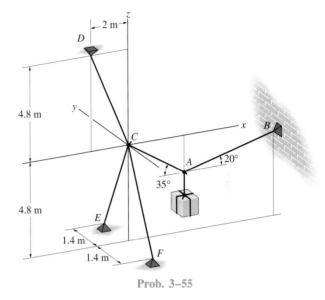

Prob. 3–55

*3–56. The 9500-N crucible is supported by three cables. Determine the force in each cable for equilibrium of the hook at *A*.

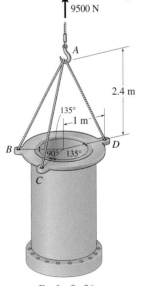

Prob. 3–56

3–57. Determine the force in each cable needed to support the 3500-N platform. Set $d = 1$ m.

3–58. Determine the force in each cable needed to support the 3500-N platform. Set $d = 2$ m.

*3–60. The 80-N chandelier is supported by three wires as shown. Determine the force in each wire for equilibrium.

3–61. If each wire can sustain a maximum tension of 120 N before it fails, determine the greatest weight of the chandelier the wires will support in the position shown.

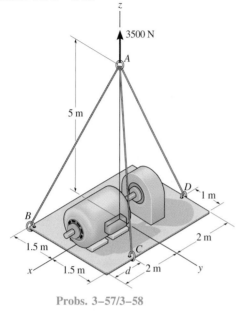

Probs. 3–57/3–58

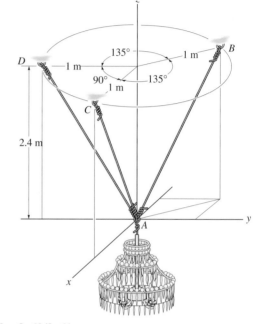

Probs. 3–60/3–61

3–59. The 800-N cylinder is supported by three chains as shown. Determine the force in each chain for equilibrium. Take $d = 1$ m.

■3–62. The 80-N ball is suspended from the horizontal ring using three springs each having an unstretched length of 1.5 m and stiffness of $k = 50$ N/m. Determine the vertical distance h from the ring to point A for equilibrium.

3–63. The ball is suspended from the horizontal ring using three springs each having a stiffness of $k = 50$ N/m and an unstretched length of 1.5 m. If $h = 2$ m, determine the weight of the ball.

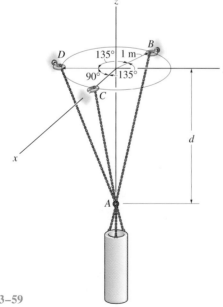

Prob. 3–59

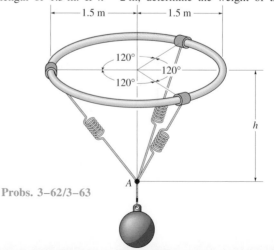

Probs. 3–62/3–63

REVIEW PROBLEMS

*3-64.   The man attempts to pull the log at C by using the three ropes. Determine the direction θ in which he should pull on his rope with a force of 80 N, so that he exerts a maximum force on the log. What is the force on the log for this case? Also, determine the direction in which he should pull in order to maximize the force in the rope attached to B. What is this maximum force?

3-66.   Romeo tries to reach Juliet by climbing with constant velocity up a rope which is knotted at point A. Any of the three segments of the rope can sustain a maximum force of 2 kN before it breaks. Determine if Romeo, who has a mass of 65 kg, can climb the rope, and if so, can he along with his Juliet, who has a mass of 60 kg, climb down with constant velocity?

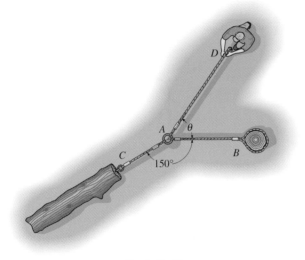

Prob. 3-64

3-65.   Determine the tension in cables AB, AC, and AD, required to hold the 60-N crate in equilibrium.

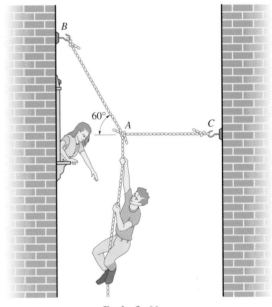

Prob. 3-66

3-67.   The 30-kg block is supported by two springs having the stiffness shown. Determine the unstretched length of each spring after the block is removed.

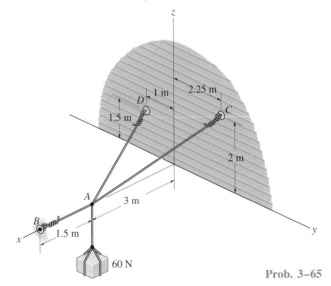

Prob. 3-65

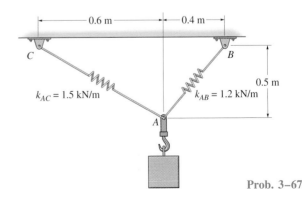

Prob. 3-67

*■**3–68.**   Determine the magnitudes of forces F_1, F_2, and F_3 necessary to hold the force $\mathbf{F} = \{-9\mathbf{i} - 8\mathbf{j} - 5\mathbf{k}\}$ kN in equilibrium.

3–70.   The 2-m-long cord AB is attached to a spring BC having an unstretched length of 2 m. If the cord sags downward an amount $\theta = 30°$ as shown, determine the vertical force F applied. The spring has a stiffness of $k = 50$ N/m.

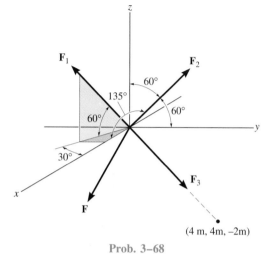

Prob. 3–68

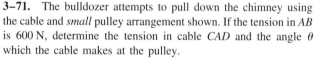
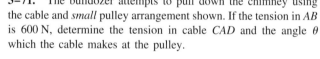

Prob. 3–70

3–69.   The boy has a mass of 60 kg and attempts to cross the creek by using the pulley and 15-m-long rope shown. If he leaves the shore at A, determine how close s he comes to the shore at B once he reaches a state of equilibrium.

3–71.   The bulldozer attempts to pull down the chimney using the cable and *small* pulley arrangement shown. If the tension in AB is 600 N, determine the tension in cable CAD and the angle θ which the cable makes at the pulley.

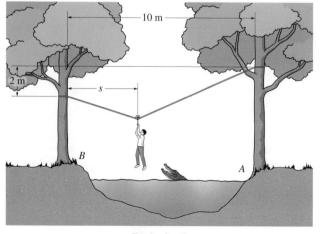

Prob. 3–69

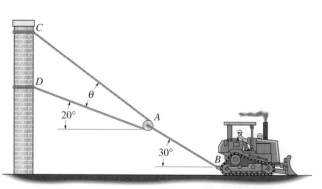

Prob. 3–71

*3–72.   Determine the magnitudes of $\mathbf{F}_1$, $\mathbf{F}_2$, and $\mathbf{F}_3$ for equilibrium.

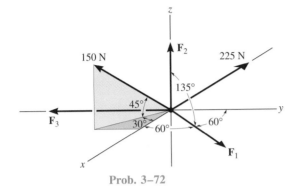

Prob. 3–72

3–73.   Determine the maximum weight of the engine that can be supported without exceeding a tension of 450 N in chain AB and 480 N in chain AC.

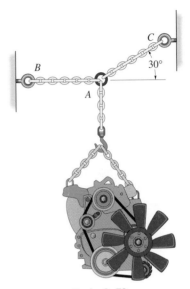

Prob. 3–73

Oftentimes a structure is subjected to a system of forces. Here the many forces of the cable-stayed bridge support the deck; the cables of the crane exert forces on the yellow spreader beam, which in turn supports the metal spool. In this chapter we will study ways of representing these force systems by their resultants.

4

Force System Resultants

In Chapter 3 it was shown that the condition for the equilibrium of a particle or a concurrent force system simply requires that the resultant of the force system be equal to zero, i.e., $\Sigma \mathbf{F} = \mathbf{0}$. In Chapter 5 it will be shown that such a restriction is necessary but not sufficient for the equilibrium of a rigid body. Since a body has a physical size, a further restriction must be made with regard to the nonconcurrency of the applied force system, giving rise to the concept of *moment*. A moment tends to turn a body, and equilibrium requires the body to have no rotation.

In this chapter a formal definition of a moment will be presented and ways of finding the moment of a force about a point or axis will be discussed. We will also present methods for determining the resultants of nonconcurrent force systems. This development is important since application of the equations for force-system simplification is similar to applying the equations of equilibrium for a rigid body. Furthermore, the resultants of a force system will influence the state of equilibrium or motion of a rigid body in the same way as the force system, and therefore we can study rigid-body behavior in a simpler manner by using the resultants.

4.1 Cross Product

The moment of a force will be formulated using Cartesian vectors in the next section. Before doing this, however, it is first necessary to expand our knowledge of vector algebra and introduce the cross-product method of vector multiplication.

The *cross product* of two vectors **A** and **B** yields the vector **C**, which is written

$$\mathbf{C} = \mathbf{A} \times \mathbf{B}$$

and is read ''**C** equals **A** cross **B**.''

Magnitude. The *magnitude* of **C** is defined as the product of the magnitudes of **A** and **B** and the sine of the angle θ between their tails ($0° \leq \theta \leq 180°$). Thus, $C = AB \sin \theta$.

Direction. Vector **C** has a *direction* that is perpendicular to the plane containing **A** and **B** such that the direction of **C** is specified by the right-hand rule; i.e., curling the fingers of the right hand from vector **A** (cross) to vector **B**, the thumb then points in the direction of **C**, as shown in Fig. 4–1.

Knowing both the magnitude and direction of **C**, we can write

$$\mathbf{C} = \mathbf{A} \times \mathbf{B} = (AB \sin \theta)\mathbf{u}_C \tag{4–1}$$

where the scalar $AB \sin \theta$ defines the *magnitude* of **C** and the unit vector $\mathbf{u}_C$ defines the *direction* of **C**. The terms of Eq. 4–1 are illustrated graphically in Fig. 4–2.

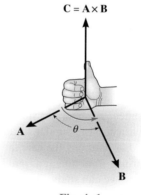

$\mathbf{C} = \mathbf{A} \times \mathbf{B}$

Fig. 4–1

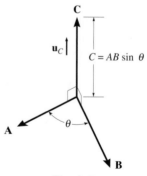

$C = AB \sin \theta$

Fig. 4–2

Laws of Operation

1. The commutative law is *not* valid; i.e.,

$$\mathbf{A} \times \mathbf{B} \neq \mathbf{B} \times \mathbf{A}$$

Rather,

$$\mathbf{A} \times \mathbf{B} = -\mathbf{B} \times \mathbf{A}$$

This is shown in Fig. 4–3 by using the right-hand rule. The cross product $\mathbf{B} \times \mathbf{A}$ yields a vector that acts in the opposite direction to $\mathbf{C}$; i.e., $\mathbf{B} \times \mathbf{A} = -\mathbf{C}.$

2. Multiplication by a scalar:

$$a(\mathbf{A} \times \mathbf{B}) = (a\mathbf{A}) \times \mathbf{B} = \mathbf{A} \times (a\mathbf{B}) = (\mathbf{A} \times \mathbf{B})a$$

This property is easily shown, since the magnitude of the resultant vector ($|a|AB \sin \theta$) and its direction are the same in each case.

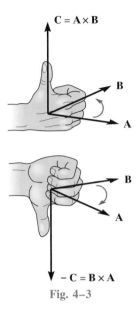

$\mathbf{C} = \mathbf{A} \times \mathbf{B}$

$-\mathbf{C} = \mathbf{B} \times \mathbf{A}$

Fig. 4–3

3. The distributive law:

$$\mathbf{A} \times (\mathbf{B} + \mathbf{D}) = (\mathbf{A} \times \mathbf{B}) + (\mathbf{A} \times \mathbf{D})$$

The proof of this identity is left as an exercise (see Prob. 4–1). It is important to note that *proper order* of the cross products must be maintained, since they are not commutative.

Cartesian Vector Formulation. Equation 4–1 may be used to find the cross product of a pair of Cartesian unit vectors. For example, to find $\mathbf{i} \times \mathbf{j}$, the *magnitude* of the resultant vector is $(i)(j)(\sin 90°) = (1)(1)(1) = 1$, and its *direction* is determined using the right-hand rule. As shown in Fig. 4–4, the resultant vector points in the $+\mathbf{k}$ direction. Thus, $\mathbf{i} \times \mathbf{j} = (1)\mathbf{k}$. In a similar manner,

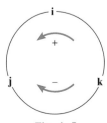

$\mathbf{k} = \mathbf{i} \times \mathbf{j}$

Fig. 4–4

$$\begin{array}{ccc}
\mathbf{i} \times \mathbf{j} = \mathbf{k} & \mathbf{i} \times \mathbf{k} = -\mathbf{j} & \mathbf{i} \times \mathbf{i} = 0 \\
\mathbf{j} \times \mathbf{k} = \mathbf{i} & \mathbf{j} \times \mathbf{i} = -\mathbf{k} & \mathbf{j} \times \mathbf{j} = 0 \\
\mathbf{k} \times \mathbf{i} = \mathbf{j} & \mathbf{k} \times \mathbf{j} = -\mathbf{i} & \mathbf{k} \times \mathbf{k} = 0
\end{array}$$

These results should *not* be memorized; rather, it should be clearly understood how each is obtained by using the right-hand rule and the definition of the cross product. A simple scheme shown in Fig. 4–5 is helpful for obtaining the same results when the need arises. If the circle is constructed as shown, then "crossing" two unit vectors in a *counterclockwise* fashion around the circle yields the *positive* third unit vector; e.g., $\mathbf{k} \times \mathbf{i} = \mathbf{j}.$ Moving *clockwise*, a *negative* unit vector is obtained; e.g., $\mathbf{i} \times \mathbf{k} = -\mathbf{j}.$

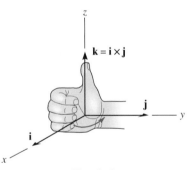

Fig. 4–5

Consider now the cross product of two general vectors **A** and **B** which are expressed in Cartesian vector form. We have

$$\mathbf{A} \times \mathbf{B} = (A_x\mathbf{i} + A_y\mathbf{j} + A_z\mathbf{k}) \times (B_x\mathbf{i} + B_y\mathbf{j} + B_z\mathbf{k})$$
$$= A_xB_x(\mathbf{i} \times \mathbf{i}) + A_xB_y(\mathbf{i} \times \mathbf{j}) + A_xB_z(\mathbf{i} \times \mathbf{k})$$
$$+ A_yB_x(\mathbf{j} \times \mathbf{i}) + A_yB_y(\mathbf{j} \times \mathbf{j}) + A_yB_z(\mathbf{j} \times \mathbf{k})$$
$$+ A_zB_x(\mathbf{k} \times \mathbf{i}) + A_zB_y(\mathbf{k} \times \mathbf{j}) + A_zB_z(\mathbf{k} \times \mathbf{k})$$

Carrying out the cross-product operations and combining terms yields

$$\mathbf{A} \times \mathbf{B} = (A_yB_z - A_zB_y)\mathbf{i} - (A_xB_z - A_zB_x)\mathbf{j} + (A_xB_y - A_yB_x)\mathbf{k} \qquad (4\text{--}2)$$

This equation may also be written in a more compact determinant form as

$$\mathbf{A} \times \mathbf{B} = \begin{vmatrix} \mathbf{i} & \mathbf{j} & \mathbf{k} \\ A_x & A_y & A_z \\ B_x & B_y & B_z \end{vmatrix} \qquad (4\text{--}3)$$

Thus, to find the cross product of any two Cartesian vectors **A** and **B,** it is necessary to expand a determinant whose first row of elements consists of the unit vectors **i, j,** and **k** and whose second and third rows represent the x, y, z components of the two vectors **A** and **B,** respectively.*

---

*A determinant having three rows and three columns can be expanded using three minors, each of which is multiplied by one of the three terms in the first row. There are four elements in each minor, e.g.,

By *definition,* this notation represents the terms $(A_{11}A_{22} - A_{12}A_{21})$, which is simply the product of the two elements of the arrow slanting downward to the right $(A_{11}A_{22})$ *minus* the product of the two elements intersected by the arrow slanting downward to the left $(A_{12}A_{21})$. For a 3×3 determinant, such as Eq. 4–3, the three minors can be generated in accordance with the following scheme:

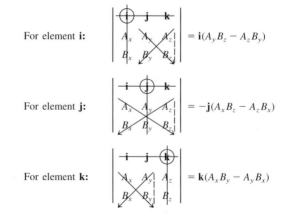

Adding the results and noting that the **j** element *must include the minus sign* yields the expanded form of **A** × **B** given by Eq. 4–2.

4.2 Moment of a Force—Scalar Formulation

The *moment* of a force about a point or axis provides a measure of the tendency of the force to cause a body to rotate about the point or axis. For example, consider the horizontal force $\mathbf{F}_x$, which acts perpendicular to the handle of the wrench and is located a distance d_y from point O, Fig. 4–6a. It is seen that this force tends to cause the pipe to turn about the z axis. The larger the force or the length d_y, the greater the turning effect. This tendency for rotation caused by $\mathbf{F}_x$ is sometimes called a *torque,* but most often it is called the *moment of a force* or simply the *moment* $(\mathbf{M}_O)_z$. In particular, note that the *moment axis* (z) is perpendicular to the shaded plane $(x–y)$ which contains both $\mathbf{F}_x$ and d_y and that this axis intersects the plane at point O. Now consider applying the force $\mathbf{F}_z$ to the wrench, Fig. 4–6b. This force will *not* rotate the pipe about the z axis. Instead, it tends to rotate it about the x axis. Keep in mind that although it may not be possible actually to "rotate" or turn the pipe in this manner, $\mathbf{F}_z$ still creates the *tendency* for rotation and so the moment $(\mathbf{M}_O)_x$ is produced. As before, the force and distance d_y lie in the shaded plane $(y–z)$ which is perpendicular to the moment axis (x). Lastly, if a force $\mathbf{F}_y$ is applied to the wrench, Fig. 4–6c, no moment is produced about point O. This lack of turning effect results, since the line of action of the force passes through O and therefore no tendency for rotation is possible.

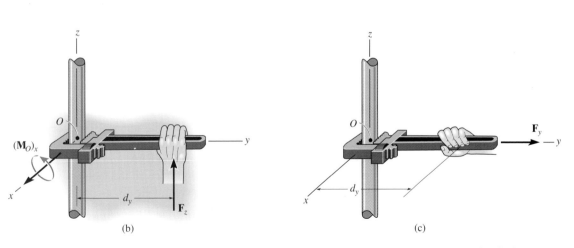

Fig. 4–6

We will now generalize the above discussion and consider the force $\mathbf{F}$ and point O which lie in a shaded plane as shown in Fig. 4–7a. The moment $\mathbf{M}_O$ about point O, or about an axis passing through O and perpendicular to the plane, is a *vector quantity* since it has a specified magnitude and direction.

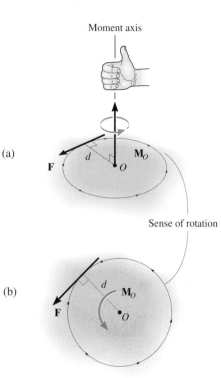

(a)

(b)

Fig. 4–7

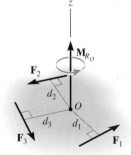

Fig. 4–8

Magnitude. The magnitude of $\mathbf{M}_O$ is

$$M_O = Fd \qquad (4\text{–}4)$$

where d is referred to as the *moment arm* or perpendicular distance from the axis at point O to the line of action of the force. Units of moment magnitude consist of force times distance, e.g., $N \cdot m$ or $lb \cdot ft$.

Direction. The direction of $\mathbf{M}_O$ will be specified by using the "right-hand rule." To do this, the fingers of the right hand are curled such that they follow the sense of rotation, which would occur if the force could rotate about point O, Fig. 4–7a. The *thumb* then *points* along the *moment axis* so that it gives the direction and sense of the moment vector, which is *upward* and *perpendicular* to the shaded plane containing $\mathbf{F}$ and d. By this definition, the moment $\mathbf{M}_O$ can be considered as a *sliding vector* and therefore acts at any point along the moment axis.

In three dimensions, $\mathbf{M}_O$ is illustrated by a vector arrow with a curl on it to *distinguish* it from a force vector, Fig. 4–7a. Many problems in mechanics, however, involve coplanar force systems that may be conveniently viewed in two dimensions. For example, a two-dimensional view of Fig. 4–7a is given in Fig. 4–7b. Here $\mathbf{M}_O$ is simply represented by the (counterclockwise) curl, which indicates the action of $\mathbf{F}$. The arrowhead on this curl is used to show the *sense of rotation* caused by $\mathbf{F}$. Using the right-hand rule, however, realize that the direction and sense of the moment vector in Fig. 4–7b are specified by the thumb, which points *out* of the page, since the fingers follow the curl. In particular, notice that *this curl or sense of rotation can always be determined by observing in which direction the force would "orbit" about point O* (counterclockwise in Fig. 4–7b). In two dimensions we will often refer to finding the moment of a force "about a point" (O). Keep in mind, however, that the moment *always acts about an axis* which is perpendicular to the plane containing $\mathbf{F}$ and d, and this axis intersects the plane at the point (O), Fig. 4–7a.

Resultant Moment of a System of Coplanar Forces. If a system of forces all lie in an x–y plane, then the moment produced by each force about point O will be directed along the z axis, Fig. 4–8. Consequently, the resultant moment $\mathbf{M}_{R_O}$ of the system can be determined by simply adding the moments of all forces *algebraically,* since all the moment vectors are collinear. We can write this vector sum symbolically as

$$\zeta + M_{R_O} = \Sigma Fd \qquad (4\text{–}5)$$

Here the counterclockwise curl written alongside the equation indicates that by the scalar sign convention, the moment of any force will be positive if it is directed along the $+z$ axis, whereas a negative moment is directed along the $-z$ axis.

The following examples illustrate numerical application of Eqs. 4–4 and 4–5.

Example 4–1

Determine the moment of the 800-N force acting on the frame in Fig. 4–9 about points A, B, C, and D.

SOLUTION (*SCALAR ANALYSIS*)

In general, $M = Fd$, where d is the moment arm or *perpendicular distance* from the *point* on the moment axis to the *line of action* of the force. Hence,

$M_A = 800 \text{ N}(2.5 \text{ m}) = 2000 \text{ N} \cdot \text{m} \downarrow$ *Ans.*

$M_B = 800 \text{ N}(1.5 \text{ m}) = 1200 \text{ N} \cdot \text{m} \downarrow$ *Ans.*

$M_C = 800 \text{ N}(0) = 0$ (line of action of **F** passes through C)   *Ans.*

$M_D = 800 \text{ N}(0.5 \text{ m}) = 400 \text{ N} \cdot \text{m} \uparrow$ *Ans.*

The curls indicate the sense of rotation of the moment, which is defined by the direction the force orbits about each point.

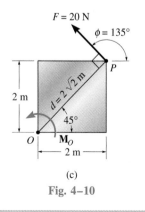

Fig. 4–9

Example 4–2

Determine the location of the point of application P and the direction of a 20-N force that lies in the plane of the square plate shown in Fig. 4–10a, so that this force creates the greatest counterclockwise moment about point O. What is this moment?

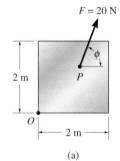

(a)

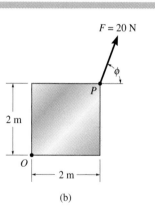

(b)

SOLUTION (*SCALAR ANALYSIS*)

Since the maximum moment created by the force is required, the force must act on the plate at a distance *farthest* from point O. As shown in Fig. 4–10b, the point of application must therefore be at the diagonal corner. In order to produce *counterclockwise* rotation of the plate about O, **F** must act at an angle $45° < \phi < 225°$. The greatest moment is produced when the line of action of **F** is *perpendicular* to d, i.e., $\phi = 135°$, Fig. 4–10c. The maximum moment is therefore

$$M_O = Fd = (20 \text{ N})(2\sqrt{2} \text{ m}) = 56.6 \text{ N} \cdot \text{m} \uparrow \qquad \textit{Ans.}$$

By the right-hand rule, $\mathbf{M}_O$ is directed out of the page.

(c)

Fig. 4–10

Example 4–3

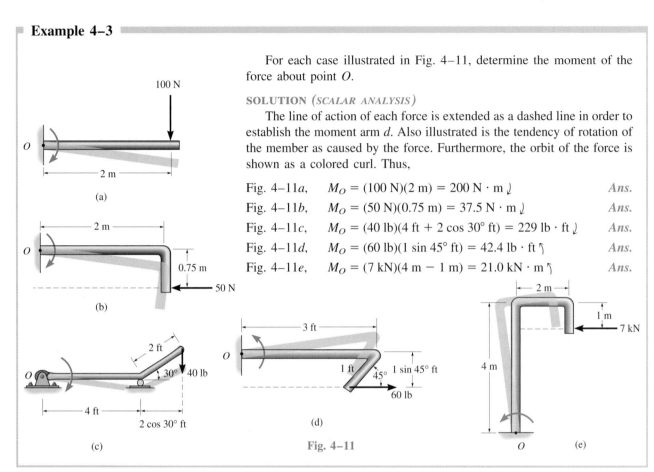

For each case illustrated in Fig. 4–11, determine the moment of the force about point O.

SOLUTION (*SCALAR ANALYSIS*)

The line of action of each force is extended as a dashed line in order to establish the moment arm d. Also illustrated is the tendency of rotation of the member as caused by the force. Furthermore, the orbit of the force is shown as a colored curl. Thus,

Fig. 4–11a, $M_O = (100 \text{ N})(2 \text{ m}) = 200 \text{ N} \cdot \text{m}$ ↴   *Ans.*

Fig. 4–11b, $M_O = (50 \text{ N})(0.75 \text{ m}) = 37.5 \text{ N} \cdot \text{m}$ ↴   *Ans.*

Fig. 4–11c, $M_O = (40 \text{ lb})(4 \text{ ft} + 2 \cos 30° \text{ ft}) = 229 \text{ lb} \cdot \text{ft}$ ↴   *Ans.*

Fig. 4–11d, $M_O = (60 \text{ lb})(1 \sin 45° \text{ ft}) = 42.4 \text{ lb} \cdot \text{ft}$ ↰   *Ans.*

Fig. 4–11e, $M_O = (7 \text{ kN})(4 \text{ m} - 1 \text{ m}) = 21.0 \text{ kN} \cdot \text{m}$ ↰   *Ans.*

Fig. 4–11

Example 4–4

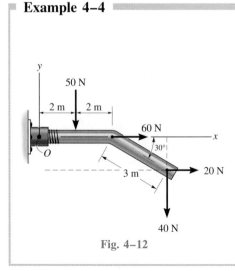

Fig. 4–12

Determine the resultant moment of the four forces acting on the rod shown in Fig. 4–12 about point O.

SOLUTION

Here it is necessary to apply Eq. 4–5. Assuming that positive moments act in the $+\mathbf{k}$ direction, i.e., counterclockwise, we have

$$\curvearrowright + M_{R_O} = \Sigma Fd;$$
$$M_{R_O} = -50 \text{ N}(2 \text{ m}) + 60 \text{ N}(0) + 20 \text{ N}(3 \sin 30° \text{ m})$$
$$-40 \text{ N}(4 \text{ m} + 3 \cos 30° \text{ m})$$
$$M_{R_O} = -334 \text{ N} \cdot \text{m} = 334 \text{ N} \cdot \text{m} \; ↴ \qquad Ans.$$

For this calculation, note how the moment-arm distances for the 20-N and 40-N forces are established from the extended (dashed) lines of action of each of these forces.

4.3 Moment of a Force—Vector Formulation

The moment of a force **F** about point O, or actually about the moment axis passing through O and perpendicular to the plane containing O and **F**, Fig. 4–13a, can also be expressed using the vector cross product, namely,

$$\mathbf{M}_O = \mathbf{r} \times \mathbf{F} \qquad\qquad (4\text{–}6)$$

Here **r** represents a position vector drawn from O to *any point* lying on the line of action of **F**. It will now be shown that indeed the moment $\mathbf{M}_O$, when determined by this cross product, has the proper magnitude and direction.

Magnitude. The magnitude of the above cross product is defined from Eq. 4–1 as $M_O = rF \sin \theta$. Here, the angle θ is measured between the *tails* of **r** and **F**. To establish this angle, **r** must be treated as a sliding vector so that θ can be constructed properly, Fig. 4–13b. Since the moment arm $d = r \sin \theta$, then

$$M_O = rF \sin \theta = F (r \sin \theta) = Fd$$

which agrees with Eq. 4–4.

Direction. The direction and sense of $\mathbf{M}_O$ in Eq. 4–6 are determined by the right-hand rule as it applies to the cross product. Thus, extending **r** to the dashed position and curling the right-hand fingers from **r** toward **F**, "**r** cross **F**," the thumb is directed upward or perpendicular to the plane containing **r** and **F** and this is in the *same direction* as $\mathbf{M}_O$, the moment of the force about point O, Fig. 4–13b. Note that the "curl" of the fingers, like the curl around the moment vector, indicates the sense of rotation caused by the force. Since the cross product is not commutative, it is important that the *proper order* of **r** and **F** be maintained in Eq. 4–6.

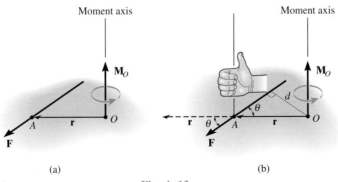

(a)   (b)

Fig. 4–13

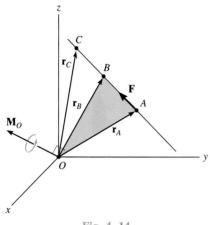

Fig. 4–14

Transmissibility of a Force.

Consider the force $\mathbf{F}$ applied at point A in Fig. 4–14. The moment created by $\mathbf{F}$ about point O is $\mathbf{M}_O = \mathbf{r}_A \times \mathbf{F}$; however, it was shown that the position vector "$\mathbf{r}$" can extend from O to *any point* on the line of action of $\mathbf{F}$. Consequently, $\mathbf{F}$ may be applied at point B or C and the same moment $\mathbf{M}_O = \mathbf{r}_B \times \mathbf{F} = \mathbf{r}_C \times \mathbf{F}$ will be computed. As a result, $\mathbf{F}$ has the properties of a *sliding vector* and can therefore act at *any point along its line of action and still create the same moment about point O*. We refer to $\mathbf{F}$ in this regard as being "transmissible," and we will discuss this property further in Sec. 4.7.

Cartesian Vector Formulation.

If we establish x, y, z coordinate axes, then the position vector $\mathbf{r}$ and force $\mathbf{F}$ can be expressed as Cartesian vectors, Fig. 4–15. Applying Eq. 4–6 we have

$$\mathbf{M}_O = \mathbf{r} \times \mathbf{F} = \begin{vmatrix} \mathbf{i} & \mathbf{j} & \mathbf{k} \\ r_x & r_y & r_z \\ F_x & F_y & F_z \end{vmatrix} \tag{4–7}$$

where r_x, r_y, r_z represent the x, y, z components of the position vector drawn from point O to *any point* on the line of action of the force

F_x, F_y, F_z represent the x, y, z components of the force vector

If the determinant is expanded, then like Eq. 4–2 we have

$$\mathbf{M}_O = (r_y F_z - r_z F_y)\mathbf{i} - (r_x F_z - r_z F_x)\mathbf{j} + (r_x F_y - r_y F_x)\mathbf{k} \tag{4–8}$$

The physical meaning of these three moment components becomes evident by studying Fig. 4–15a. For example, the $\mathbf{i}$ component of $\mathbf{M}_O$ is determined from the moments of $\mathbf{F}_x$, $\mathbf{F}_y$, and $\mathbf{F}_z$ about the x axis. In particular, note that $\mathbf{F}_x$ does *not* create a moment or tendency to cause turning about the x axis, since this

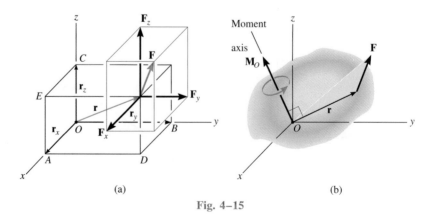

(a)   (b)

Fig. 4–15

force is *parallel* to the x axis. The line of action of $\mathbf{F}_y$ passes through point E, and so the magnitude of the moment of $\mathbf{F}_y$ about point A on the x axis is $r_z F_y$. By the right-hand rule this component acts in the negative $\mathbf{i}$ direction. Likewise, $\mathbf{F}_z$ contributes a moment component of $r_y F_z \mathbf{i}$. Thus, $(M_O)_x = (r_y F_z - r_z F_y)$ as shown in Eq. 4–8. As an exercise, establish the $\mathbf{j}$ and $\mathbf{k}$ components of $\mathbf{M}_O$ in this manner, and show that indeed the expanded form of the determinant, Eq. 4–8, represents the moment of $\mathbf{F}$ about point O. Once determined, realize that $\mathbf{M}_O$ will always be *perpendicular* to the shaded plane containing vectors $\mathbf{r}$ and $\mathbf{F}$, Fig. 4–15b.

It will be shown in Example 4–5 that the computation of the moment using the cross product has a distinct advantage over the scalar formulation when solving problems in *three dimensions*. This is because it is generally easier to establish the position vector $\mathbf{r}$ to the force, rather than determining the moment-arm distance d that must be directed *perpendicular* to the line of action of the force.

Resultant Moment of a System of Forces. The resultant moment of a system of forces about point O can be determined by vector addition resulting from successive applications of Eq. 4–7. This resultant can be written symbolically as

$$\mathbf{M}_{R_O} = \Sigma(\mathbf{r} \times \mathbf{F}) \qquad (4\text{–}9)$$

and is shown in Fig. 4–16.

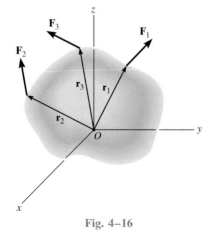

Fig. 4–16

Example 4–5

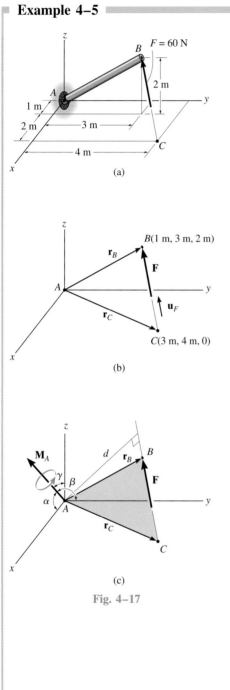

F = 60 N

2 m

1 m

2 m   3 m

4 m

(a)

B(1 m, 3 m, 2 m)

$\mathbf{r}_B$

F

$\mathbf{r}_C$

$\mathbf{u}_F$

C(3 m, 4 m, 0)

(b)

$\mathbf{M}_A$

d $\mathbf{r}_B$ B

γ β **F**

α A

$\mathbf{r}_C$

C

(c)

Fig. 4–17

The pole in Fig. 4–17a is subjected to a 60-N force that is directed from C to B. Determine the magnitude of the moment created by this force about the support at A.

SOLUTION (*VECTOR ANALYSIS*)

As shown in Fig. 4–17b, either one of two position vectors can be used for the solution, since $\mathbf{M}_A = \mathbf{r}_B \times \mathbf{F}$ or $\mathbf{M}_A = \mathbf{r}_C \times \mathbf{F}$. The position vectors are represented as

$$\mathbf{r}_B = \{1\mathbf{i} + 3\mathbf{j} + 2\mathbf{k}\} \text{ m} \quad \text{and} \quad \mathbf{r}_C = \{3\mathbf{i} + 4\mathbf{j}\} \text{ m}$$

The force has a magnitude of 60 N and a direction specified by the unit vector $\mathbf{u}_F$, directed from C to B. Thus,

$$\mathbf{F} = (60 \text{ N})\mathbf{u}_F = (60 \text{ N}) \left[\frac{(1-3)\mathbf{i} + (3-4)\mathbf{j} + (2-0)\mathbf{k}}{\sqrt{(-2)^2 + (-1)^2 + (2)^2}} \right]$$

$$= \{-40\mathbf{i} - 20\mathbf{j} + 40\mathbf{k}\} \text{ N}$$

Substituting into the determinant formulation, Eq. 4–7, and following the scheme for determinant expansion as stated in the footnote on page 110, we have

$$\mathbf{M}_A = \mathbf{r}_B \times \mathbf{F} = \begin{vmatrix} \mathbf{i} & \mathbf{j} & \mathbf{k} \\ 1 & 3 & 2 \\ -40 & -20 & 40 \end{vmatrix}$$

$$= [3(40) - 2(-20)]\mathbf{i} - [1(40) - 2(-40)]\mathbf{j} + [1(-20) - 3(-40)]\mathbf{k}$$

or

$$\mathbf{M}_A = \mathbf{r}_C \times \mathbf{F} = \begin{vmatrix} \mathbf{i} & \mathbf{j} & \mathbf{k} \\ 3 & 4 & 0 \\ -40 & -20 & 40 \end{vmatrix}$$

$$= [4(40) - 0(-20)]\mathbf{i} - [3(40) - 0(-40)]\mathbf{j} + [3(-20) - 4(-40)]\mathbf{k}$$

In both cases,

$$\mathbf{M}_A = \{160\mathbf{i} - 120\mathbf{j} + 100\mathbf{k}\} \text{ N} \cdot \text{m}$$

The *magnitude* of $\mathbf{M}_A$ is therefore

$$M_A = \sqrt{(160)^2 + (-120)^2 + (100)^2} = 224 \text{ N} \cdot \text{m} \qquad Ans.$$

As expected, $\mathbf{M}_A$ acts perpendicular to the shaded plane containing vectors **F**, $\mathbf{r}_B$, and $\mathbf{r}_C$, Fig. 4–17c. (How would you find its coordinate direction angles $\alpha = 44.3°$, $\beta = 122°$, $\gamma = 63.4°$?) Had this problem been worked using a scalar approach, where $M_A = Fd$, notice the difficulty that might arise in obtaining the moment arm d.

Example 4-6

Three forces act on the rod shown in Fig. 4–18a. Determine the resultant moment they create about the flange at O, and determine the direction of the moment axis.

SOLUTION

Here we must apply Eq. 4–9. Position vectors are directed from point O to each force as shown in Fig. 4–18b. These vectors are

$$\mathbf{r}_A = \{5\mathbf{j}\} \text{ m}$$
$$\mathbf{r}_B = \{4\mathbf{i} + 5\mathbf{j} - 2\mathbf{k}\} \text{ m}$$

Since $\mathbf{F}_2 = \{50\mathbf{j}\}$ lb, and the Cartesian components of the other forces are given, the resultant moment about O is therefore

$$\mathbf{M}_{R_O} = \Sigma(\mathbf{r} \times \mathbf{F})$$
$$= \mathbf{r}_A \times \mathbf{F}_1 + \mathbf{r}_A \times \mathbf{F}_2 + \mathbf{r}_B \times \mathbf{F}_3$$

$$= \begin{vmatrix} \mathbf{i} & \mathbf{j} & \mathbf{k} \\ 0 & 5 & 0 \\ -60 & 40 & 20 \end{vmatrix} + \begin{vmatrix} \mathbf{i} & \mathbf{j} & \mathbf{k} \\ 0 & 5 & 0 \\ 0 & 50 & 0 \end{vmatrix} + \begin{vmatrix} \mathbf{i} & \mathbf{j} & \mathbf{k} \\ 4 & 5 & -2 \\ 80 & 40 & -30 \end{vmatrix}$$

$$= [5(20) - 40(0)]\mathbf{i} - [0]\mathbf{j} + [0(40) - (-60)(5)]\mathbf{k} + [0\mathbf{i} - 0\mathbf{j} + 0\mathbf{k}]$$
$$+ [5(-30) - (40)(-2)]\mathbf{i} - [4(-30) - 80(-2)]\mathbf{j} + [4(40) - 80(5)]\mathbf{k}$$
$$= \{30\mathbf{i} - 40\mathbf{j} + 60\mathbf{k}\} \text{ N} \cdot \text{m} \qquad \qquad Ans.$$

The moment axis is directed along the line of action of $\mathbf{M}_{R_O}$. Since the magnitude of this moment is

$$M_{R_O} = \sqrt{(30)^2 + (-40)^2 + (60)^2} = 78.10 \text{ N} \cdot \text{m}$$

the unit vector which defines the direction of the moment axis is

$$\mathbf{u} = \frac{\mathbf{M}_{R_O}}{M_{R_O}} = \frac{30\mathbf{i} - 40\mathbf{j} + 60\mathbf{k}}{78.10} = 0.3841\mathbf{i} - 0.5121\mathbf{j} + 0.7682\mathbf{k}$$

Therefore, the coordinate direction angles of the moment axis are

$$\cos \alpha = 0.3841; \qquad \alpha = 67.4° \qquad \qquad Ans.$$
$$\cos \beta = -0.5121; \qquad \beta = 121° \qquad \qquad Ans.$$
$$\cos \gamma = 0.7682; \qquad \gamma = 39.8° \qquad \qquad Ans.$$

These results are shown in Fig. 4–18c. Realize that the three forces tend to cause the rod to rotate about this axis in the manner shown by the curl indicated on the moment vector.

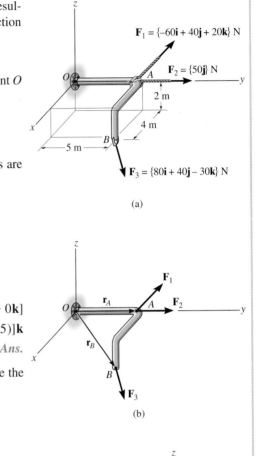

$$\mathbf{F}_1 = \{-60\mathbf{i} + 40\mathbf{j} + 20\mathbf{k}\} \text{ N}$$
$$\mathbf{F}_2 = \{50\mathbf{j}\} \text{ N}$$
$$\mathbf{F}_3 = \{80\mathbf{i} + 40\mathbf{j} - 30\mathbf{k}\} \text{ N}$$

(a)

(b)

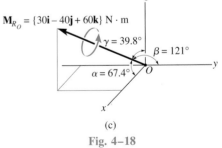

$$\mathbf{M}_{R_O} = \{30\mathbf{i} - 40\mathbf{j} + 60\mathbf{k}\} \text{ N} \cdot \text{m}$$
$$\gamma = 39.8°$$
$$\beta = 121°$$
$$\alpha = 67.4°$$

(c)

Fig. 4–18

4.4 Principle of Moments

A concept often used in mechanics is the *principle of moments,* which is sometimes referred to as *Varignon's theorem* since it was originally developed by the French mathematician Varignon (1654–1722). It states that *the moment of a force about a point is equal to the sum of the moments of the force's components about the point.* The proof follows directly from the distributive law of the vector cross product. To show this, consider the force $\mathbf{F}$ and two of its components, where $\mathbf{F} = \mathbf{F}_1 + \mathbf{F}_2$, Fig. 4–19. We have

$$\mathbf{M}_O = \mathbf{r} \times \mathbf{F}_1 + \mathbf{r} \times \mathbf{F}_2 = \mathbf{r} \times (\mathbf{F}_1 + \mathbf{F}_2) = \mathbf{r} \times \mathbf{F}$$

This concept has important applications to the solution of problems and proofs of theorems that follow, since it is often easier to determine the moments of a force's components rather than the moment of the force itself.

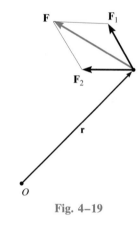

Fig. 4–19

Example 4-7

A 200-N force acts on the bracket shown in Fig. 4–20a. Determine the moment of the force about point A.

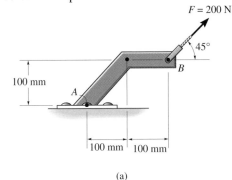

F = 200 N

45°

B

100 mm

A

100 mm   100 mm

(a)

SOLUTION I

The moment arm d can be found by trigonometry, using the construction shown in Fig. 4–20b. From the right triangle BCD,

$$CB = d = 100 \cos 45° = 70.71 \text{ mm} = 0.070\ 71 \text{ m}$$

Thus,

$$M_A = Fd = 200 \text{ N}(0.070\ 71 \text{ m}) = 14.1 \text{ N} \cdot \text{m} \,\rceil$$

According to the right-hand rule, $\mathbf{M}_A$ is directed in the $+\mathbf{k}$ direction since the force tends to rotate or orbit *counterclockwise* about point A. Hence, reporting the moment as a Cartesian vector, we have

$$\mathbf{M}_A = \{14.1\mathbf{k}\} \text{ N} \cdot \text{m} \qquad Ans.$$

SOLUTION II

The 200-N force may be resolved into x and y components, as shown in Fig. 4–20c. In accordance with the principle of moments, the moment of $\mathbf{F}$ computed about point A is equivalent to the sum of the moments produced by the two force components. Assuming counterclockwise rotation as positive, i.e., in the $+\mathbf{k}$ direction, we can apply Eq. 4–5 ($M_A = \Sigma Fd$), in which case

$$\langle +M_A = (200 \sin 45° \text{ N})(0.20 \text{ m}) - (200 \cos 45° \text{ N})(0.10 \text{ m})$$
$$= 14.1 \text{ N} \cdot \text{m} \,\rceil$$

Thus

$$\mathbf{M}_A = \{14.1\mathbf{k}\} \text{ N} \cdot \text{m} \qquad Ans.$$

By comparison, it is seen that Solution II provides a more *convenient method* for analysis than Solution I, since the moment arm for each component force is easier to establish.

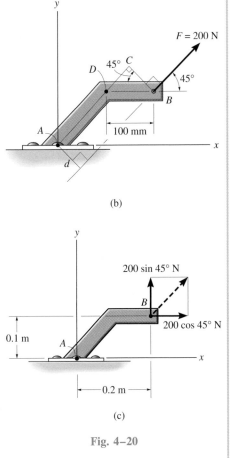

y

F = 200 N

45° C

D

45°

B

A   100 mm

x

d

(b)

y

200 sin 45° N

B

200 cos 45° N

0.1 m

A

x

0.2 m

(c)

Fig. 4–20

Example 4–8

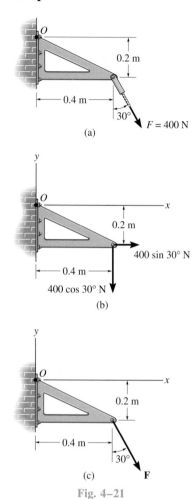

(a)

(b)

(c)

Fig. 4–21

The force **F** acts at the end of the angle bracket shown in Fig. 4–21a. Determine the moment of the force about point O.

SOLUTION I (*SCALAR ANALYSIS*)

The force is resolved into its x and y components as shown in Fig. 4–21b and the moments of the components are computed about point O. Taking positive moments as counterclockwise, i.e., in the $+\mathbf{k}$ direction, we have

$$\zeta+ M_O = 400 \sin 30° \,\text{N}(0.2 \text{ m}) - 400 \cos 30° \,\text{N}(0.4 \text{ m})$$
$$= -98.6 \text{ N} \cdot \text{m} = 98.6 \text{ N} \cdot \text{m} \,\downarrow$$

or

$$\mathbf{M}_O = \{-98.6\mathbf{k}\} \text{ N} \cdot \text{m} \qquad\qquad Ans.$$

SOLUTION II (*VECTOR ANALYSIS*)

Using a Cartesian vector approach, the force and position vectors shown in Fig. 4–21c can be represented as

$$\mathbf{r} = \{0.4\mathbf{i} - 0.2\mathbf{j}\} \text{ m}$$
$$\mathbf{F} = \{400 \sin 30°\mathbf{i} - 400 \cos 30°\mathbf{j}\} \text{ N}$$
$$= \{200.0\mathbf{i} - 346.4\mathbf{j}\} \text{ N}$$

The moment is therefore

$$\mathbf{M}_O = \mathbf{r} \times \mathbf{F} = \begin{vmatrix} \mathbf{i} & \mathbf{j} & \mathbf{k} \\ 0.4 & -0.2 & 0 \\ 200.0 & -346.4 & 0 \end{vmatrix}$$
$$= 0\mathbf{i} - 0\mathbf{j} + [0.4(-346.4) - (-0.2)(200.0)]\mathbf{k}$$
$$= \{-98.6\mathbf{k}\} \text{ N} \cdot \text{m} \qquad\qquad Ans.$$

By comparison, it is seen that the scalar analysis (Solution I) provides a more *convenient method* for analysis than Solution II, since the direction of the moment and the moment arm for each component force are easy to establish. Hence, this method is generally recommended for solving problems displayed in two dimensions. On the other hand, Cartesian vector analysis is generally recommended only for solving three-dimensional problems, where the moment arms and force components are often more difficult to determine.

PROBLEMS

4–1 If **A, B,** and **D** are given vectors, prove the distributive law for the vector cross product, i.e., **A** × (**B** + **D**) = (**A** × **B**) + (**A** × **D**).

4–2. Prove the triple scalar product identity **A** · **B** × **C** = **A** × **B** · **C.**

4–3. Given the three nonzero vectors **A**, **B**, and **C**, show that if **A** · (**B** × **C**) = 0, the three vectors *must* lie in the same plane.

*4–4. Determine the magnitude and directional sense of the moment of the force at *A* about point *O.*

4–5. Determine the magnitude and directional sense of the moment of the force at *A* about point *P.*

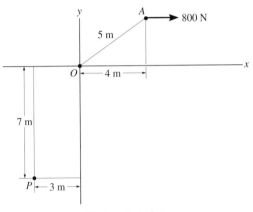

Probs. 4–4/4–5

4–6. Determine the magnitude and directional sense of the moment of the force at *A* about point *P.*

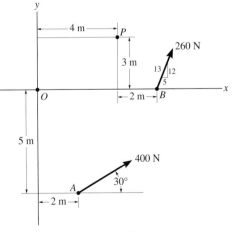

Prob. 4–6

4–7. Determine the magnitude and directional sense of the resultant moment of the forces about point *O.*

*4–8. Determine the magnitude and directional sense of the resultant moment of the forces about point *P.*

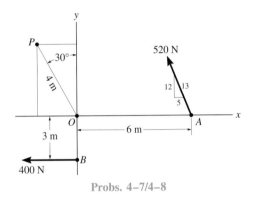

Probs. 4–7/4–8

4–9. Determine the magnitude and directional sense of the resultant moment of the forces about point *O.*

4–10. Determine the magnitude and directional sense of the resultant moment of the forces about point *P.*

Probs. 4–9/4–10

4–11. Determine the magnitude and directional sense of the resultant moment of the forces about point O.

*4–12. Determine the magnitude and directional sense of the resultant moment of the forces about point P.

4–15. Determine the moment of each force about the bolt located at A. Take $F_B = 40$ N, $F_C = 50$ N.

*4–16. If $F_B = 30$ N and $F_C = 45$ N, determine the resultant moment about the bolt located at A.

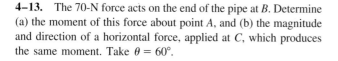

Probs. 4–11/4–12

Probs. 4–15/4–16

4–13. The 70-N force acts on the end of the pipe at B. Determine (a) the moment of this force about point A, and (b) the magnitude and direction of a horizontal force, applied at C, which produces the same moment. Take $\theta = 60°$.

4–14. The 70-N force acts on the end of the pipe at B. Determine the angles θ ($0° \le \theta \le 180°$) of the force that will produce maximum and minimum moments about point A. What are the magnitudes of these moments?

4–17. The torque wrench ABC is used to measure the moment or torque applied to a bolt when the bolt is located at A and a force is applied to the handle at C. The mechanic reads the torque on the scale at B. If an extension AO of length d is used on the wrench, determine the required scale reading if the desired torque on the bolt at O is to be M.

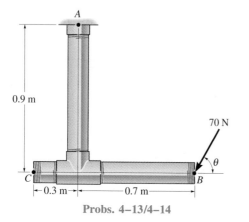

Probs. 4–13/4–14

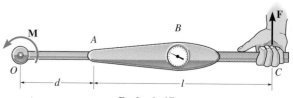

Prob. 4–17

4–18. Determine the resultant moment of the forces about point A. Solve the problem first by considering each force as a whole, and then by using the principle of moments. Take $F_1 = 250$ N, $F_2 = 300$ N, $F_3 = 500$ N.

4–19. If the resultant moment about point A is 4800 N · m clockwise, determine the magnitude of $\mathbf{F}_3$ if $F_1 = 300$ N and $F_2 = 400$ N.

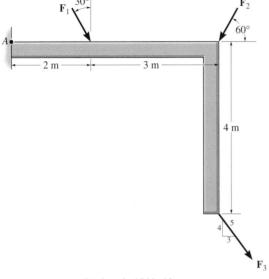

Probs. 4–18/4–19

4–20. The lift has a boom that has a length of 9.15 m, a weight of 3.56 kN, and mass center at G. If the bucket is designed to hold $W = 1.56$ kN, with mass center at G', determine the moment $\mathbf{M}$ that must be supplied by the motor at A to overcome the moment of the two forces of 3.56 kN and 1.56 kN. Take $\theta = 30°$.

4–21. The boom has a length of 9.15 m, a weight of 3.56 kN, and mass center at G. If the maximum moment that can be developed by the motor at A is $M = 27.1$ kN · m, determine the maximum load W, having a mass center at G', that can be lifted. Take $\theta = 30°$.

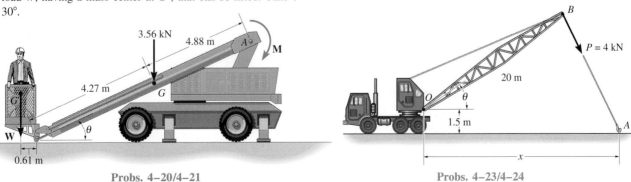

Probs. 4–20/4–21

4–22. Determine the direction $\theta(0° \le \theta \le 180°)$ of the force $F = 40$ N so that it produces (a) the maximum moment about point A and (b) the minimum moment about point A. Compute the moment in each case.

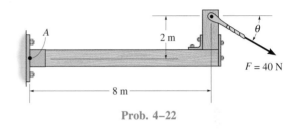

Prob. 4–22

4–23. The towline exerts a force of $P = 4$ kN at the end of the 20-m-long crane boom. If $\theta = 30°$, determine the placement x of the hook at A so that this force creates a maximum moment about point O. What is this moment?

4–24. The towline exerts a force of $P = 4$ kN at the end of the 20-m-long crane boom. If $x = 25$ m, determine the position θ of the boom so that this force creates a maximum moment about point O. What is this moment?

Probs. 4–23/4–24

4–25. The tool at *A* is used to hold a power lawnmower blade stationary while the nut is being loosened with the wrench. If a force of 50 N is applied to the wrench at *B* in the direction shown, determine the moment it creates about the nut at *C*. What is the magnitude of force **F** at *A* so that it creates the opposite moment about *C*?

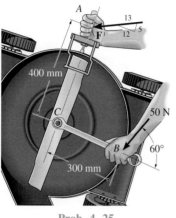

Prob. 4–25

4–26. The bucket boom carries a worker who has a weight of 1.02 kN and mass center at *G*. Determine the moment of this force about (a) point *A* and (b) point *B*.

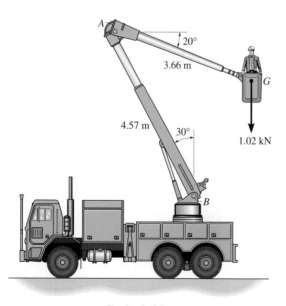

Prob. 4–26

4–27. The worker is using the bar to pull two pipes together in order to complete the connection. If he applies a horizontal force of 360 N to the handle of the lever, determine the moment of this force about the end *A*. What would be the tension *T* in the cable needed to cause the opposite moment about point *A*?

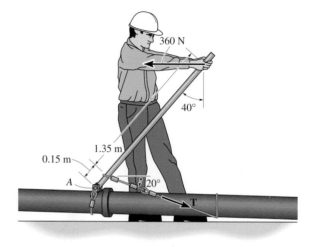

Prob. 4–27

*4–28. The crowbar is subjected to a vertical force of *P* = 110 N at the grip, whereas it takes a force of *F* = 690 N at the claw to pull the nail out. Find the moment of each force about point *A* and determine if **P** is sufficient to pull out the nail. The crowbar contacts the board at point *A*.

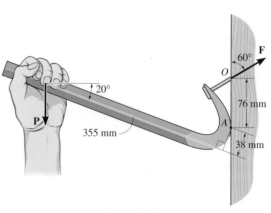

Prob. 4–28

4–29. If it takes a force of $F = 550$ N to pull the nail out, determine the smallest vertical force **P** that must be applied to the handle of the crowbar. *Hint:* This requires the moment of **F** about point A to be equal to the moment of **P** about A. Why?

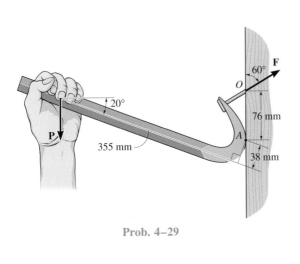

Prob. 4–29

4–30. Two forces act on the skew caster. Determine the resultant moment of these forces about point A and about point B.

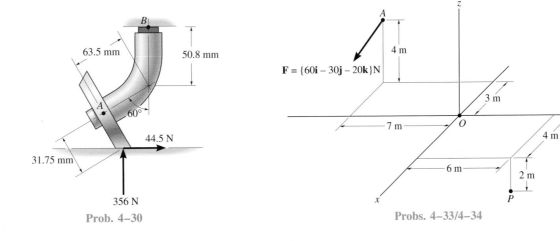

Prob. 4–30

4–31. Determine the moment of the force **F** at A about point O. Express the result as a Cartesian vector.

*4–32. Determine the moment of the force **F** at A about point P. Express the result as a Cartesian vector.

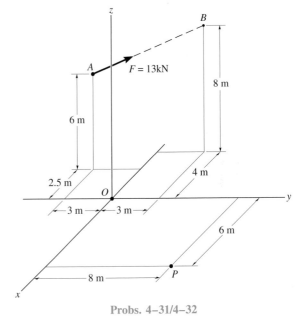

Probs. 4–31/4–32

4–33. Determine the moment of the force at A about point O. Express the result as a Cartesian vector.

4–34. Determine the moment of the force at A about point P. Express the result as a Cartesian vector.

$$F = \{60\mathbf{i} - 30\mathbf{j} - 20\mathbf{k}\}\,N$$

Probs. 4–33/4–34

4–35. The pole supports a 10-kg traffic light. Using Cartesian vectors, determine the moment of the weight of the traffic light about the base of the pole at A.

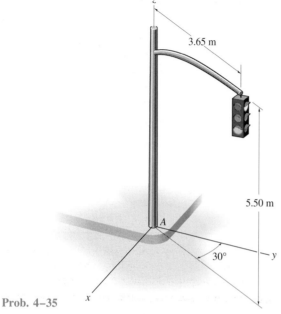

3.65 m

5.50 m

30°

Prob. 4–35

*4–36.** Using Cartesian vector analysis, determine the moment of each of the three forces acting on the column about point A. Take $\mathbf{F}_1 = \{300\mathbf{i} + 200\mathbf{j} + 80\mathbf{k}\}$ N.

4–37. Using Cartesian vector analysis, determine the resultant moment of the three forces about the base of the column at A. Take $\mathbf{F}_1 = \{400\mathbf{i} + 300\mathbf{j} + 120\mathbf{k}\}$ N.

$\mathbf{F}_2 = \{100\mathbf{i} - 100\mathbf{j} - 60\mathbf{k}\}$N

B $\mathbf{F}_1$

4 m

$\mathbf{F}_3 = \{-500\mathbf{k}\}$N

8 m

1 m

E

A

Probs. 4–36/4–37

4–38. The man pulls on the rope with a force of $F = 20$ N. Determine the moment that this force exerts about the base of the pole at O. Solve the problem two ways, i.e., by using a position vector from O to A, then O to B.

4–39. Determine the smallest force F that must be applied to the rope in order to cause the pole to break at its base O. This requires a moment of $M = 900$ N · m to be developed at O.

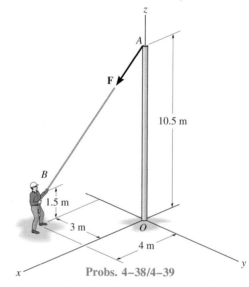

A

F

10.5 m

B

1.5 m

3 m

O

4 m

Probs. 4–38/4–39

*4–40.** The curved rod lies in the x-y plane and has a radius of 3 m. If a force of $F = 80$ N acts at its end as shown, determine the moment of this force about point O.

4–41. The curved rod lies in the x-y plane and has a radius of 3 m. If a force of $F = 80$ N acts at its end as shown, determine the moment of this force about point B.

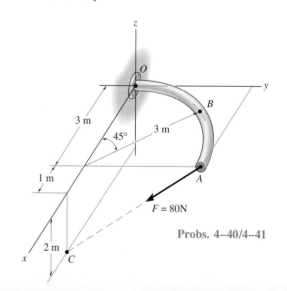

O

B

3 m

3 m

45°

1 m

A

$F = 80$N

2 m

C

Probs. 4–40/4–41

4–42. The force $\mathbf{F} = \{600\mathbf{i} + 300\mathbf{j} - 600\mathbf{k}\}$ N acts at the end B of the beam. Determine the moment of this force about point O.

4–43. The force $\mathbf{F} = \{600\mathbf{i} + 300\mathbf{j} - 600\mathbf{k}\}$ N acts at the end of the beam. Determine the moment of the force about point A.

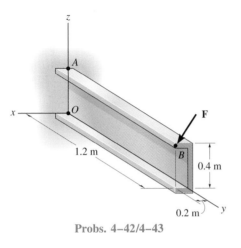

Probs. 4–42/4–43

4–46. A force of $\mathbf{F} = \{6\mathbf{i} - 2\mathbf{j} + 1\mathbf{k}\}$ kN produces a moment of $\mathbf{M}_O = \{4\mathbf{i} + 5\mathbf{j} - 14\mathbf{k}\}$ kN · m about the origin of coordinates, point O. If the force acts at a point having an x coordinate of $x = 1$ m, determine the y and z coordinates.

4–47. The force $\mathbf{F} = \{6\mathbf{i} + 8\mathbf{j} + 10\mathbf{k}\}$ N creates a moment about point O of $\mathbf{M}_O = \{-14\mathbf{i} + 8\mathbf{j} + 2\mathbf{k}\}$ N · m. If the force passes through a point having an x coordinate of 1 m, determine the y and z coordinates of the point. Also, realizing that $M_O = Fd$, determine the perpendicular distance d from point O to the line of action of $\mathbf{F}$.

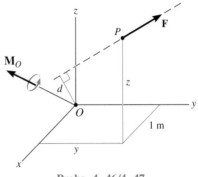

Probs. 4–46/4–47

4–44. The curved rod has a radius of 5 m. If a force of $F = 60$ N acts at its end as shown, determine the moment of this force about point C.

4–45. Determine the smallest force F that must be applied along the rope in order to cause the curved rod, which has a radius of 5 m, to fail at the support C. This requires a moment of $M = 80$ N · m to be developed at C.

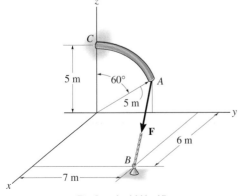

Probs. 4–44/4–45

4–48. Segments of drill pipe D for an oil well are tightened a prescribed amount by using a set of tongs T, which grip the pipe, and a hydraulic cylinder (not shown) to regulate the force $\mathbf{F}$ applied to the tongs. This force acts along the cable which passes around the small pulley P. If the cable is originally perpendicular to the tongs as shown, determine the magnitude of force $\mathbf{F}$ which must be applied so that the moment about the pipe is $M = 2000$ N · m. In order to maintain this same moment what magnitude of $\mathbf{F}$ is required when the tongs rotate 30° to the dashed position?

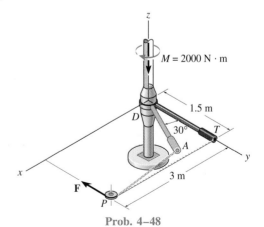

Prob. 4–48

4.5 Moment of a Force About a Specified Axis

Recall that when the moment of a force is computed about a point, the moment and its axis are *always* perpendicular to the plane containing the force and the moment arm. In some problems it is important to find the *component* of this moment along a *specified axis* that passes through the point. To solve this problem either a scalar or vector analysis can be used.

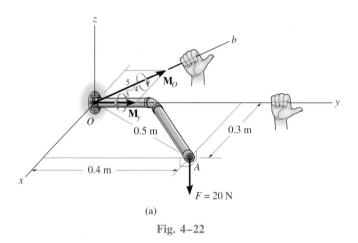

(a)

Fig. 4–22

Scalar Analysis. As a numerical example of this problem, consider the pipe assembly shown in Fig. 4–22*a*, which lies in the horizontal plane and is subjected to the vertical force of $F = 20$ N applied at point A. The moment of this force about point O has a *magnitude* of $M_O = (20 \text{ N})(0.5 \text{ m}) = 10 \text{ N} \cdot \text{m}$ and a *direction* defined by the right-hand rule, as shown in Fig. 4–22*a*. This moment tends to turn the pipe about the *Ob* axis. For practical reasons, however, it may be necessary to determine the *component* of M_O about the *y* axis, M_y, since this component tends to unscrew the pipe from the flange at O. From Fig. 4–22*a*, M_y has a magnitude of $M_y = \frac{3}{5}(10 \text{ N} \cdot \text{m}) = 6 \text{ N} \cdot \text{m}$ and a sense of direction shown by the vector resolution. Rather than performing this *two-step* process of first finding the moment of the force about point O and then resolving the moment along the *y* axis, it is also possible to solve this problem *directly*. To do so, it is necessary to determine the perpendicular or moment-arm distance from the line of action of **F** to the *y* axis. From Fig. 4–22*a* this distance is 0.3 m. Thus the *magnitude* of the moment of the force about the *y* axis is again $M_y = 0.3(20 \text{ N}) = 6 \text{ N} \cdot \text{m}$, and the *direction* is determined by the right-hand rule as shown.

In general, then, *if the line of action of a force* **F** *is perpendicular to any specified axis aa*, the magnitude of the moment of **F** about the axis can be determined from the equation

$$M_a = F d_a \tag{4-10}$$

Here d_a is the *perpendicular or shortest distance* from the force line of action to the axis. The direction is determined from the thumb of the right hand when the fingers are curled in accordance with the direction of rotation as produced by the force. In particular, realize that a *force will not contribute a moment about a specified axis if the force line of action is parallel to the axis or its line of action passes through the axis.*

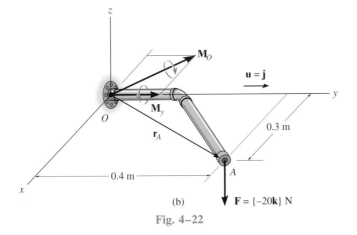

(b)

Fig. 4–22

Vector Analysis. The previous two-step solution of first finding the moment of the force about a point on the axis and then finding the projected component of the moment about the axis can also be performed using a vector analysis, Fig. 4–22b. Here the moment about point O is first determined from $\mathbf{M}_O = \mathbf{r}_A \times \mathbf{F} = (0.3\mathbf{i} + 0.4\mathbf{j}) \times (-20\mathbf{k}) = \{-8\mathbf{i} + 6\mathbf{j}\}\,\text{N} \cdot \text{m}$. The component or projection of this moment along the y axis is then determined from the dot product (Sec. 2.9). Since the unit vector for this axis (or line) is $\mathbf{u} = \mathbf{j}$, then $M_y = \mathbf{M}_O \cdot \mathbf{u} = (-8\mathbf{i} + 6\mathbf{j}) \cdot \mathbf{j} = 6\,\text{N} \cdot \text{m}$. This result, of course, is to be expected, since it represents the $\mathbf{j}$ component of $\mathbf{M}_O$.

A vector analysis such as this is particularly advantageous for finding the moment of a force about an axis when the force components or the appropriate moment arms are difficult to determine. For this reason, the above two-step process will now be generalized and applied to a body of arbitrary shape. To do so, consider the body in Fig. 4–23, which is subjected to the force **F** acting at point A. Here we wish to determine the effect of **F** in tending to rotate the body about the aa' axis. This tendency for rotation is measured by the moment component $\mathbf{M}_a$. To determine $\mathbf{M}_a$ we first compute the moment of **F** about any *arbitrary point* O that lies on the aa' axis. In this case, $\mathbf{M}_O$ is expressed by the

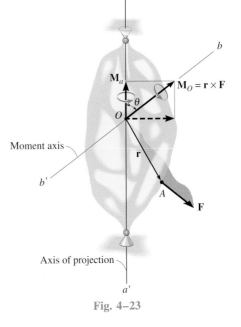

Fig. 4–23

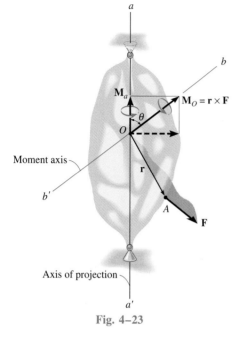

Fig. 4–23

cross product $\mathbf{M}_O = \mathbf{r} \times \mathbf{F}$, where $\mathbf{r}$ is directed from O to A. Since $\mathbf{M}_O$ acts along the moment axis bb', which is perpendicular to the plane containing $\mathbf{r}$ and $\mathbf{F}$, the component or projection of $\mathbf{M}_O$ onto the aa' axis is then represented by $\mathbf{M}_a$. The *magnitude* of $\mathbf{M}_a$ is determined by the dot product, $M_a = M_O \cos \theta = \mathbf{M}_O \cdot \mathbf{u}_a$, where $\mathbf{u}_a$ is a unit vector that defines the direction of the aa' axis. Combining these two steps as a general expression, we have $M_a = (\mathbf{r} \times \mathbf{F}) \cdot \mathbf{u}_a$. Since the dot product is commutative, we can also write

$$M_a = \mathbf{u}_a \cdot (\mathbf{r} \times \mathbf{F})$$

In vector algebra, this combination of dot and cross product yielding the scalar M_a is called the *triple scalar product.* Provided x, y, z axes are established and the Cartesian components of each of the vectors can be determined, then the triple scalar product may be written in determinant form as

$$M_a = (u_{a_x}\mathbf{i} + u_{a_y}\mathbf{j} + u_{a_z}\mathbf{k}) \cdot \begin{vmatrix} \mathbf{i} & \mathbf{j} & \mathbf{k} \\ r_x & r_y & r_z \\ F_x & F_y & F_z \end{vmatrix}$$

or simply

$$M_a = \mathbf{u}_a \cdot (\mathbf{r} \times \mathbf{F}) = \begin{vmatrix} u_{a_x} & u_{a_y} & u_{a_z} \\ r_x & r_y & r_z \\ F_x & F_y & F_z \end{vmatrix} \tag{4–11}$$

where $\quad u_{a_x}, u_{a_y}, u_{a_z}$ represent the x, y, z components of the unit vector defining the direction of the aa' axis

r_x, r_y, r_z represent the x, y, z components of the position vector drawn from any point O on the aa' axis to any point A on the line of action of the force

F_x, F_y, F_z represent the x, y, z components of the force vector.

When M_a is evaluated from Eq. 4–11, it will yield a positive or negative scalar. The sign of this scalar indicates the sense of direction of $\mathbf{M}_a$ along the aa' axis. If it is positive, then $\mathbf{M}_a$ will have the same sense as $\mathbf{u}_a$, whereas if it is negative, then $\mathbf{M}_a$ will act opposite to $\mathbf{u}_a$.

Once M_a is determined, we can then express $\mathbf{M}_a$ as a Cartesian vector, namely,

$$\mathbf{M}_a = M_a\mathbf{u}_a = [\mathbf{u}_a \cdot (\mathbf{r} \times \mathbf{F})]\mathbf{u}_a \tag{4–12}$$

Finally, if the resultant moment of a series of forces is to be computed about the axis, then the moment components of each force are added together *algebraically,* since each component lies along the same axis, i.e., its magnitude is

$$M_A = \Sigma [\mathbf{u}_a \cdot (\mathbf{r} \times \mathbf{F})] = \mathbf{u}_a \cdot \Sigma (\mathbf{r} \times \mathbf{F})$$

The following examples illustrate a numerical application of the above concepts.

Example 4–9

The force $\mathbf{F} = \{-40\mathbf{i} + 20\mathbf{j} + 10\mathbf{k}\}\,\mathrm{N}$ acts at point A shown in Fig. 4–24a. Determine the moments of this force about the x and Oa axes.

SOLUTION I (*VECTOR ANALYSIS*)

We can solve this problem by using the position vector $\mathbf{r}_A$. Why? Since $\mathbf{r}_A = \{-3\mathbf{i} + 4\mathbf{j} + 6\mathbf{k}\}\,\mathrm{m}$, and $\mathbf{u}_x = \mathbf{i}$, then applying Eq. 4–11,

$$M_x = \mathbf{i} \cdot (\mathbf{r}_A \times \mathbf{F}) = \begin{vmatrix} 1 & 0 & 0 \\ -3 & 4 & 6 \\ -40 & 20 & 10 \end{vmatrix}$$

$$= 1[4(10) - 6(20)] - 0[(-3)(10) - 6(-40)] + 0[(-3)(20) - 4(-40)]$$

$$= -80\,\mathrm{N} \cdot \mathrm{m} \qquad \qquad \textit{Ans.}$$

The negative sign indicates that the sense of $\mathbf{M}_x$ is opposite to $\mathbf{i}$.

We can compute M_{Oa} also using $\mathbf{r}_A$ because $\mathbf{r}_A$ extends from a point on the Oa axis to the force. Also, $\mathbf{u}_{Oa} = -\frac{3}{5}\mathbf{i} + \frac{4}{5}\mathbf{j}$. Thus,

$$M_{Oa} = \mathbf{u}_{Oa} \cdot (\mathbf{r}_A \times \mathbf{F}) = \begin{vmatrix} -\frac{3}{5} & \frac{4}{5} & 0 \\ -3 & 4 & 6 \\ -40 & 20 & 10 \end{vmatrix}$$

$$= -\tfrac{3}{5}[4(10) - 6(20)] - \tfrac{4}{5}[(-3)(10) - 6(-40)] + 0[(-3)(20) - 4(-40)]$$

$$= -120\,\mathrm{N} \cdot \mathrm{m} \qquad \qquad \textit{Ans.}$$

What does the negative sign indicate?

The moment components are shown in Fig. 4–24b.

SOLUTION II (*SCALAR ANALYSIS*)

Since the force components and moment arms are easy to determine for computing M_x a scalar analysis can be used to solve this problem. Referring to Fig. 4–24c, only the 10-N and 20-N forces contribute moments about the x axis. (The line of action of the 40-N force is *parallel* to this axis and hence its moment about the x axis is zero.) Using the right-hand rule, the algebraic sum of the moment components about the x axis is therefore

$$M_x = (10\,\mathrm{N})(4\,\mathrm{m}) - (20\,\mathrm{N})(6\,\mathrm{m}) = -80\,\mathrm{N} \cdot \mathrm{m} \qquad \textit{Ans.}$$

Although not required here, note also that

$$M_y = (10\,\mathrm{N})(3\,\mathrm{m}) - (40\,\mathrm{N})(6\,\mathrm{m}) = -210\,\mathrm{N} \cdot \mathrm{m}$$

$$M_z = (40\,\mathrm{N})(4\,\mathrm{m}) - (20\,\mathrm{N})(3\,\mathrm{m}) = 100\,\mathrm{N} \cdot \mathrm{m}$$

If we were to determine M_{Oa} by this scalar method it would require much more effort, since the force components of 40 N and 20 N are *not* perpendicular to the direction of Oa. The vector analysis yields a more direct solution.

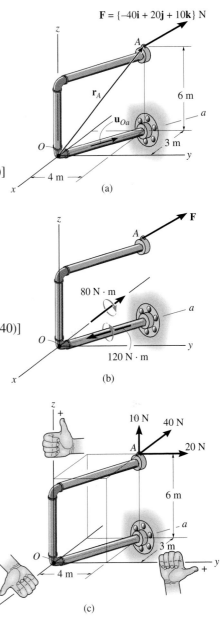

Fig. 4–24

Example 4–10

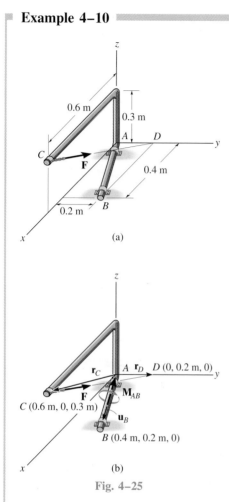

(a)

(b)

Fig. 4–25

The rod shown in Fig. 4–25*a* is supported by two brackets at *A* and *B*. Determine the moment $\mathbf{M}_{AB}$ produced by $\mathbf{F} = \{-600\mathbf{i} + 200\mathbf{j} - 300\mathbf{k}\}$ N, which tends to rotate the rod about the *AB* axis.

SOLUTION

A vector analysis using $M_{AB} = \mathbf{u}_B \cdot (\mathbf{r} \times \mathbf{F})$ will be considered for the solution since the moment arm or perpendicular distance from the line of action of $\mathbf{F}$ to the *AB* axis is difficult to determine. Each of the terms in the equation will now be identified.

Unit vector $\mathbf{u}_B$ defines the direction of the *AB* axis of the rod, Fig. 4–25*b*, where

$$\mathbf{u}_B = \frac{\mathbf{r}_B}{r_B} = \frac{0.4\mathbf{i} + 0.2\mathbf{j}}{\sqrt{(0.4)^2 + (0.2)^2}} = 0.894\mathbf{i} + 0.447\mathbf{j}$$

Vector $\mathbf{r}$ is directed from *any point* on the *AB* axis to *any point* on the line of action of the force. For example, position vectors $\mathbf{r}_C$ and $\mathbf{r}_D$ are suitable, Fig. 4–25*b*. (Although not shown, $\mathbf{r}_{BC}$ or $\mathbf{r}_{BD}$ can also be used.) For simplicity, we choose $\mathbf{r}_D$, where

$$\mathbf{r}_D = \{0.2\mathbf{j}\} \text{ m}$$

The force is

$$\mathbf{F} = \{-600\mathbf{i} + 200\mathbf{j} - 300\mathbf{k}\} \text{ N}$$

Substituting these vectors into the determinant form and expanding, we have

$$M_{AB} = \mathbf{u}_B \cdot (\mathbf{r}_D \times \mathbf{F}) = \begin{vmatrix} 0.894 & 0.447 & 0 \\ 0 & 0.2 & 0 \\ -600 & 200 & -300 \end{vmatrix}$$

$$= 0.894[0.2(-300) - 0(200)] - 0.447[0(-300) - 0(-600)] + $$
$$0[0(200) - 0.2(-600)]$$

$$= -53.67 \text{ N} \cdot \text{m}$$

The negative sign indicates that the sense of $\mathbf{M}_{AB}$ is opposite to that of $\mathbf{u}_B$. Expressing $\mathbf{M}_{AB}$ as a Cartesian vector yields

$$\mathbf{M}_{AB} = M_{AB}\mathbf{u}_B = (-53.67 \text{ N} \cdot \text{m})(0.894\mathbf{i} + 0.447\mathbf{j})$$
$$= \{-48.0\mathbf{i} - 24.0\mathbf{j}\} \text{ N} \cdot \text{m} \qquad \textit{Ans.}$$

The result is shown in Fig. 4–25*b*.

Note that if axis *AB* is defined using a unit vector directed from *B* toward *A*, then in the above formulation $-\mathbf{u}_B$ would have to be used. This would lead to $M_{AB} = +53.67$ N · m. Consequently, $\mathbf{M}_{AB} = M_{AB}(-\mathbf{u}_B)$, and the above result would again be determined.

PROBLEMS

4–49. Determine the moment of the force **F** about the *Oa* axis. Express the result as a Cartesian vector.

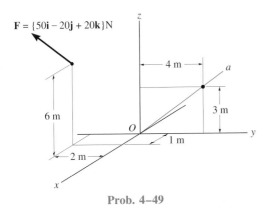

Prob. 4–49

4–50. Determine the moment of the force **F** about the *Aa* axis. Express the result as a Cartesian vector.

4–51. Determine the resultant moment of the two forces about the *Oa* axis. Express the result as a Cartesian vector.

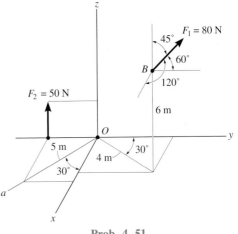

Prob. 4–51

*4–52. Determine the resultant moment of the two forces about the *aa* axis. Express the result as a Cartesian vector.

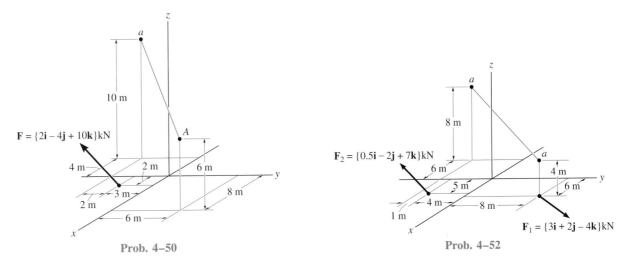

Prob. 4–50

Prob. 4–52

4–53. The cutting tool on the lathe exerts a force **F** on the shaft in the direction shown. Determine the moment of this force about the *y* axis of the shaft.

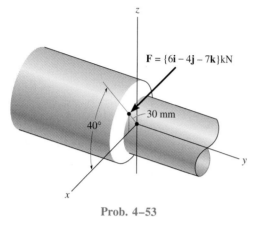

$$\mathbf{F} = \{6\mathbf{i} - 4\mathbf{j} - 7\mathbf{k}\}\text{kN}$$

30 mm

40°

Prob. 4–53

4–54. The hood of the automobile is supported by the strut *AB*, which exerts a force of $F = 24$ N on the hood. Determine the moment of this force about the hinged axis *y*.

4–55. When the strut *AB* is removed, a moment of required about the *y* axis to hold the hood of the automobile in the position shown. Determine the magnitude of the *smallest force* **F′** that should be applied at *B* to do this.

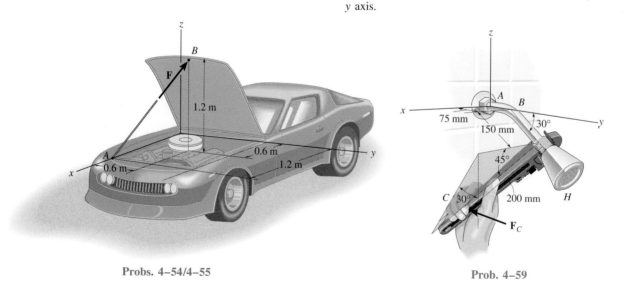

Probs. 4–54/4–55

4–56. Determine the magnitude of the moment of the force $\mathbf{F} = \{50\mathbf{i} - 20\mathbf{j} - 80\mathbf{k}\}$ N about the base line *AB* of the tripod.

4–57. Determine the magnitude of the moment of the force $\mathbf{F} = \{50\mathbf{i} - 20\mathbf{j} - 80\mathbf{k}\}$ N about the base line *BC* of the tripod.

4–58. Determine the magnitude of the moment of the force $\mathbf{F} = \{50\mathbf{i} - 20\mathbf{j} - 80\mathbf{k}\}$ N about the base line *CA* of the tripod.

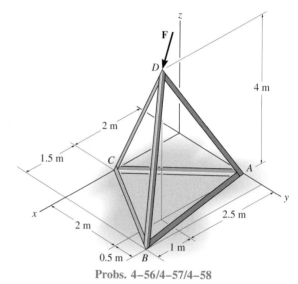

4 m

2 m

1.5 m

2.5 m

2 m

0.5 m   1 m

Probs. 4–56/4–57/4–58

4–59. The showerhead *H* is installed into the shower arm using a wrench. Determine the moment that the force $\mathbf{F}_C = \{8\mathbf{i} - 24\mathbf{j} + 16\mathbf{k}\}$ N applied to the wrench develops about the *y* axis.

75 mm

150 mm

30°

45°

200 mm

30°

Prob. 4–59

*4–60. The flex-headed ratchet wrench is subjected to a force of $P = 8$ N, applied perpendicular to the handle as shown. Determine the moment or torque this imparts along the vertical axis of the bolt at A.

4–61. If a torque or moment of 1000 N · mm is required to loosen the bolt at A, determine the force P that must be applied perpendicular to the handle of the flex-headed ratchet wrench.

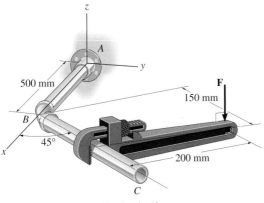

Probs. 4–60/4–61

4–62. A vertical force of $F = 60$ N is applied to the handle of the pipe wrench. Determine the moment that this force exerts along the axis AB (x axis) of the pipe assembly. Both the wrench and pipe assembly ABC lie in the x-y plane. *Suggestion:* Use a scalar analysis.

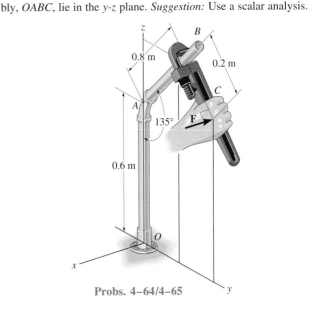

Prob. 4–62

4–63. Determine the magnitude of the vertical force $\mathbf{F}$ acting on the handle of the wrench so that this force produces a component of moment along the AB axis (x axis) of the pipe assembly of $(\mathbf{M}_A)_x = \{-5\mathbf{i}\}$ N · m. Both the pipe assembly ABC and the wrench lie in the x-y plane. *Suggestion:* Use a scalar analysis.

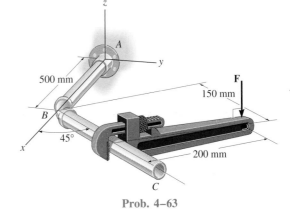

Prob. 4–63

*4–64. A horizontal force of $\mathbf{F} = \{-50\mathbf{i}\}$ N is applied perpendicular to the handle of the pipe wrench. Determine the moment that this force exerts along the axis OA (z axis) of the pipe assembly. Both the wrench and pipe assembly, $OABC$, lie in the y-z plane. *Suggestion:* Use a scalar analysis.

4–65. Determine the magnitude of the horizontal force $\mathbf{F} = -F\mathbf{i}$ acting on the handle of the wrench so that this force produces a component of moment along the OA axis (z axis) of the pipe assembly of $\mathbf{M}_z = \{4\mathbf{k}\}$ N · m. Both the wrench and the pipe assembly, $OABC$, lie in the y-z plane. *Suggestion:* Use a scalar analysis.

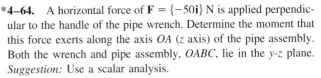

Probs. 4–64/4–65

4.6 Moment of a Couple

A *couple* is defined as two parallel forces that have the same magnitude, opposite directions, and are separated by a perpendicular distance d, Fig. 4–26. Since the resultant force of the two forces composing the couple is zero, the only effect of a couple is to produce a rotation or tendency of rotation in a specified direction. As a practical example, a couple is produced on the steering wheel of an automobile when turning the wheels.

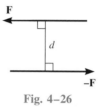

Fig. 4–26

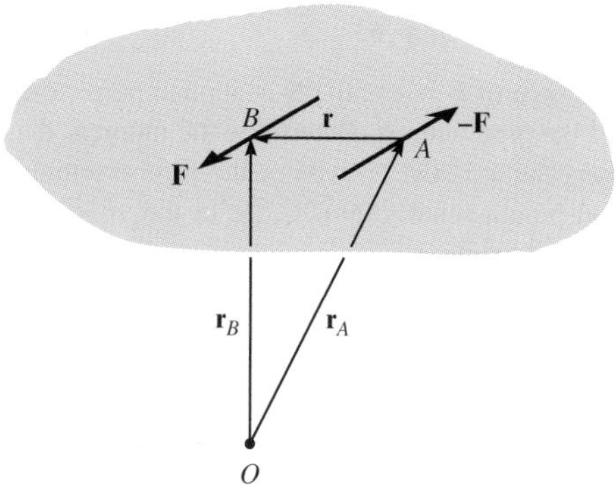

The moment produced by a couple, called a *couple moment*, is equivalent to the sum of the moments of both couple forces, determined about *any* arbitrary point O in space. To show this, consider position vectors $\mathbf{r}_A$ and $\mathbf{r}_B$, directed from O to points A and B lying on the line of action of $-\mathbf{F}$ and $\mathbf{F}$, Fig. 4–27. The couple moment computed about O is therefore

$$\mathbf{M} = \mathbf{r}_A \times (-\mathbf{F}) + \mathbf{r}_B \times (\mathbf{F})$$
$$= (\mathbf{r}_B - \mathbf{r}_A) \times \mathbf{F}$$

By the triangle law of vector addition, $\mathbf{r}_A + \mathbf{r} = \mathbf{r}_B$ or $\mathbf{r} = \mathbf{r}_B - \mathbf{r}_A$, so that

$$\mathbf{M} = \mathbf{r} \times \mathbf{F} \qquad (4\text{–}13)$$

This result indicates that a couple moment is a *free vector*, i.e., it can act at *any point*, since $\mathbf{M}$ depends *only* upon the position vector directed *between* the forces and *not* the position vectors $\mathbf{r}_A$ and $\mathbf{r}_B$, directed from point O to the forces. This concept is therefore unlike the moment of a force, which requires a definite point (or axis) about which moments are determined.

Fig. 4–27

Scalar Formulation. The moment of a couple, $\mathbf{M}$, Fig. 4–28, is defined as having a *magnitude* of

$$M = Fd \qquad (4\text{–}14)$$

where F is the magnitude of one of the forces and d is the perpendicular distance or moment arm between the forces. The *direction* and sense of the couple moment are determined by the right-hand rule, where the thumb indicates the direction when the fingers are curled with the sense of rotation caused by the two forces. In all cases, $\mathbf{M}$ acts perpendicular to the plane containing these forces.

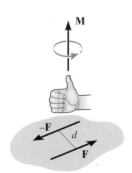

Fig. 4–28

Vector Formulation.

The moment of a couple can also be expressed by the vector cross product using Eq. 4–13, i.e.,

$$\mathbf{M} = \mathbf{r} \times \mathbf{F} \qquad (4\text{–}15)$$

Application of this equation is easily remembered if one thinks of taking the moments of both forces about a point lying on the line of action of one of the forces. For example, if moments are taken about point *A* in Fig. 4–27 the moment of $-\mathbf{F}$ is zero about this point and the moment of $\mathbf{F}$ is defined from Eq. 4–15. Therefore, in the formulation $\mathbf{r}$ is crossed with the force $\mathbf{F}$ to which it is directed.

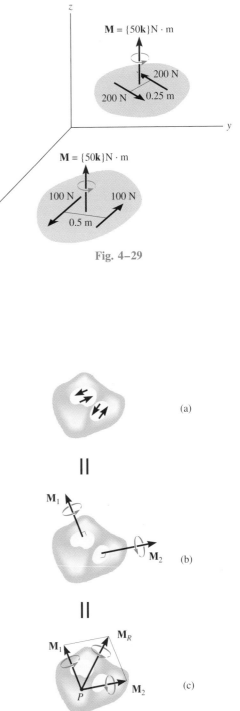

Fig. 4–29

Equivalent Couples.

Two couples are said to be equivalent if they produce the same moment. Since the moment produced by a couple is always perpendicular to the plane containing the couple forces, it is therefore necessary that the forces of equal couples lie either in the same plane or in planes that are *parallel* to one another. In this way, the direction of each couple moment will be the same, that is, perpendicular to the parallel planes. For example, the two couples shown in Fig. 4–29 are equivalent. One couple is produced by a pair of 100-N forces separated by a distance of $d = 0.5$ m, and the other is produced by a pair of 200-N forces separated by a distance of 0.25 m. Since the planes in which the forces act are parallel to the *x–y* plane, the moment produced by each of the couples may be expressed as $\mathbf{M} = \{50\mathbf{k}\}$ N · m.

Resultant Couple Moment.

Since couple moments are free vectors, they may be applied at any point *P* on a body and added vectorially. For example, the two couples acting on different planes of the rigid body in Fig. 4–30*a* may be replaced by their corresponding couple moments $\mathbf{M}_1$ and $\mathbf{M}_2$, Fig. 4–30*b*, and then these free vectors may be moved to the *arbitrary point P* and added to obtain the resultant couple moment $\mathbf{M}_R = \mathbf{M}_1 + \mathbf{M}_2$, shown in Fig. 4–30*c*.

If more than two couple moments act on the body, we may generalize this concept and write the vector resultant as

$$\mathbf{M}_R = \Sigma\,(\mathbf{r} \times \mathbf{F}) \qquad (4\text{–}16)$$

where each couple moment is computed in accordance with Eq. 4–15.

The following examples illustrate these concepts numerically. In general, problems projected in two dimensions should be solved using a scalar analysis, since the moment arms and force components are easy to compute.

Fig. 4–30

Example 4–11

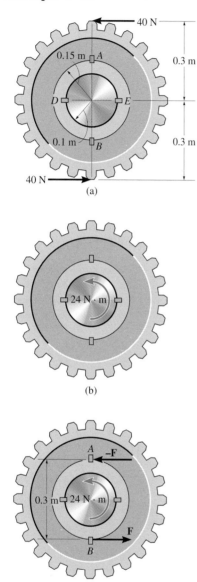

(a)

(b)

(c)

Fig. 4–31

A couple acts on the gear teeth as shown in Fig. 4–31*a*. Replace it by an equivalent couple having a pair of forces that act through (a) points *A* and *B*, and (b) points *D* and *E*.

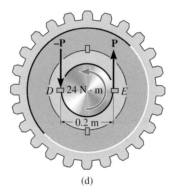

(d)

SOLUTION (*SCALAR ANALYSIS*)

The couple has a magnitude of $M = Fd = 40(0.6) = 24$ N · m and a direction that is out of the page since the forces tend to rotate counterclockwise. **M** is a free vector so that it can be placed at any point on the gear, Fig. 4–31*b*.

Part (a). To preserve the counterclockwise rotation of **M,** *horizontal* forces acting through points *A* and *B* must be directed as shown in Fig. 4–31*c*. The magnitude of each force is

$$M = Fd$$
$$24 \text{ N} \cdot \text{m} = F(0.3 \text{ m})$$
$$F = 80 \text{ N} \qquad \qquad \textit{Ans.}$$

Part (b). Likewise, for counterclockwise rotation, forces acting through points *D* and *E* must be *vertical* and directed as shown in Fig. 4–31*d*. The magnitude of each force is

$$M = Pd$$
$$24 \text{ N} \cdot \text{m} = P(0.2 \text{ m})$$
$$P = 120 \text{ N} \qquad \qquad \textit{Ans.}$$

Example 4–12

Determine the moment of the couple acting on the machine member shown in Fig. 4–32a.

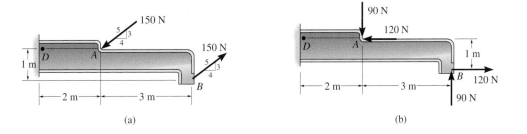

(a)   (b)

SOLUTION (*SCALAR ANALYSIS*)

Here it is somewhat difficult to determine the perpendicular distance between the forces and compute the couple moment as $M = Fd$. Instead, we can resolve each force into its horizontal and vertical components, $F_x = \frac{4}{5}(150 \text{ N}) = 120 \text{ N}$ and $F_y = \frac{3}{5}(150 \text{ N}) = 90 \text{ N}$, Fig. 4–32b, and then use the principle of moments. The couple moment can be determined about *any point*. For example, if point D is chosen, we have for all four forces,

$$\zeta+M = 120 \text{ N}(0 \text{ m}) - 90 \text{ N}(2 \text{ m}) + 90 \text{ N}(5 \text{ m}) + 120 \text{ N}(1 \text{ m})$$
$$= 390 \text{ N} \cdot \text{m} \uparrow \qquad\qquad\qquad\qquad\qquad\qquad Ans.$$

It is easier, however, to determine the moments about point A or B in order to *eliminate* the moment of the forces acting at the moment point. For point A, Fig. 4–32b, we have

$$\zeta+M = 90 \text{ N}(3 \text{ m}) + 120 \text{ N}(1 \text{ m})$$
$$= 390 \text{ N} \cdot \text{m} \uparrow \qquad\qquad\qquad\qquad\qquad\qquad Ans.$$

Show that one obtains this same result if moments are summed about point B. Notice also that the couple in Fig. 4–32a can be replaced by *two* couples in Fig. 4–32b. Using $M = Fd$, one couple has a moment of $M_1 = 90 \text{ N}(3 \text{ m}) = 270 \text{ N} \cdot \text{m}$ and the other has a moment of $M_2 = 120 \text{ N}(1 \text{ m}) = 120 \text{ N} \cdot \text{m}$. By the right-hand rule, both couple moments are counterclockwise and are therefore directed out of the page. Since these couples are free vectors, they can be moved to any point and added, which yields $M = 270 \text{ N} \cdot \text{m} + 120 \text{ N} \cdot \text{m} = 390 \text{ N} \cdot \text{m} \uparrow$, the same result determined above. **M** is a free vector and can therefore act at any point on the member, Fig. 4–32c. Also, realize that the external effect, such as the support reactions on the member, will be the *same* if the member supports the couple, Fig. 4–32a, or the couple moment, Fig. 4–32c.

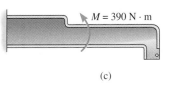

$M = 390 \text{ N} \cdot \text{m}$

(c)

Fig. 4–32

Example 4–13

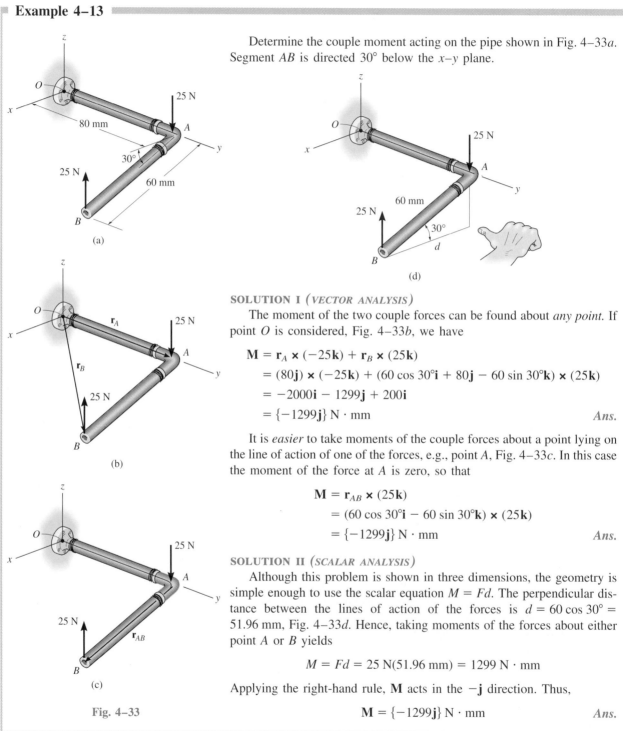

Determine the couple moment acting on the pipe shown in Fig. 4–33a. Segment AB is directed 30° below the x–y plane.

SOLUTION I (*VECTOR ANALYSIS*)

The moment of the two couple forces can be found about *any point*. If point O is considered, Fig. 4–33b, we have

$$\mathbf{M} = \mathbf{r}_A \times (-25\mathbf{k}) + \mathbf{r}_B \times (25\mathbf{k})$$
$$= (80\mathbf{j}) \times (-25\mathbf{k}) + (60 \cos 30°\mathbf{i} + 80\mathbf{j} - 60 \sin 30°\mathbf{k}) \times (25\mathbf{k})$$
$$= -2000\mathbf{i} - 1299\mathbf{j} + 200\mathbf{i}$$
$$= \{-1299\mathbf{j}\} \, \text{N} \cdot \text{mm} \qquad\qquad \textit{Ans.}$$

It is *easier* to take moments of the couple forces about a point lying on the line of action of one of the forces, e.g., point A, Fig. 4–33c. In this case the moment of the force at A is zero, so that

$$\mathbf{M} = \mathbf{r}_{AB} \times (25\mathbf{k})$$
$$= (60 \cos 30°\mathbf{i} - 60 \sin 30°\mathbf{k}) \times (25\mathbf{k})$$
$$= \{-1299\mathbf{j}\} \, \text{N} \cdot \text{mm} \qquad\qquad \textit{Ans.}$$

SOLUTION II (*SCALAR ANALYSIS*)

Although this problem is shown in three dimensions, the geometry is simple enough to use the scalar equation $M = Fd$. The perpendicular distance between the lines of action of the forces is $d = 60 \cos 30° = 51.96$ mm, Fig. 4–33d. Hence, taking moments of the forces about either point A or B yields

$$M = Fd = 25 \, \text{N}(51.96 \, \text{mm}) = 1299 \, \text{N} \cdot \text{mm}$$

Applying the right-hand rule, $\mathbf{M}$ acts in the $-\mathbf{j}$ direction. Thus,

$$\mathbf{M} = \{-1299\mathbf{j}\} \, \text{N} \cdot \text{mm} \qquad\qquad \textit{Ans.}$$

Fig. 4–33

Example 4–14

Replace the two couples acting on the pipe column in Fig. 4–34a by a resultant couple moment.

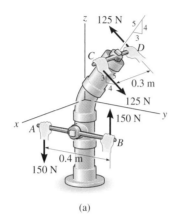

(a)

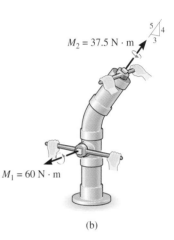

(b)

(c)

Fig. 4–34

SOLUTION (*VECTOR ANALYSIS*)

The couple moment $\mathbf{M}_1$, developed by the forces at A and B, can easily be determined from a scalar formulation.

$$M_1 = Fd = 150 \text{ N}(0.4 \text{ m}) = 60 \text{ N} \cdot \text{m}$$

By the right-hand rule, $\mathbf{M}_1$ acts in the $+\mathbf{i}$ direction, Fig. 4–34b. Hence,

$$\mathbf{M}_1 = \{60\mathbf{i}\} \text{ N} \cdot \text{m}$$

Vector analysis will be used to determine $\mathbf{M}_2$, caused by forces at C and D. If moments are computed about point D, Fig. 4–34a, $\mathbf{M}_2 = \mathbf{r}_{DC} \times \mathbf{F}_C$, then

$$\mathbf{M}_2 = \mathbf{r}_{DC} \times \mathbf{F}_C = (0.3\mathbf{i}) \times [125(\tfrac{4}{5})\mathbf{j} - 125(\tfrac{3}{5})\mathbf{k}]$$
$$= (0.3\mathbf{i}) \times [100\mathbf{j} - 75\mathbf{k}] = 30(\mathbf{i} \times \mathbf{j}) - 22.5(\mathbf{i} \times \mathbf{k})$$
$$= \{22.5\mathbf{j} + 30\mathbf{k}\} \text{ N} \cdot \text{m}$$

Try to establish $\mathbf{M}_2$ by using a scalar formulation, Fig. 4–34b.

Since $\mathbf{M}_1$ and $\mathbf{M}_2$ are free vectors, they may be moved to some arbitrary point P and added vectorially, Fig. 4–34c. The resultant couple moment becomes

$$\mathbf{M}_R = \mathbf{M}_1 + \mathbf{M}_2 = \{60\mathbf{i} + 22.5\mathbf{j} + 30\mathbf{k}\} \text{ N} \cdot \text{m} \qquad \textit{Ans.}$$

PROBLEMS

4–66. Determine the magnitude and sense of the couple moment.

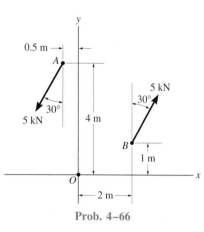

Prob. 4–66

4–67. Determine the magnitude and sense of the couple moment. Each force has a magnitude of $F = 8$ kN.

*4–68. If the couple moment has a magnitude of 250 N · m, determine the magnitude F of the couple forces.

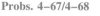

Probs. 4–67/4–68

4–69. Determine the magnitude and sense of the couple moment. Take $F = 260$ N.

4–70. If the couple moment has a magnitude of 300 N · m, determine the magnitude F of the couple forces.

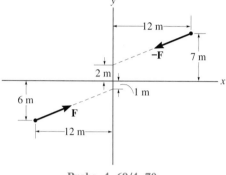

Probs. 4–69/4–70

4–71. Three couple moments act on the pipe assembly. Determine the magnitude of the resultant couple moment if $M_2 = 50$ N · m and $M_3 = 35$ N · m.

*4–72. Three couple moments act on the pipe assembly. Determine the magnitudes of $\mathbf{M}_2$ and $\mathbf{M}_3$ so that the resultant couple moment is zero.

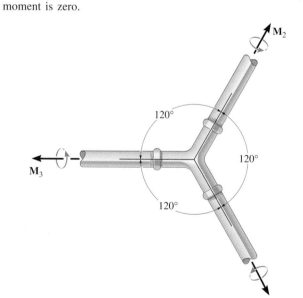

Probs. 4–71/4–72

4–73. Two couples act on the beam as shown. Determine the magnitude of **F** so that the resultant couple moment is 300 kN · m counterclockwise. Where on the beam does the resultant couple act?

4–74. If $F = 180$ kN, determine the magnitude and sense of the resultant couple moment. Where on the beam does the resultant couple act?

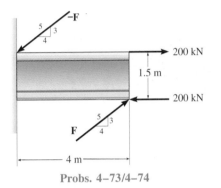

Probs. 4–73/4–74

4–75. Two couples act on the frame. If the resultant couple moment is to be zero, determine the distance d between the 80-N couple forces.

*4–76. Two couples act on the frame. If $d = 4$ m, determine the resultant couple moment. Compute the result by resolving each force into x and y components and (a) finding the moment of each couple (Eq. 4–13) and (b) summing the moments of all the force components about point A.

4–77. Two couples act on the frame. If $d = 4$ m, determine the resultant couple moment. Compute the result by resolving each force into x and y components and (a) finding the moment of each couple (Eq. 4–13) and (b) summing the moments of all the force components about point B.

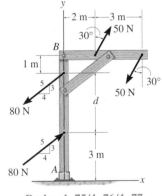

Probs. 4–75/4–76/4–77

4–78. The resultant couple moment created by the two couples acting on the disk is $\mathbf{M}_R = \{10\mathbf{k}\}$ kN · m. Determine the magnitude of force **T**.

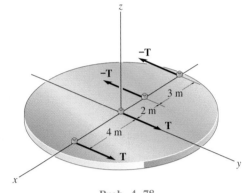

Prob. 4–78

4–79. Two couples act on the beam. Determine the magnitude of **F** so that the resultant couple moment is 450 kN · m, counterclockwise. Where on the beam does the resultant couple moment act?

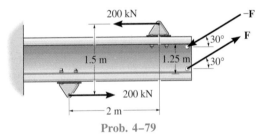

Prob. 4–79

*4–80. Determine the couple moment. Express the result as a Cartesian vector.

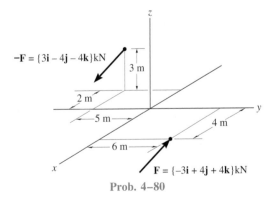

Prob. 4–80

4–81. Determine the couple moment. Express the result as a Cartesian vector.

*4–84. A couple acts on each of the handles of the minidual valve. Determine the magnitude and coordinate direction angles of the resultant couple moment.

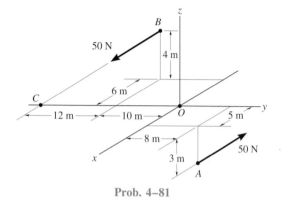

Prob. 4–81

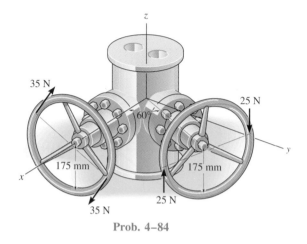

Prob. 4–84

4–82. Express the moment of the couple acting on the pipe in Cartesian vector form. What is the magnitude of the couple moment? Take $F = 125$ N.

4–83. If the couple moment acting on the pipe has a magnitude of 300 N · m, determine the magnitude F of the forces applied to the wrenches.

4–85. Express the moment of the couple acting on the pipe assembly in Cartesian vector form. Solve the problem (a) using Eq. 4–13, and (b) summing the moment of each force about point O. Take $\mathbf{F} = \{25\mathbf{k}\}$ N.

4–86. If the couple moment acting on the pipe has a magnitude of 400 N · m, determine the magnitude F of the vertical force applied to each wrench.

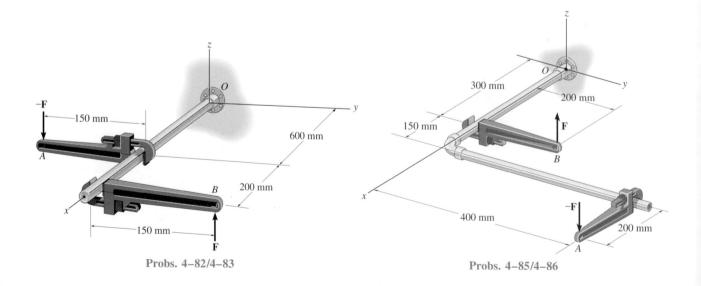

Probs. 4–82/4–83   Probs. 4–85/4–86

4–87. The meshed gears are subjected to the couple moments shown. Determine the magnitude and coordinate direction angles of the resultant couple moment.

4–89. Express the moment of the couple acting on the pipe assembly in Cartesian vector form. What is the magnitude of the couple moment?

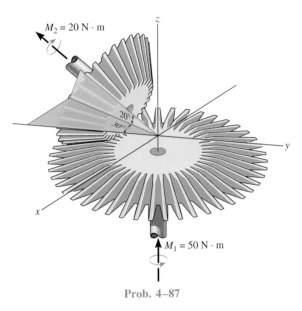

Prob. 4–87

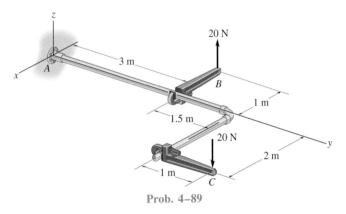

Prob. 4–89

4–88. Determine the resultant couple moment of the two couples that act on the assembly. Member *OB* lies in the *x-z* plane.

4–90. If $\mathbf{F} = \{100\ \mathbf{k}\}$ N, determine the couple moment that acts on the assembly. Express the result as a Cartesian vector. Member *BA* lies in the *x-y* plane.

4–91. If the magnitude of the resultant couple moment is 15 N · m, determine the magnitude *F* of the forces applied to the wrenches.

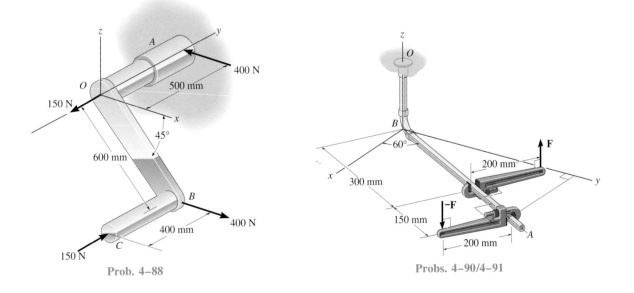

Prob. 4–88

Probs. 4–90/4–91

4.7 Movement of a Force on a Rigid Body

Many problems in statics, including the reduction of a force system to its simplest possible form, require moving a force from one point to another on a rigid body. Since a force tends to both *translate* and *rotate* a body, it is important that these two "external" effects remain the same if the force is moved from one point to another on the body. Two cases for the location of the point O to which the force is moved will now be considered.

Point O Is On the Line of Action of the Force. Consider the rigid body shown in Fig. 4–35a, which is subjected to the force **F** applied to point A. In order to move the force to point O without altering the external effects on the body, we will first apply equal but opposite forces **F** and $-$**F** at O, as shown in Fig. 4–35b. The two forces indicated by the slash across them can be canceled, leaving the force at point O as required, Fig. 4–35c. By using this construction procedure, an *equivalent system* has been maintained between each of the diagrams, as shown by the equal sign. Note, however, that the force has simply been "transmitted" along its line of action, from point A, Fig. 4–35a, to point O, Fig. 4–35c. In other words, the force can be considered as a *sliding vector* since it can act at any point O along its line of action. In Sec. 4.3 we referred to this concept as the *principle of transmissibility*. It can be formally stated as follows: *The external effects on a rigid body remain unchanged when a force, acting at a given point on the body, is applied to another point lying on the line of action of the force.* It is important to realize that only the *external effects*, such as the body's motion or the forces needed to support the body if it is stationary, remain *unchanged* after **F** is moved. Certainly the *internal effects* depend on where **F** is located. For example, when **F** acts at A, the internal forces in the body have a high intensity around A; whereas movement of **F** away from this point will cause these internal forces to decrease.

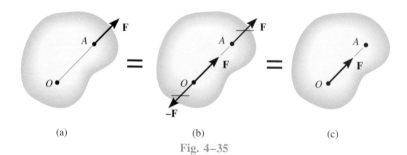

(a)   (b)   (c)

Fig. 4–35

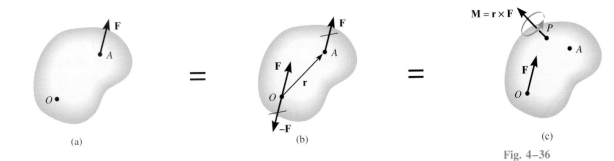

Fig. 4–36

Point *O* Is Not On the Line of Action of the Force. This case is shown in Fig. 4–36*a*, where **F** is to be moved to point *O* without altering the external effects on the body. Following the same procedure as before, we first apply equal but opposite forces **F** and −**F** at point *O*, Fig. 4–36*b*. Here the two forces indicated by a slash across them form a couple which has a moment that is perpendicular to **F** and is defined by the cross product **M** = **r** × **F**. Since the couple moment is a *free vector,* it may be applied at *any point P* on the body as shown in Fig. 4–36*c*. In addition to this couple moment, **F** now acts at point *O* as required.

As a physical illustration of the above two cases, consider the effect on the hand when holding the end *O* of a stick of negligible weight. If a vertical force **F** is applied at the other end *A* and the stick is held in the vertical position, Fig. 4–37*a*, then only the force is felt at the grip, regardless of where **F** is applied along its line of action *OA*. This is a consequence of the principle of transmissibility. When the stick is held in the horizontal position, Fig. 4–37*b*, the force at *A* has the effect of producing *both* a downward force at the grip *O* and a clockwise twist. Conversely, these same effects are felt at the grip if the force **F** is applied at the grip and the couple moment $M = Fd$ is applied to the stick.

Fig. 4–37

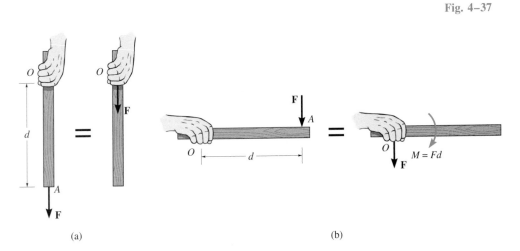

Example 4–15

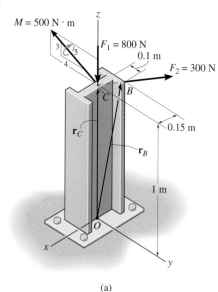

(a)

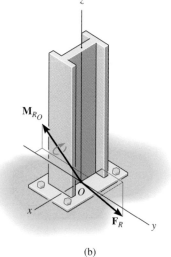

(b)

Fig. 4–38

A structural member is subjected to a couple moment $\mathbf{M}$ and forces $\mathbf{F}_1$ and $\mathbf{F}_2$ as shown in Fig. 4–38a. Replace this system by an equivalent resultant force and couple moment acting at its base, point O.

SOLUTION (*VECTOR ANALYSIS*)

The three-dimensional aspects of the problem can be simplified by using a Cartesian vector analysis. Expressing the forces and couple moment as Cartesian vectors, we have

$$\mathbf{F}_1 = \{-800\mathbf{k}\}\,\text{N}$$

$$\mathbf{F}_2 = (300\text{ N})\mathbf{u}_{CB} = (300\text{ N})\left(\frac{\mathbf{r}_{CB}}{r_{CB}}\right)$$

$$= 300\left[\frac{-0.15\mathbf{i} + 0.1\mathbf{j}}{\sqrt{(-0.15)^2 + (0.1)^2}}\right] = \{-249.6\mathbf{i} + 166.4\mathbf{j}\}\,\text{N}$$

$$\mathbf{M} = -500(\tfrac{4}{5})\mathbf{j} + 500(\tfrac{3}{5})\mathbf{k} = \{-400\mathbf{j} + 300\mathbf{k}\}\,\text{N}\cdot\text{m}$$

Force Summation

$$\mathbf{F}_R = \Sigma\mathbf{F}; \qquad \mathbf{F}_R = \mathbf{F}_1 + \mathbf{F}_2 = -800\mathbf{k} - 249.6\mathbf{i} + 166.4\mathbf{j}$$

$$= \{-249.6\mathbf{i} + 166.4\mathbf{j} - 800\mathbf{k}\}\,\text{N} \qquad\qquad Ans.$$

Moment Summation

$$\mathbf{M}_{R_O} = \Sigma\mathbf{M}_O;$$

$$\mathbf{M}_{R_O} = \mathbf{M} + \mathbf{r}_C \times \mathbf{F}_1 + \mathbf{r}_B \times \mathbf{F}_2$$

$$= (-400\mathbf{j} + 300\mathbf{k}) + (1\mathbf{k}) \times (-800\mathbf{k}) + \begin{vmatrix} \mathbf{i} & \mathbf{j} & \mathbf{k} \\ -0.15 & 0.1 & 1 \\ -249.6 & 166.4 & 0 \end{vmatrix}$$

$$= (-400\mathbf{j} + 300\mathbf{k}) + (\mathbf{0}) + (-166.4\mathbf{i} - 249.6\mathbf{j})$$

$$= \{-166.4\mathbf{i} - 649.6\mathbf{j} + 300\mathbf{k}\}\,\text{N}\cdot\text{m} \qquad\qquad Ans.$$

The results are shown in Fig. 4–38b.

4.8 Resultants of a Force and Couple System

When a rigid body is subjected to a system of forces and couple moments, it is often simpler to study the external effects on the body by using the force and couple moment resultants, rather than the force and couple moment system. To show how to simplify a system of forces and couple moments to their resultants, consider the rigid body in Fig. 4–39a. The force and couple moment system acting on it will be simplified by moving the forces and couple moments to the arbitrary point O. In this regard, the couple moment $\mathbf{M}$ is simply moved to O, since it is a free vector. Forces $\mathbf{F}_1$ and $\mathbf{F}_2$ are sliding vectors, and since O does not lie on the line of action of these forces, each must be moved to O in accordance with the procedure stated in Sec. 4.7. For example, when $\mathbf{F}_1$ is applied at O, a corresponding couple moment $\mathbf{M}_1 = \mathbf{r}_1 \times \mathbf{F}_1$ must also be applied to the body, Fig. 4–39b. By vector addition, the force and couple moment system shown in Fig. 4–39b can now be reduced to an *equivalent* resultant force $\mathbf{F}_R = \mathbf{F}_1 + \mathbf{F}_2$ and resultant couple moment $\mathbf{M}_{R_O} = \mathbf{M} + \mathbf{M}_1 + \mathbf{M}_2$ as shown in Fig. 4–39c. Note that both the magnitude and direction of $\mathbf{F}_R$ are independent of the location of point O; however, $\mathbf{M}_{R_O}$ depends upon this location, since the moments $\mathbf{M}_1$ and $\mathbf{M}_2$ are computed using the position vectors $\mathbf{r}_1$ and $\mathbf{r}_2$. Realize also that $\mathbf{M}_{R_O}$ is a free vector and can act at *any point* on the body, although point O is generally chosen as its point of application.

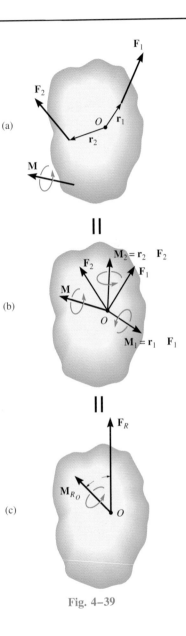

Fig. 4–39

PROCEDURE FOR ANALYSIS

The above method for simplifying any force and couple moment system to a resultant force acting at point O and a resultant couple moment will now be stated in general terms. To apply this method it is first necessary to establish an x, y, z coordinate system.

Three-Dimensional Systems. A Cartesian vector analysis is generally used to solve problems involving three-dimensional force and couple systems for which the force components and moment arms are difficult to determine.

Force Summation. The resultant force is equivalent to the vector sum of all the forces in the system; i.e.,

$$\mathbf{F}_R = \Sigma \mathbf{F} \qquad (4\text{--}17)$$

Moment Summation. The resultant couple moment is equivalent to the vector sum of all the couple moments in the system plus the moments about point O of all the forces in the system; i.e.,

$$\mathbf{M}_{R_O} = \Sigma\mathbf{M} + \Sigma\mathbf{M}_O \qquad (4\text{--}18)$$

Coplanar Force Systems.   Since force components and moment arms are easy to determine in two dimensions, a scalar analysis provides the most convenient solution to problems involving coplanar force systems. Assuming that the forces lie in the x–y plane and any couple moments are perpendicular to this plane (along the z axis), then the resultants are determined as follows:

Force Summation.   The *resultant force* $\mathbf{F}_R$ is equivalent to the vector sum of its two components $\mathbf{F}_{R_x}$ and $\mathbf{F}_{R_y}$. Each component is found from the scalar (algebraic) sum of the components of all the forces in the system that act in the same direction; i.e.,

$$\begin{aligned} F_{R_x} &= \Sigma F_x \\ F_{R_y} &= \Sigma F_y \end{aligned} \qquad (4\text{--}19)$$

Moment Summation.   The *resultant couple moment* $\mathbf{M}_{R_O}$ is perpendicular to the plane containing the forces and is equivalent to the scalar (algebraic) sum of all the couple moments in the system *plus* the moments about point O of all the forces in the system; i.e.,

$$M_{R_O} = \Sigma M + \Sigma M_O \qquad (4\text{--}20)$$

When determining the moments of the forces about O, it is generally advantageous to use the *principle of moments;* i.e., determine the moments of the *components* of each force rather than the moment of the force itself.

It is important to remember that, when applying any of these equations, attention should be paid to the sense of direction of the force components and the moments of the forces. If they are along the positive coordinate axes, they represent positive scalars; whereas if these components have a directional sense along the negative coordinate axes, they are negative scalars. By following this convention, a positive result, for example, indicates that the resultant vector has a sense of direction along the positive coordinate axis.

The following example illustrates these procedures numerically.

Example 4–16

Replace the forces acting on the brace shown in Fig. 4–40*a* by an equivalent resultant force and couple moment acting at point *A*.

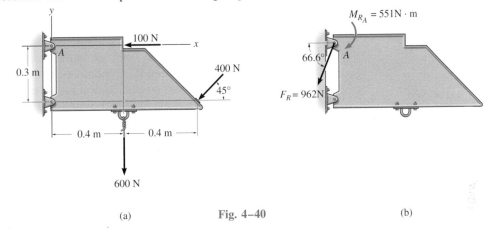

(a)   **Fig. 4–40**   (b)

SOLUTION (*SCALAR ANALYSIS*)

The principle of moments will be applied to the 400-N force, whereby the moments of its two rectangular components will be considered.

Force Summation.   The resultant force has *x* and *y* components of

$$\xrightarrow{+} F_{R_x} = \Sigma F_x; \quad F_{R_x} = -100 \text{ N} - 400 \cos 45° \text{ N} = -382.8 \text{ N} = 382.8 \text{ N} \leftarrow$$

$$+ \uparrow F_{R_y} = \Sigma F_y; \quad F_{R_y} = -600 \text{ N} - 400 \sin 45° \text{ N} = -882.8 \text{ N} = 882.8 \text{ N} \downarrow$$

As shown in Fig. 4–40*b*, $\mathbf{F}_R$ has a magnitude of

$$F_R = \sqrt{(F_{R_x})^2 + (F_{R_y})^2} = \sqrt{(382.8)^2 + (882.8)^2} = 962 \text{ N} \quad \textit{Ans.}$$

and a direction defined from the vector sketch of

$$\theta = \tan^{-1}\left(\frac{F_{R_y}}{F_{R_x}}\right) = \tan^{-1}\left(\frac{882.8}{382.8}\right) = 66.6° \quad _\theta\nearrow \quad \textit{Ans.}$$

Moment Summation.   The resultant couple moment $\mathbf{M}_{R_A}$ is determined by summing the moments of the forces about point *A*. Assuming that positive moments act counterclockwise, i.e., in the $+\mathbf{k}$ direction, we have

$$\zeta + M_{R_A} = \Sigma M_A;$$

$$M_{R_A} = 100 \text{ N}(0) - 600 \text{ N}(0.4 \text{ m}) - (400 \sin 45° \text{ N})(0.8 \text{ m})$$
$$- (400 \cos 45° \text{ N})(0.3 \text{ m})$$

$$= -551 \text{ N} \cdot \text{m} = 551 \text{ N} \cdot \text{m} \downarrow \quad \textit{Ans.}$$

In conclusion, when $\mathbf{M}_{R_A}$ and $\mathbf{F}_R$ act on the brace at point *A*, Fig. 4–40*b*, they will produce the *same* external effect or reactions at the supports as that produced by the force system in Fig. 4–40*a*.

4.9 Further Reduction of a Force and Couple System

Simplification to a Single Resultant Force. Consider now a special case for which the system of forces and couple moments acting on a rigid body, Fig. 4–41a, reduces at point O to a resultant force $\mathbf{F}_R = \Sigma\mathbf{F}$ and resultant couple moment $\mathbf{M}_{R_O} = \Sigma\mathbf{M}_O$, which are *perpendicular* to one another, Fig. 4–41b. Whenever this occurs, we can further simplify the force and couple moment system by moving $\mathbf{F}_R$ to another point P, located either on or off the body so that no resultant couple moment has to be applied to the body, Fig. 4–41c. In other words, if the force and couple moment system in Fig. 4–41a is reduced to a resultant system at point P, only the force resultant will have to be applied to the body, Fig. 4–41c.

The location of point P, measured from point O, can always be determined provided $\mathbf{F}_R$ and $\mathbf{M}_{R_O}$ are known, Fig. 4–41b. As shown in Fig. 4–41c, P must lie on the bb axis, which is perpendicular to the line of action of $\mathbf{F}_R$ and the aa axis. This point is chosen such that the distance d satisfies the scalar equation $M_{R_O} = F_R d$ or $d = M_{R_O}/F_R$. With $\mathbf{F}_R$ so located, it will produce the same external effects on the body as the force and couple moment system in Fig. 4–41a, or the force and couple moment resultants in Fig. 4–41b. We refer to the force and couple moment system in Fig. 4–41a as being *equivalent or equipollent* to the single force "system" in Fig. 4–41c, because each system produces the *same* resultant force and resultant moment when replaced at point O.

If a system of forces is either concurrent, coplanar, or parallel, it can always be reduced, as in the above case, to a single resultant force $\mathbf{F}_R$ acting through a unique point P. This is because in each of these cases $\mathbf{F}_R$ and $\mathbf{M}_{R_O}$ will always be perpendicular to each other when the force system is simplified at *any* point O.

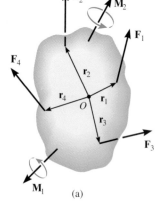

(a)

(b)

(c)

Concurrent Force Systems. A concurrent force system has been treated in detail in Chapter 2. Obviously, all the forces act at a point for which there is no resultant couple moment, so the point P is automatically specified, Fig. 4–42.

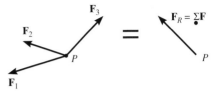

Fig. 4–42

Coplanar Force Systems. Coplanar force systems, which may include couple moments directed perpendicular to the plane of the forces, as shown in Fig. 4–43a, can be reduced to a single resultant force, because when each force in the system is moved to any point O in the x–y plane, it produces a couple moment that is *perpendicular* to the plane, i.e., in the $\pm\mathbf{k}$ direction. The resultant moment $\mathbf{M}_{R_O} = \Sigma\mathbf{M} + \Sigma\mathbf{r} \times \mathbf{F}$ is thus perpendicular to the resultant force $\mathbf{F}_R$, Fig. 4–43b; and so $\mathbf{F}_R$ can be positioned a distance d from O so as to create this same moment $\mathbf{M}_{R_O}$ about O, Fig. 4–43c.

(a)   (b)   (c)

Fig. 4–43

Parallel Force Systems. Parallel force systems, which can include couple moments that are perpendicular to the forces, as shown in Fig. 4–44a, can be reduced to a single resultant force, because when each force is moved to any point O in the x–y plane, it produces a couple moment that has components only about the x and y axes. The resultant moment $\mathbf{M}_{R_O} = \Sigma\mathbf{M} + \Sigma\mathbf{r} \times \mathbf{F}$ is thus perpendicular to the resultant force $\mathbf{F}_R$, Fig. 4–44b; and so $\mathbf{F}_R$ can be moved to a point a distance d away so that it produces the same moment about O.

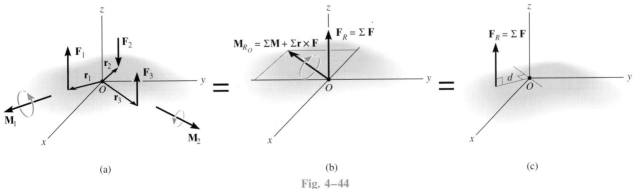

(a)   (b)   (c)

Fig. 4–44

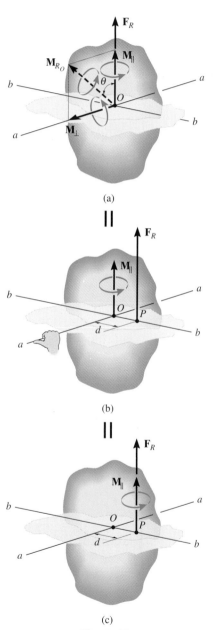

(a)

‖

(b)

‖

(c)

Fig. 4–45

PROCEDURE FOR ANALYSIS

The technique used to reduce a coplanar or parallel force system to a single resultant force follows the general procedure outlined in the previous section. First establish an x, y, z coordinate system. Then simplification requires the following two steps:

Force Summation. The resultant force $\mathbf{F}_R$ equals the sum of all the forces of the system, Fig. 4–41a and 4–41c; i.e.,

$$\mathbf{F}_R = \Sigma \mathbf{F} \qquad (4\text{--}21)$$

Moment Summation. The distance d from the arbitrary point O to the line of action of $\mathbf{F}_R$ is determined by equating the moment of $\mathbf{F}_R$ about O, $\mathbf{M}_{R_O}$, Fig. 4–41c, to the sum of the moments about point O of all the couple moments and forces in the system, $\Sigma \mathbf{M}_O$, Fig. 4–41a; i.e.,

$$\mathbf{M}_{R_O} = \Sigma \mathbf{M}_O \qquad (4\text{--}22)$$

Most often a scalar analysis can be used to apply these equations, since the force components and the moment arms are easily determined for either coplanar or parallel force systems.

Reduction to a Wrench. In the general case, the force and couple moment system acting on a body, Fig. 4–39a, will reduce to a single resultant force $\mathbf{F}_R$ and couple moment $\mathbf{M}_{R_O}$ at O which are *not* perpendicular. Instead, $\mathbf{F}_R$ will act at an angle θ from $\mathbf{M}_{R_O}$, Fig. 4–39c. As shown in Fig. 4–45a, however, $\mathbf{M}_{R_O}$ may be resolved into two components: one perpendicular, $\mathbf{M}_\perp$, and the other parallel, $\mathbf{M}_\|$, to the line of action of $\mathbf{F}_R$. As in the previous discussion, the perpendicular component $\mathbf{M}_\perp$ may be *eliminated* by moving $\mathbf{F}_R$ to point P, as shown in Fig. 4–45b. This point lies on axis bb, which is perpendicular to both $\mathbf{M}_{R_O}$ and $\mathbf{F}_R$. In order to maintain an equivalency of loading, the distance from O to P is $d = M_\perp / F_R$. Furthermore, when $\mathbf{F}_R$ is applied at P, the moment of $\mathbf{F}_R$ tending to cause rotation of the body *about O* is in the *same direction* as $\mathbf{M}_\perp$, Fig. 4–45a. Finally, since $\mathbf{M}_\|$ is a free vector, it may be moved to P so that it is collinear with $\mathbf{F}_R$, Fig. 4–45c. This combination of a collinear force and couple moment is called a *wrench* or *screw*. The *axis of the wrench* has the same line of action as the force. Hence, the wrench tends to cause both a translation along and a rotation about this axis. Comparing Fig. 4–45a to Fig. 4–45c, it is seen that a general force and couple moment system acting on a body can be reduced to a wrench. The axis of the wrench and a point through which this axis passes are unique and can always be determined.

Example 4–17

Replace the system of forces acting on the aircraft wing shown in Fig. 4–46a by an equivalent resultant force. Specify the distance the force acts from point A on the fuselage.

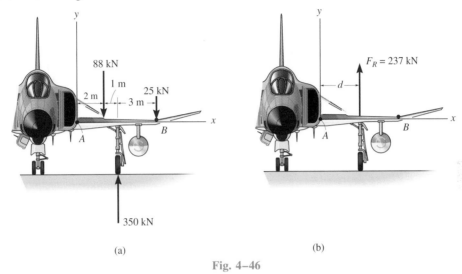

(a)   (b)

Fig. 4–46

SOLUTION

Here the force system is parallel and coplanar. We will use the established x, y, z axes to solve this problem.

Force Summation.   From Fig. 4–46a the force resultant F_R is

$$+\uparrow F_R = \Sigma F; \quad F_R = -88 \text{ kN} + 350 \text{ kN} - 25 \text{ kN} = 237 \text{ kN} \uparrow \qquad Ans.$$

Moment Summation.   Moments will be summed about point A. Considering counterclockwise rotations as positive, i.e., positive moment vectors act in the $+\mathbf{k}$ direction, then from Figs. 4–46a and 4–46b we require the moment of $\mathbf{F}_R$ about A to equal the moments of the force system about A; i.e.,

$$\zeta +M_{R_A} = \Sigma M_A;$$
$$237 \text{ kN}(d) = -(88 \text{ kN})(2 \text{ m}) + (350 \text{ kN})(3 \text{ m}) - (25 \text{ kN})(6 \text{ m})$$
$$237 \text{ kN}(d) = 724 \text{ kN} \cdot \text{m}$$
$$d = 3.05 \text{ m} \qquad\qquad Ans.$$

Note that using a clockwise sign convention would yield the same result. Since d is *positive*, $\mathbf{F}_R$ acts to the right of A as shown in Fig. 4–46b. Try and solve this problem by summing moments about point B and show that $d' = 2.95$ m, measured to the left of B.

Example 4–18

The beam AE in Fig. 4–47a is subjected to a system of coplanar forces. Determine the magnitude, direction, and location on the beam of a resultant force which is equivalent to the given system of forces.

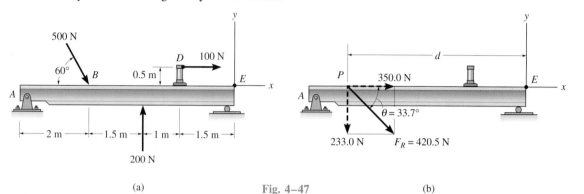

(a)   Fig. 4–47   (b)

SOLUTION

The origin of coordinates is arbitrarily located at point E as shown in Fig. 4–47a.

Force Summation.   Resolving the 500-N force into x and y components, and summing the force components, yields

$$\xrightarrow{+} F_{R_x} = \Sigma F_x; \qquad F_{R_x} = 500 \cos 60° \text{ N} + 100 \text{ N} = 350.0 \text{ N} \rightarrow$$

$$+\uparrow F_{R_y} = \Sigma F_y; \qquad F_{R_y} = -500 \sin 60° \text{ N} + 200 \text{ N} = -233.0 \text{ N}$$

$$= 233.0 \text{ N} \downarrow$$

The magnitude and direction of the resultant force are established from the vector addition shown in Fig. 4–47b. We have

$$F_R = \sqrt{(350.0)^2 + (233.0)^2} = 420.5 \text{ N} \qquad \qquad Ans.$$

$$\theta = \tan^{-1}\left(\frac{233.0}{350.0}\right) = 33.7° \quad \diagdown_\theta \qquad \qquad Ans.$$

Moment Summation.   Moments will be summed about point E. Hence, from Figs. 4–47a and 4–47b, we require the moments of the components of $\mathbf{F}_R$ (or the moment of $\mathbf{F}_R$) about point E to equal the moments of the force system about E; i.e.,

$$\zeta + M_{R_E} = \Sigma M_E;$$

$$233.0 \text{ N}(d) + 350.0 \text{ N}(0) = (500 \sin 60° \text{ N})(4 \text{ m}) + (500 \cos 60° \text{ N})(0)$$

$$-(100 \text{ N})(0.5 \text{ m}) - (200 \text{ N})(2.5 \text{ m})$$

$$d = \frac{1182.1}{233.0} = 5.07 \text{ m} \qquad \qquad Ans.$$

Example 4–19

The jib crane shown in Fig. 4–48*a* is subjected to three coplanar forces. Replace this loading by an equivalent resultant force and specify where the resultant's line of action intersects the column *AB* and boom *BC*.

SOLUTION

Force Summation. Resolving the 250-N force into *x* and *y* components and summing the force components yields

$$\xrightarrow{+} F_{R_x} = \Sigma F_x; \qquad F_{R_x} = -250 \text{ N}(\tfrac{3}{5}) - 175 \text{ N} = -325 \text{ N} = 325 \text{ N} \leftarrow$$
$$+ \uparrow F_{R_y} = \Sigma F_y; \qquad F_{R_y} = -250 \text{ N}(\tfrac{4}{5}) - 60 \text{ N} = -260 \text{ N} = 260 \text{ N} \downarrow$$

As shown by the vector addition in Fig. 4–48*b*,

$$F_R = \sqrt{(325)^2 + (260)^2} = 416 \text{ N} \qquad\qquad Ans.$$

$$\theta = \tan^{-1}\left(\frac{260}{325}\right) = 38.7° \quad {}^{\theta}\!\nearrow \qquad\qquad Ans.$$

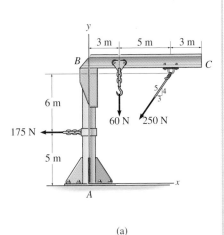

(a)

Moment Summation. Moments will be summed about point *A*. If the line of action of $\mathbf{F}_R$ intersects *AB*, Fig. 4–48*b*, we require the moment of the components of $\mathbf{F}_R$ in Fig. 4–48*b* about *A* to equal the moments of the force system in Fig. 4–48*a* about *A*; i.e.,

$$\zeta + M_{R_A} = \Sigma M_A; \qquad 325 \text{ N}(y) + 260 \text{ N}(0)$$
$$= 175 \text{ N}(5 \text{ m}) - 60 \text{ N}(3 \text{ m}) + 250 \text{ N}(\tfrac{3}{5})(11 \text{ m}) - 250 \text{ N}(\tfrac{4}{5})(8 \text{ m})$$
$$y = 2.29 \text{ m} \qquad\qquad Ans.$$

By the principle of transmissibility, $\mathbf{F}_R$ can also intersect *BC*, Fig. 4–48*b*, in which case we have

$$\zeta + M_{R_A} = \Sigma M_A; \qquad 325 \text{ N}(11 \text{ m}) - 260 \text{ N}(x)$$
$$= 175 \text{ N}(5 \text{ m}) - 60 \text{ N}(3 \text{ m}) + 250 \text{ N}(\tfrac{3}{5})(11 \text{ m}) - 250 \text{ N}(\tfrac{4}{5})(8 \text{ m})$$
$$x = 10.9 \text{ m} \qquad\qquad Ans.$$

We can also solve for these positions by assuming $\mathbf{F}_R$ acts at the arbitrary point (x, y) on its line of action, Fig. 4–48*b*. Summing moments about point *A* yields

$$\zeta + M_{R_A} = \Sigma M_A; \qquad 325 \text{ N}(y) - 260 \text{ N}(x)$$
$$= 175 \text{ N}(5 \text{ m}) - 60 \text{ N}(3 \text{ m}) + 250 \text{ N}(\tfrac{3}{5})(11 \text{ m}) - 250 \text{ N}(\tfrac{4}{5})(8 \text{ m})$$
$$325y - 260x = 745$$

which is the equation of the colored dashed line in Fig. 4–48*b*. To find the points of intersection with the crane, set $x = 0$, then $y = 2.29$ m, and set $y = 11$ m, then $x = 10.9$ m.

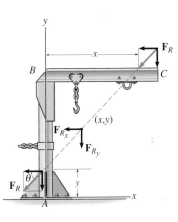

(b)

Fig. 4–48

Example 4–20

The slab in Fig. 4–49a is subjected to four parallel forces. Determine the magnitude and direction of a resultant force equivalent to the given force system, and locate its point of application on the slab.

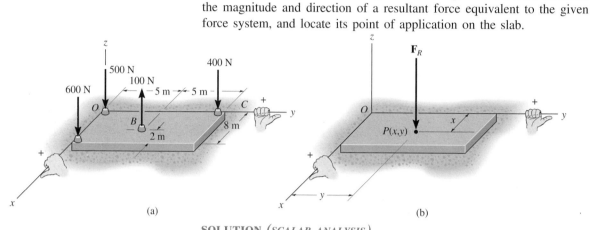

Fig. 4–49

SOLUTION (*SCALAR ANALYSIS*)

Force Summation. From Fig. 4–49a, the resultant force is

$$+ \uparrow F_R = \Sigma F; \quad F_R = -600 \text{ N} + 100 \text{ N} - 400 \text{ N} - 500 \text{ N}$$
$$= -1400 \text{ N} = 1400 \text{ N} \downarrow \qquad \textit{Ans.}$$

Moment Summation. We require the moment about the x axis of the resultant force, Fig. 4–49b, to be equal to the sum of the moments about the x axis of all the forces in the system, Fig. 4–49a. The moment arms are determined from the y coordinates since these coordinates represent the *perpendicular distances* from the x axis to the lines of action of the forces. Using the right-hand rule, where positive moments act in the $+\mathbf{i}$ direction, we have

$$M_{R_x} = \Sigma M_x;$$
$$-(1400 \text{ N})y = 600 \text{ N}(0) + 100 \text{ N}(5 \text{ m}) - 400 \text{ N}(10 \text{ m}) + 500 \text{ N}(0)$$
$$-1400y = -3500 \qquad y = 2.50 \text{ m} \qquad \textit{Ans.}$$

In a similar manner, assuming that positive moments act in the $+\mathbf{j}$ direction, a moment equation can be written about the y axis using moment arms defined by the x coordinates of each force.

$$M_{R_y} = \Sigma M_y;$$
$$(1400 \text{ N})x = 600 \text{ N}(8 \text{ m}) - 100 \text{ N}(6 \text{ m}) + 400 \text{ N}(0) + 500 \text{ N}(0)$$
$$1400x = 4200 \qquad x = 3.00 \text{ m} \qquad \textit{Ans.}$$

Hence, a force of $F_R = 1400$ N placed at point $P(3.00 \text{ m}, 2.50 \text{ m})$ on the slab, Fig. 4–49b, is equivalent to the parallel force system acting on the slab in Fig. 4–49a.

Example 4–21

Three parallel bolting forces act on the rim of the circular cover plate in Fig. 4–50*a*. Determine the magnitude and direction of a resultant force equivalent to the given force system and locate its point of application, *P*, on the cover plate.

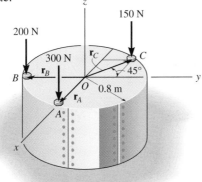

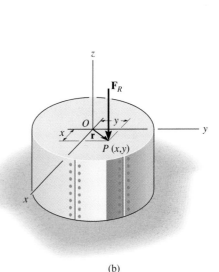

SOLUTION (*VECTOR ANALYSIS*)   (a)

(b)

Force Summation.   From Fig. 4–50*a*, the force resultant $\mathbf{F}_R$ is

Fig. 4–50

$$\mathbf{F}_R = \Sigma\mathbf{F}; \qquad \mathbf{F}_R = -300\mathbf{k} - 200\mathbf{k} - 150\mathbf{k}$$
$$= \{-650\mathbf{k}\}\,\text{N} \qquad\qquad Ans.$$

Moment Summation.   Choosing point *O* as a reference for computing moments and assuming that $\mathbf{F}_R$ acts at a point $P(x, y)$, Fig. 4–50*b*, we require

$$\mathbf{M}_{R_O} = \Sigma\mathbf{M}_O; \quad \mathbf{r} \times \mathbf{F}_R = \mathbf{r}_A \times (-300\mathbf{k}) + \mathbf{r}_B \times (-200\mathbf{k}) + \mathbf{r}_C \times (-150\mathbf{k})$$
$$(x\mathbf{i} + y\mathbf{j}) \times (-650\mathbf{k}) = (0.8\mathbf{i}) \times (-300\mathbf{k}) + (-0.8\mathbf{j}) \times (-200\mathbf{k})$$
$$+ (-0.8 \sin 45°\mathbf{i} + 0.8 \cos 45°\mathbf{j}) \times (-150\mathbf{k})$$
$$650x\mathbf{j} - 650y\mathbf{i} = 240\mathbf{j} + 160\mathbf{i} - 84.85\mathbf{j} - 84.85\mathbf{i}$$

Equating the corresponding **j** and **i** components yields

$$650x = 240 - 84.85 \qquad\qquad (1)$$
$$-650y = 160 - 84.85 \qquad\qquad (2)$$

Solving these equations, we obtain the coordinates of point *P*,

$$x = 0.239 \text{ m} \qquad y = -0.116 \text{ m} \qquad\qquad Ans.$$

The negative sign indicates that it was wrong to have assumed a $+y$ position for $\mathbf{F}_R$ as shown in Fig. 4–50*b*.

As a review, try to establish Eqs. 1 and 2 by using a scalar analysis; i.e., apply the sum of moments about the *x* and *y* axes, respectively.

PROBLEMS

*4–92. Replace the force at A by an equivalent force and couple moment at point O.

4–93. Replace the force at A by an equivalent force and couple moment at point P.

*4–96. Replace the force system by an equivalent force and couple moment at point O.

4–97. Replace the force system by an equivalent force and couple moment at point P.

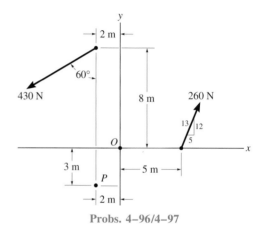

Probs. 4–92/4–93

Probs. 4–96/4–97

4–94. Replace the force at A by an equivalent force and couple moment at point O.

4–95. Replace the force at A by an equivalent force and couple moment at point P.

4–98. Replace the force and couple system by an equivalent force and couple moment at point O.

4–99. Replace the force and couple system by an equivalent force and couple moment at point P.

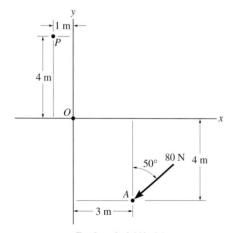

Probs. 4–94/4–95

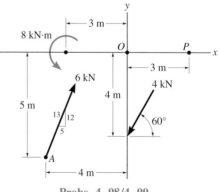

Probs. 4–98/4–99

*4–100. Replace the force system by a single force resultant and specify its point of application, measured along the *x* axis from point *O*.

4–101. Replace the force system by a single force resultant and specify its point of application, measured along the *x* axis from point *P*.

4–103. Replace the three forces acting on the shaft by a single resultant force. Specify where the force acts, measured from end *A*.

*4–104. Replace the three forces acting on the shaft by a single resultant force. Specify where the force acts, measured from end *B*.

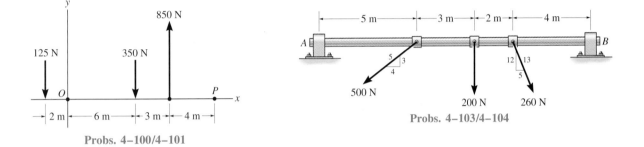

Probs. 4–100/4–101

Probs. 4–103/4–104

4–102. The forces $\mathbf{F}_1 = \{-4\mathbf{i} + 2\mathbf{j} - 3\mathbf{k}\}$ kN and $\mathbf{F}_2 = \{3\mathbf{i} - 4\mathbf{j} - 2\mathbf{k}\}$ kN act on the end of the beam. Replace these forces by an equivalent force and couple moment acting at point *O*.

4–105. Replace the three forces acting on the beam by a single resultant force. Specify where the force acts, measured from end *A*.

4–106. Replace the three forces acting on the beam by a single resultant force. Specify where the force acts, measured from *B*.

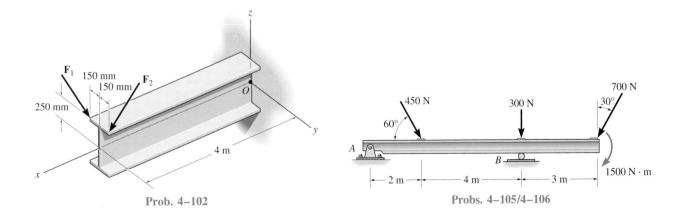

Prob. 4–102

Probs. 4–105/4–106

4–107. Replace the loading on the frame by a single resultant force. Specify where its line of action intersects member *AB*, measured from *A*.

*4–108. Replace the loading on the frame by a single resultant force. Specify where its line of action intersects member *AB*, measured from *B*.

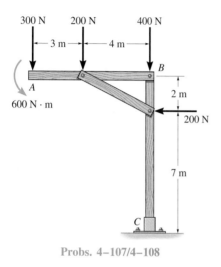

Probs. 4–107/4–108

4–109. Replace the loading on the frame by a single resultant force. Specify where its line of action intersects member *AB*, measured from *A*.

4–110. Replace the loading on the frame by a single resultant force. Specify where its line of action intersects member *CD*, measured from end *C*.

4–111. The resultant force of the wind and the weights of the various components act on the sign. Determine the equivalent force and moment acting at its base, *A*.

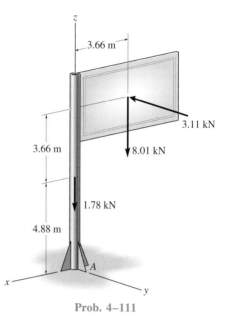

Prob. 4–111

*4–112. The forces and couple moments which are exerted on the toe and heel plates of a snow ski are $\mathbf{F}_t = \{-50\mathbf{i} + 80\mathbf{j} - 158\mathbf{k}\}$ N, $\mathbf{M}_t = \{-6\mathbf{i} + 4\mathbf{j} + 2\mathbf{k}\}$ N · m, and $\mathbf{F}_h = \{-20\mathbf{i} + 60\mathbf{j} - 250\mathbf{k}\}$ N, $\mathbf{M}_h = \{-20\mathbf{i} + 8\mathbf{j} + 3\mathbf{k}\}$ N · m, respectively. Replace this system by an equivalent force and couple moment acting at point *O*. Express the results in Cartesian vector form.

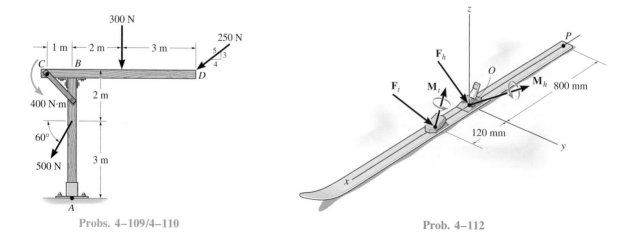

Probs. 4–109/4–110   **Prob. 4–112**

4–113. The forces and couple moments that are exerted on the toe and heel plates of a snow ski are $\mathbf{F}_t = \{-50\mathbf{i} + 80\mathbf{j} - 158\mathbf{k}\}\,\text{N}$, $\mathbf{M}_t = \{-6\mathbf{i} + 4\mathbf{j} + 2\mathbf{k}\}\,\text{N} \cdot \text{m}$, and $\mathbf{F}_h = \{-20\mathbf{i} + 60\mathbf{j} - 250\mathbf{k}\}\,\text{N}$, $\mathbf{M}_h = \{-20\mathbf{i} + 8\mathbf{j} + 3\mathbf{k}\}\,\text{N} \cdot \text{m}$, respectively. Replace this system by an equivalent force and couple moment acting at point P. Express the results in Cartesian vector form.

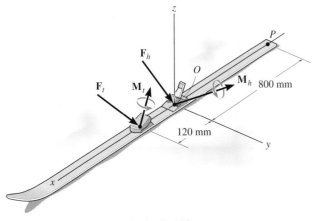

Prob. 4–113

4–114. The belt passing over the pulley is subjected to forces $\mathbf{F}_1$ and $\mathbf{F}_2$, each having a magnitude of 40 N. $\mathbf{F}_1$ acts in the $-\mathbf{k}$ direction. Replace these forces by an equivalent force and couple moment at point A. Express the result in Cartesian vector form. Set $\theta = 0°$ so that $\mathbf{F}_2$ acts in the $-\mathbf{j}$ direction.

4–115. The belt passing over the pulley is subjected to two forces $\mathbf{F}_1$ and $\mathbf{F}_2$, each having a magnitude of 40 N. $\mathbf{F}_1$ acts in the $-\mathbf{k}$ direction. Replace these forces by an equivalent force and couple moment at point A. Express the result in Cartesian vector form. Take $\theta = 45°$.

Probs. 4–114/4–115

*4–116.** Three parallel bolting forces act on the circular plate. Determine the resultant force, and specify its location (x, z) on the plate. $F_A = 200\,\text{N}$, $F_B = 100\,\text{N}$, and $F_C = 400\,\text{N}$.

4–117. The three parallel bolting forces act on the circular plate. If the force at A has a magnitude of $F_A = 200\,\text{N}$, determine the magnitudes of $\mathbf{F}_B$ and $\mathbf{F}_C$ so that the resultant force $\mathbf{F}_R$ of the system has a line of action that coincides with the y axis. *Hint:* This requires $\Sigma M_x = 0$ and $\Sigma M_z = 0$. Why?

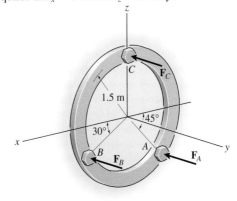

Probs. 4–116/4–117

4–118. A biomechanical model of the lumbar region of the human trunk is shown. The forces acting in the four muscle groups consist of $F_R = 35\,\text{N}$ for the rectus, $F_O = 45\,\text{N}$ for the oblique, $F_L = 23\,\text{N}$ for the lumbar latissimus dorsi, and $F_E = 32\,\text{N}$ for the erector spinae. These loadings are symmetric with respect to the y-z plane. Replace this system of parallel forces by an equivalent force and couple moment acting at the spine, point O. Express the results in Cartesian vector form.

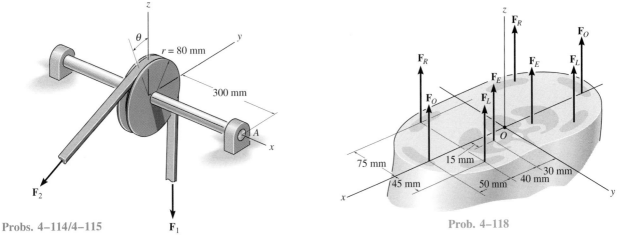

Prob. 4–118

4–119. A biomechanical model of the lumbar region of the human trunk is shown. The forces acting in the four muscle groups consist of $F_R = 35$ N for the rectus, $F_O = 45$ N for the oblique, $F_L = 23$ N for the lumbar latissimus dorsi, and $F_E = 32$ N for the erector spinae. These loadings are symmetric with respect to the y-z plane. Determine the resultant force equivalent to the given parallel force system, and locate its point of application (x, y) on the trunk.

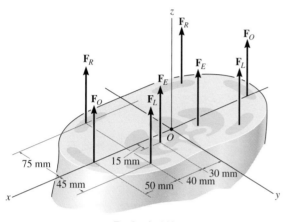

Prob. 4–119

***4–120.** The building slab is subjected to four parallel column loadings. Determine the equivalent resultant force and specify its location (x, y) on the slab. Take $F_1 = 30$ kN, $F_2 = 40$ kN.

4–121. The building slab is subjected to four parallel column loadings. Determine the equivalent resultant force and specify its location (x, y) on the slab. Take $F_1 = 20$ kN, $F_2 = 50$ kN.

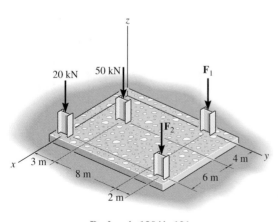

Probs. 4–120/4–121

4–122. A force and couple act on the pipe assembly. Replace this system by an equivalent resultant force. Specify the point where the line of action of the resultant force intersects the x axis. The pipe lies in the x-y plane. Take $F_1 = F_2 = 45$ N.

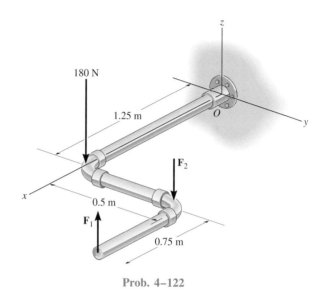

Prob. 4–122

4–123. A force and couple act on the pipe assembly. If $F_1 = 50$ N and $F_2 = 80$ N, replace this system by an equivalent resultant force and couple moment acting at O. Express the results in Cartesian vector form.

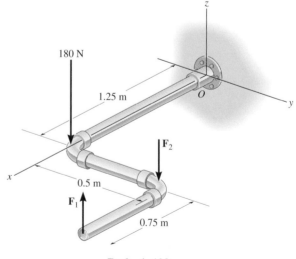

Prob. 4–123

*4–124. The three forces acting on the block each have a magnitude of 10 N. Replace this system by a wrench and specify the point where the wrench intersects the z axis, measured from point O.

4–126. The pipe assembly is subjected to the action of a wrench at B and a couple at A. Determine the magnitude F of the couple forces so that the system can be simplified to a wrench acting at point C.

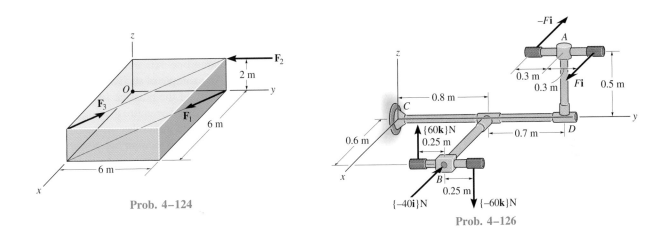

Prob. 4–124

Prob. 4–126

4–125. The pipe assembly is subjected to the action of a wrench at B and a couple at A. Simplify this system to a resultant wrench and specify the location of the wrench along the axis of pipe CD, measured from point C. Set F = 40 N.

4–127. Replace the three forces acting on the plate by a wrench. Specify the magnitude of the force and couple moment for the wrench and the point P(x, y) where its line of action intersects the plate.

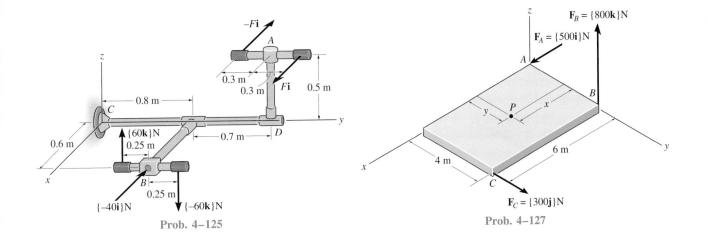

Prob. 4–125

Prob. 4–127

4.10 Reduction of a Simple Distributed Loading

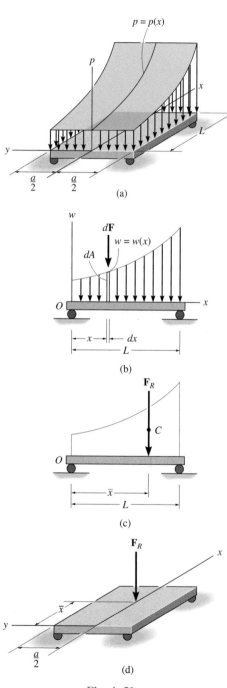

Fig. 4–51

In many situations a very large surface area of a body may be subjected to *distributed loadings* such as those caused by wind, fluids, or simply the weight of material supported over the body's surface. The *intensity* of these loadings at each point on the surface is defined as the *pressure p* (force per unit area), which can be measured in units of lb/ft² or pascals (Pa), where 1 Pa = 1 N/m².

In this section we will consider the most common case of a distributed pressure loading, which is *uniform* along one axis of a flat rectangular body upon which the loading is applied.* An example of such a loading is shown in Fig. 4–51a. The direction of the intensity of the pressure load is indicated by arrows shown on the *load-intensity diagram*. The entire loading on the plate is therefore a system of parallel forces, infinite in number and each acting on a separate differential area of the plate. Here the *loading function, p = p(x)* Pa, is only a function of x since the pressure is uniform along the y axis. If we multiply $p = p(x)$ by the *width a* m of the plate, we obtain $w = [p(x) \text{ N/m}^2]a \text{ m} = w(x)$ N/m. This loading function, shown in Fig. 4–51b, is a measure of load distribution along the line y = 0 which is in the plane of symmetry of the loading, Fig. 4–51a. As noted, it is measured as a force per unit length, rather than a force per unit area. Consequently, the load-intensity diagram for $w = w(x)$ can be represented by a system of *coplanar* parallel forces, shown in two dimensions in Fig. 4–51b. Using the methods of Sec. 4.9, this system of forces can be simplified to a single resultant force $\mathbf{F}_R$ and its location $\bar{x}$ specified, Fig. 4–51c.

Magnitude of Resultant Force. From Eq. 4–21 ($F_R = \Sigma F$), the magnitude of $\mathbf{F}_R$ is equivalent to the sum of all the forces in the system. In this case integration must be used, since there is an infinite number of parallel forces $d\mathbf{F}$ acting along the plate, Fig. 4–51b. Since $d\mathbf{F}$ is acting on an element of length dx, and $w(x)$ is a force per unit length, then at the location x, $dF = w(x) \, dx = dA$. In other words, the magnitude of $d\mathbf{F}$ is determined from the colored differential *area dA* under the loading curve. For the entire plate length,

$+\downarrow F_R = \Sigma F;$

$$F_R = \int_L w(x) \, dx = \int_A dA = A \qquad (4\text{–}23)$$

Hence, the magnitude of the resultant force is equal to the total area under the loading diagram $w = w(x)$.

*The more general case of a nonuniform surface loading acting on a body is considered in Sec. 9.5.

Location of Resultant Force. Applying Eq. 4–22 ($M_{R_O} = \Sigma M_O$), the location $\bar{x}$ of the line of action of $\mathbf{F}_R$ can be determined by equating the moments of the force resultant and the force distribution about point O (the y axis). Since $d\mathbf{F}$ produces a moment of $x \, dF = x \, w(x) \, dx$ about O, Fig. 4–51b, then for the entire plate, Fig. 4–51c,

$$\curvearrowright + M_{R_O} = \Sigma M_O; \qquad \bar{x} F_R = \int_L x \, w(x) \, dx$$

Solving for $\bar{x}$, using Eq. 4–23, we can write

$$\bar{x} = \frac{\displaystyle\int_L x \, w(x) \, dx}{\displaystyle\int_L w(x) \, dx} = \frac{\displaystyle\int_A x \, dA}{\displaystyle\int_A dA} \qquad\qquad (4\text{–}24)$$

This equation represents the x coordinate for the geometric center or *centroid* of the *area* under the distributed-loading diagram $w(x)$. *Therefore, the resultant force has a line of action which passes through the centroid C (geometric center) of the area defined by the distributed-loading diagram $w(x)$,* Fig. 4–51c.

Once $\bar{x}$ is determined, $\mathbf{F}_R$ by symmetry passes through point $(\bar{x}, 0)$ on the surface of the plate, Fig. 4–51d. If we now consider the three-dimensional pressure loading $p(x)$, Fig. 4–51a, we can therefore conclude that *the resultant force has a magnitude equal to the volume under the distributed-loading curve $p = p(x)$ and a line of action which passes through the centroid (geometric center) of this volume.* Detailed treatment of the integration techniques for computing the centroids of volumes or areas is given in Chapter 9. In many cases, however, the distributed-loading diagram is in the shape of a rectangle, triangle, or other simple geometric form. The centroids for such common shapes do not have to be determined from Eq. 4–24; rather, they can be obtained directly from the tabulation given on the inside back cover.

Example 4–22

In each case, determine the magnitude and location of the resultant of the distributed load acting on the beams in Fig. 4–52.

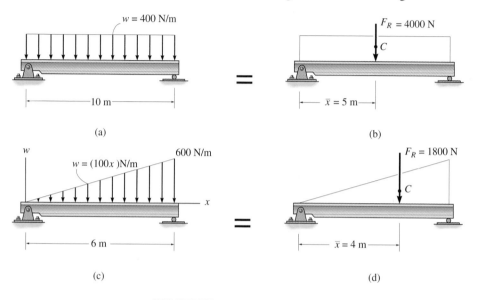

(a)   (b)

(c)   (d)

Fig. 4–52

SOLUTION

Uniform Loading. As indicated $w = 400$ N/m, which is constant over the entire beam, Fig. 4–52a. This loading forms a rectangle, the area of which is equal to the resultant force, Fig. 4–52b; i.e.,

$$F_R = (400 \text{ N/m})(10 \text{ m}) = 4000 \text{ N} \qquad \textit{Ans.}$$

The location of $\mathbf{F}_R$ passes through the geometric center or centroid C of this rectangular area, so

$$\bar{x} = 5 \text{ m} \qquad \textit{Ans.}$$

Triangular Loading. Here the loading varies uniformly in intensity from 0 to 600 N/m, Fig. 4–52c. These values can be verified by substitution of $x = 0$ and $x = 6$ m into the loading function $w = 100x$ N/m. The area of this triangular loading is equal to $\mathbf{F}_R$, Fig. 4–52d. From the table on the inside back cover, $A = \frac{1}{2}bh$, so that

$$F_R = \tfrac{1}{2}(6 \text{ m})(600 \text{ N/m}) = 1800 \text{ N} \qquad \textit{Ans.}$$

The line of action of $\mathbf{F}_R$ passes through the centroid C of the triangle. Using the table on the inside back cover, this point lies at a distance of one third the length of the beam, measured from the right side. Hence,

$$\bar{x} = 6 \text{ m} - \tfrac{1}{3}(6 \text{ m}) = 4 \text{ m} \qquad \textit{Ans.}$$

Example 4–23

The granular material exerts the distributed loading on the beam as shown in Fig. 4–53a. Determine the magnitude and location of the resultant of this load.

SOLUTION

The area of the loading diagram is a *trapezoid,* and therefore the solution can be obtained directly from the area and centroid formulas for a trapezoid listed on the inside back cover. Since these formulas are not easily remembered, instead we will solve this problem by using "composite" areas. In this regard, we can divide the trapezoidal loading into a rectangular and triangular loading as shown in Fig. 4–53b. The magnitude of the force represented by each of these loadings is equal to its associated *area,*

$$F_1 = \tfrac{1}{2}(9 \text{ m})(50 \text{ N/m}) = 225 \text{ N}$$
$$F_2 = (9 \text{ m})(50 \text{ N/m}) = 450 \text{ N}$$

The lines of action of these parallel forces act through the *centroid* of their associated areas and therefore intersect the beam at

$$\bar{x}_1 = \tfrac{1}{3}(9 \text{ m}) = 3 \text{ m}$$
$$\bar{x}_2 = \tfrac{1}{2}(9 \text{ m}) = 4.5 \text{ m}$$

The two parallel forces F_1 and F_2 can be reduced to a single resultant F_R. The magnitude of F_R is

$$+\downarrow F_R = \Sigma F; \qquad F_R = 225 + 450 = 675 \text{ N} \qquad\qquad Ans.$$

With reference to point A, Fig. 4–53b and c, we can define the location of F_R. We require

$$\curvearrowright + M_{R_A} = \Sigma M_A; \qquad \bar{x}(675) = 3(225) + 4.5(450)$$
$$\bar{x} = 4 \text{ m} \qquad\qquad Ans.$$

Note: The trapezoidal area in Fig. 4–53a can also be divided into two triangular areas as shown in Fig. 4–53d. In this case

$$F_1 = \tfrac{1}{2}(9 \text{ m})(100 \text{ N/m}) = 450 \text{ N}$$
$$F_2 = \tfrac{1}{2}(9 \text{ m})(50 \text{ N/m}) = 225 \text{ N}$$

and

$$\bar{x}_1 = \tfrac{1}{3}(9 \text{ m}) = 3 \text{ m}$$
$$\bar{x}_2 = \tfrac{1}{3}(9 \text{ m}) = 3 \text{ m}$$

Using these results, show that again $F_R = 675$ N and $\bar{x} = 4$ m.

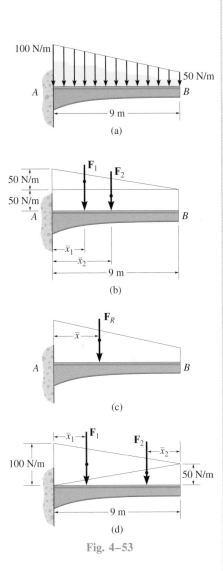

Fig. 4–53

Example 4–24

Determine the magnitude and location of the resultant force acting on the shaft in Fig. 4–54a.

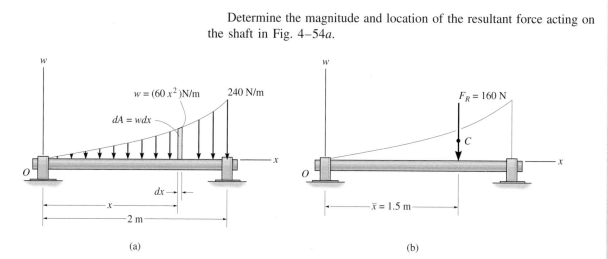

Fig. 4–54

SOLUTION

Since $w = w(x)$ is given, this problem will be solved by integration. The colored differential area element $dA = w\,dx = 60x^2\,dx$. Applying Eq. 4–23, by summing these elements from $x = 0$ to $x = 2$ m, we obtain the resultant force $\mathbf{F}_R$,

$F_R = \Sigma F;$

$$F_R = \int_A dA = \int_0^2 60x^2\,dx = 60\left[\frac{x^3}{3}\right]_0^2 = 60\left[\frac{2^3}{3} - \frac{0^3}{3}\right]$$

$$= 160\ \text{N} \qquad\qquad\qquad\qquad Ans.$$

Since the element of area dA is located an arbitrary distance x from O, the location $\bar{x}$ of $\mathbf{F}_R$ *measured from O,* Fig. 4–54b, is determined from Eq. 4–24.

$$\bar{x} = \frac{\displaystyle\int_A x\,dA}{\displaystyle\int_A dA} = \frac{\displaystyle\int_0^2 x(60x^2)\,dx}{160} = \frac{60\left[\dfrac{x^4}{4}\right]_0^2}{160} = \frac{60\left[\dfrac{2^4}{4} - \dfrac{0^4}{4}\right]}{160}$$

$$= 1.5\ \text{m} \qquad\qquad\qquad\qquad Ans.$$

These results may be checked by using the table on the inside back cover, where it is shown that for an exparabolic area of length a, height b, and shape shown in Fig. 4–54a,

$$A = \frac{ab}{3} = \frac{2(240)}{3} = 160\ \text{N} \quad\text{and}\quad \bar{x} = \frac{3}{4}a = \frac{3}{4}(2) = 1.5\ \text{m}$$

Example 4–25

A distributed loading of $p = 800x$ Pa acts over the top surface of the beam shown in Fig. 4–55a. Determine the magnitude and location of the resultant force.

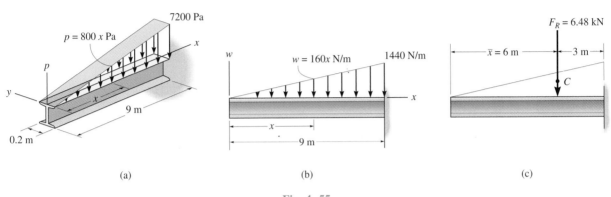

Fig. 4–55

SOLUTION

The loading function $p = 800x$ Pa indicates that the load intensity varies uniformly from $p = 0$ at $x = 0$ to $p = 7200$ Pa at $x = 9$ m. Since the intensity is uniform along the width of the beam (the y axis), the loading may be viewed in two dimensions as shown in Fig. 4–55b. Here

$$w = (800x \text{ N/m}^2)(0.2 \text{ m})$$
$$= (160x) \text{ N/m}$$

At $x = 9$ m, note that $w = 1440$ N/m. Although we may again apply Eqs. 4–23 and 4–24 as in Example 4–24, it is simpler to use the table on the inside back cover.

The magnitude of the resultant force is

$$F_R = \tfrac{1}{2}(9 \text{ m})(1440 \text{ N/m}) = 6480 \text{ N} = 6.48 \text{ kN} \qquad \textit{Ans.}$$

The line of action of $\mathbf{F}_R$ passes through the *centroid* C of the triangle. Hence,

$$\bar{x} = 9 \text{ m} - \tfrac{1}{3}(9 \text{ m}) = 6 \text{ m} \qquad \textit{Ans.}$$

The results are shown in Fig. 4–55c.

We may also view the resultant $\mathbf{F}_R$ as *acting* through the *centroid* of the *volume* of the loading diagram $p = p(x)$ in Fig. 4–55a. Hence $\mathbf{F}_R$ intersects the x–y plane at the point (6 m, 0). Furthermore, the *magnitude* of $\mathbf{F}_R$ is equal to the *volume* under the loading diagram; i.e.,

$$F_R = V = \tfrac{1}{2}(7200 \text{ N/m}^2)(9 \text{ m})(0.2 \text{ m}) = 6.48 \text{ kN} \qquad \textit{Ans.}$$

PROBLEMS

*4–128.   The loading on the bookshelf is distributed as shown. Determine the magnitude of the equivalent resultant force and its location, measured from point O.

4–129.   The loading on the bookshelf is distributed as shown. Determine the equivalent resultant force and its location, measured from point A.

*4–132.   The column is used to support the floor which exerts a force of 13.34 kN on the top of the column. The effect of soil pressure along its side is distributed as shown. Replace this loading by an equivalent resultant force and specify where it acts along the column, measured from its base A.

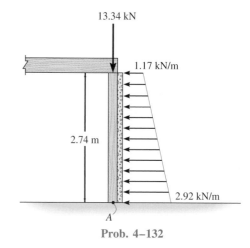

Prob. 4–132

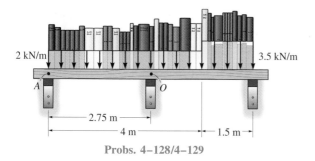

Probs. 4–128/4–129

4–130.   Replace the loading by an equivalent resultant force and couple moment acting at point O.

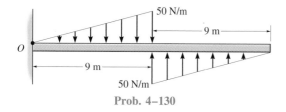

Prob. 4–130

4–131.   Replace the distributed loading by an equivalent resultant force, and specify its location on the beam, measured from the pin at C.

4–133.   Replace the loading by an equivalent force and couple moment acting at point O.

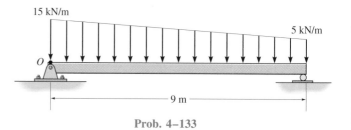

Prob. 4–133

4–134.   The masonry support creates the loading distribution acting on the end of the beam. Replace this load by an equivalent resultant force and specify its location, measured from point O.

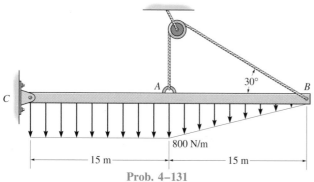

Prob. 4–131

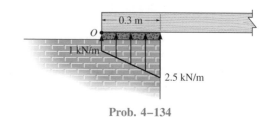

Prob. 4–134

4–135. Replace the loading by an equivalent resultant force and specify its location on the beam, measured from point B.

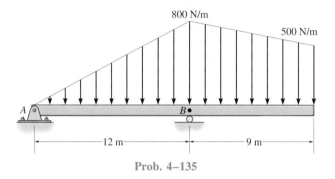

Prob. 4–135

***4–136.** Replace the distributed loading by an equivalent resultant force and specify its location, measured from point A.

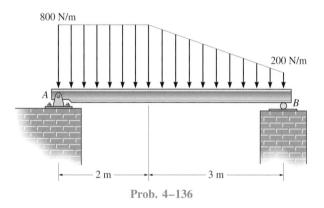

Prob. 4–136

4–137. The beam is subjected to the parabolic loading. Replace this loading by an equivalent force and couple moment at point O.

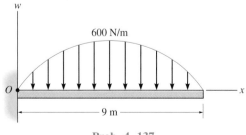

Prob. 4–137

4–138. Replace the distributed loading by an equivalent resultant force and specify where its line of action intersects member AB, measured from A.

4–139. Replace the distributed loading by an equivalent resultant force and specify where its line of action intersects member BC, measured from C.

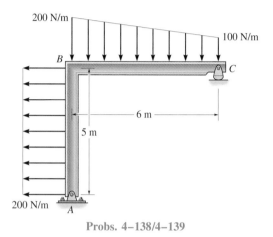

Probs. 4–138/4–139

***4–140.** Replace the loading by an equivalent resultant force and couple moment acting at point O.

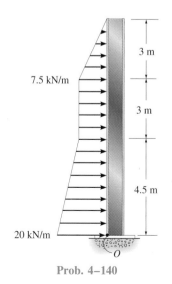

Prob. 4–140

4–141. The distribution of soil loading on the bottom of a building slab is shown. Replace this loading by an equivalent resultant force and specify its location, measured from point O.

4–143. The distribution of soil loading on the bottom of a building slab is shown. The center portion of the loading is parabolic. Simplify this loading to an equivalent resultant force and specify its location, measured from point O.

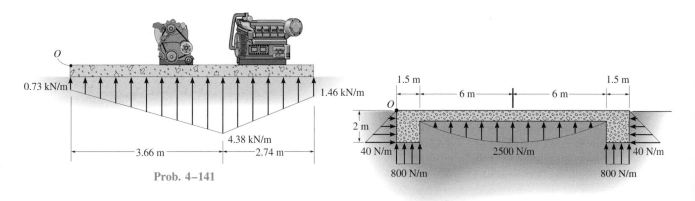

Prob. 4–141

Prob. 4–143

4–142. The bricks on top of the beam and the supports at the bottom create the distributed loading shown in the second figure. Determine the required intensity w and dimension d of the right support so that the equivalent resultant force and couple moment about point A of the system are both zero.

*__4–144.__ Wind has blown sand over a platform such that the intensity of the load can be approximated by the function $w = (0.5x^3)$ N/m. Simplify this distributed loading to an equivalent resultant force and specify the magnitude and location of the force, measured from A.

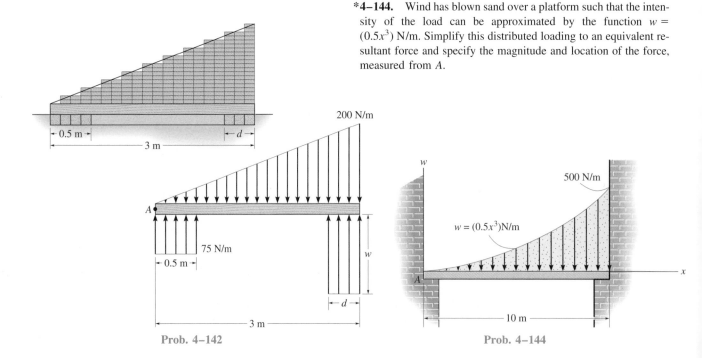

Prob. 4–142

Prob. 4–144

4–145. The form is used to cast a concrete wall having a width of 5 m. Determine the equivalent resultant force the wet concrete exerts on the form AB if the pressure distribution due to the concrete can be approximated as shown. Specify the location of the resultant force, measured from point B.

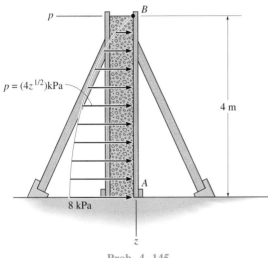

Prob. 4–145

4–147. Determine the magnitude of the equivalent resultant force and its location, measured from point O.

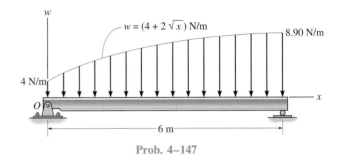

Prob. 4–147

*4–148.** The distributed load acts on the beam as shown. Determine the magnitude of the equivalent resultant force and specify where it acts, measured from the support, A.

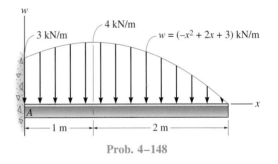

Prob. 4–148

■**4–146.** Determine the equivalent resultant force of the distributed loading and its location, measured from point A. Evaluate the integrals using Simpson's rule.

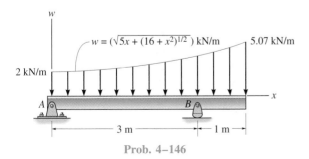

Prob. 4–146

4–149. The distributed load acts on the shaft as shown. Determine the magnitude of the equivalent resultant force and specify its location, measured from the support, A.

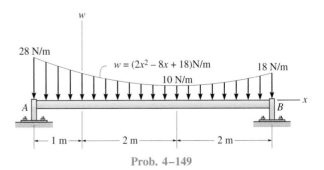

Prob. 4–149

REVIEW PROBLEMS

4–150.   Determine the equivalent resultant force of the distributed loading and its location, measured from point A.

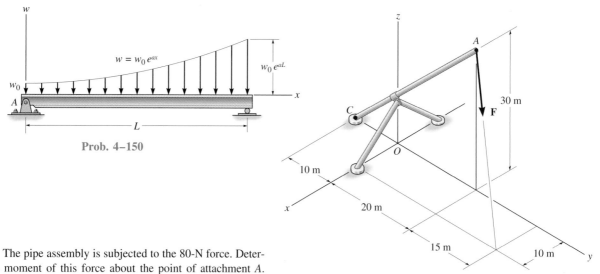

$$w = w_0 e^{ax}$$

w_0

$w_0 e^{aL}$

A

L

Prob. 4–150

4–151.   The pipe assembly is subjected to the 80-N force. Determine the moment of this force about the point of attachment A.

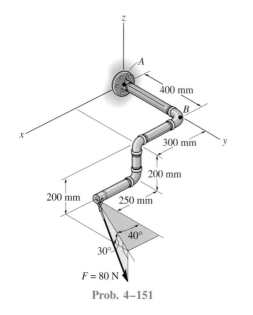

400 mm

B

300 mm

200 mm

200 mm

250 mm

40°

30°

$F = 80$ N

Prob. 4–151

*4–152.   Replace the force $\mathbf{F}$ having a magnitude of $F = 50$ N and acting at point A by an equivalent force and couple moment at point C.

z

A

30 m

C

$\mathbf{F}$

O

10 m

20 m

15 m

10 m

y

x

Prob. 4–152

4–153.   Determine the resultant moment about point A of the forces acting on the beam.

4–154.   Determine the resultant moment about point B of the forces acting on the beam.

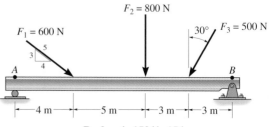

$F_2 = 800$ N

$F_1 = 600$ N

$30°$ $F_3 = 500$ N

$3\ \ \ 5$

4

A

B

4 m   5 m   3 m   3 m

Probs. 4–153/4–154

4–155. The gear reducer is subjected to the two couple moments shown. Express the resultant couple moment in Cartesian vector form and specify its magnitude and coordinate direction angles. The coordinate direction angles for M_1 are $\alpha_1 = 120°$, $\beta_1 = 45°$, $\gamma_1 = 60°$.

*4–156. The gear reducer is subjected to the two couple moments shown. Determine the coordinate direction angles of M_1 so that the resultant couple moment acts in the $+\mathbf{k}$ direction. What is the magnitude of the resultant couple moment?

4–159. Determine the moments of the force $\mathbf{F}$ about the x, y, and z axes. Solve the problem (a) using a Cartesian vector approach and (b) using a scalar approach. Express each result as a Cartesian vector.

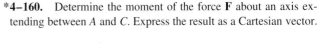

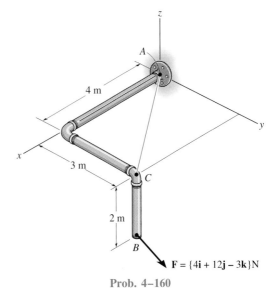

Prob. 4–159

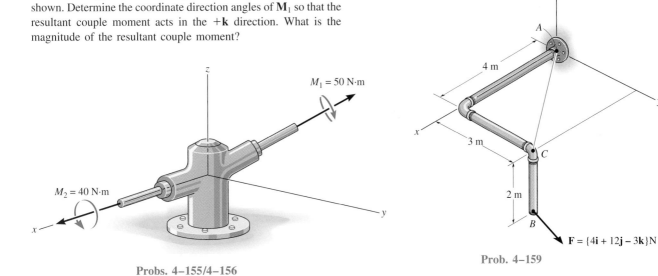

Probs. 4–155/4–156

4–157. The force of $F = 80$ N acts along the edge DB of the tetrahedron. Determine the magnitude of the moment of this force about the edge AC.

4–158. If the moment of the force $\mathbf{F}$ about the edge AC of the tetrahedron has a magnitude of $M = 200$ N · m and is directed from C towards A, determine the magnitude of $\mathbf{F}$.

*4–160. Determine the moment of the force $\mathbf{F}$ about an axis extending between A and C. Express the result as a Cartesian vector.

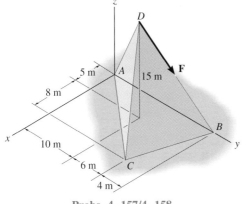

Probs. 4–157/4–158

Prob. 4–160

Utility poles are often subjected to the forces of cables and the weight of transformers. The equilibrium analysis, as explained in this chapter, must account for these loadings.

5

Equilibrium of a Rigid Body

In this chapter the fundamental concepts of rigid-body equilibrium will be discussed. It will be shown that equilibrium requires both a *balance of forces,* to prevent the body from translating with accelerated motion, and a *balance of moments,* to prevent the body from rotating.

Many types of engineering problems involve symmetric loadings and can be solved by projecting all the forces acting on a body onto a single plane. Hence, in the first part of this chapter, the equilibrium of a body subjected to a *coplanar* or *two-dimensional force system* will be considered. Ordinarily the geometry of such problems is not very complex, so a scalar solution is suitable for analysis. The more general discussion of rigid bodies subjected to *three-dimensional force systems* is given in the second part of this chapter. It will be seen that many of these types of problems can best be solved by using vector analysis.

5.1 Conditions for Rigid-Body Equilibrium

In Chapter 3 it was stated that a particle is in equilibrium if it remains at rest or moves with constant velocity. For this to be the case, it is both necessary and sufficient to require the resultant force acting on the particle to be equal to zero. Using this fact, we will now develop the conditions required to maintain equilibrium for a rigid body. To do this, consider the rigid body in Fig. 5–1a, which is fixed in the x, y, z reference and is either at rest or moves with the reference at constant velocity. A free-body diagram of the arbitrary ith particle of the body is shown in Fig. 5–1b. There are two types of forces which act on

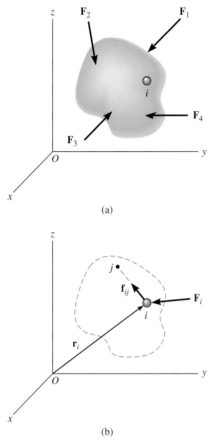

(a)

(b)

Fig. 5–1

it. The *internal forces,* represented symbolically as

$$\sum_{\substack{j=1 \\ (j \neq 1)}}^{n} \mathbf{f}_{ij} = \mathbf{f}_i$$

are forces which all the other particles exert on the ith particle and produce the resultant $\mathbf{f}_i$. Although only one of these particles is shown in Fig. 5–1b, the summation extends over all n particles composing the body. For this summation note that it is meaningless for $i = j$ since the ith particle cannot exert a force on itself. The resultant *external force* $\mathbf{F}_i$ represents, for example, the effects of gravitational, electrical, magnetic, or contact forces between the ith particle and adjacent bodies or particles *not* included within the body. If the particle is in equilibrium, then applying Newton's first law we have

$$\mathbf{F}_i + \mathbf{f}_i = \mathbf{0}$$

When the equation of equilibrium is applied to each of the other particles of the body, similar equations will result. If all these equations are added together *vectorially,* we obtain

$$\Sigma\mathbf{F}_i + \Sigma\mathbf{f}_i = \mathbf{0}$$

The summation of the internal forces if carried out will equal zero since the internal forces between particles within the body will occur in equal but opposite collinear pairs, Newton's third law. Consequently, only the sum of the *external forces* will remain; and therefore, letting $\Sigma\mathbf{F}_i = \Sigma\mathbf{F},$ the above equation can be written as

$$\Sigma\mathbf{F} = \mathbf{0}$$

Let us now consider the moments of the forces acting on the ith particle about the arbitrary point O, Fig. 5–1b. Using the particle equilibrium equation and the distributive law of the vector cross product yields

$$\mathbf{r}_i \times (\mathbf{F}_i + \mathbf{f}_i) = \mathbf{r}_i \times \mathbf{F}_i + \mathbf{r}_i \times \mathbf{f}_i = \mathbf{0}$$

Similar equations can be written for the other particles of the body, and adding them together vectorially, we obtain

$$\Sigma\mathbf{r}_i \times \mathbf{F}_i + \Sigma\mathbf{r}_i \times \mathbf{f}_i = \mathbf{0}$$

The second term is zero since, as stated above, the internal forces occur in equal but opposite collinear pairs, and by the transmissibility of a force, as discussed in Sec. 4.4, the moment of each pair of forces about point O is therefore zero. Hence, using the notation $\Sigma\mathbf{M}_O = \Sigma\mathbf{r}_i \times \mathbf{F}_i,$ we can write the previous equation as

$$\Sigma\mathbf{M}_O = \mathbf{0}$$

Hence the *equations of equilibrium* for a rigid body can be summarized as follows:

$$\Sigma \mathbf{F} = \mathbf{0}$$
$$\Sigma \mathbf{M}_O = \mathbf{0}$$

(5–1)

These equations require that a rigid body will be in equilibrium provided the sum of all the *external forces* acting on the body is equal to zero and the sum of the moments of the external forces about a point is equal to zero. The fact that these conditions are *necessary* for equilibrium has now been proven. They are also *sufficient* conditions. To show this, let us assume that the body is *not* in equilibrium, and yet the force system acting on it satisfies Eqs. 5–1. Suppose that an *additional force* $\mathbf{F}'$ is required to hold the body in equilibrium. As a result, the equilibrium equations become

$$\Sigma \mathbf{F} + \mathbf{F}' = \mathbf{0}$$
$$\Sigma \mathbf{M}_O + \mathbf{M}'_O = \mathbf{0}$$

where $\mathbf{M}'_O$ is the moment of $\mathbf{F}'$ about O. Since $\Sigma \mathbf{F} = \mathbf{0}$ and $\Sigma \mathbf{M}_O = \mathbf{0}$, then we require $\mathbf{F}' = \mathbf{0}$ (also $\mathbf{M}'_O = \mathbf{0}$). Consequently, the additional force $\mathbf{F}'$ is not required for holding the body, and indeed Eqs. 5–1 are also sufficient conditions for equilibrium.

Equilibrium in Two Dimensions

5.2 Free-Body Diagrams

Successful application of the equations of equilibrium requires a complete specification of *all* the known and unknown external forces that act *on* the body. The best way to account for these forces is to draw the body's free-body diagram. This diagram is a sketch of the outlined shape of the body, which represents it as being *isolated* or "free" from its surroundings, i.e., a "free body". On this sketch it is necessary to show *all* the forces and couple moments that the surroundings exert *on the body*. By using this diagram the effects of all the applied forces and couple moments acting on the body can be accounted for when the equations of equilibrium are applied. For this reason, *a thorough understanding of how to draw a free-body diagram is of primary importance for solving problems in mechanics.*

Table 5–1   Supports for Rigid Bodies Subjected to Two-Dimensional Force Systems

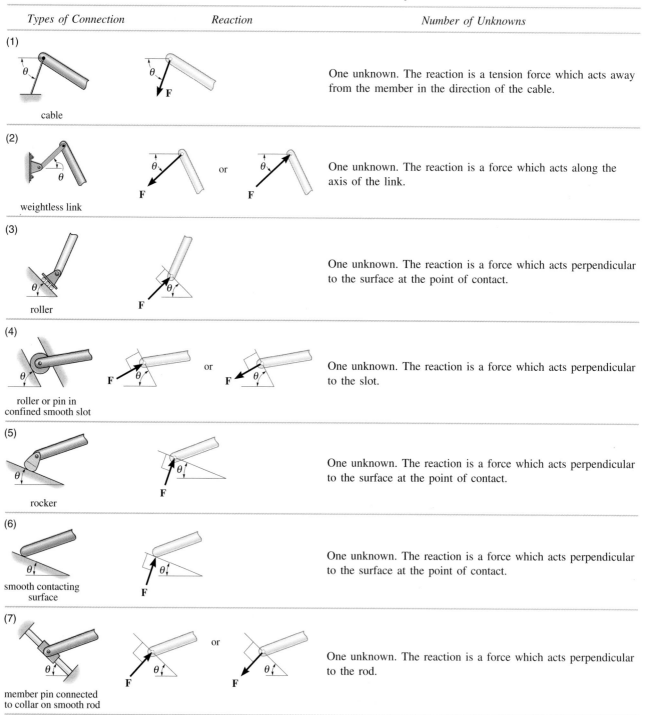

Types of Connection	Reaction	Number of Unknowns
(1) cable		One unknown. The reaction is a tension force which acts away from the member in the direction of the cable.
(2) weightless link	or	One unknown. The reaction is a force which acts along the axis of the link.
(3) roller		One unknown. The reaction is a force which acts perpendicular to the surface at the point of contact.
(4) roller or pin in confined smooth slot	or	One unknown. The reaction is a force which acts perpendicular to the slot.
(5) rocker		One unknown. The reaction is a force which acts perpendicular to the surface at the point of contact.
(6) smooth contacting surface		One unknown. The reaction is a force which acts perpendicular to the surface at the point of contact.
(7) member pin connected to collar on smooth rod	or	One unknown. The reaction is a force which acts perpendicular to the rod.

Table 5–1 (Contd.)

Types of Connection	Reaction	Number of Unknowns
(8) 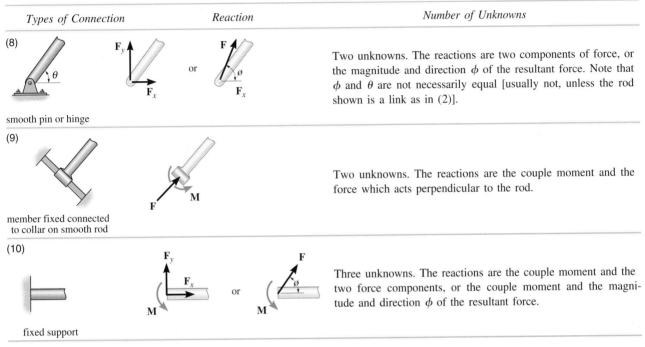 smooth pin or hinge		Two unknowns. The reactions are two components of force, or the magnitude and direction ϕ of the resultant force. Note that ϕ and θ are not necessarily equal [usually not, unless the rod shown is a link as in (2)].
(9) member fixed connected to collar on smooth rod		Two unknowns. The reactions are the couple moment and the force which acts perpendicular to the rod.
(10) fixed support		Three unknowns. The reactions are the couple moment and the two force components, or the couple moment and the magnitude and direction ϕ of the resultant force.

Support Reactions. Before presenting a formal procedure as to how to draw a free-body diagram, we will first consider the various types of reactions that occur at supports and points of support between bodies subjected to co-planar force systems. *As a general rule, if a support prevents the translation of a body in a given direction, then a force is developed on the body in that direction. Likewise, if rotation is prevented, a couple moment is exerted on the body.*

For example, let us consider three ways in which a horizontal member, such as a beam, is commonly supported at its end. The first method of support consists of a *roller* or cylinder, Fig. 5–2a. Since this type of support only prevents the beam from translating in the vertical direction, the roller can only exert a force on the beam in this direction, Fig. 5–2b.

The beam can be supported in a more restrictive manner by using a *pin* as shown in Fig. 5–3a. The pin passes through holes in the beam and two leaves which are fixed to the ground. Here the pin will prevent translation of the beam in *any direction* ϕ, Fig. 5–3b, and so the pin must exert a force **F** on the beam in this direction. For purposes of analysis, it is generally easier to represent this effect by its two components $\mathbf{F}_x$ and $\mathbf{F}_y$, Fig. 5–3c. If F_x and F_y are known, then F and ϕ can be calculated.

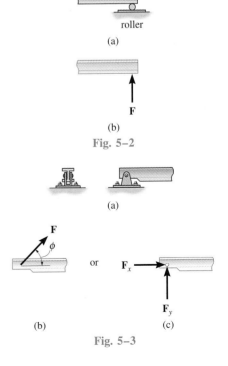

Fig. 5–2

Fig. 5–3

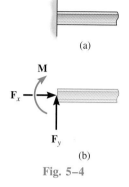

(a)

(b)

Fig. 5–4

The most restrictive way to support the beam would be to use a *fixed support* as shown in Fig. 5–4*a*. This support will prevent both translation and rotation of the beam, and so to do this a force and couple moment must be developed on the beam at its point of connection, Fig. 5–4*b*. As in the case of the pin, the force is usually represented by its components F_x and F_y.

Table 5–1 lists other common types of supports for bodies subjected to coplanar force systems. (In all cases the angle θ is assumed to be known.) Carefully study each of the symbols used to represent these supports and the types of reactions they exert on their contacting members. Although concentrated forces and couple moments are shown in this table, they actually represent the *resultants* of *distributed surface loads* that exist between each support and its contacting member. It is these *resultants* which will be determined from the equations of equilibrium. It is generally not important to determine the actual distribution of the load, since the surface area over which it acts is considerably *smaller* than the *total surface area* of the connected member.

External and Internal Forces. Since a rigid body is a composition of particles, both *external* and *internal* loadings may act on it. It is important to realize, however, that if the free-body diagram for the body is drawn, the forces that are *internal* to the body are *not represented* on the free-body diagram. As discussed in Sec. 5.1, these forces always occur in equal but opposite collinear pairs, and therefore their *net effect* on the body is zero.

In some problems, a free-body diagram for a "system" of connected bodies may be used for an analysis. An example would be the free-body diagram of an entire automobile (system) composed of its many parts. Obviously, the connecting forces between its parts would represent *internal forces* which would *not* be included on the free-body diagram of the automobile. To summarize, then, internal forces act between particles which are located *within* a specified system which is contained within the boundary of the free-body diagram. Particles or bodies outside this boundary exert external forces on the system, and these alone must be shown on the free-body diagram.

Weight and the Center of Gravity. When a body is subjected to a gravitational field, each of its particles has a specified weight as defined by Newton's law of gravitation, $F = Gm_1m_2/r^2$, Eq. 1–2. If we assume the size of the body to be "small" in relation to the size of the earth, then it is appropriate to consider these gravitational forces to be represented as a *system of parallel forces* acting on the particles contained within the boundary of the body. It was shown in Sec. 4.9 that such a system can be reduced to a single resultant force acting through a specified point. We refer to this force resultant as the *weight* **W** of the body, and to the location of its point of application as the *center of gravity G*. The methods used for its calculation will be developed in Chapter 9.

In the examples and problems that follow, if the weight of the body is important for the analysis, this force will then be reported in the problem statement. Also, when the body is *uniform* or made of homogeneous material, the center of gravity will be located at the body's *geometric center* or *centroid;* however, if the body is nonhomogeneous or has an unusual shape, then its center of gravity will be given.

PROCEDURE FOR DRAWING
A FREE-BODY DIAGRAM

To construct a free-body diagram for a rigid body or group of bodies considered as a single system, the following steps should be performed:

Step 1.   Imagine the body to be *isolated* or cut "free" from its constraints and connections, and draw (sketch) its outlined shape.

Step 2.   Identify all the external forces and couple moments that act on the body. Those generally encountered are due to (1) applied loadings, (2) reactions occurring at the supports or at points of contact with other bodies (see Table 5–1), and (3) the weight of the body. To account for all these effects, it may help to trace over the boundary, carefully noting each force or couple moment acting on it.

Step 3.   Indicate the dimensions of the body necessary for computing the moments of forces. The forces and couple moments that are known should be labeled with their proper magnitudes and directions. Letters are used to represent the magnitudes and direction angles of forces and couple moments that are *unknown*. Establish an x, y coordinate system so that these unknowns, A_x, B_y, etc., can be identified. In particular, if a force or couple moment has a known line of action but unknown magnitude, the arrowhead which defines the sense of the vector can be assumed. The correctness of the assumed sense will become apparent after solving the equilibrium equations for the unknown magnitude. By definition, the *magnitude* of a vector is *always positive,* so that if the solution yields a "negative" scalar, the *minus sign* indicates that the vector's sense is *opposite* to that which was originally assumed.

Before proceeding, review this section; then carefully study the following examples. Afterward, attempt to draw the free-body diagrams for the objects in Figs. 5–5 through 5–9 without "looking" at the solutions. Further practice in drawing free-body diagrams should be gained by solving *all* the problems given at the end of this section.

Example 5–1

Draw the free-body diagram of the uniform beam shown in Fig. 5–5a. The beam has a mass of 100 kg.

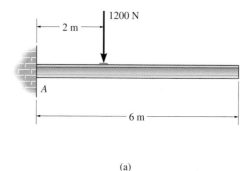

(a)

SOLUTION

The free-body diagram of the beam is shown in Fig. 5–5b. Since the support at A is a fixed wall, there are three reactions acting *on the beam* at A, denoted as $\mathbf{A}_x$, $\mathbf{A}_y$, and $\mathbf{M}_A$. The magnitudes of these vectors are *unknown*, and their sense has been *assumed*. (How does one obtain the *correct* sense of these vectors?) The weight of the beam, $W = 100(9.81) = 981$ N, acts through the beam's center of gravity G, 3 m from A since the beam is uniform.

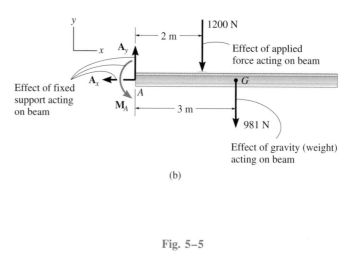

(b)

Fig. 5–5

Example 5–2

Draw the free-body diagram for the bell crank *ABC* shown in Fig. 5–6*a*.

SOLUTION

The free-body diagram is shown in Fig. 5–6*b*. The pin support at *B* exerts force components $\mathbf{B}_x$ and $\mathbf{B}_y$ *on the bell crank,* each having a known line of action but unknown magnitude. The link at *C* exerts a force $\mathbf{F}_C$ acting in the direction of the link and having an unknown magnitude. The dimensions of the crank are also labeled on the free-body diagram, since this information will be useful in computing the moments of the forces. As usual, the sense of the three unknown forces has been assumed. The correct sense will become apparent after solving the equilibrium equations.

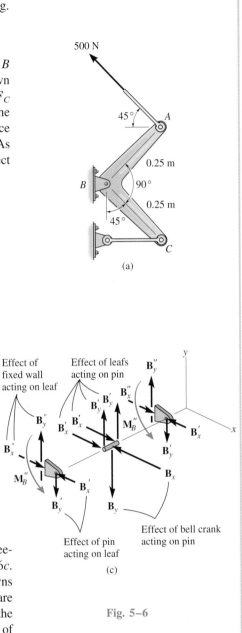

(a)

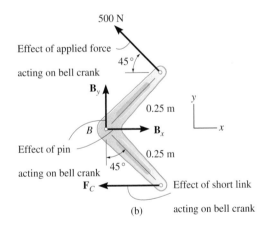

(b)

Although not part of this problem, three-dimensional views of the free-body diagrams of the pin and two pin leaves at *B* are shown in Fig. 5–6*c*. Since the leaves are *fixed-connected* to the wall, there are three unknowns that the wall exerts on each leaf; namely, $\mathbf{B}''_x$, $\mathbf{B}''_y$, $\mathbf{M}''_B$. These reactions are shown to be equal in magnitude and direction on each leaf due to the symmetry of the loading and geometry. Note carefully how the principle of action—equal but opposite collinear reaction—is used when applying the forces $\mathbf{B}'_x$ and $\mathbf{B}'_y$ to each leaf and the pin. All of these unknowns can be obtained from the equations of equilibrium once $\mathbf{B}_x$ and $\mathbf{B}_y$ are obtained.

Fig. 5–6

Example 5–3

Two smooth tubes A and B, each having a mass of 2 kg, rest between the inclined plates shown in Fig. 5–7a. Draw the free-body diagrams for tube A, tube B, and tubes A and B together.

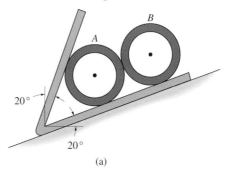

(a)

SOLUTION

The free-body diagram for tube A is shown in Fig. 5–7b. Its weight is $W = 2(9.81) = 19.62$ N. Since all contacting surfaces are *smooth,* the reactive forces **T, F, R** act in a direction *normal* to the tangent at their surfaces of contact.

The free-body diagram of tube B is shown in Fig. 5–7c. Can you identify each of the three forces acting *on the tube?* In particular, note that **R,** representing the force of tube A on tube B, Fig. 5–7c, is equal and opposite to **R** representing the force of tube B on tube A, Fig. 5–7b. This is a consequence of Newton's third law of motion.

The free-body diagram of both tubes combined ("system") is shown in Fig. 5–7d. Here the contact force **R,** which acts between A and B, is considered as an *internal* force and hence is not shown on the free-body diagram. That is, it represents a pair of equal but opposite collinear forces which cancel each other.

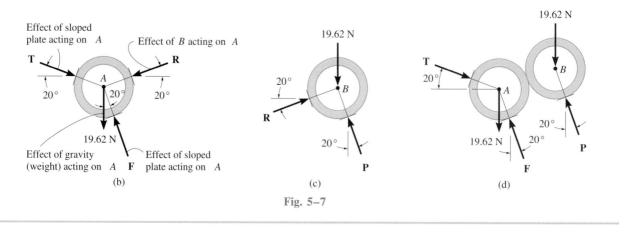

Fig. 5–7

Example 5–4

The free-body diagram of each object in Fig. 5–8 is drawn. Carefully study each solution and identify what each loading represents, as was done in Fig. 5–7b. The weight of the objects is neglected except where indicated.

SOLUTION

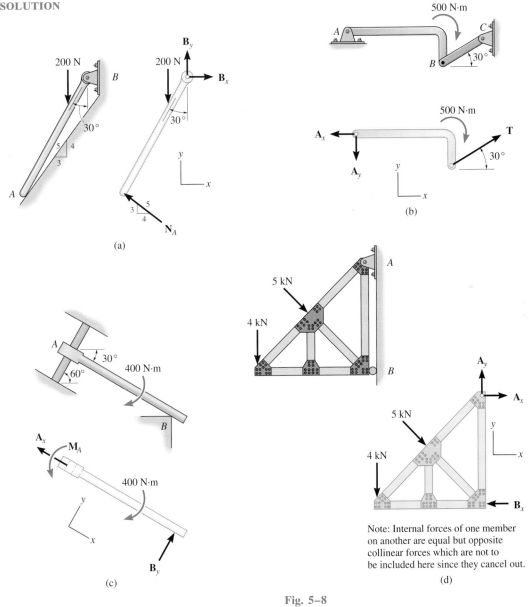

Note: Internal forces of one member on another are equal but opposite collinear forces which are not to be included here since they cancel out.

(a)

(b)

(c)

(d)

Fig. 5–8

Example 5–5

The highway sign shown in Fig. 5–9a has a mass of 100 kg with a center of gravity at G. It is supported by pins at C and D and a cable AB. Draw a free-body diagram of the sign and the supporting frame. Neglect the weight of the frame.

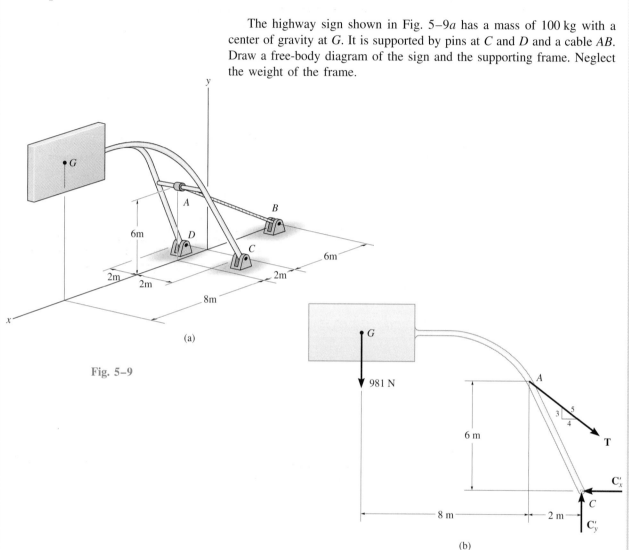

Fig. 5–9

(a)

(b)

SOLUTION

By observation, the frame, sign, and the loading are all symmetrical about the vertical x–y plane, hence the problem may be analyzed using a system of *coplanar forces*. The free-body diagram is shown in Fig. 5–9b. Note that the force **T** that the cable exerts on the frame has a known line of action indicated by the 3–4–5 slope triangle. The force components C'_x and C'_y represent the horizontal and vertical reactions that *both* pins C and D exert on the frame. Consequently, after the solution for these reactions is obtained, *half* their magnitude is developed at C and half at D.

PROBLEMS

5–1. Draw the free-body diagram of the 50-kg uniform pipe, which is supported by the smooth contacts at *A* and *B*.

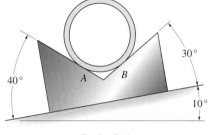

Prob. 5–1

5–2. Draw the free-body diagram of the hand punch, which is pinned at *A* and bears down on the smooth surface at *B*.

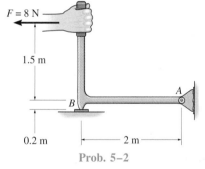

Prob. 5–2

5–3. Draw the free-body diagram of the smooth bar, which has points of contact at *A*, *B*, and *C*.

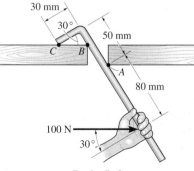

Prob. 5–3

*5–4. Draw the free-body diagram of the jib crane *AB*, which is pin-connected at *A* and supported by member (link) *BC*.

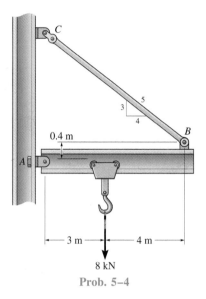

Prob. 5–4

5–5. Draw the free-body diagram of the dumpster *D* of the truck, which has a weight of 5000 N and a center of gravity at *G*. It is supported by a pin at *A* and a pin-connected hydraulic cylinder *BC* (short link).

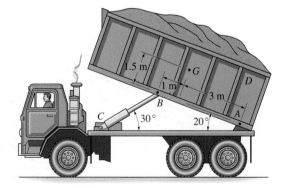

Prob. 5–5

5–6. Draw the free-body diagram of the link *CAB*, which is pin-connected at *A* and rests on the smooth cam at *B*.

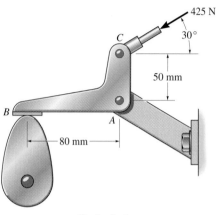

Prob. 5–6

5–7. Draw the free-body diagram of the uniform pipe, which has a mass of 100 kg and a center of mass at *G*. The supports *A*, *B*, and *C* are smooth.

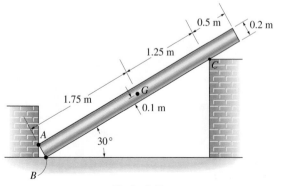

Prob. 5–7

*5–8. Draw the free-body diagram of the beam, which is pin-connected at *A* and rocker-supported at *B*.

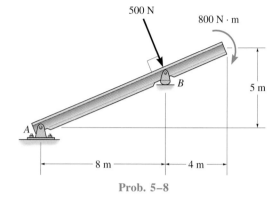

Prob. 5–8

5–9. Draw the free-body diagram of the beam, which is pin-supported at *A* and rests on the smooth incline at *B*.

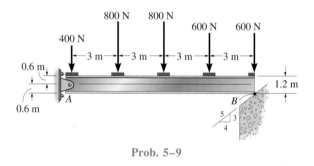

Prob. 5–9

5–10. Draw the free-body diagram of member *ABC*, which is supported by a pin at *A* and a horizontal short link *BD*.

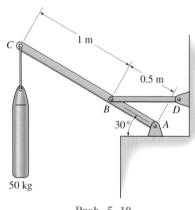

Prob. 5–10

5.3 Equations of Equilibrium

In Sec. 5.1 we developed the two equations which are both necessary and sufficient for the equilibrium of a rigid body, namely, $\Sigma \mathbf{F} = \mathbf{0}$ and $\Sigma \mathbf{M}_O = \mathbf{0}$. When the body is subjected to a system of forces, which all lie in the x–y plane, then the forces can be resolved into their x and y components. Consequently, the conditions for equilibrium in two dimensions are

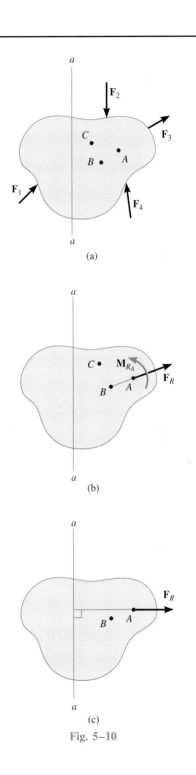

$$\Sigma F_x = 0$$
$$\Sigma F_y = 0 \qquad (5\text{–}2)$$
$$\Sigma M_O = 0$$

Here ΣF_x and ΣF_y represent, respectively, the algebraic sums of the x and y components of all the forces acting on the body, and ΣM_O represents the algebraic sum of the couple moments and the moments of all the force components about an axis perpendicular to the x–y plane and passing through the arbitrary point O, which may lie either on or off the body.

Alternative Sets of Equilibrium Equations. Although Eqs. 5–2 are *most often* used for solving equilibrium problems involving coplanar force systems, two *alternative* sets of three independent equilibrium equations may also be used. One such set is

$$\Sigma F_a = 0$$
$$\Sigma M_A = 0 \qquad (5\text{–}3)$$
$$\Sigma M_B = 0$$

When using these equations it is required that the moment points A and B do *not* lie on a line that is *perpendicular* to the a axis. To prove that Eqs. 5–3 provide the *conditions* for equilibrium, consider the free-body diagram of an arbitrarily shaped body shown in Fig. 5–10a. Using the methods of Sec. 4.8, the loading on the free-body diagram may be replaced by a single resultant force $\mathbf{F}_R = \Sigma \mathbf{F}$, acting at point A, and a resultant couple moment $\mathbf{M}_{R_A} = \Sigma \mathbf{M}_A$, Fig. 5–10b. If $\Sigma M_A = 0$ is satisfied, it is necessary that $\mathbf{M}_{R_A} = \mathbf{0}$. Furthermore, in order that $\mathbf{F}_R$ satisfy $\Sigma F_a = 0$, it must have *no component* along the a axis, and therefore its line of action must be perpendicular to the a axis, Fig. 5–10c. Finally, if it is required that $\Sigma M_B = 0$, where B does not lie on the line of action of $\mathbf{F}_R$, then $\mathbf{F}_R = \mathbf{0}$, and indeed the body shown in Fig. 5–10a must be in equilibrium.

(a)

(b)

(c)

Fig. 5–10

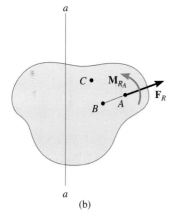

(b)

A second alternative set of equilibrium equations is

$$\Sigma M_A = 0$$
$$\Sigma M_B = 0 \qquad (5\text{--}4)$$
$$\Sigma M_C = 0$$

Here it is necessary that points A, B, and C do not lie on the same line. To prove that these equations when satisfied ensure equilibrium, consider again the free-body diagram in Fig. 5–10b. If $\Sigma M_A = 0$ is to be satisfied, then $\mathbf{M}_{R_A} = \mathbf{0}$. $\Sigma M_B = 0$ is satisfied if the line of action of $\mathbf{F}_R$ passes through point B as shown, and finally, if we require $\Sigma M_C = 0$, where C does not lie on line AB, Fig. 5–10b, it is necessary that $\mathbf{F}_R = \mathbf{0}$, and the body in Fig. 5–10a must then be in equilibrium.

PROCEDURE FOR ANALYSIS

The following procedure provides a method for solving coplanar force equilibrium problems:

Free-Body Diagram.   Draw a free-body diagram of the body as discussed in Sec. 5.2. Briefly, this requires showing all the external forces and couple moments acting *on the body*. The magnitudes of these vectors must be labeled and their directions specified relative to an established set of x, y axes. Dimensions of the body, necessary for computing the moments of forces, are also included on the free-body diagram. Identify the unknowns. The sense of a force or couple moment having an *unknown* magnitude but known line of action can be *assumed*.

Equations of Equilibrium.   Using the established x, y axes, apply the equations of equilibrium: $\Sigma F_x = 0$, $\Sigma F_y = 0$, $\Sigma M_O = 0$ (or the alternative sets of Eqs. 5–3 or 5–4). To *avoid* having to solve simultaneous equations, apply the moment equation $\Sigma M_O = 0$ about a point (O) *that lies at the intersection of the lines of action of two unknown forces.* In this way, the moments of these unknowns are *zero* about O, and one can obtain a *direct solution* for the third unknown. When applying the force equations $\Sigma F_x = 0$ and $\Sigma F_y = 0$, orient the x and y axes along lines that will provide the simplest resolution of the forces into their x and y components. If the solution of the equilibrium equations yields a *negative* scalar for an unknown force or couple moment, it indicates that the sense is *opposite* to that which was assumed on the free-body diagram.

The following example problems illustrate this procedure numerically.

Example 5–6

Determine the horizontal and vertical components of reaction for the beam loaded as shown in Fig. 5–11a. Neglect the weight of the beam in the calculations.

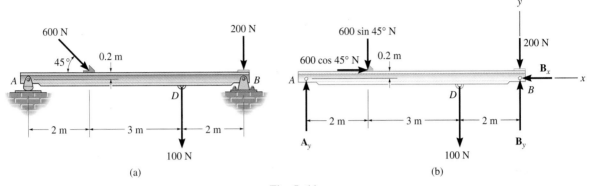

Fig. 5–11

SOLUTION

Free-Body Diagram.   Can you identify each of the forces shown on the free-body diagram of the beam, Fig. 5–11b? For simplicity in applying the equilibrium equations, the 600-N force is represented by its x and y components as shown. Also, note that a 200-N force acts on the beam at B, and is independent of the force components $\mathbf{B}_x$ and $\mathbf{B}_y$ which represent the effect of the pin on the beam.

Equations of Equilibrium.   Summing forces in the x direction yields

$$\xrightarrow{+} \Sigma F_x = 0; \qquad 600 \cos 45° \text{ N} - B_x = 0$$
$$B_x = 424 \text{ N} \qquad\qquad Ans.$$

A direct solution for $\mathbf{A}_y$ can be obtained by applying the moment equation $\Sigma M_B = 0$ about point B. For the calculation, it should be apparent that forces 200 N, $\mathbf{B}_x$, and $\mathbf{B}_y$ all create zero moment about B. Assuming counterclockwise rotation about B to be positive (in the $+\mathbf{k}$ direction), Fig. 5–11b, we have

$$\zeta + \Sigma M_B = 0; \quad 100 \text{ N}(2 \text{ m}) + (600 \sin 45° \text{ N})(5 \text{ m})$$
$$- (600 \cos 45° \text{ N})(0.2 \text{ m}) - A_y(7 \text{ m}) = 0$$
$$A_y = 319 \text{ N} \qquad\qquad Ans.$$

Summing forces in the y direction, using this result, gives

$$+ \uparrow \Sigma F = 0; \quad 319 \text{ N} - 600 \sin 45° \text{ N} - 100 \text{ N} - 200 \text{ N} + B = 0$$
$$B_y = 405 \text{ N} \qquad\qquad Ans.$$

Example 5–7

The cord shown in Fig. 5–12a supports a force of 100 N and wraps over the frictionless pulley. Determine the tension in the cord at *C* and the horizontal and vertical components of reaction at pin *A*.

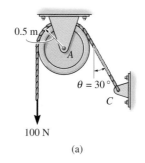

$\theta = 30°$

100 N

(a)

SOLUTION

Free-Body Diagrams. The free-body diagrams of the cord and pulley are shown in Fig. 5–12b. Note that the principle of action, equal but opposite reaction, must be carefully observed when drawing each of these diagrams: the cord exerts an unknown load distribution *p* along part of the pulley's surface, whereas the pulley exerts an equal but opposite effect on the cord. For the solution, however, it is simpler to *combine* the free-body diagrams of the pulley and the contacting portion of the cord, so that the distributed load becomes *internal* to the system and is therefore eliminated from the analysis, Fig. 5–12c.

Equations of Equilibrium. Summing moments about point *A* to eliminate $\mathbf{A}_x$ and $\mathbf{A}_y$, Fig. 5–12c, we have

$$\zeta + \Sigma M_A = 0; \qquad 100 \text{ N}(0.5 \text{ m}) - T(0.5 \text{ m}) = 0$$
$$T = 100 \text{ N} \qquad \qquad \textit{Ans.}$$

It is seen that the tension remains *constant* as the cord passes over the pulley. (This of course is true for *any angle* θ at which the cord is directed and for *any radius r* of the pulley.) Using the result for *T*, a force summation is applied to determine the components of reaction at pin *A*.

$$\xrightarrow{+} \Sigma F_x = 0; \qquad -A_x + 100 \sin 30° \text{ N} = 0$$
$$A_x = 50.0 \text{ N} \qquad \qquad \textit{Ans.}$$

$$+ \uparrow \Sigma F_y = 0; \qquad A_y - 100 \text{ N} - 100 \cos 30° \text{ N} = 0$$
$$A_y = 187 \text{ N} \qquad \qquad \textit{Ans.}$$

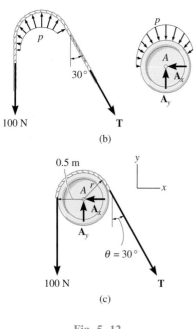

(b)

(c)

Fig. 5–12

Example 5–8

The open-ended box wrench in Fig. 5–13a is used to tighten the bolt at A. If the wrench does not turn when the load is applied to the handle, determine the torque or moment applied to the bolt and the force of the wrench on the bolt.

SOLUTION

Free-Body Diagram. The free-body diagram for the wrench is shown in Fig. 5–13b. Since the bolt acts as a "fixed support," it exerts force components A_x and A_y and a torque M_A on the wrench at A.

Equations of Equilibrium

$$\xrightarrow{+} \Sigma F_x = 0; \qquad A_x - 52(\tfrac{5}{13}) \text{ N} + 30 \cos 60° \text{ N} = 0$$

$$A_x = 5.00 \text{ N} \qquad\qquad Ans.$$

$$+\uparrow \Sigma F_y = 0; \qquad A_y - 52(\tfrac{12}{13}) \text{ N} - 30 \sin 60° \text{ N} = 0$$

$$A_y = 74.0 \text{ N} \qquad\qquad Ans.$$

$$\downarrow{+}\Sigma M_A = 0; \quad M_A - 52(\tfrac{12}{13}) \text{ N} (0.3 \text{ m}) - (30 \sin 60° \text{ N})(0.7 \text{ m}) = 0$$

$$M_A = 32.6 \text{ N} \cdot \text{m} \qquad\qquad Ans.$$

Point A was chosen for summing moments because the lines of action of the *unknown* forces A_x and A_y pass through this point, and therefore these forces were not included in the moment summation. Realize, however, that M_A must be *included* in this moment summation. This couple moment is a free vector and represents the twisting resistance of the bolt on the wrench. By Newton's third law, the wrench exerts an equal but opposite moment or torque on the bolt. Furthermore, the resultant force on the wrench or bolt is

$$F_A = \sqrt{(5.00)^2 + (74.0)^2} = 74.1 \text{ N} \qquad\qquad Ans.$$

Because the force components A_x and A_y were calculated as positive quantities, their directional sense is shown correctly on the free-body diagram in Fig. 5–13b. Hence

$$\theta = \tan^{-1} \frac{74.0 \text{ N}}{5.00 \text{ N}} = 86.1° \quad \angle$$

Realize that F_A acts in the opposite direction on the bolt. Why?

Although only *three* independent equilibrium equations can be written for a rigid body, it is a good practice to *check* the calculations using a fourth equilibrium equation. For example, the above computations may be verified in part by summing moments about point C:

$$\downarrow{+}\Sigma M_C = 0; \quad 52(\tfrac{12}{13}) \text{ N} (0.4 \text{ m}) + 32.6 \text{ N} \cdot \text{m} - 74.0 \text{ N}(0.7 \text{ m}) = 0$$

$$19.2 \text{ N} \cdot \text{m} + 32.6 \text{ N} \cdot \text{m} - 51.8 \text{ N} \cdot \text{m} = 0$$

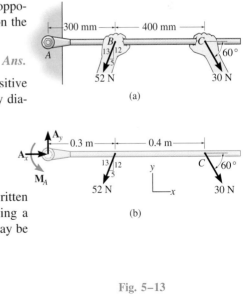

(a)

(b)

Fig. 5–13

Example 5–9

The uniform smooth rod shown in Fig. 5–14a is subjected to a force and couple moment. If the rod is supported at A by a smooth wall and at B and C either at the top or bottom by rollers, determine the reactions at these supports. Neglect the weight of the rod.

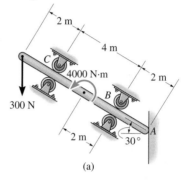

(a)

SOLUTION

Free-Body Diagram. As shown in Fig. 5–14b, all the support reactions act normal to the surface of contact since the contacting surfaces are smooth. The reactions at B and C are shown acting in the positive y' direction. This assumes that only the rollers located on the bottom of the rod are used for support.

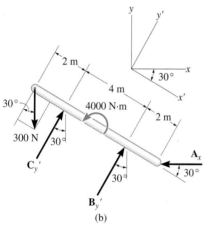

(b)

Fig. 5–14

Equations of Equilibrium. Using the x, y coordinate system in Fig. 5–14b, we have

$$\xrightarrow{+}\Sigma F_x = 0; \qquad C_{y'} \sin 30° + B_{y'} \sin 30° - A_x = 0 \qquad (1)$$

$$+\uparrow\Sigma F_y = 0; \quad -300\text{ N} + C_{y'} \cos 30° + B_{y'} \cos 30° = 0 \qquad (2)$$

$$\zeta+\Sigma M_A = 0; \qquad -B_{y'}(2\text{ m}) + 4000\text{ N}\cdot\text{m} - C_{y'}(6\text{ m})$$
$$+ (300 \cos 30°\text{ N})(8\text{ m}) = 0 \qquad (3)$$

When writing the moment equation, it should be noticed that the line of action of the force component 300 sin 30° N passes through point A, and therefore this force is not included in the moment equation.

Solving Eqs. 2 and 3 simultaneously, we obtain

$$B_{y'} = -1000.0\text{ N} \qquad\qquad Ans.$$
$$C_{y'} = 1346.4\text{ N} \qquad\qquad Ans.$$

Since $B_{y'}$ is a negative scalar, the sense of $\mathbf{B}_{y'}$ is opposite to that shown on the free-body diagram in Fig. 5–14b. Therefore, the top roller at B serves as the support rather than the bottom one. Retaining the negative sign for $B_{y'}$ (Why?) and substituting the results into Eq. 1, we obtain

$$1346.4 \sin 30°\text{ N} - 1000.0 \sin 30°\text{ N} - A_x = 0$$
$$A_x = 173.2\text{ N} \qquad\qquad Ans.$$

Example 5–10

The link shown in Fig. 5–15a is pin-connected at A and rests against a smooth support at B. Compute the horizontal and vertical components of reaction at the pin A.

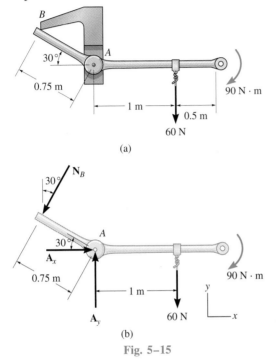

(a)

(b)

Fig. 5–15

SOLUTION

Free-Body Diagram.   As shown in Fig. 5–15b, the reaction $\mathbf{N}_B$ is perpendicular to the link at B. Also, horizontal and vertical components of reaction are represented at A, even though the base of the pin support is tilted.

Equations of Equilibrium.   Summing moments about A, we obtain a direct solution for N_B,

$$\zeta + \Sigma M_A = 0; \quad -90 \text{ N} \cdot \text{m} - 60 \text{ N}(1 \text{ m}) + N_B(0.75 \text{ m}) = 0$$
$$N_B = 200 \text{ N}$$

Using this result,

$$\xrightarrow{+} \Sigma F_x = 0; \qquad A_x - 200 \sin 30° \text{ N} = 0$$
$$A_x = 100 \text{ N} \qquad\qquad Ans.$$

$$+ \uparrow \Sigma F_y = 0; \qquad A_y - 200 \cos 30° \text{ N} - 60 \text{ N} = 0$$
$$A_y = 233 \text{ N} \qquad\qquad Ans.$$

Example 5–11

A force of 150 N acts on the end of the beam shown in Fig. 5–16a. Determine the magnitude and direction of the reaction at the pin A and the tension in the cable.

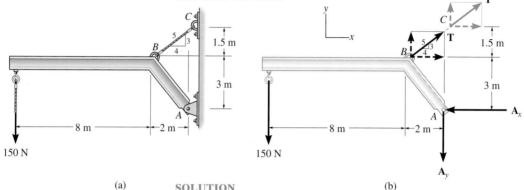

(a)   (b)

Fig. 5–16

SOLUTION

Free-Body Diagram.   The forces acting on the beam are shown in Fig. 5–16b.

Equations of Equilibrium.   Summing moments about point A to obtain a direct solution for the cable tension yields

$$\zeta +\Sigma M_A = 0; \quad -(\tfrac{3}{5}T)(2 \text{ m}) - (\tfrac{4}{5}T)(3 \text{ m}) + 150 \text{ N}(10 \text{ m}) = 0$$

$$-3.6T + 150 \text{ N}(10 \text{ m}) = 0 \qquad\qquad (1)$$

$$T = 416.7 \text{ N} \qquad\qquad\qquad Ans.$$

Using the principle of transmissibility it is also possible to locate **T** at C, even though this point is not on the beam, Fig. 5–16b. In this case, the vertical component of **T** creates *zero moment* about A and the moment arm of the horizontal component ($\tfrac{4}{5}T$) becomes 4.5 ft. Hence, $\Sigma M_A = 0$ yields Eq. 1 directly since $(\tfrac{4}{5}T)(4.5) = 3.6T$.

Summing forces to obtain A_x and A_y, using the result for T, we have

$$\xrightarrow{+}\Sigma F_x = 0; \qquad\qquad -A_x + (\tfrac{4}{5})(416.7 \text{ N}) = 0$$

$$A_x = 333.3 \text{ N} \leftarrow$$

$$+\uparrow \Sigma F_y = 0; \qquad\qquad (\tfrac{3}{5})416.7 \text{ N} - 150 \text{ N} - A_y = 0$$

$$A_y = 100 \text{ N} \downarrow$$

Thus,

$$F_A = \sqrt{(333.3 \text{ N})^2 + (100 \text{ N})^2}$$

$$= 348.0 \text{ N} \qquad\qquad\qquad Ans.$$

$$\theta = \tan^{-1}\frac{100 \text{ N}}{333.3 \text{ N}} = 16.7° \quad \triangleright \qquad Ans.$$

Example 5–12

The oil-drilling rig shown in Fig. 5–17a has a mass of 24 Mg and mass center at G. If the rig is pin-connected at its base, determine the tension in the hoisting cable and the magnitude of the resultant force at A when the rig is in the position shown.

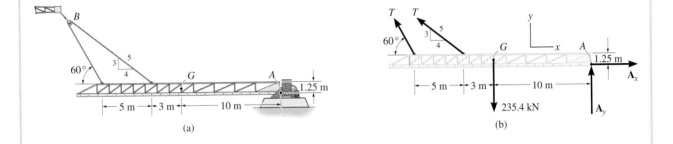

(a)

(b)

Fig. 5–17

SOLUTION

Free-Body Diagram. Because the hoisting cable is continuous and passes over the pulley, the cable is subjected to the same tension T throughout its length. Hence the cable exerts a force T on the rig at its points of attachment, Fig. 5–17b. The weight of the rig is $(24(10^3) \text{ kg})(9.81 \text{ m/s}^2) = 235.4 \text{ kN}$.

Equations of Equilibrium. By summing moments about point A, it is possible to obtain a direct solution for T. Why?

$$\zeta + \Sigma M_A = 0; \quad (235.4 \text{ kN})(10 \text{ m}) - T(\tfrac{3}{5})(13 \text{ m}) + T(\tfrac{4}{5})(1.25 \text{ m})$$
$$-(T \sin 60°)(18 \text{ m}) + (T \cos 60°)(1.25 \text{ m}) = 0$$
$$T = 108.2 \text{ kN} \qquad \textit{Ans.}$$

$$\xrightarrow{+} \Sigma F_x = 0; \quad A_x - 108.2(\tfrac{4}{5}) \text{ kN} - 108.2 \cos 60° \text{ kN} = 0$$
$$A_x = 140.6 \text{ kN}$$

$$+\uparrow \Sigma F_y = 0; \quad A_y - 235.4 \text{ kN} + 108.2(\tfrac{3}{5}) \text{ kN} + 108.2 \sin 60° \text{ kN} = 0$$
$$A_y = 76.8 \text{ kN}$$

Thus,

$$F_A = \sqrt{(140.6)^2 + (76.8)^2} = 160 \text{ kN} \qquad \textit{Ans.}$$

5.4 Two- and Three-Force Members

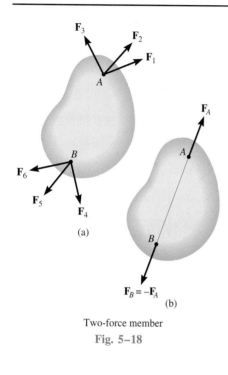

(a)

$F_B = -F_A$

(b)

Two-force member

Fig. 5–18

The solution to some equilibrium problems can be simplified if one is able to recognize members that are subjected to only two or three forces.

Two-Force Members. When a member is subject to *no couple moments* and forces are applied at only two points on a member, the member is called a *two-force member*. An example of this situation is shown in Fig. 5–18a. The forces at A and B are first summed to obtain their respective *resultants* F_A and F_B, Fig. 5–18b. These two forces will maintain *translational or force equilibrium* ($\Sigma F = 0$) provided F_A is of equal magnitude and opposite direction to F_B. Furthermore, *rotational or moment equilibrium* ($\Sigma M_O = 0$) is satisfied if F_A is *collinear* with F_B. As a result, the line of action of both forces is known, since it always passes through A and B. Hence, only the force magnitude must be determined or stated. Other examples of two-force members held in equilibrium are shown in Fig. 5–19.

Three-Force Members. If a member is subjected to only three forces, then it is necessary that the forces be either *concurrent* or *parallel* if the member is to be in equilibrium. To show the concurrency requirement, consider the body in Fig. 5–20a and suppose that any two of the three forces acting on the body have lines of action that intersect at point O. To satisfy moment equilibrium about O, i.e., $\Sigma M_O = 0$, the third force must also pass through O, which then makes the force system *concurrent*. If two of the three forces are parallel, Fig. 5–20b, the point of concurrency, O, is considered to be at "infinity" and the third force must be parallel to the other two forces to intersect at this "point."

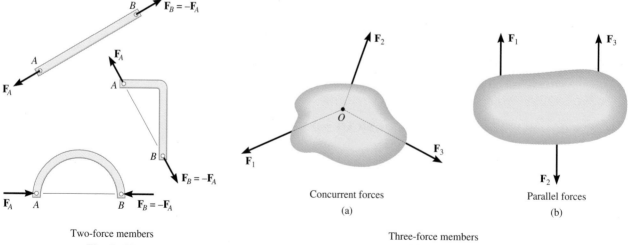

Two-force members

Fig. 5–19

Concurrent forces

(a)

Parallel forces

(b)

Three-force members

Fig. 5–20

Example 5–13

The lever *ABC* is pin-supported at *A* and connected to a short link *BD* as shown in Fig. 5–21*a*. If the weight of the members is negligible, determine the force of the pin on the lever at *A*.

SOLUTION

Free-Body Diagrams. As shown by the free-body diagram, Fig. 5–21*b*, the short link *BD* is a *two-force member,* so the *resultant forces* at pins *D* and *B* must be equal, opposite, and collinear. Although the magnitude of the force is unknown, the line of action is known, since it passes through *B* and *D*.

Lever *ABC* is a *three-force member,* and therefore, in order to satisfy moment equilibrium, the three nonparallel forces acting on it must be concurrent at *O*, Fig. 5–21*c*. In particular, note that the force **F** on the lever is equal but opposite to **F** acting at *B* on the link. Why? The distance *CO* must be 0.5 m, since the lines of action of **F** and the 400-N force are known.

Equations of Equilibrium. By requiring the force system to be concurrent at *O*, so that $\Sigma M_O = 0$, the angle θ which defines the line of action of $\mathbf{F}_A$ can be determined from trigonometry,

$$\theta = \tan^{-1}\left(\frac{0.7}{0.4}\right) = 60.3° \quad \measuredangle^\theta \qquad \qquad Ans.$$

Using the *x*, *y* axes and applying the force equilibrium equations, we can obtain *F* and F_A.

$\xrightarrow{+}\Sigma F_x = 0;$ $F_A \cos 60.3° - F \cos 45° + 400 \text{ N} = 0$

$+\uparrow \Sigma F_y = 0;$ $F_A \sin 60.3° - F \sin 45° = 0$

Solving, we get

$$F_A = 1075 \text{ N} \qquad \qquad Ans.$$
$$F = 1320 \text{ N}$$

Note: We can also solve this problem by representing the force at *A* by its two components $\mathbf{A}_x$ and $\mathbf{A}_y$ and applying $\Sigma M_A = 0$, $\Sigma F_x = 0$, $\Sigma F_y = 0$ to the lever. Once A_x and A_y are determined, how would you find F_A and θ?

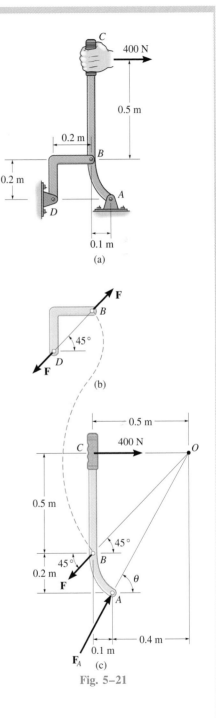

400 N

0.5 m

0.2 m

B

0.2 m

A

D

0.1 m

(a)

F

B

45°

D

F

(b)

0.5 m

400 N

C

O

0.5 m

45°

45°   *B*

0.2 m

F

θ

A

0.1 m

0.4 m

$\mathbf{F}_A$ (c)

Fig. 5–21

PROBLEMS

5–11. Determine the magnitude of the resultant force acting at pin A of the hand punch in Prob. 5–2.

*5–12. Determine the reactions at the supports A and B of the jib crane in Prob. 5–4.

5–13. Determine the reactions at the supports for the beam in Prob. 5–8.

5–14. Determine the reactions at the pin A and the force on the hydraulic cylinder of the truck dumpster in Prob. 5–5.

5–15. Determine the support reactions on the beam in Prob. 5–9.

*5–16. Determine the reactions on the uniform pipe at A, B, and C in Prob. 5–7.

5–17. Determine the reactions at the points of contact at A, B, and C of the bar in Prob. 5–3.

5–18. Determine the magnitude of the reactions on the beam at A and B. Neglect the thickness of the beam.

5–19. The forces acting on the plane while it is flying at constant velocity are shown. If the engine thrust is $F_T = 480$ kN and the plane's weight is $W = 750$ kN, determine the atmospheric drag $\mathbf{F}_D$ and the wing lift $\mathbf{F}_L$. Also, determine the distance s to the line of action of the drag force.

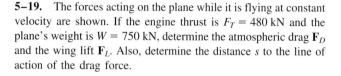
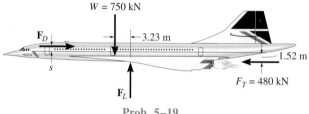

Prob. 5–19

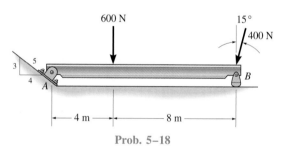

Prob. 5–18

*5–20. When holding the 2.3 kg stone in equilibrium, the humerus H, assumed to be smooth, exerts normal forces $\mathbf{F}_C$ and $\mathbf{F}_A$ on the radius C and ulna A as shown. Determine these forces and the force $\mathbf{F}_B$ that the biceps B exerts on the radius for equilibrium. The stone has a center of mass at G. Neglect the weight of the arm.

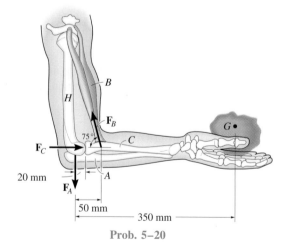

Prob. 5–20

5–21. Determine the reactions at the roller B, the rocker C, and where the beam contacts the smooth plane at A. Neglect the thickness of the beam.

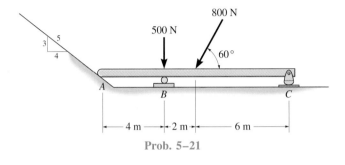

Prob. 5–21

5–22. The ramp of a ship has a weight of 2000 N and a center of gravity at G. Determine the cable force in CD needed to just start lifting the ramp, i.e., so the reaction at B becomes zero. Also, determine the horizontal and vertical components of force at the hinge (pin) at A.

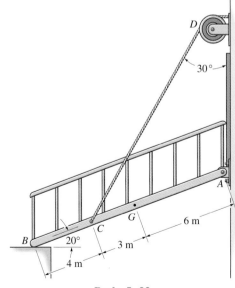

Prob. 5–22

5–23. The man is pulling a load of 35 N with one arm held as shown. Determine the force $\mathbf{F}_H$ this exerts on the humerus bone H, and the tension developed in the biceps muscle B. Neglect the weight of the man's arm.

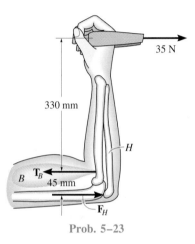

Prob. 5–23

*5–24. Determine the reactions at the pin A and the tension in cord BC. Set $F = 40$ kN. Neglect the thickness of the beam.

5–25. If rope BC will fail when the tension becomes 50 kN, determine the greatest vertical load F that can be applied to the beam at B. What is the magnitude of the reaction at A for this loading? Neglect the thickness of the beam.

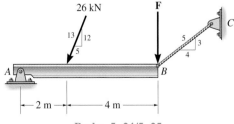

Probs. 5–24/5–25

5–26. The platform assembly has a weight of 250 N and center of gravity at G_1. If it is intended to support a maximum load of 400 N placed at point G_2, determine the smallest counterweight W that should be placed at B in order to prevent the platform from tipping over.

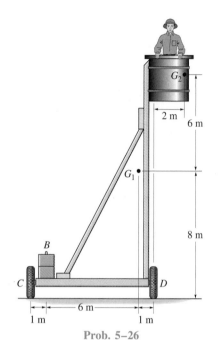

Prob. 5–26

5–27. The device is used to hold an elevator door open. If the spring has a stiffness of $k = 40$ N/m and it is compressed 0.2 m, determine the horizontal and vertical components of reaction at the pin A and the resultant force at the wheel bearing B.

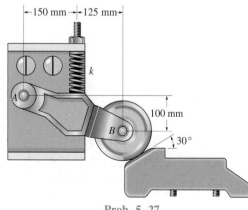

Prob. 5–27

*5–28. The linkage rides along the top and bottom flanges of the crane rail. If the load it supports is 5 kN, determine the force of each roller on the flange.

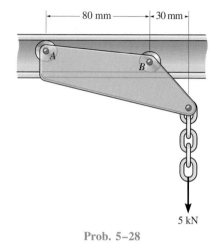

Prob. 5–28

5–29. The wall crane is supported by the journal bearing (smooth collar) at B and thrust bearing at A. Determine the horizontal and vertical components of force at A and the horizontal force at B if $P = 8$ kN.

5–30. The wall crane is supported by the journal bearing (smooth collar) at B and thrust bearing at A. If the journal bearing can support a force of 12 kN before it fails, determine the maximum load P that can be suspended from the crane. The thrust bearing at A can support both horizontal and vertical components of force.

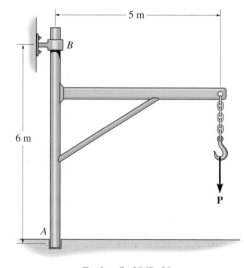

Probs. 5–29/5–30

5–31. The cantilevered jib crane is used to support the load of 780 N. If the trolley T can be placed anywhere between $1.5 \text{ m} \leq x \leq 7.5 \text{ m}$, determine the maximum magnitude of reaction at the supports A and B. Note that the supports are collars that allow the crane to rotate freely about the vertical axis. The collar at B supports a force in the vertical direction, whereas the one at A does not.

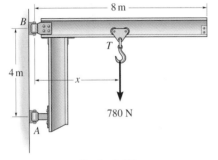

Prob. 5–31

5–33. The power pole supports the three lines, each line exerting a vertical force on the pole due to its weight as shown. Determine the reactions at the fixed support D. If it is possible for wind or ice to snap the lines, determine which line(s) when removed create(s) a condition for the greatest moment reaction at D.

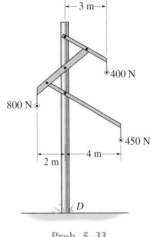

Prob. 5–33

5–34. The framework is supported by the member AB which rests on the smooth floor. When loaded, the pressure distribution on AB is linear as shown. Determine the smallest size d of member AB so that it will not cause the frame to tip over. What is the intensity w for this case?

*5–32. Determine the reactions on the beam at A and B.

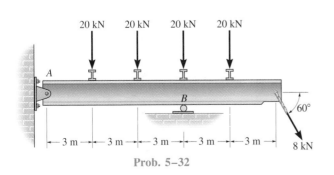

Prob. 5–32

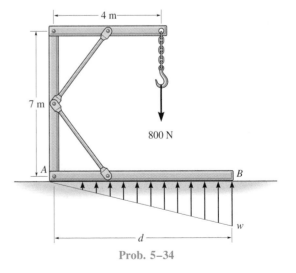

Prob. 5–34

5–35. If the wheelbarrow and its contents have a mass of 60 kg and center of mass at *G*, determine the magnitude of the resultant force which the man must exert on *each* of the two handles in order to hold the wheelbarrow in equilibrium.

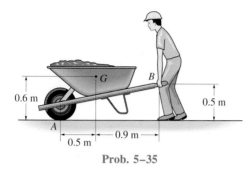

0.6 m

G   *B*

0.5 m

A

0.5 m   0.9 m

Prob. 5–35

*5–36. Determine the resultant normal force acting on *each* set of wheels of the airplane. There is a set of wheels in the front, *A*, and a set of wheels under each wing, *B*. Both wings have a total weight of 200 kN and center of gravity at G_w, the fuselage has a weight of 800 kN and center of gravity at G_f, and both engines (one on each side) have a total weight of 100 kN and center of gravity at G_e.

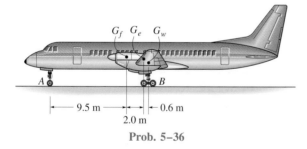

G_f G_e G_w

A   *B*

9.5 m   0.6 m

2.0 m

Prob. 5–36

5–37. The telephone pole of negligible thickness is subjected to the cable force of 360 N directed as shown. It is supported by the cable *BCD* and the pole can be assumed pinned at its base *A*. In order to provide clearance for a sidewalk right of way, where *D* is located, a strut *CE* is attached to the pole at its midheight *h* = 4.5 m and two cables, *BC* and *CD′*, are attached to the strut at *C*, as shown by the dashed line (cable segment *CD* is removed). Determine the tension in cable *CD′* and the resultant force at *A*.

5–38. The telephone pole of negligible thickness is subjected to the force of 360 N directed as shown. It is supported by the cable *BCD* and can be assumed pinned at its base *A*. In order to provide clearance for a sidewalk right of way, where *D* is located, a strut *CE* is attached to the pole and two cables, *BC* and *CD′*, are attached to the strut at *C*, as shown by the dashed lines (cable segment *CD* is removed). If the tension in *CD′* is to be twice the tension in *BCD*, determine the height *h* for placement of the strut *CE*.

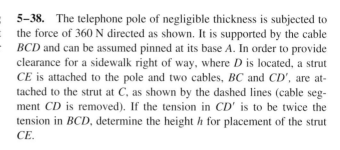
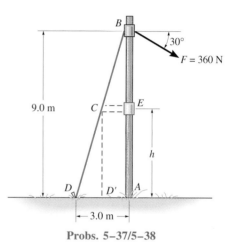

B

30°

F = 360 N

9.0 m   *C*   *E*

h

D   *D′*   *A*

3.0 m

Probs. 5–37/5–38

5–39. The mechanism shown was thought by its inventor to be a perpetual-motion machine. It consists of the stand *A*, two smooth idler wheels *B* and *C*, and in between a uniform hollow cylindrical ring *D* suspended in the manner shown. The ring has a weight *W* and it was expected to revolve in the direction indicated by the arrow. Draw a free-body diagram of the ring and using an appropriate equation of equilibrium show that it will not rotate.

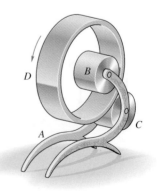

D   *B*

A   *C*

Prob. 5–39

*5–40   The shelf supports the electric motor which has a mass of 15 kg and mass center at G_m. The platform upon which it rests has a mass of 4 kg and mass center at G_p. Assuming that a single bolt B holds the shelf up and the bracket bears against the smooth wall at A, determine this normal force at A and the horizontal and vertical components of reaction of the bolt on the bracket.

5–43.   Determine the reactions at the smooth collar A, the rocker B, and the short link CD.

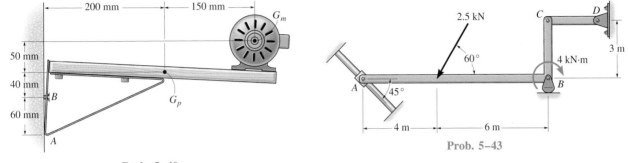

Prob. 5–40

Prob. 5–43

5–41.   The boom supports the two vertical loads. Neglect the size of the collars at D and B and the thickness of the boom, and compute the horizontal and vertical components of force at the pin A and the force in cable CB. Set $F_1 = 800$ N and $F_2 = 350$ N.

5–42.   The boom is intended to support two vertical loads, F_1 and F_2. If the cable CB can sustain a maximum load of 1500 N before it fails, determine the critical loads if $F_1 = 2F_2$. Also, what is the magnitude of the maximum reaction at pin A?

*5–44.   The worker uses the hand truck to move material down the ramp. If the truck and its contents are held in the position shown and have a weight of 450 N with center of gravity at G, determine the resultant normal force of both wheels on the ground A and the magnitude of the force required at the grip B.

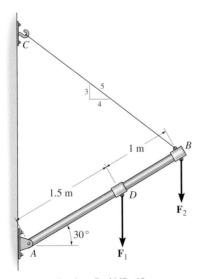

Probs. 5–41/5–42

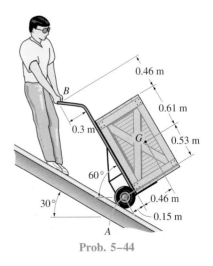

Prob. 5–44

5–45. The upper portion of the crane boom consists of the jib *AB*, which is supported by the pin at *A*, the guy line *BC*, and the backstay *CD*, each cable being separately attached to the mast at *C*. If the 5-kN load is supported by the hoist line, which passes over the pulley at *B*, determine the magnitude of the resultant force the pin exerts on the jib at *A* for equilibrium, the tension in the guy line *BC*, and the tension *T* in the hoist line. Neglect the weight of the jib. The pulley at *B* has a radius of 0.1 m.

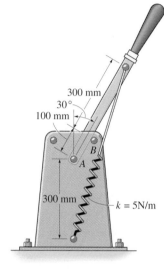

Prob. 5–45

5–46. The mobile crane has a weight of 600 000 N and center of gravity at G_1; the boom has a weight of 150 000 N and center of gravity at G_2. Determine the smallest angle of tilt θ of the boom, without causing the crane to overturn if the suspended load is $W = 200\,000$ N. Neglect the thickness of the tracks at *A* and *B*.

5–47. The mobile crane has a weight of 600 000 N and center of gravity at G_1; the boom has a weight of 150 000 N and center of gravity at G_2. If the suspended load has a weight of $W = 80\,000$ N, determine the normal reactions at the tracks *A* and *B*. For the calculation, neglect the thickness of the tracks and take $\theta = 30°$.

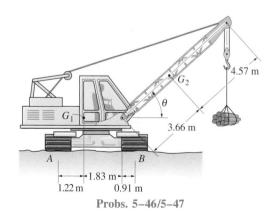

Probs. 5–46/5–47

*5–48. The toggle switch consists of a cocking lever that is pinned to a fixed frame at *A* and held in place by the spring which has an unstretched length of 200 mm. Determine the magnitude of the resultant force at *A* and the normal force on the peg at *B* when the lever is in the position shown.

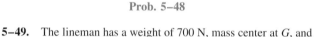

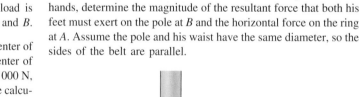

Prob. 5–48

5–49. The lineman has a weight of 700 N, mass center at *G*, and stands in the position shown. If he lets go of the pole with his hands, determine the magnitude of the resultant force that both his feet must exert on the pole at *B* and the horizontal force on the ring at *A*. Assume the pole and his waist have the same diameter, so the sides of the belt are parallel.

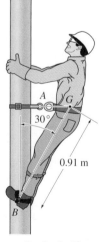

Prob. 5–49

5–50. The wheel support on a cart is attached to the frame of the cart by a smooth collar. Since it is not a snug fit, the shaft on the wheel support bears on the collar at the smooth points A, B, and C. If the wheel loading is 900 N, determine the reactive forces on the shaft at its points of contact.

*5–52. The file cabinet contains four uniform drawers, each 0.76 m long and weighing 400 N/m. If the empty cabinet (without drawers) has a weight of 160 N and a center of mass at G, determine how many of the drawers can be fully pulled out and the extension x of the last drawer that will cause the assembly to be on the verge of tipping over.

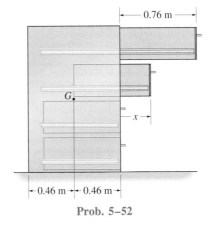

Prob. 5–52

Prob. 5–50

5–51. The rigid beam of negligible weight is supported horizontally by two springs and a pin. If the springs are uncompressed when the load is removed, determine the force in each spring when the load **P** is applied. Also, compute the vertical deflection of end C. Assume the spring stiffness k is large enough so that only small deflections occur. *Hint:* The beam rotates about A so the deflections in the springs can be related.

5–53. The smooth pipe rests against the wall at the points of contact A, B, and C. Determine the reactions at these points needed to support the vertical force of 200 N. Neglect the pipe's thickness in the calculation.

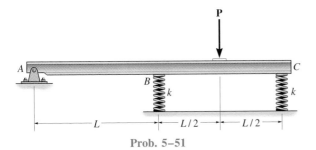

Prob. 5–51

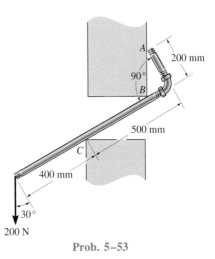

Prob. 5–53

5–54. Determine the distance d for placement of the load **P** for equilibrium of the smooth bar in the position θ as shown. Neglect the weight of the bar.

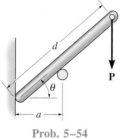

Prob. 5–54

5–55. The assembly is made from two boards. The board on the left has a weight of 10 N and center of gravity at G_1, and the board on the right has a weight of 7 N and center of gravity at G_2. Determine the force that the two smooth pipes exert on it at A, B, and C.

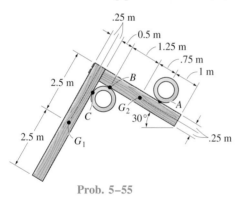

Prob. 5–55

***5–56.** The disk has a mass of 20 kg and is supported on the smooth cylindrical surface by a spring having a stiffness of $k =$ 400 N/m and unstretched length of $l_0 = 1$ m. The spring remains in the horizontal position since its end A is attached to the small roller guide which has negligible weight. Determine the angle θ to the nearest degree for equilibrium of the roller.

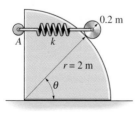

Prob. 5–56

5–57. The wheelbarrow and its contents have a mass m and center of mass at G. Determine the greatest angle of tilt θ without causing the wheelbarrow to tip over.

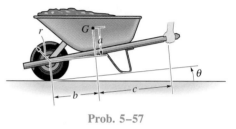

Prob. 5–57

5–58. The smooth uniform rod has a mass m and is placed on the semicircular arch and against the wall. Show that for equilibrium the angle θ must satisfy $\sin \theta = \dfrac{1}{r}(\sqrt{1 + 3 \cos^2 \theta})(d - l \sin \theta)$.

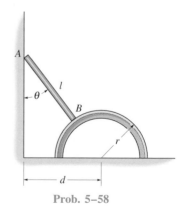

Prob. 5–58

5–59. The uniform ladder has a mass of 60 kg and is placed against the smooth step A. It is lowered to the horizontal position by a man who applies a normal force to it always from a height of 2.5 m. Determine the largest length L at which it can be so that he can let it down slowly without causing the ladder to slip at A.

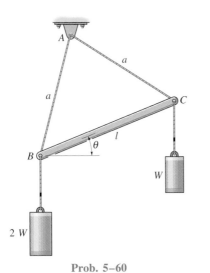

Prob. 5–59

5–61. The disk has a mass of 20 kg and rests on the smooth inclined surface. One end of a spring is attached to the center of the disk and the other end is attached to a roller at A. Consequently, the spring remains in the horizontal position when the disk is in equilibrium. If the unstretched length of the spring is 200 mm, determine its stretched length when the disk is in equilibrium. Neglect the size and weight of the roller.

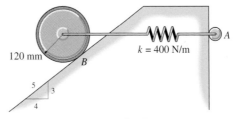

Prob. 5–61

***5–60.** The rod BC is supported by two cords, each of length a, which are attached to the pin at A. If weights W and 2W are suspended from the ends of the rod, determine the angle θ for equilibrium, measured from the horizontal. Express the answer in terms of a and l. Neglect the weight of the rod.

■**5–62.** A linear *torsional spring* deforms such that an applied couple moment M is related to the spring's rotation θ in radians by the equation $M = (20\,\theta)$ N · m. If such a spring is attached to the end of a pin-connected uniform 10-kg rod, determine the angle θ for equilibrium. The spring is undeformed when $\theta = 0°$.

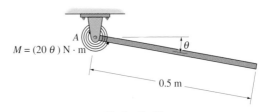

Prob. 5–62

Prob. 5–60

Equilibrium in Three Dimensions

5.5 Free-Body Diagrams

The first step in solving three-dimensional equilibrium problems, as in the case of two dimensions, is to draw a free-body diagram of the body (or group of bodies considered as a system). Before we show this, however, it is necessary to discuss the types of reactions that can occur at the supports.

Support Reactions. The reactive forces and couple moments acting at various types of supports and connections, when the members are viewed in three dimensions, are listed in Table 5–2. It is important to recognize the symbols used to represent each of these supports and to understand clearly how the forces and couple moments are developed by each support. As in the two-dimensional case, *a force is developed by a support that restricts the translation of the attached member, whereas a couple moment is developed when rotation of the attached member is prevented.* For example, in Table 5–2, the ball-and-socket joint (4) prevents any translation of the connecting member; therefore, a force must act on the member at the point of connection. This force has three components having unknown magnitudes, F_x, F_y, F_z. Provided these components are known, one can obtain the magnitude of force, $F = \sqrt{F_x^2 + F_y^2 + F_z^2}$, and the force's orientation defined by the coordinate direction angles α, β, γ, Eqs. 2–7.* Since the connecting member is allowed to rotate freely about *any* axis, no couple moment is resisted by a ball-and-socket joint.

It should be noted that the *single* bearing supports (5) and (7), the *single* pin (8), and the *single* hinge (9) are shown to support both force and couple-moment components. If, however, these supports are used in conjunction with *other* bearings, pins, or hinges to hold the body in equilibrium, and provided the physical body maintains its *rigidity* when loaded and the supports are *properly aligned* when connected to the body, then the *force reactions* at these supports may *alone* be adequate for supporting the body. In other words, the couple moments become redundant and may be neglected on the free-body diagram. The reason for this will be clear after studying the examples which follow, but essentially the couple moments will not be developed at these supports since the rotation of the body is prevented by the reactions developed at the other supports and not by the supporting couple moments.

---

*The three unknowns may also be represented as an unknown force magnitude F and two unknown coordinate direction angles. The third direction angle is obtained using the identity $\cos^2 \alpha + \cos^2 \beta + \cos^2 \gamma = 1$, Eq. 2–10.

Table 5–2   **Supports for Rigid Bodies Subjected to Three-Dimensional Force Systems**

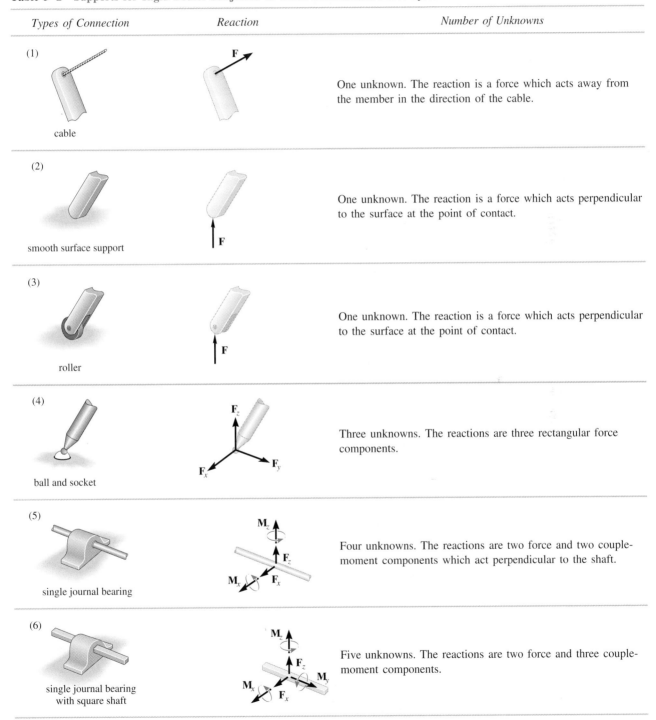

Types of Connection	*Reaction*	*Number of Unknowns*
(1) cable	**F**	One unknown. The reaction is a force which acts away from the member in the direction of the cable.
(2) smooth surface support	**F**	One unknown. The reaction is a force which acts perpendicular to the surface at the point of contact.
(3) roller	**F**	One unknown. The reaction is a force which acts perpendicular to the surface at the point of contact.
(4) ball and socket	F_z F_y F_x	Three unknowns. The reactions are three rectangular force components.
(5) single journal bearing	M_z F_z M_x F_x	Four unknowns. The reactions are two force and two couple-moment components which act perpendicular to the shaft.
(6) single journal bearing with square shaft	M_z F_z M_y M_x F_x	Five unknowns. The reactions are two force and three couple-moment components.

Table 5–2 (Contd.)

Types of Connection	Reaction	Number of Unknowns

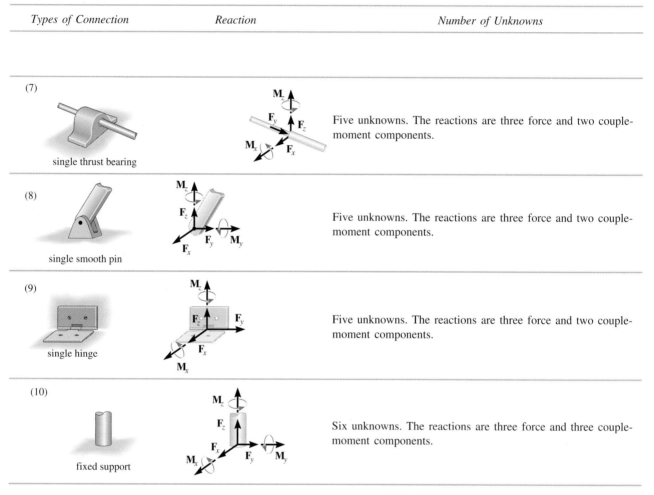

(7) single thrust bearing		Five unknowns. The reactions are three force and two couple-moment components.
(8) single smooth pin		Five unknowns. The reactions are three force and two couple-moment components.
(9) single hinge		Five unknowns. The reactions are three force and two couple-moment components.
(10) fixed support		Six unknowns. The reactions are three force and three couple-moment components.

Free-Body Diagrams. The general procedure for establishing the free-body diagram of a rigid body has been outlined in Sec. 5.2. Essentially it requires first "isolating" the body by drawing its outlined shape. This is followed by a careful *labeling* of *all* the forces and couple moments in reference to an established x, y, z coordinate system. As a general rule, *components of reaction* having an *unknown magnitude* are shown acting on the free-body diagram in the *positive sense*. In this way, if any negative values are obtained, they will indicate that the components act in the negative coordinate directions.

Example 5–14

Several examples of objects along with their associated free-body diagrams are shown in Fig. 5–22. In all cases, the x, y, z axes are established and the unknown reaction components are indicated in the positive sense. The weight of the objects is neglected.

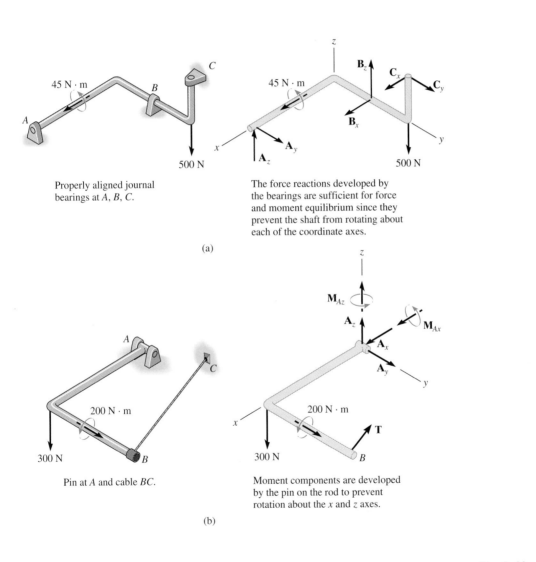

Properly aligned journal bearings at A, B, C.

The force reactions developed by the bearings are sufficient for force and moment equilibrium since they prevent the shaft from rotating about each of the coordinate axes.

(a)

Pin at A and cable BC.

Moment components are developed by the pin on the rod to prevent rotation about the x and z axes.

(b)

Fig. 5–22*a* and *b*
Fig. 5–22 (*continued*)

Fig. 5–22 (*continued*)

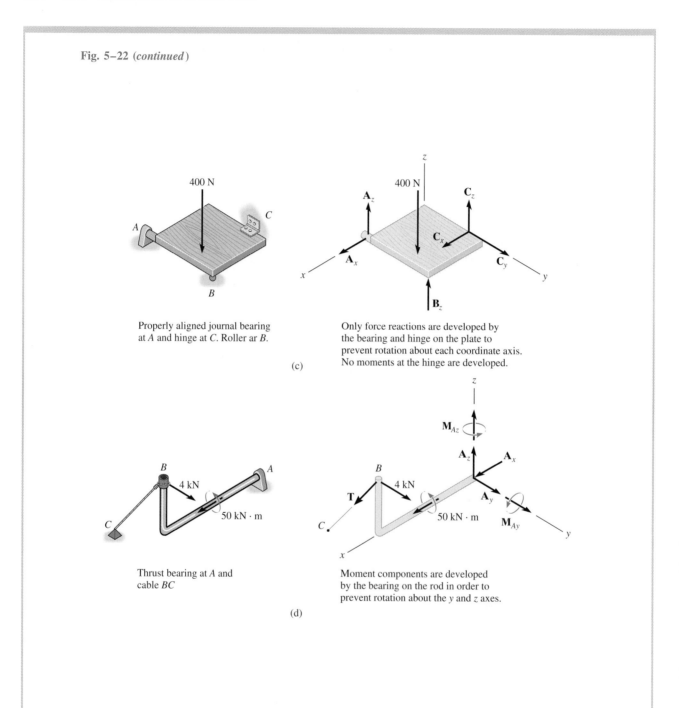

400 N

A

C

B

Properly aligned journal bearing
at *A* and hinge at *C*. Roller ar *B*.

(c)

400 N

Only force reactions are developed by
the bearing and hinge on the plate to
prevent rotation about each coordinate axis.
No moments at the hinge are developed.

B   *A*

4 kN

50 kN · m

C

Thrust bearing at *A* and
cable *BC*

(d)

B   4 kN

T

50 kN · m

C

Moment components are developed
by the bearing on the rod in order to
prevent rotation about the *y* and *z* axes.

Fig. **5–22c and d**

5.6 Equations of Equilibrium

As stated in Sec. 5.1, the conditions for equilibrium of a rigid body subjected to a three-dimensional force system require that both the *resultant* force and *resultant* couple moment acting on the body be equal to *zero*.

Vector Equations of Equilibrium. The two conditions for equilibrium of a rigid body may be expressed mathematically in vector form as

$$\Sigma \mathbf{F} = \mathbf{0}$$
$$\Sigma \mathbf{M}_O = \mathbf{0}$$

(5–5)

where $\Sigma \mathbf{F}$ is the vector sum of all the external forces acting on the body and $\Sigma \mathbf{M}_O$ is the sum of the couple moments and the moments of all the forces about any point O located either on or off the body.

Scalar Equations of Equilibrium. If all the applied external forces and couple moments are expressed in Cartesian vector form and substituted into Eqs. 5–5, we have

$$\Sigma \mathbf{F} = \Sigma F_x \mathbf{i} + \Sigma F_y \mathbf{j} + \Sigma F_z \mathbf{k} = \mathbf{0}$$
$$\Sigma \mathbf{M}_O = \Sigma M_x \mathbf{i} + \Sigma M_y \mathbf{j} + \Sigma M_z \mathbf{k} = \mathbf{0}$$

Since the $\mathbf{i}, \mathbf{j},$ and $\mathbf{k}$ components are independent from one another, the above equations are satisfied provided

$$\Sigma F_x = 0$$
$$\Sigma F_y = 0$$
$$\Sigma F_z = 0$$

(5–6a)

and

$$\Sigma M_x = 0$$
$$\Sigma M_y = 0$$
$$\Sigma M_z = 0$$

(5–6b)

These *six scalar equilibrium equations* may be used to solve for at most six unknowns shown on the free-body diagram. Equations 5–6a express the fact that the sum of the external force components acting in the x, y, and z directions must be zero, and Eqs. 5–6b require the sum of the moment components about the x, y, and z axes to be zero.

PROCEDURE FOR ANALYSIS

The following procedure provides a method for solving three-dimensional equilibrium problems.

Free-Body Diagram.   Construct the free-body diagram for the body. Be sure to include *all* the forces and couple moments that act *on* the body. These interactions are commonly caused by the externally applied loadings, contact forces exerted by adjacent bodies, support reactions, and the weight of the body if it is significant compared to the magnitudes of the other applied forces. Establish the origin of the x, y, z axes at a convenient point, and orient the axes so that they are parallel to as many of the external forces and moments as possible. Identify the unknowns, and in general show all the unknown components having a positive sense if the sense cannot be determined. Dimensions of the body, necessary for calculating the moments of forces, are also included on the free-body diagram.

Equations of Equilibrium.   Apply the equations of equilibrium. In many cases, problems can be solved by *direct application* of the six scalar equations $\Sigma F_x = 0$, $\Sigma F_y = 0$, $\Sigma F_z = 0$, $\Sigma M_x = 0$, $\Sigma M_y = 0$, $\Sigma M_z = 0$, Eqs. 5–6; however, if the force components or moment arms seem difficult to determine, it is recommended that the solution be obtained by using vector equations: $\Sigma \mathbf{F} = \mathbf{0}, \Sigma \mathbf{M}_O = \mathbf{0}$, Eqs. 5–5. In any case, it is *not necessary* that the set of axes chosen for force summation *coincide* with the set of axes chosen for moment summation. Instead, it is recommended that one *choose the direction of an axis for moment summation such that it intersects the lines of action of as many unknown forces as possible.* The moments of forces passing through points on this axis or forces which are parallel to the axis will then be zero. Furthermore, *any set of three nonorthogonal axes* may be chosen for either the force or moment summations. By the proper choice of axes, it may be possible to solve directly for an unknown quantity, or at least reduce the need for solving a large number of simultaneous equations for the unknowns.

5.7 Constraints for a Rigid Body

To ensure the equilibrium of a rigid body, it is not only necessary to satisfy the equations of equilibrium, but the body must also be properly held or constrained by its supports. Some bodies may have more supports than are necessary for equilibrium, whereas others may not have enough or the supports may be arranged in a particular manner that could cause the body to collapse. Each of these cases will now be discussed.

Redundant Constraints. When a body has redundant supports, that is, more supports than are necessary to hold it in equilibrium, it becomes statically indeterminate. *Statically indeterminate* means that there will be more unknown loadings on the body than equations of equilibrium available for their solution. For example, the two-dimensional problem, Fig. 5–23a, and the three-dimensional problem, Fig. 5–23b, shown together with their free-body diagrams, are both statically indeterminate because of additional support reactions. In the two-dimensional case, there are five unknowns, that is, M_A, A_x, A_y, B_y, and C_y, for which only three equilibrium equations can be written ($\Sigma F_x = 0$, $\Sigma F_y = 0$, and $\Sigma M_O = 0$, Eqs. 5–2). The three-dimensional problem has eight unknowns, for which only six equilibrium equations can be written, Eqs. 5–6. The additional equations needed to solve indeterminate problems of the type shown in Fig. 5–23 are generally obtained from the deformation conditions at the points of support. These equations involve the physical properties of the body which are studied in subjects dealing with the mechanics of deformation, such as "mechanics of materials."*

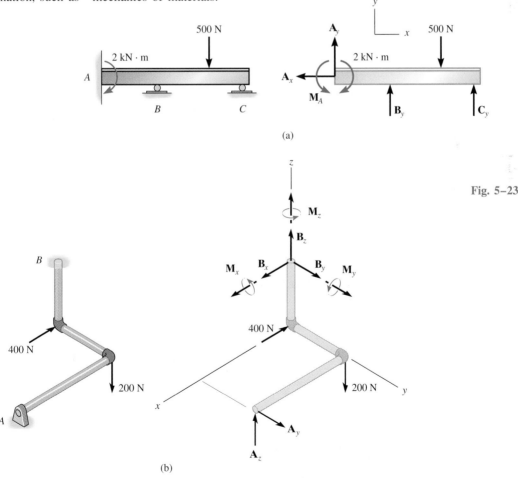

(a)

(b)

Fig. 5–23

*See R. C. Hibbeler, *Mechanics of Materials*, 2nd edition (New York: Macmillan, 1994).

Improper Constraints. In some cases, there may be as many unknown forces on the body as there are equations of equilibrium; however, *instability* of the body can develop because of *improper constraining* by the supports. In the case of three-dimensional problems, the body is improperly constrained if the support reactions *all intersect a common axis.* For two-dimensional problems, this axis is *perpendicular* to the plane of the forces and therefore appears as a point. Hence, when all the reactive forces are *concurrent* at this point, the body is improperly constrained. Examples of both cases are given in Fig. 5–24. From the free-body diagrams it is seen that the summation of moments about the x axis, Fig. 5–24a, or point O, Fig. 5–24b, will *not* be equal to zero; thus rotation about the x axis or point O will take place.* Furthermore, in both cases, it becomes *impossible* to solve *completely* for all the unknowns, since one can write a moment equation that *does not* involve any of the unknown support reactions, and as a result, this reduces the number of available equilibrium equations by one.

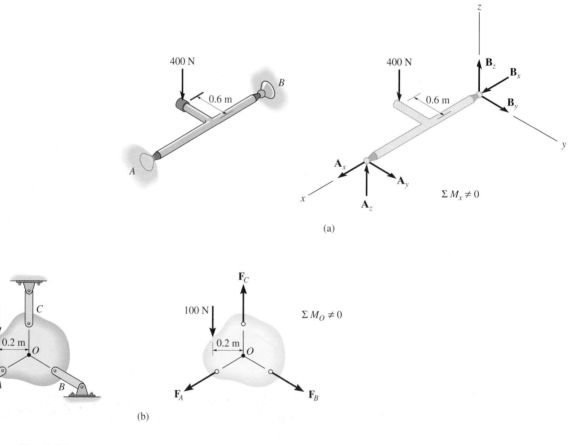

(a)

(b)

Fig. 5–24

---

*For the three-dimensional problem, $\Sigma M_x = (400 \text{ N})(0.6 \text{ m}) \neq 0$, and for the two-dimensional problem, $\Sigma M_O = (100 \text{ N})(0.2 \text{ m}) \neq 0$.

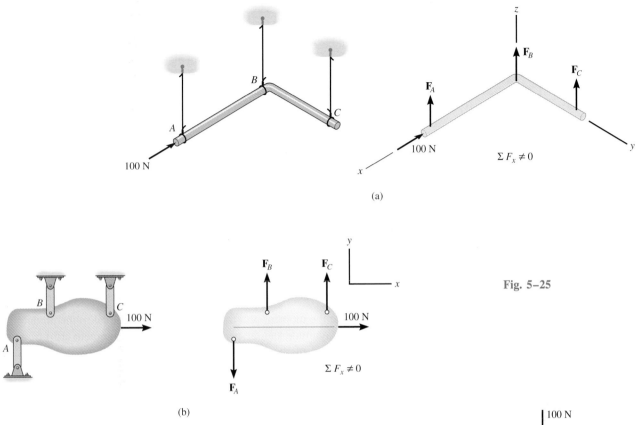

(a)

(b)

Fig. 5–25

Another way in which improper constraining leads to instability occurs when the *reactive forces* are all *parallel.* Three- and two-dimensional examples of this are shown in Fig. 5–25. In both cases, the summation of forces along the x axis will not equal zero.

In some cases, a body may have *fewer* reactive forces than equations of equilibrium that must be satisfied. The body then becomes only *partially constrained.* For example, consider the body shown in Fig. 5–26a with its corresponding free-body diagram in Fig. 5–26b. If O is a point not located on the line AB, the equations $\Sigma F_x = 0$ and $\Sigma M_O = 0$ will be satisfied by proper choice of the reactions $\mathbf{F}_A$ and $\mathbf{F}_B$. The equation $\Sigma F_y = 0$, however, will not be satisfied for the loading conditions and therefore equilibrium will not be maintained.

Proper constraining therefore requires that (1) the lines of action of the reactive forces do not intersect points on a common axis, and (2) the reactive forces must not all be parallel to one another. When the number of reactive forces needed to properly constrain the body in question is a *minimum,* the problem will be statically determinate, and therefore the equations of equilibrium can be used to determine *all* the reactive forces.

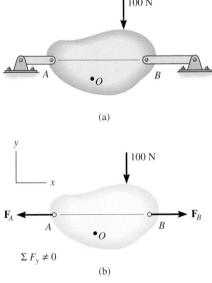

(a)

(b)

Fig. 5–26

Example 5–15

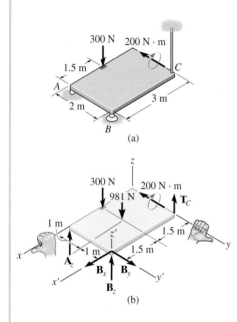

(a)

(b)

Fig. 5–27

The homogeneous plate shown in Fig. 5–27a has a mass of 100 kg and is subjected to a force and couple moment along its edges. If it is supported in the horizontal plane by means of a roller at A, a ball-and-socket joint at B, and a cord at C, determine the components of reaction at the supports.

SOLUTION (*SCALAR ANALYSIS*)

Free-Body Diagram. There are five unknown reactions acting on the plate, as shown in Fig. 5-27b. Each of these reactions is assumed to act in a positive coordinate direction.

Equations of Equilibrium. Since the three-dimensional geometry is rather simple, a *scalar analysis* provides a *direct solution* to this problem. A force summation along each axis yields

$$\Sigma F_x = 0; \qquad B_x = 0 \qquad\qquad\qquad Ans.$$
$$\Sigma F_y = 0; \qquad B_y = 0 \qquad\qquad\qquad Ans.$$
$$\Sigma F_z = 0; \qquad A_z + B_z + T_C - 300\text{ N} - 981\text{ N} = 0 \qquad (1)$$

Recall that the moment of a force about an axis is equal to the product of the force magnitude and the perpendicular distance (moment arm) from the line of action of the force to the axis. The sense of the moment is determined by the right-hand rule. Hence, summing moments of the forces on the free-body diagram, with positive moments acting along the positive x or y axis, we have

$$\Sigma M_x = 0; \qquad T_C(2\text{ m}) - 981\text{ N}(1\text{ m}) + B_z(2\text{ m}) = 0 \qquad (2)$$
$$\Sigma M_y = 0;$$
$$300\text{ N}(1.5\text{ m}) + 981\text{ N}(1.5\text{ m}) - B_z(3\text{ m}) - A_z(3\text{ m}) - 200\text{ N} \cdot \text{m} = 0$$
$$(3)$$

The components of force at B can be eliminated if the x', y', z' axes are used. We obtain

$$\Sigma M_{x'} = 0; \qquad 981\text{ N}(1\text{ m}) + 300\text{ N}(2\text{ m}) - A_z(2\text{ m}) = 0 \qquad (4)$$
$$\Sigma M_{y'} = 0;$$
$$-300\text{ N}(1.5\text{ m}) - 981\text{ N}(1.5\text{ m}) - 200\text{ N} \cdot \text{m} + T_C(3\text{ m}) = 0 \qquad (5)$$

Solving Eqs. 1 through 3 or the more convenient Eqs. 1, 4, and 5 yields

$$A_z = 790\text{ N} \qquad B_z = -217\text{ N} \qquad T_C = 707\text{ N} \qquad Ans.$$

The negative sign indicates that $\mathbf{B}_z$ acts downward.

Note that the solution of this problem does not require the use of a summation of moments about the z axis. The plate is partially constrained since the supports will not prevent it from turning about the z axis if a force is applied to it in the x–y plane.

Example 5–16

The windlass shown in Fig. 5–28a is supported by a thrust bearing at A and a smooth journal bearing at B, which are properly aligned on the shaft. Determine the magnitude of the vertical force **P** that must be applied to the handle to maintain equilibrium of the 100-kg crate. Also calculate the reactions at the bearings.

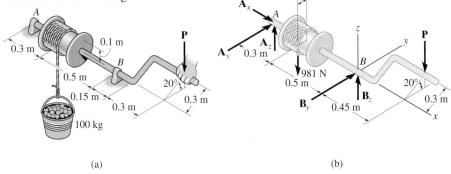

(a)   (b)

SOLUTION (*SCALAR ANALYSIS*)   Fig. 5–28

Free-Body Diagram.   Since the bearings at A and B are aligned correctly, *only* force reactions occur at these supports, Fig. 5–28b. Why are there no moment reactions?

Equations of Equilibrium.   Summing moments about the x axis yields a direct solution for **P.** Why? For a scalar moment summation, it is necessary to determine the moment of each force as the product of the force magnitude and the *perpendicular distance* from the x axis to the line of action of the force. Using the right-hand rule and assuming positive moments act in the $+\mathbf{i}$ direction, we have

$\Sigma M_x = 0;$ $981 \text{ N}(0.1 \text{ m}) - P(0.3 \cos 20° \text{ m}) = 0$

$$P = 348.0 \text{ N} \qquad \qquad Ans.$$

Using this result and summing moments about the y and z axes yields

$\Sigma M_y = 0;$

$$-981 \text{ N}(0.5 \text{ m}) + A_z(0.8 \text{ m}) + (348.0 \text{ N})(0.45 \text{ m}) = 0$$
$$A_z = 417.4 \text{ N} \qquad \qquad Ans.$$

$\Sigma M_z = 0;$ $-A_y(0.8 \text{ m}) = 0 \qquad A_y = 0$ *Ans.*

The reactions at B are determined by a force summation, using the results obtained above.

$\Sigma F_x = 0;$ $A_x = 0$ *Ans.*

$\Sigma F_y = 0;$ $0 + B_y = 0 \qquad B_y = 0$ *Ans.*

$\Sigma F_z = 0;$ $417.4 - 981 + B_z - 348.0 = 0 \qquad B_z = 911.6 \text{ N}$ *Ans.*

Example 5–17

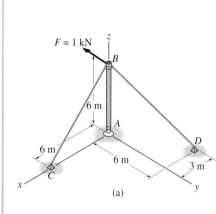

$F = 1$ kN

B

6 m

A

6 m

C

6 m

D

3 m

(a)

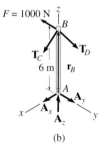

$F = 1000$ N

B

T_C

T_D

6 m r_B

A

A_x A_y

A_z

(b)

Fig. 5–29

Determine the tension in cables BC and BD and the reactions at the ball-and-socket joint A for the mast shown in Fig. 5–29a.

SOLUTION (*VECTOR ANALYSIS*)

Free-Body Diagram. There are five unknown force magnitudes shown on the free-body diagram, Fig. 5–29b.

Equations of Equilibrium. Expressing each force in Cartesian vector form, we have

$$\mathbf{F} = \{-1000\mathbf{j}\} \text{ N}$$

$$\mathbf{F}_A = A_x\mathbf{i} + A_y\mathbf{j} + A_z\mathbf{k}$$

$$\mathbf{T}_C = 0.707T_C\mathbf{i} - 0.707T_C\mathbf{k}$$

$$\mathbf{T}_D = T_D\left(\frac{\mathbf{r}_{BD}}{r_{BD}}\right) = -0.333T_D\mathbf{i} + 0.667T_D\mathbf{j} - 0.667T_D\mathbf{k}$$

Applying the force equation of equilibrium gives

$$\Sigma\mathbf{F} = \mathbf{0}; \qquad \mathbf{F} + \mathbf{F}_A + \mathbf{T}_C + \mathbf{T}_D = \mathbf{0}$$
$$(A_x + 0.707T_C - 0.333T_D)\mathbf{i} + (-1000 + A_y + 0.667T_D)\mathbf{j}$$
$$+ (A_z - 0.707T_C - 0.667T_D)\mathbf{k} = \mathbf{0}$$

$$\Sigma F_x = 0; \qquad A_x + 0.707T_C - 0.333T_D = 0 \tag{1}$$
$$\Sigma F_y = 0; \qquad A_y + 0.667T_D - 1000 = 0 \tag{2}$$
$$\Sigma F_z = 0; \qquad A_z - 0.707T_C - 0.667T_D = 0 \tag{3}$$

Summing moments about point A, we have

$$\Sigma\mathbf{M}_A = \mathbf{0}; \qquad \mathbf{r}_B \times (\mathbf{F} + \mathbf{T}_C + \mathbf{T}_D) = \mathbf{0}$$
$$6\mathbf{k} \times (-1000\mathbf{j} + 0.707T_C\mathbf{i} - 0.707T_C\mathbf{k}$$
$$-0.333T_D\mathbf{i} + 0.667T_D\mathbf{j} - 0.667T_D\mathbf{k}) = \mathbf{0}$$

Evaluating the cross product and combining terms yields

$$(-4T_D + 6000)\mathbf{i} + (4.24T_C - 2T_D)\mathbf{j} = \mathbf{0}$$
$$\Sigma M_x = 0; \qquad -4T_D + 6000 = 0 \tag{4}$$
$$\Sigma M_y = 0; \qquad 4.24T_C - 2T_D = 0 \tag{5}$$

The moment equation about the z axis, $\Sigma M_z = 0$, is automatically satisfied. Why? Solving Eqs. 1 through 5 we have

$$T_C = 707 \text{ N} \qquad T_D = 1500 \text{ N} \qquad \qquad \textit{Ans.}$$
$$A_x = 0 \text{ N} \qquad A_y = 0 \text{ N} \qquad A_z = 1500 \text{ N} \qquad \textit{Ans.}$$

Since the mast is a two-force member, note that the value $A_x = A_y = 0$ could have been determined *by inspection*.

Example 5–18

Rod AB shown in Fig. 5–30a is subjected to the 200-N force. Determine the reactions at the ball-and-socket joint A and the tension in cables BD and BE.

SOLUTION (*VECTOR ANALYSIS*)

Free-Body Diagram.   Fig. 5–30b.

Equations of Equilibrium.   Representing each force on the free-body diagram in Cartesian vector form, we have

$$\mathbf{F}_A = A_x\mathbf{i} + A_y\mathbf{j} + A_z\mathbf{k}$$
$$\mathbf{T}_E = T_E\mathbf{i}$$
$$\mathbf{T}_D = T_D\mathbf{j}$$
$$\mathbf{F} = \{-200\mathbf{k}\} \text{ N}$$

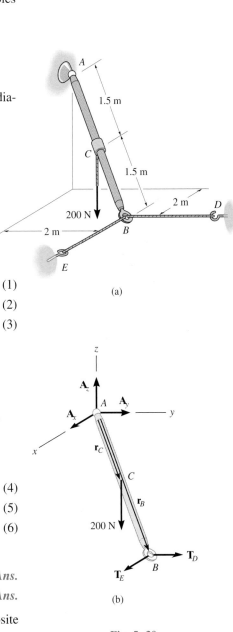

Applying the force equation of equilibrium,

$\Sigma\mathbf{F} = \mathbf{0};$ $\qquad\qquad \mathbf{F}_A + \mathbf{T}_E + \mathbf{T}_D + \mathbf{F} = \mathbf{0}$

$\qquad\qquad (A_x + T_E)\mathbf{i} + (A_y + T_D)\mathbf{j} + (A_z - 200)\mathbf{k} = \mathbf{0}$

$\Sigma F_x = 0;$ $\qquad\qquad A_x + T_E = 0$ $\qquad\qquad$ (1)

$\Sigma F_y = 0;$ $\qquad\qquad A_y + T_D = 0$ $\qquad\qquad$ (2)

$\Sigma F_z = 0;$ $\qquad\qquad A_z - 200 = 0$ $\qquad\qquad$ (3)

Summing moments about point A yields

$\Sigma\mathbf{M}_A = \mathbf{0};$ $\qquad \mathbf{r}_C \times \mathbf{F} + \mathbf{r}_B \times (\mathbf{T}_E + \mathbf{T}_D) = \mathbf{0}$

Since $\mathbf{r}_C = \frac{1}{2}\mathbf{r}_B$, then

$\quad (1\mathbf{i} + 1\mathbf{j} - 0.5\mathbf{k}) \times (-200\mathbf{k}) + (2\mathbf{i} + 2\mathbf{j} - 1\mathbf{k}) \times (T_E\mathbf{i} + T_D\mathbf{j}) = \mathbf{0}$

Expanding and rearranging terms gives

$\qquad (T_D - 200)\mathbf{i} + (-T_E + 200)\mathbf{j} + (2T_D - 2T_E)\mathbf{k} = \mathbf{0}$

$\Sigma M_x = 0;$ $\qquad\qquad T_D - 200 = 0$ $\qquad\qquad$ (4)

$\Sigma M_y = 0;$ $\qquad\qquad -T_E + 200 = 0$ $\qquad\qquad$ (5)

$\Sigma M_z = 0;$ $\qquad\qquad 2T_D - 2T_E = 0$ $\qquad\qquad$ (6)

Solving Eqs. 1 through 6, we get

$$A_x = A_y = -200 \text{ N} \qquad\qquad \textit{Ans.}$$
$$A_z = T_E = T_D = 200 \text{ N} \qquad\qquad \textit{Ans.}$$

The negative sign indicates that $\mathbf{A}_x$ and $\mathbf{A}_y$ have a sense which is opposite to that shown on the free-body diagram, Fig. 5–30b.

Fig. 5–30

Example 5–19

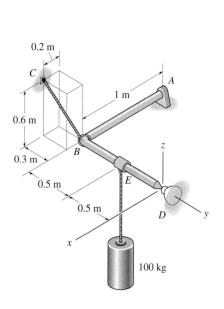

0.2 m

C

A

1 m

0.6 m

B

z

0.3 m

E

0.5 m

0.5 m

D

y

x

100 kg

(a)

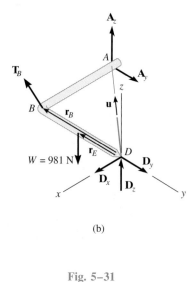

A_z

T_B

A

A_y

z

B

r_B

u

r_E D D_y

W = 981 N

D_x D_z

x   y

(b)

Fig. 5–31

The bent rod in Fig. 5–31*a* is supported at *A* by a journal bearing, at *D* by a ball-and-socket joint, and at *B* by means of cable *BC*. Using only *one equilibrium equation*, obtain a direct solution for the tension in cable *BC*. The bearing at *A* is capable of exerting force components only in the *z* and *y* directions, since it is properly aligned on the shaft.

SOLUTION (*VECTOR ANALYSIS*)

Free-Body Diagram.   As shown in Fig. 5–31*b*, there are six unknowns: three force components caused by the ball-and-socket joint, two caused by the bearing, and one caused by the cable.

Equations of Equilibrium.   The cable tension $\mathbf{T}_B$ may be obtained *directly* by summing moments about an axis passing through points *D* and *A*. Why? The direction of the axis is defined by the unit vector **u,** where

$$\mathbf{u} = \frac{\mathbf{r}_{DA}}{r_{DA}} = -\frac{1}{\sqrt{2}}\mathbf{i} - \frac{1}{\sqrt{2}}\mathbf{j}$$
$$= -0.707\mathbf{i} - 0.707\mathbf{j}$$

Hence, the sum of the moments about this axis is zero provided

$$\Sigma M_{DA} = \mathbf{u} \cdot \Sigma(\mathbf{r} \times \mathbf{F}) = 0$$

Here **r** represents a position vector drawn from *any point* on the axis *DA* to any point on the line of action of force **F** (see Eq. 4–11). With reference to Fig. 5–31*b*, we can therefore write

$$\mathbf{u} \cdot (\mathbf{r}_B \times \mathbf{T}_B + \mathbf{r}_E \times \mathbf{W}) = 0$$

$$(-0.707\mathbf{i} - 0.707\mathbf{j}) \cdot [(-1\mathbf{j}) \times \left(\tfrac{0.2}{0.7}T_B\mathbf{i} - \tfrac{0.3}{0.7}T_B\mathbf{j} + \tfrac{0.6}{0.7}T_B\mathbf{k}\right)$$
$$+ (-0.5\mathbf{j}) \times (-981\mathbf{k})] = 0$$
$$(-0.707\mathbf{i} - 0.707\mathbf{j}) \cdot [(-0.857T_B + 490.5)\mathbf{i} + 0.286T_B\mathbf{k}] = 0$$
$$-0.707(-0.857T_B + 490.5) + 0 + 0 = 0$$

$$T_B = \frac{490.5}{0.857} = 572 \text{ N} \qquad\qquad Ans.$$

The advantage of using Cartesian vectors for this solution should be noted. It would be especially tedious to determine the perpendicular distance from the *DA* axis to the line of action of $\mathbf{T}_B$ using scalar methods.

Note: In Example 5–17, a direct solution for A_z is possible by summing moments about an axis passing through the supports at *C* and *D*, Fig. 5–29*a*. If this is done only the moment of **F** and A_z must be considered. Go back to that example and try to apply the above technique to determine the result $A_z = 1500$ N.

PROBLEMS

5–63. Determine the x, y, z components of reaction at the fixed wall A. The 150-N force is parallel to the z axis and the 200-N force is parallel to the y axis.

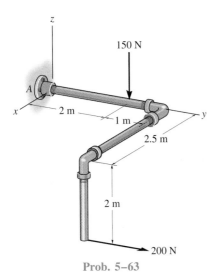

150 N

2 m

1 m

2.5 m

2 m

200 N

Prob. 5–63

*5–64. The wing of the jet aircraft is subjected to a thrust of $T = 8$ kN from its engine and the resultant lift force $L = 45$ kN. If the mass of the wing is 2.1 Mg and the mass center is at G, determine the x, y, z components of reaction where the wing is fixed to the fuselage at A.

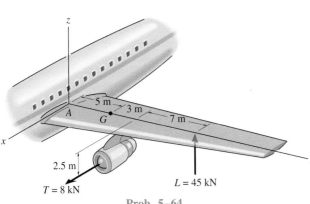

5 m

3 m

7 m

A

G

2.5 m

$T = 8$ kN

$L = 45$ kN

Prob. 5–64

5–65. The uniform concrete slab has a weight of 5500 N. Determine the tension in each of the three parallel supporting cables when the slab is held in the horizontal plane as shown.

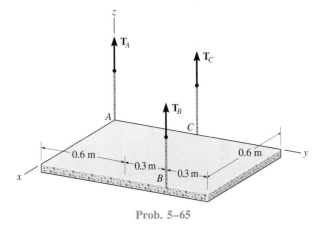

$\mathbf{T}_A$

$\mathbf{T}_C$

$\mathbf{T}_B$

A

C

0.6 m

0.3 m

0.6 m

B

0.3 m

Prob. 5–65

5–66. The air-conditioning unit is hoisted to the roof of a building using the three cables. If the unit has a weight of 4.0 kN and a center of gravity G, located at $x = 1.68$ m, $y = 1.83$ m, determine the tension in each of the cables for equilibrium.

5–67. The air-conditioning unit is hoisted to the roof of a building using the three cables. If the tensions in the cables are $T_A = 1.00$ kN, $T_B = 1.20$ kN, and $T_C = 800$ N, determine the weight of the unit and the location (x, y) of its center of gravity G.

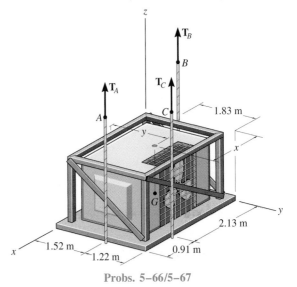

$\mathbf{T}_B$

B

$\mathbf{T}_A$

$\mathbf{T}_C$

A

C

1.83 m

x

y

G

2.13 m

1.52 m

1.22 m

0.91 m

Probs. 5–66/5–67

*5–68. The platform truck supports the three loadings shown. Determine the normal reactions on each of its three wheels.

5–70. The pole for a power line is subjected to the two cable forces of 60 N, each force lying in a plane parallel to the *x-y* plane. If the tension in the guy wire *AB* is 80 N, determine the *x, y, z* components of reaction at the base of the pole, *O*, assuming it to be a ball-and-socket joint.

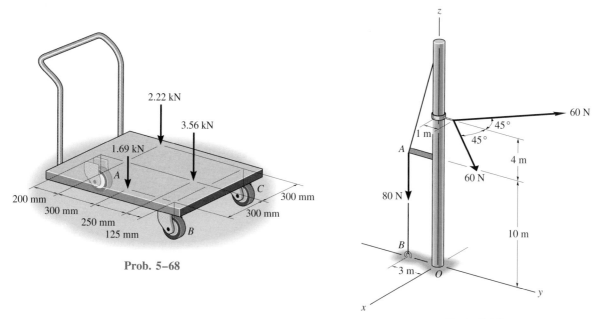

2.22 kN

3.56 kN

1.69 kN

A

C   300 mm

200 mm

300 mm

250 mm

125 mm   *B*

300 mm

Prob. 5–68

z

60 N

45°

1 m

45°

A

4 m

60 N

80 N

10 m

B

3 m   *O*

y

x

Prob. 5–70

5–69. Determine the force components acting on the ball-and-socket at *A*, the reaction at the roller *B* and the tension in the cord *CD* needed for equilibrium of the quarter circular plate.

5–71. The windlass is subjected to a load of 150 N. Determine the horizontal force **P** needed to hold the handle in the position shown, and the components of reaction at the ball-and-socket joint *A* and the smooth journal bearing *B*. The bearing at *B* is in proper alignment and exerts only force reactions on the windlass.

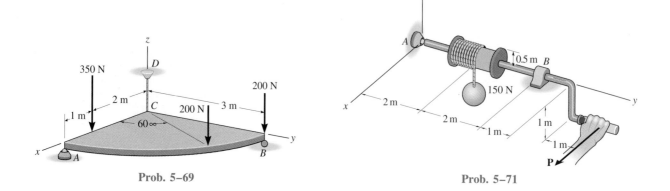

z

D

350 N

200 N

2 m

C

200 N   3 m

1 m

60∞

y

x

A

B

Prob. 5–69

z

A

0.5 m *B*

150 N

x

2 m

2 m

1 m

1 m

1 m

P

y

Prob. 5–71

*5–72. The stiff-leg derrick used on ships is supported by a ball-and-socket joint at *D* and two cables *BA* and *BC*. The cables are attached to a smooth collar ring at *B*, which allows rotation of the derrick about the *z* axis. If the derrick supports a crate having a mass of 200 kg, determine the tension in the cables and the *x*, *y*, *z* components of reaction at *D*.

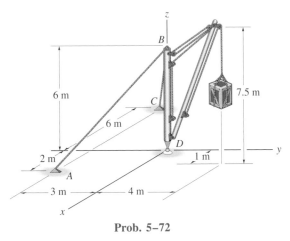

Prob. 5–72

5–73. The pole is subjected to the two forces shown. Determine the components of reaction at *A* assuming it to be a ball-and-socket joint. Also, compute the tension in each of the guy wires, *BC* and *ED*.

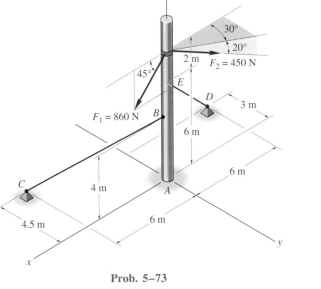

Prob. 5–73

5–74. The boom *AB* is held in equilibrium by a ball-and-socket joint *A* and a pulley and cord system as shown. Determine the *x*, *y*, *z* components of reaction at *A* and the tension in cable *DEC* if $F = \{-1500k\}$ N.

5–75. The cable *CED* can sustain a maximum tension of 800 N before it fails. Determine the greatest vertical force *F* that can be applied to the boom. Also, what are the *x*, *y*, *z* components of reaction at the ball-and-socket joint *A*?

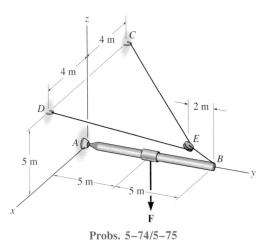

Probs. 5–74/5–75

*5–76. The power line is subjected to a tension of 17 000 N. If the insulator *AB* weighs 500 N with center of gravity at *G*, determine the angle θ it makes with the pole.

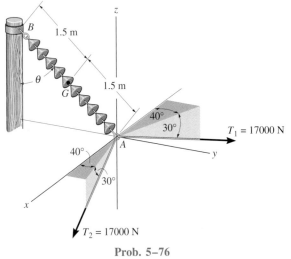

Prob. 5–76

5–77. The boom is supported by a ball-and-socket joint at A and a guy wire at B. If the loads in the cables are each 5 kN and they lie in a plane which is parallel to the x-z plane, determine the components of reaction at A for equilibrium.

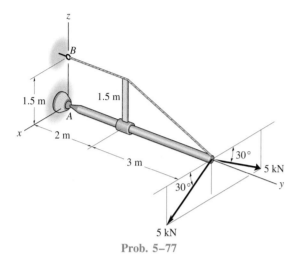

Prob. 5–77

5–78. The pipe assembly supports the vertical loads shown. Determine the components of reaction at the ball-and-socket joint A and the tension in the supporting cables BC and BD.

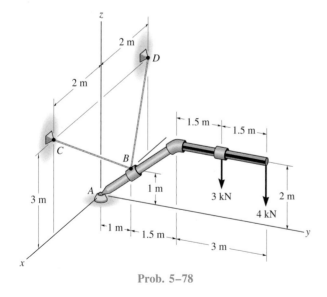

Prob. 5–78

5–79. The boom supports a crate having a weight of 850 N. Determine the x, y, z components of reaction at the ball-and-socket joint A and the tension in cables BC and DE.

*5–80. Cable BC or DE can support a maximum tension of 9000 N before it fails. Determine the greatest weight W of the crate that can be suspended from the end of the boom. Also, determine the x, y, z components of reaction at the ball-and-socket joint A.

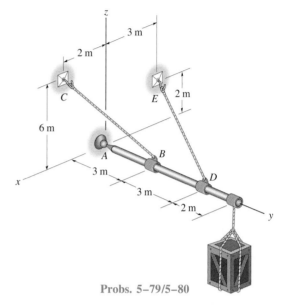

Probs. 5–79/5–80

5–81. Both pulleys are fixed to the shaft and as the shaft turns with constant angular velocity, the power of pulley A is transmitted to pulley B. Determine the horizontal tension **T** in the belt on pulley B and the x, y, z components of reaction at the journal bearing C and thrust bearing D if $\theta = 0°$. The bearings are in proper alignment and exert only force reactions on the shaft.

5–82. Both pulleys are fixed to the shaft and as the shaft turns with constant angular velocity, the power of pulley A is transmitted to pulley B. Determine the horizontal tension **T** in the belt on pulley B and the x, y, z components of reaction at the journal bearing C and thrust bearing D if $\theta = 45°$. The bearings are in proper alignment and exert only force reactions on the shaft.

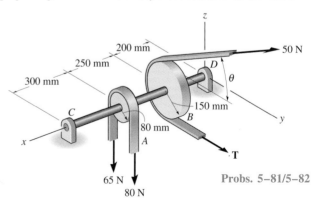

Probs. 5–81/5–82

5–83. The platform has a mass of 3 Mg and center of mass located at G. If it is lifted with constant velocity using the three cables, determine the force in each of the cables.

*5–84. The platform has a mass of 2 Mg and center of mass located at G. If it is lifted using the three cables, determine the force in each of these cables. Solve for each force by using a single moment equation of equilibrium.

5–86. The silo has a weight of 35 000 N and a center of gravity at G. Determine the vertical component of force that each of the three struts at A, B, and C exerts on the silo if it is subjected to a resultant wind loading of 2500 N which acts in the direction shown.

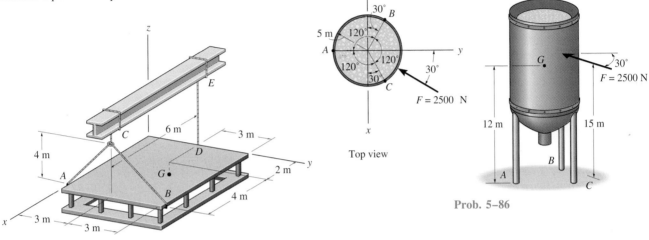

Top view

Prob. 5–86

Probs. 5–83/5–84

5–85. The cables exert the forces shown on the pole. Assuming the pole is supported by a ball-and-socket joint at its base, determine the tension in each cable for equilibrium and the x, y, z components of reaction at A. The forces of 560 N and 300 N lie in a horizontal plane.

5–87. Member AB is supported by a cable BC and at A by a *square* rod which fits loosely through the square hole at the end joint of the member as shown. Determine the components of reaction at A and the tension in the cable needed to hold the 800-N cylinder in equilibrium.

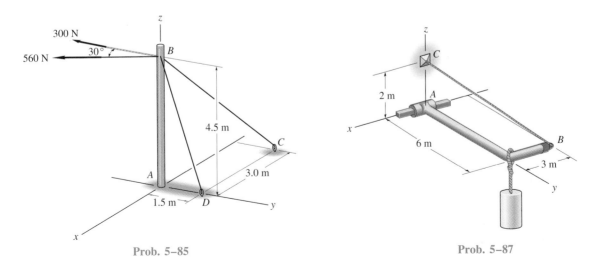

Prob. 5–85

Prob. 5–87

*5–88. Determine the tensions in the cables and the components of reaction acting on the smooth collar at A necessary to hold the 500-N sign in equilibrium. The center of gravity for the sign is at G.

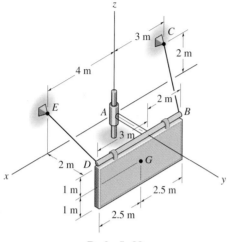

Prob. 5–88

5–89. The boom AC is supported at A by a ball-and-socket joint and by two cables BDC and CE. Cable BDC is continuous and passes over a pulley at D. Calculate the tension in the cables and the x, y, z components of reaction at A if a crate has a weight of 800 N.

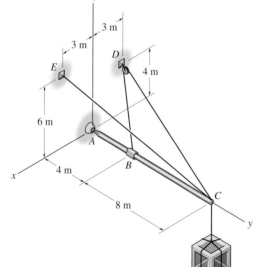

Prob. 5–89

5–90. The bent rod is supported at A, B, and C by smooth journal bearings. Compute the x, y, z components of reaction at the bearings if the rod is subjected to forces F_1 = 300 N and F_2 = 250 N. F_1 lies in the y-z plane. The bearings are in proper alignment and exert only force reactions on the rod.

5–91. The bent rod is supported at A, B, and C by smooth journal bearings. Determine the magnitude of F_2 which will cause the reaction C_y at the bearing C to be equal to zero. The bearings are in proper alignment and exert only force reactions on the rod. Set F_1 = 300 N.

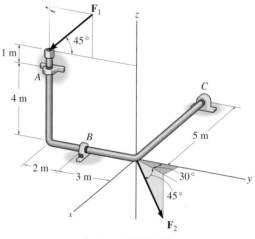

Probs. 5–90/5–91

*5–92. The bar AB is supported by two smooth collars. At A the connection is with a ball-and-socket joint and at B it is a rigid attachment. If a 50-N load is applied to the bar, determine the x, y, z components of reaction at A and B.

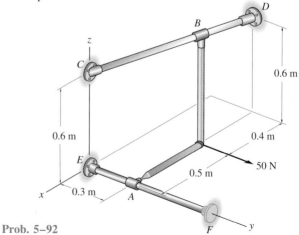

Prob. 5–92

5–93. The left shaft of the universal joint (or Hooke's joint) is subjected to a torque (or couple moment) of 50 N · m. Determine the required equilibrium couple moment M' on the connected shaft when the shafts are in the position $\theta = 30°$, $\phi = 60°$ as shown. Axles AB and CD are perpendicular to one another and are free to turn in their bearings.

5–94. The rod has a weight of 60 N/m. If it is supported by a ball-and-socket joint at C and a journal bearing at D, determine the x, y, z components of reaction at these supports and the moment M that must be applied along the axis of the rod to hold it in the position shown.

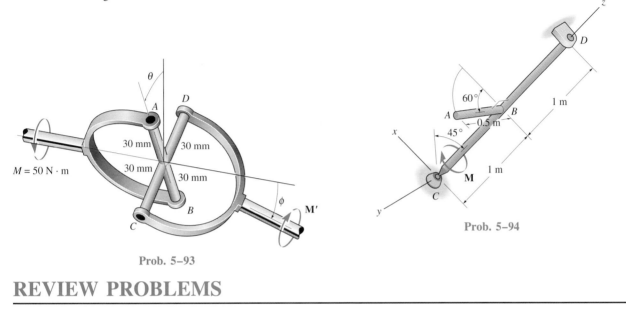

Prob. 5–94

Prob. 5–93

REVIEW PROBLEMS

5–95. Determine the x and z components of reaction at the journal bearing A and the tension in cords BC and BD necessary for equilibrium of the rod.

*5–96. The shaft assembly is supported by two smooth journal bearings A and B and a short link DC. If a couple moment is applied to the shaft as shown, determine the components of force reaction at the bearings and the force in the link. The link lies in a plane parallel to the y-z plane and the bearings are properly aligned on the shaft.

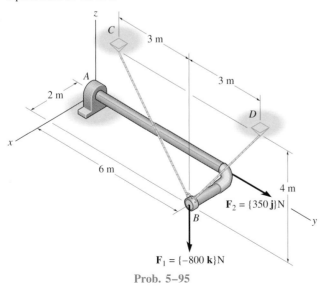

$F_2 = \{350\,\mathbf{j}\}\,N$

$F_1 = \{-800\,\mathbf{k}\}\,N$

Prob. 5–95

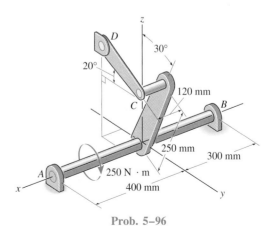

Prob. 5–96

5–97.   Determine the reactions at the roller *A* and pin *B*.

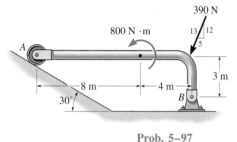

Prob. 5–97

5–98.   The wheel of radius *R* has a spring attached to its central hub. If the spring has a stiffness *k* and unstretched length *R*, show that the *horizontal* force **F** needed to pull the wheel forward so that the spring makes an angle θ with the horizontal is $F = kR(\cot \theta - \cos \theta)$.

Prob. 5–98

5–99.   Determine the reactions at the roller *A* and pin *B*.

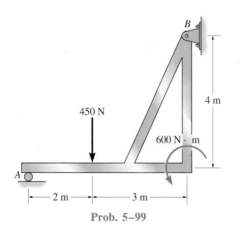

Prob. 5–99

*5–100.   The member is supported by cable *BC* and at *A* by a smooth fixed *square* rod which fits loosely through the square hole of the collar. Determine the *x, y, z* components of reaction at *A* and the tension in the cable needed to hold the member in equilibrium.

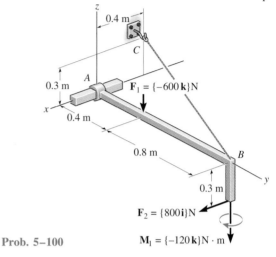

Prob. 5–100

5–101.   Member *AB* is supported at *B* by a cable and at *A* by a smooth fixed *square* rod which fits loosely through the square hole of the collar. If $F = \{20\mathbf{i} - 40\mathbf{j} - 75\mathbf{k}\}$ N, determine the *x, y, z* components of reaction at *A* and the tension in the cable.

5–102.   Member *AB* is supported at *B* by a cable and at *A* by a smooth fixed *square* rod which fits loosely through the square hole of the collar. Determine the tension in cable *BC* if the force $\mathbf{F} = \{-45\mathbf{k}\}$ N.

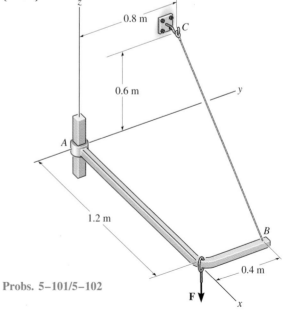

Probs. 5–101/5–102

5–103. The wall footing is used to support the column load of 12,000 N. Determine the intensities w_1 and w_2 of the distributed loading acting on the base of the footing for equilibrium.

5–105. The uniform rod AB has a mass of 5 kg and is supported by a ball-and-socket joint at A, a cord BC, and a smooth wall at B. Determine the x, y, z components of reaction at the supports.

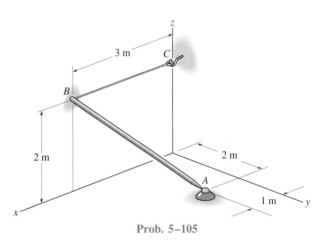

12 000 N

125 mm 225 mm 225 mm

w_1 w_2

875 mm

Prob. 5–103

3 m

C

B

2 m

2 m

A

1 m

y

x

z

Prob. 5–105

5–106. Determine the reactions at roller A and pin B for equilibrium of the member.

***5–104.** Compute the horizontal and vertical components of force at pin B. The belt is subjected to a tension of $T = 100$ N and passes over each of the three pulleys.

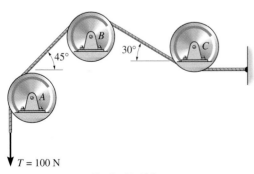

B

45°

30°

C

A

T = 100 N

Prob. 5–104

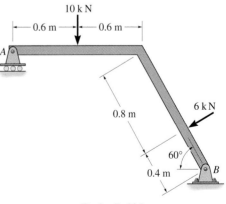

10 k N

0.6 m   0.6 m

A

0.8 m

6 k N

60°

0.4 m   B

Prob. 5–106

The forces in the truss members of the booms and in the supporting cables of these tower cranes can be determined using the principles discussed in this chapter.

6

Structural Analysis

In this chapter we will use the equations of equilibrium to analyze structures composed of pin-connected members. The analysis is based on the principle that if a structure is in equilibrium, then each of its members is also in equilibrium. By applying the equations of equilibrium to the various parts of a simple truss, frame, or machine, we will be able to determine all the forces acting at the connections.

The topics in this chapter are very important since they provide practice in drawing free-body diagrams, using the principle of action, equal but opposite collinear force reaction, and applying the equations of equilibrium.

6.1 Simple Trusses

A *truss* is a structure composed of slender members joined together at their end points. The members commonly used in construction consist of wooden struts or metal bars. The joint connections are usually formed by bolting or welding the ends of the members to a common plate, called a *gusset plate,* as shown in Fig. 6–1a, or by simply passing a large bolt or pin through each of the members, Fig. 6–1b.

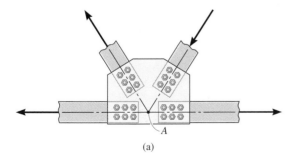

(a)

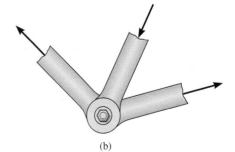

(b)

Fig. 6–1

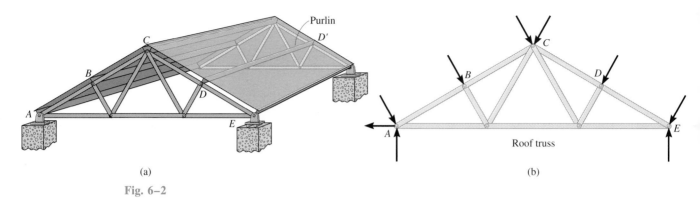

(a)

Fig. 6–2

(b)

Planar Trusses.

Planar trusses lie in a single plane and are often used to support roofs and bridges. The truss *ABCDE,* shown in Fig. 6–2a, is an example of a typical roof-supporting truss. In this figure, the roof load is transmitted to the truss *at the joints* by means of a series of *purlins,* such as *DD′.* Since the imposed loading acts in the same plane as the truss, Fig. 6–2b, the analysis of the forces developed in the truss members is two-dimensional.

In the case of a bridge, such as shown in Fig. 6–3a, the load on the deck is first transmitted to *stringers,* then to *floor beams,* and finally to the *joints B, C,* and *D* of the two supporting side trusses. Like the roof truss, the bridge truss loading is also coplanar, Fig. 6–3b.

When bridge or roof trusses extend over large distances, a rocker or roller is commonly used for supporting one end, joint *E* in Figs. 6–2a and 6–3a. This type of support allows freedom for expansion or contraction of the members due to temperature or application of loads.

Fig. 6–3

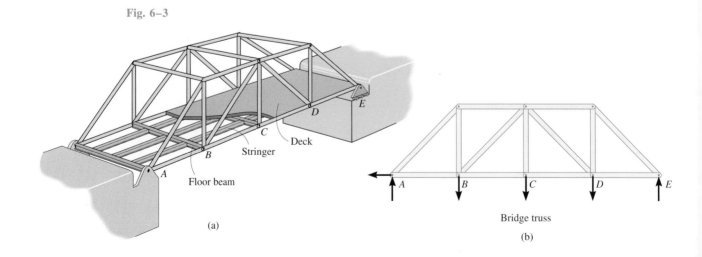

(a)

(b)

Assumptions for Design. To design both the members and the connections of a truss, it is first necessary to determine the *force* developed in each member when the truss is subjected to a given loading. In this regard, two important assumptions will be made:

1. *All loadings are applied at the joints.* In most situations, such as for bridge and roof trusses, this assumption is true. Frequently in the force analysis the weights of the members are neglected, since the forces supported by the members are usually large in comparison with their weights. If the member's weight is to be included in the analysis, it is generally satisfactory to apply it as a vertical force, half of its magnitude applied at each end of the member.
2. *The members are joined together by smooth pins.* In cases where bolted or welded joint connections are used, this assumption is satisfactory provided the center lines of the joining members are *concurrent,* as in the case of point A in Fig. 6–1a.

Because of these two assumptions, *each truss member acts as a two-force member,* and therefore the forces at the ends of the member must be directed along the axis of the member. If the force tends to *elongate* the member, it is a *tensile force* (**T**), Fig. 6–4a; whereas if it tends to *shorten* the member, it is a *compressive force* (**C**), Fig. 6–4b. In the actual design of a truss it is important to state whether the nature of the force is tensile or compressive. Often, compression members must be made *thicker* than tension members, because of the buckling or column effect that occurs when a member is in compression.

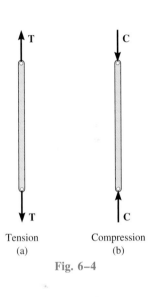

Tension
(a)

Compression
(b)

Fig. 6–4

Simple Truss. To prevent collapse, the form of a truss must be rigid. Obviously, the four-bar shape *ABCD* in Fig. 6–5 will collapse unless a diagonal, such as *AC*, is added for support. The simplest form that is rigid or stable is a *triangle.* Consequently, a *simple truss* is constructed by *starting* with a basic triangular element, such as *ABC* in Fig. 6–6, and connecting two members (*AD* and *BD*) to form an additional element. Thus it is seen that as each additional element of two members is placed on the truss, the number of joints for a simple truss is increased by one.

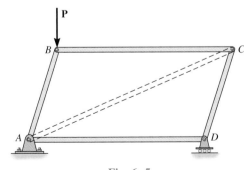

Fig. 6–5

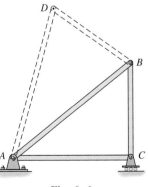

Fig. 6–6

6.2 The Method of Joints

If a truss is in equilibrium, then each of its joints must also be in equilibrium. The method of joints is based on this fact, since it consists of satisfying the equilibrium conditions for the forces exerted *on the pin* at each joint of the truss. Because the truss members are all straight two-force members lying in the same plane, the force system acting at each pin is *coplanar and concurrent.* Consequently, rotational or moment equilibrium is automatically satisfied at the joint (or pin), and it is only necessary to satisfy $\Sigma F_x = 0$ and $\Sigma F_y = 0$ to ensure translational or force equilibrium.

When using the method of joints, it is *first* necessary to draw the joint's free-body diagram before applying the equilibrium equations. To do this, recall that the *line of action* of each member force acting on the joint is *specified* from the geometry of the truss, since the force in a member passes along the axis of the member. As an example, consider the pin at joint *B* of the truss in Fig. 6–7*a*. Three forces act on the pin, namely, the 500-N force and the forces exerted by members *BA* and *BC*. The free-body diagram is shown in Fig. 6–7*b*. As shown, $\mathbf{F}_{BA}$ is "pulling" on the pin, which means that member *BA* is in *tension;* whereas $\mathbf{F}_{BC}$ is "pushing" on the pin, and consequently member *BC* is in *compression.* These effects are clearly demonstrated by isolating the joint with small segments of the member connected to the pin, Fig. 6–7*c*. Notice that pushing or pulling on these small segments indicates the effect of the member being either in compression or tension.

In all cases, the analysis should start at a joint having at least one known force and at most two unknown forces, as in Fig. 6–7*b*. In this way, application of $\Sigma F_x = 0$ and $\Sigma F_y = 0$ yields two algebraic equations which can be solved for the two unknowns. When applying these equations, the correct sense of an unknown member force can be determined using one of two possible methods.

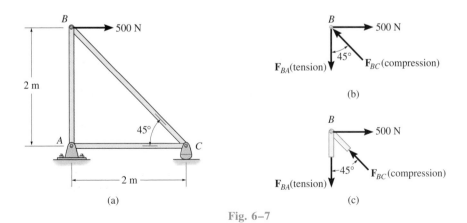

Fig. 6–7

1. *Always assume* the *unknown member forces* acting on the joint's free-body diagram to be in *tension,* i.e., "pulling" on the pin. If this is done, then numerical solution of the equilibrium equations will yield *positive scalars for members in tension and negative scalars for members in compression.* Once an unknown member force is found, use its *correct* magnitude and sense (T or C) on subsequent joint free-body diagrams.
2. The *correct* sense of direction of an unknown member force can, in many cases, be determined "by inspection." For example, $\mathbf{F}_{BC}$ in Fig. 6–7b must push on the pin (compression) since its horizontal component, $F_{BC} \sin 45°$, must balance the 500-N force ($\Sigma F_x = 0$). Likewise, $\mathbf{F}_{BA}$ is a tensile force since it balances the vertical component, $F_{BC} \cos 45°$ ($\Sigma F_y = 0$). In more complicated cases, the sense of an unknown member force can be *assumed;* then, after applying the equilibrium equations, the assumed sense can be verified from the numerical results. A *positive* answer indicates that the sense is *correct,* whereas a *negative* answer indicates that the sense shown on the free-body diagram must be *reversed.* This is the method we will use in the example problems which follow.

PROCEDURE FOR ANALYSIS

The following procedure provides a typical means for analyzing a truss using the method of joints.

Draw the free-body diagram of a joint having at least one known force and at most two unknown forces. (If this joint is at one of the supports, it generally will be necessary to know the external reactions at the truss support.) Use one of the two methods described above for establishing the sense of an unknown force. Orient the x and y axes such that the forces on the free-body diagram can be easily resolved into their x and y components and then apply the two force equilibrium equations $\Sigma F_x = 0$ and $\Sigma F_y = 0$. Solve for the two unknown member forces and verify their correct sense.

Continue to analyze each of the other joints, where again it is necessary to choose a joint having at most two unknowns and at least one known force. Realize that once the force in a member is found from the analysis of a joint at one of its ends, the result can be used to analyze the forces acting on the joint at its other end. Strict adherence to the principle of action, equal but opposite reaction must, of course, be observed. Remember, a member in *compression* "pushes" on the joint and a member in *tension* "pulls" on the joint.

Once the force analysis of the truss has been completed, the size of the members and their connections can be determined using the theory of mechanics of materials along with information given in engineering design codes.

Example 6–1

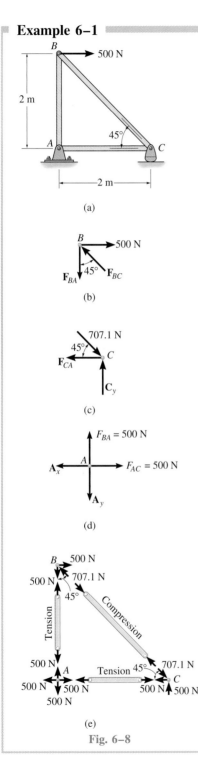

(a)

(b)

(c)

(d)

(e)

Fig. 6–8

Determine the force in each member of the truss shown in Fig. 6–8a and indicate whether the members are in tension or compression.

SOLUTION

By inspection of Fig. 6–8a, there are two unknown member forces at joint B, two unknown member forces and an unknown reaction force at joint C, and two unknown member forces and two unknown reaction forces at joint A. Since we must have no more than two unknowns at the joint and at least one known force acting there, we must begin the analysis at joint B.

Joint B. The free-body diagram of the pin at B is shown in Fig. 6–8b. Three forces act on the pin: the external force of 500 N and the *two* unknown forces developed by members BA and BC. Applying the equations of joint equilibrium, we have

$$\xrightarrow{+}\Sigma F_x = 0; \quad 500\text{ N} - F_{BC}\sin 45° = 0 \quad F_{BC} = 707.1\text{ N} \quad (C) \quad \textit{Ans.}$$
$$+\uparrow \Sigma F_y = 0; \quad F_{BC}\cos 45° - F_{BA} = 0 \quad F_{BA} = 500\text{ N} \quad (T) \quad \textit{Ans.}$$

Since the force in member BC has been calculated, we can proceed to analyze joint C in order to determine the force in member CA and the support reaction at the rocker.

Joint C. From the free-body diagram of joint C, Fig. 6–8c, we have

$$\xrightarrow{+}\Sigma F_x = 0; \quad -F_{CA} + 707.1\cos 45°\text{ N} = 0 \quad F_{CA} = 500\text{ N} \quad (T) \; \textit{Ans.}$$
$$+\uparrow \Sigma F_y = 0; \quad C_y - 707.1\sin 45°\text{ N} = 0 \quad C_y = 500\text{ N} \quad \textit{Ans.}$$

Joint A. Although not necessary, we can determine the support reactions at joint A using the results of $F_{AC} = 500$ N and $F_{AB} = 500$ N. From the free-body diagram, Fig. 6–8d, we have

$$\xrightarrow{+}\Sigma F_x = 0; \quad 500\text{ N} - A_x = 0 \quad A_x = 500\text{ N}$$
$$+\uparrow \Sigma F_y = 0; \quad 500\text{ N} - A_y = 0 \quad A_y = 500\text{ N}$$

The results of the analysis are summarized in Fig. 6–8e. Note that the free-body diagram of each pin shows the effects of all the connected members and external forces applied to the pin, whereas the free-body diagram of each member shows only the effects of the end pins on the member.

Example 6–2

Determine the forces acting in all the members of the roof truss shown in Fig. 6–9*a*.

SOLUTION

By inspection, there are more than two unknowns at each joint. Consequently, the support reactions on the truss must first be determined. Show that they have been correctly calculated on the free-body diagram in Fig. 6–9*b*. We can now begin the analysis at joint *C*. Why?

Joint C. From the free-body diagram, Fig. 6–9*c*,

$$\xrightarrow{+}\Sigma F_x = 0; \qquad -F_{CD}\cos 30° + F_{CB}\sin 45° = 0$$
$$+\uparrow \Sigma F_y = 0; \quad 1.5\text{ kN} + F_{CD}\sin 30° - F_{CB}\cos 45° = 0$$

These two equations must be solved *simultaneously* for each of the two unknowns. Note, however, that a *direct solution* for one of the unknown forces may be obtained by applying a force summation along an axis that is *perpendicular* to the direction of the other unknown force. For example, summing forces along the *y'* axis, which is perpendicular to the direction of $\mathbf{F}_{CD}$, Fig. 6–9*d*, yields a direct solution for F_{CB}.

$$+\nearrow\Sigma F_{y'} = 0;$$
$$1.5\cos 30°\text{ kN} - F_{CB}\sin 15° = 0 \qquad F_{CB} = 5.02\text{ kN} \quad (C) \quad Ans.$$

In a similar fashion, summing forces along the *y"* axis, Fig. 6–9*e*, yields a direct solution for F_{CD}.

$$+\nearrow\Sigma F_{y''} = 0;$$
$$1.5\cos 45°\text{ kN} - F_{CD}\sin 15° = 0 \qquad F_{CD} = 4.10\text{ kN} \quad (T) \quad Ans.$$

Joint D. We can now proceed to analyze joint *D*. The free-body diagram is shown in Fig. 6–9*f*.

$$\xrightarrow{+}\Sigma F_x = 0; \qquad -F_{DA}\cos 30° + 4.10\cos 30°\text{ kN} = 0$$
$$F_{DA} = 4.10\text{ kN} \quad (T) \qquad\qquad Ans.$$
$$+\uparrow \Sigma F_y = 0; \qquad F_{DB} - 2(4.10\sin 30°\text{ kN}) = 0$$
$$F_{DB} = 4.10\text{ kN} \quad (T) \qquad\qquad Ans.$$

The force in the last member, *BA*, can be obtained from joint *B* or joint *A*. As an exercise, draw the free-body diagram of joint *B*, sum the forces in the horizontal direction, and show that $F_{BA} = 0.776\text{ kN}$ (C).

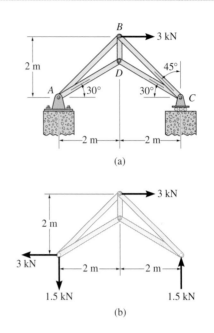

(a)

(b)

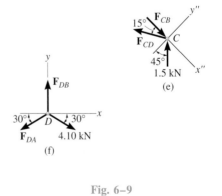

(e)

(f)

Fig. 6–9

Example 6–3

Determine the force in each member of the truss shown in Fig. 6–10a. Indicate whether the members are in tension or compression.

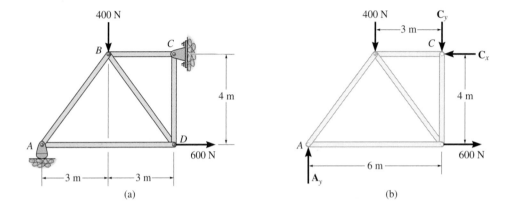

(a)   (b)

SOLUTION

Support Reactions.   No joint can be analyzed until the support reactions are determined. Why? A free-body diagram of the entire truss is given in Fig. 6–10b. Applying the equations of equilibrium, we have

$$\xrightarrow{+}\Sigma F_x = 0; \qquad\qquad\qquad 600\ \text{N} - C_x = 0 \quad C_x = 600\ \text{N}$$

$$\zeta+\Sigma M_C = 0; \quad -A_y(6\ \text{m}) + 400\ \text{N}(3\ \text{m}) + 600\ \text{N}(4\ \text{m}) = 0 \quad A_y = 600\ \text{N}$$

$$+\uparrow \Sigma F_y = 0; \qquad\qquad\qquad 600\ \text{N} - 400\ \text{N} - C_y = 0 \quad C_y = 200\ \text{N}$$

The analysis can now start at either joint A or C. The choice is arbitrary, since there are one known and two unknown member forces acting on the pin at each of these joints.

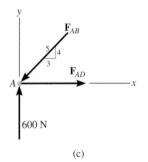

(c)

Fig. 6–10(a–f)

Joint A (Fig. 6–10c).   As shown on the free-body diagram, there are three forces that act on the pin at joint A. The inclination of $\mathbf{F}_{AB}$ is determined from the geometry of the truss. By inspection, can you see why this force is assumed to be compressive and $\mathbf{F}_{AD}$ tensile? Applying the equations of equilibrium, we have

$$+\uparrow \Sigma F_y = 0; \quad 600\ \text{N} - \tfrac{4}{5}F_{AB} = 0 \qquad F_{AB} = 750\ \text{N} \quad (\text{C}) \qquad\qquad Ans.$$

$$\xrightarrow{+}\Sigma F_x = 0; \quad F_{AD} - \tfrac{3}{5}(750\ \text{N}) = 0 \qquad F_{AD} = 450\ \text{N} \quad (\text{T}) \qquad\qquad Ans.$$

Joint D (Fig. 6–10*d*).   The pin at this joint is chosen next since, by inspection of Fig. 6–10*a*, the force in *AD* is known and the unknown forces in *DB* and *DC* can be determined. Summing forces in the horizontal direction, Fig. 6–10*d*, we have

$$\xrightarrow{+}\Sigma F_x = 0; \quad -450 \text{ N} + \tfrac{3}{5}F_{DB} + 600 \text{ N} = 0 \qquad F_{DB} = -250 \text{ N}$$

The negative sign indicates that $\mathbf{F}_{DB}$ acts in the *opposite sense* to that shown in Fig. 6–10*d*.* Hence,

$$F_{DB} = 250 \text{ N} \quad \text{(T)} \qquad\qquad Ans.$$

To determine $\mathbf{F}_{DC}$, we can either correct the sense of $\mathbf{F}_{DB}$ and then apply $\Sigma F_y = 0$, or apply this equation and retain the negative sign for F_{DB}, i.e.,

$$+\uparrow \Sigma F_y = 0; \quad -F_{DC} - \tfrac{4}{5}(-250 \text{ N}) = 0 \qquad F_{DC} = 200 \text{ N} \quad \text{(C)} \qquad Ans.$$

Joint C (Fig. 6–10*e*)

$$\xrightarrow{+}\Sigma F_x = 0; \qquad F_{CB} - 600 \text{ N} = 0 \qquad F_{CB} = 600 \text{ N} \quad \text{(C)} \qquad\qquad Ans.$$
$$+\uparrow \Sigma F_y = 0; \qquad\qquad 200 \text{ N} - 200 \text{ N} \equiv 0 \quad \text{(check)}$$

The analysis is summarized in Fig. 6–10*f*, which shows the correct free-body diagram for each pin and member.

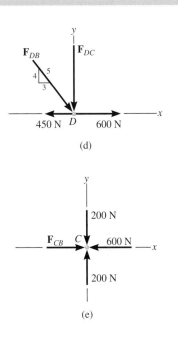

(d)

(e)

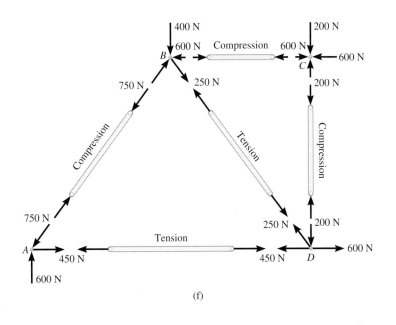

(f)

*The proper sense could have been determined by inspection, prior to applying $\Sigma F_x = 0$.

6.3 Zero-Force Members

Truss analysis using the method of joints is greatly simplified if one is able to first determine those members which support *no loading*. These *zero-force members* are used to increase the stability of the truss during construction and to provide support if the applied loading is changed.

The zero-force members of a truss can generally be determined *by inspection* of each of its joints. For example, consider the truss shown in Fig. 6–11*a*. If a free-body diagram of the pin at joint *A* is drawn, Fig. 6–11*b*, it is seen that members *AB* and *AF* are zero-force members. On the other hand, notice that we could not have come to this conclusion if we had considered the free-body diagrams of joints *F* or *B*, simply because there are five unknowns at each of these joints. In a similar manner, consider the free-body diagram of joint *D*, Fig. 6–11*c*. Here again it is seen that *DC* and *DE* are zero-force members. As a general rule, then, *if only two members form a truss joint and no external load or support reaction is applied to the joint, the members must be zero-force members.* The load on the truss in Fig. 6–11*a* is therefore supported by only five members as shown in Fig. 6–11*d*.

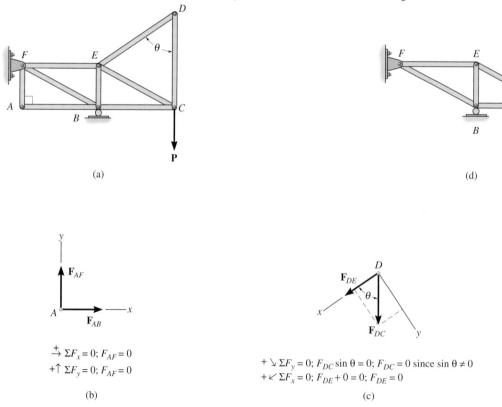

$\xrightarrow{+} \Sigma F_x = 0;\ F_{AF} = 0$
$+\uparrow \Sigma F_y = 0;\ F_{AF} = 0$

(b)

$+\searrow \Sigma F_y = 0;\ F_{DC}\sin\theta = 0;\ F_{DC} = 0$ since $\sin\theta \neq 0$
$+\swarrow \Sigma F_x = 0;\ F_{DE} + 0 = 0;\ F_{DE} = 0$

(c)

Fig. 6–11

Now consider the truss shown in Fig. 6–12a. The free-body diagram of the pin at joint D is shown in Fig. 6–12b. By orienting the y axis along members DC and DE and the x axis along member DA, it is seen that DA is a zero-force member. Note that this is also the case for member CA, Fig. 6–12c. In general, then, *if three members form a truss joint for which two of the members are collinear, the third member is a zero-force member provided no external force or support reaction is applied to the joint.* The truss shown in Fig. 6–12d is therefore suitable for supporting the load **P.**

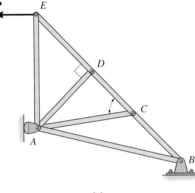

(a)

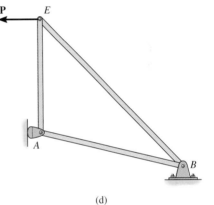

(d)

Fig. 6–12

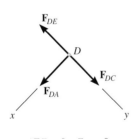

$$+\swarrow \Sigma F_x = 0; \quad F_{DA} = 0$$
$$+\searrow \Sigma F_y = 0; \quad F_{DC} = F_{DE}$$

(b)

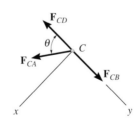

$$+\swarrow \Sigma F_x = 0; \quad F_{CA} \sin \theta = 0; \quad F_{CA} = 0 \text{ since } \sin \theta \neq 0;$$
$$+\searrow \Sigma F_y = 0; \quad F_{CB} = F_{CD}$$

(c)

Example 6–4

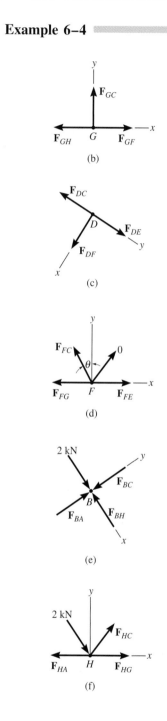

(b)

(c)

(d)

(e)

(f)

Using the method of joints, determine all the zero-force members of the *Fink roof truss* shown in Fig. 6–13a. Assume all joints are pin-connected.

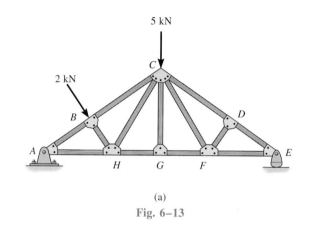

(a)

Fig. 6–13

SOLUTION

Looking for joint geometries that are similar to those outlined in Figs. 6–11 and 6–12, we have

Joint G (Fig. 6–13b)

$$+\uparrow \Sigma F_y = 0; \qquad\qquad F_{GC} = 0 \qquad\qquad \textit{Ans.}$$

Realize that we could not conclude that *GC* is a zero-force member by considering joint *C*, where there are five unknowns. The fact that *GC* is a zero-force member means that the 5-kN load at *C* must be supported by members *CB*, *CH*, *CF*, and *CD*.

Joint D (Fig. 6–13c)

$$+\swarrow \Sigma F_x = 0; \qquad\qquad F_{DF} = 0 \qquad\qquad \textit{Ans.}$$

Joint F (Fig. 6–13d)

$$+\uparrow \Sigma F_y = 0; \quad F_{FC} \cos \theta = 0 \qquad \text{Since } \theta \neq 90°, \qquad F_{FC} = 0 \qquad \textit{Ans.}$$

Note that if joint *B* is analyzed, Fig. 6–13e,

$$+\searrow \Sigma F_x = 0; \quad 2 \text{ kN} - F_{BH} = 0 \qquad F_{BH} = 2 \text{ kN} \quad \text{(C)}$$

Consequently, the numerical value of F_{HC} must satisfy $\Sigma F_y = 0$, Fig. 6–13f, and therefore *HC* is *not* a zero-force member.

PROBLEMS

6–1. The truss, used to support a balcony, is subjected to the loading shown. Approximate each joint as a pin and determine the force in each member. State whether the members are in tension or compression. Set $P_1 = 600$ N, $P_2 = 400$ N.

6–2. The truss, used to support a balcony, is subjected to the loading shown. Approximate each joint as a pin and determine the force in each member. State whether the members are in tension or compression. Set $P_1 = 800$ N, $P_2 = 0$.

6–5. Determine the force in each member of the truss and state if the members are in tension or compression. Set $P_1 = 0$, $P_2 = 4.45$ kN.

6–6. Determine the force in each member of the truss and state if the members are in tension or compression. Set $P_1 = 2.22$ kN, $P_2 = 6.66$ kN.

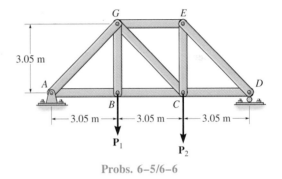

Probs. 6–5/6–6

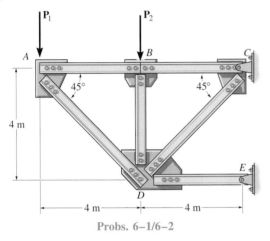

Probs. 6–1/6–2

6–3. Determine the force in each member of the truss and state if the members are in tension or compression. Set $P_1 = 7$ kN, $P_2 = 7$ kN.

*6–4.** Determine the force in each member of the truss and state if the members are in tension or compression. Set $P_1 = 8$ kN, $P_2 = 10$ kN.

6–7. Determine the force in each member of the truss and state if the members are in tension or compression. Set $P_1 = 10$ kN, $P_2 = 15$ kN.

*6–8.** Determine the force in each member of the truss and state if the members are in tension or compression. Set $P_1 = 0$, $P_2 = 20$ kN.

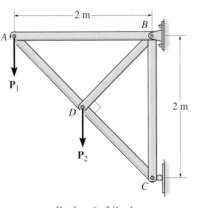

Probs. 6–3/6–4

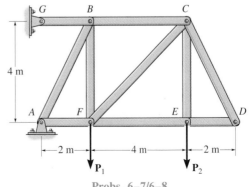

Probs. 6–7/6–8

6–9. Determine the force in each member of the truss and state if the members are in tension or compression.

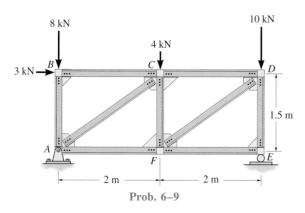

Prob. 6–9

6–10. Determine the force in each member of the truss and state if the members are in tension or compression. Set $P_1 = 10$ kN, $P_2 = 20$ kN, $P_3 = 30$ kN.

6–11. Determine the force in each member of the truss and state if the members are in tension or compression. Set $P_1 = 40$ kN, $P_2 = 40$ kN, $P_3 = 0$.

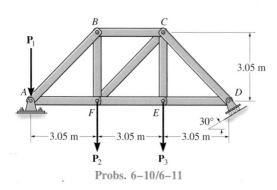

Probs. 6–10/6–11

*6–12. Determine the force in each member of the truss in terms of the load P and state if the members are in tension or compression.

6–13. Members AB and BC can support a maximum compressive force of 800 N, and members AD, DC, and BD can support a maximum tensile force of 1500 N. If $a = 10$ m, determine the greatest load P the truss can support.

6–14. Members AB and BC can support a maximum compressive force of 800 N, and members AD, DC, and BD can support a maximum tensile force of 2000 N. If $a = 6$ m, determine the greatest load P the truss can support.

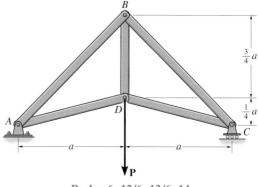

Probs. 6–12/6–13/6–14

6–15. Determine the force in each member of the truss and state if the members are in tension or compression. Approximate each joint as a pin. Set $P = 4$ kN.

*6–16. Assume that each member of the truss is made of steel having a mass per length of 4 kg/m. Set $P = 0$, determine the force in each member, and state if the members are in tension or compression. Neglect the weight of the gusset plates and approximate each joint as a pin. Solve the problem by *assuming* the weight of each member can be represented as a vertical force, half of which is applied at each end of the member.

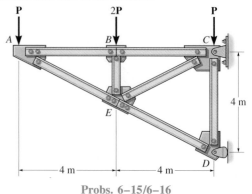

Probs. 6–15/6–16

6–17. Determine the force in each member of the truss and state if the members are in tension or compression. *Hint:* The vertical component of force at *C* must equal zero. Why?

6–18. Each member of the truss is uniform and has a mass of 8 kg/m. Remove the external loads of 6 kN and 8 kN and determine the approximate force in each member due to the weight of the truss. State if the members are in tension or compression. Solve the problem by *assuming* the weight of each member can be represented as a vertical force, half of which is applied at each end of the member.

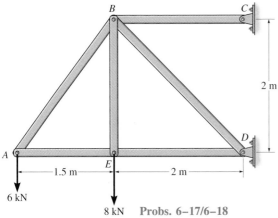

Probs. 6–17/6–18

6–19. Determine the force in each member of the truss and state if the members are in tension or compression. *Hint:* The resultant force at the pin *E* acts along member *ED*. Why?

*6–20. Each member of the truss is uniform and has a mass of 8 kg/m. Remove the external loads of 3 kN and 2 kN and determine the approximate force in each member due to the weight of the truss. State if the members are in tension or compression. Solve the problem by *assuming* the weight of each member can be represented as a vertical force, half of which is applied at each end of the member.

6–21. Determine the force in each member of the truss and state if the members are in tension or compression. *Hint:* The horizontal force component at *A* must be zero. Why?

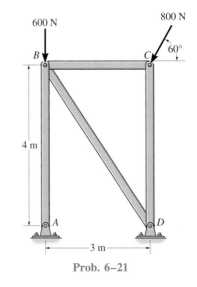

Prob. 6–21

6–22. Determine the force in each member of the double scissors truss in terms of the load *P* and state if the members are in tension or compression.

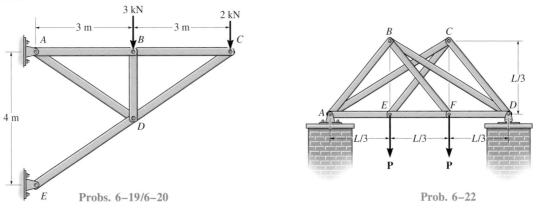

Probs. 6–19/6–20

Prob. 6–22

6–23. Determine the force in each member of the truss in terms of the external loading and state if the members are in tension or compression.

*6–24. The maximum allowable tensile force in the members of the truss is $(F_t)_{max} = 1500$ N, and the maximum allowable compressive force is $(F_c)_{max} = 800$ N. Determine the maximum magnitude P of the two loads that can be applied to the truss. Take $a = 8$ m.

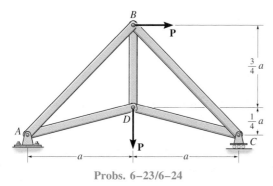

Probs. 6–23/6–24

6–25. Determine the force in each member of the truss in terms of the external loading and state if the members are in tension or compression.

6–26. The maximum allowable tensile force in the members of the truss is $(F_t)_{max} = 2$ kN, and the maximum allowable compressive force is $(F_c)_{max} = 1.2$ kN. Determine the maximum magnitude P of the two loads that can be applied to the truss. Take $L = 2$ m and $\theta = 30°$.

6–27. Determine the force in each member of the truss in terms of the load P and state if the members are in tension or compression.

*6–28. The maximum allowable tensile force in the members of the truss is $(F_t)_{max} = 3$ kN, and the maximum allowable compressive force is $(F_c)_{max} = 5$ kN. Determine the maximum magnitude of the load **P** that can be applied to the truss. Take $d = 2$ m.

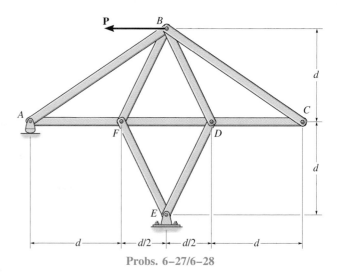

Probs. 6–27/6–28

6–29. Determine the force in each member of the truss and state if the members are in tension or compression. Set $P_1 = 4$ kN, $P_2 = 5$ kN.

6–30. Determine the force in each member of the truss and state if the members are in tension or compression. Set $P_1 = 0$, $P_2 = 8$ kN.

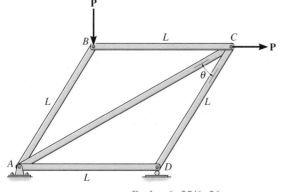

Probs. 6–25/6–26

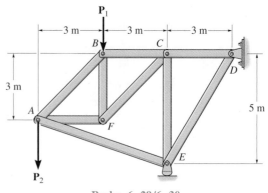

Probs. 6–29/6–30

6.4 The Method of Sections

The *method of sections* is used to determine the loadings acting within a body. It is based on the principle that if a body is in equilibrium, then any part of the body is also in equilibrium. To apply this method, one passes an *imaginary section* through the body, thus cutting it into two parts. When a free-body diagram of one of the parts is drawn, the loads acting at the section must be *included* on the free-body diagram. One then applies the equations of equilibrium to the part in order to determine the loading at the section. For example, consider the two truss members shown colored in Fig. 6–14. The internal loads at the section indicated by the blue line can be obtained using one of the free-body diagrams shown on the right. Clearly, it can be seen that equilibrium requires that the member in tension be subjected to a "pull" **T** at the section, whereas the member in compression is subjected to a "push" **C.**

The method of sections can also be used to "cut" or section several members of an entire truss. If either of the two parts of the truss is isolated as a free-body diagram, we can then apply the equations of equilibrium to that part to determine the member forces at the "cut section." Since only *three* independent equilibrium equations ($\Sigma F_x = 0$, $\Sigma F_y = 0$, $\Sigma M_O = 0$) can be applied to the isolated part of the truss, one should try to select a section that, in general, passes through not more than *three* members in which the forces are unknown. For example, consider the truss in Fig. 6–15a. If the force in member *GC* is to be determined, section *aa* would be appropriate. The free-body diagrams of the two parts are shown in Figs. 6–15b and 6–15c. In particular, note that the line of action of each cut member force is specified from the *geometry* of the truss, since the force in a member passes along its axis. Also, the member forces acting on one part of the truss are equal but opposite to those acting on the other part—Newton's third law. As noted above, members assumed to be in *tension* (*BC* and *GC*) are subjected to a "pull," whereas the member in *compression* (*GF*) is subjected to a "push."

Fig. 6–14

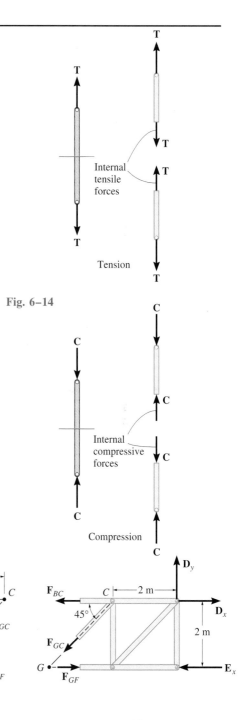

Fig. 6–15

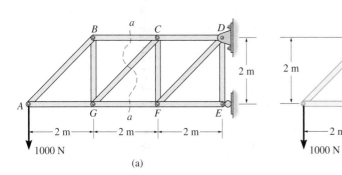

(a)   (b)   (c)

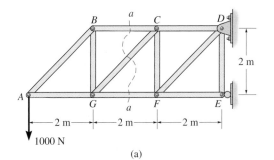

(a)

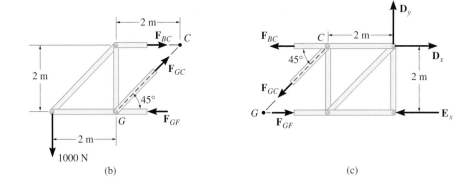

(b)   (c)

Fig. 6–15 *(Repeated)*

The three unknown member forces $\mathbf{F}_{BC}$, $\mathbf{F}_{GC}$, and $\mathbf{F}_{GF}$ can be obtained by applying the three equilibrium equations to the free-body diagram in Fig. 6–15b. If, however, the free-body diagram in Fig. 6–15c is considered, the three support reactions $\mathbf{D}_x$, $\mathbf{D}_y$, and $\mathbf{E}_x$ will have to be determined *first*. Why? (This, of course, is done in the usual manner by considering a free-body diagram of the *entire truss*.) When applying the equilibrium equations, one should consider ways of writing the equations so as to yield a *direct solution* for each of the unknowns, rather than having to solve simultaneous equations. For example, summing moments about C in Fig. 6–15b would yield a direct solution for $\mathbf{F}_{GF}$ since $\mathbf{F}_{BC}$ and $\mathbf{F}_{GC}$ create zero moment about C. Likewise, $\mathbf{F}_{BC}$ can be directly obtained by summing moments about G. Finally, $\mathbf{F}_{GC}$ can be found directly from a force summation in the vertical direction since $\mathbf{F}_{GF}$ and $\mathbf{F}_{BC}$ have no vertical components. This ability to *determine directly* the force in a particular truss member is one of the main advantages of using the method of sections.*

*By comparison, if the method of joints were used to determine, say, the force in member GC, it would be necessary to analyze joints A, B, and G in sequence.

As in the method of joints, there are two ways in which one can determine the correct sense of an unknown member force.

1. *Always assume* that the unknown member forces at the cut section are in *tension,* i.e., ''pulling'' on the member. By doing this, the numerical solution of the equilibrium equations will yield *positive scalars for members in tension and negative scalars for members in compression.*

2. The correct sense of an unknown member force can in many cases be determined ''by inspection.'' For example, $\mathbf{F}_{BC}$ is a tensile force as represented in Fig. 6–15*b*, since moment equilibrium about G requires that $\mathbf{F}_{BC}$ create a moment opposite to that of the 1000-N force. Also, $\mathbf{F}_{GC}$ is tensile since its vertical component must balance the 1000-N force acting downward. In more complicated cases, the sense of an unknown member force may be *assumed.* If the solution yields a *negative* scalar, it indicates that the force's sense is *opposite* to that shown on the free-body diagram. This is the method we will use in the example problems which follow.

PROCEDURE FOR ANALYSIS

The following procedure provides a means for applying the method of sections to determine the forces in the members of a truss.

Free-Body Diagram.   Make a decision as to how to ''cut'' or section the truss through the members where forces are to be determined. Before isolating the appropriate section, it may first be necessary to determine the truss's *external* reactions, so that the three equilibrium equations are used *only* to solve for member forces at the cut section. Draw the free-body diagram of that part of the sectioned truss which has the least number of forces acting on it. Use one of the two methods described above for establishing the sense of an unknown member force.

Equations of Equilibrium.   Try to apply the three equations of equilibrium such that simultaneous solution of equations is avoided. In this regard, moments should be summed about a point that lies at the intersection of the lines of action of two unknown forces, so that the third unknown force is determined directly from the moment equation. If two of the unknown forces are *parallel,* forces may be summed *perpendicular* to the direction of these unknowns to determine *directly* the third unknown force.

The following examples illustrate these concepts numerically.

Example 6–5

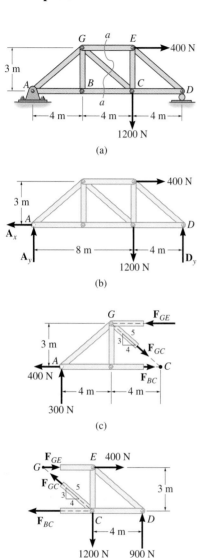

Fig. 6–16

Determine the force in members *GE*, *GC*, and *BC* of the truss shown in Fig. 6–16*a*. Indicate whether the members are in tension or compression.

SOLUTION

Section *aa* in Fig. 6–16*a* has been chosen since it cuts through the *three* members whose forces are to be determined. In order to use the method of sections, however, it is *first* necessary to determine the external reactions at *A* or *D*. Why? A free-body diagram of the entire truss is shown in Fig. 6–16*b*. Applying the equations of equilibrium, we have

$$\xrightarrow{+}\Sigma F_x = 0; \qquad 400 \text{ N} - A_x = 0 \qquad A_x = 400 \text{ N}$$
$$\zeta +\Sigma M_A = 0; \quad -1200 \text{ N(8 m)} - 400 \text{ N(3 m)} + D_y(12 \text{ m}) = 0$$
$$D_y = 900 \text{ N}$$
$$+\uparrow \Sigma F_y = 0; \quad A_y - 1200 \text{ N} + 900 \text{ N} = 0 \qquad A_y = 300 \text{ N}$$

Free-Body Diagrams. The free-body diagrams of the sectioned truss are shown in Figs. 6–16*c* and 6–16*d*. For the analysis the free-body diagram in Fig. 6–16*c* will be used since it involves the least number of forces.

Equations of Equilibrium. Summing moments about point *G* eliminates $\mathbf{F}_{GE}$ and $\mathbf{F}_{GC}$ and yields a direct solution for F_{BC}.

$$\zeta +\Sigma M_G = 0; \quad -300 \text{ N(4 m)} - 400 \text{ N(3 m)} + F_{BC}(3 \text{ m}) = 0$$
$$F_{BC} = 800 \text{ N} \quad \text{(T)} \qquad\qquad Ans.$$

In the same manner, by summing moments about point *C* we obtain a direct solution for F_{GE}.

$$\zeta +\Sigma M_C = 0; \qquad -300 \text{ N(8 m)} + F_{GE}(3 \text{ m}) = 0$$
$$F_{GE} = 800 \text{ N} \quad \text{(C)} \qquad\qquad Ans.$$

Since $\mathbf{F}_{BC}$ and $\mathbf{F}_{GE}$ have no vertical components, summing forces in the *y* direction directly yields F_{GC}, i.e.,

$$+\uparrow \Sigma F_y = 0; \qquad\qquad 300 \text{ N} - \tfrac{3}{5}F_{GC} = 0$$
$$F_{GC} = 500 \text{ N} \quad \text{(T)} \qquad\qquad Ans.$$

Obtain these results by applying the equations of equilibrium to the free-body diagram shown in Fig. 6–16*d*.

Example 6–6

Determine the force in member *CF* of the bridge truss shown in Fig. 6–17*a*. Indicate whether the member is in tension or compression. Assume each member is pin-connected.

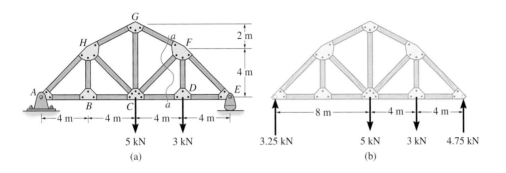

(a)   (b)

SOLUTION

Free-Body Diagram. Section *aa* in Fig. 6–17*a* will be used since this section will "expose" the internal force in member *CF* as "external" on the free-body diagram of either the right or left portion of the truss. It is first necessary, however, to determine the external reactions on either the left or right side of the truss. Verify the results shown on the free-body diagram in Fig. 6–17*b*.

The free-body diagram of the right portion of the truss, which is the easiest to analyze, is shown in Fig. 6–17*c*. There are three unknowns, F_{FG}, F_{CF}, and F_{CD}.

Equations of Equilibrium. The most direct method for solving this problem requires application of the moment equation about a point that eliminates two of the unknown forces. Hence, to obtain $\mathbf{F}_{CF}$, we will eliminate $\mathbf{F}_{FG}$ and $\mathbf{F}_{CD}$ by summing moments about point *O*, Fig. 6–17*c*. Note that the location of point *O* measured from *E* is determined from proportional triangles, i.e., $4/(4 + x) = 6/(8 + x)$, $x = 4$ m. Or, stated in another manner, the slope of member *GF* has a drop of 2 m to a horizontal distance of $CD = 4$ m. Since *FD* is 4 m, Fig. 6–17*c*, then from *D* to *O* the distance must be 8 m.

An easy way to determine the moment of $\mathbf{F}_{CF}$ about point *O* is to resolve $\mathbf{F}_{CF}$ into its two rectangular components and then use the principle of transmissibility to move $\mathbf{F}_{CF}$ to point *C*. We have

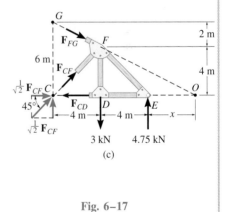

Fig. 6–17

$$\zeta + \Sigma M_O = 0; \quad -\frac{1}{\sqrt{2}} F_{CF}(12 \text{ m}) + (3 \text{ kN})(8 \text{ m}) - (4.75 \text{ kN})(4 \text{ m}) = 0$$

$$F_{CF} = 0.589 \text{ kN} \quad (C) \qquad\qquad \textit{Ans.}$$

Example 6–7

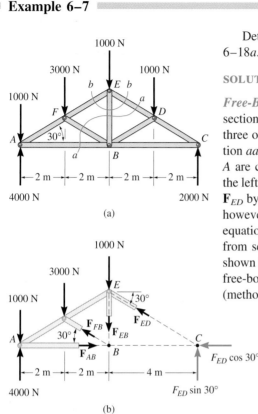

1000 N

3000 N   1000 N

1000 N

b   *E b*

a

F   *D*

A   30°   *C*

a   *B*

← 2 m → ← 2 m → ← 2 m → ← 2 m →

4000 N   2000 N

(a)

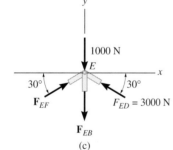

1000 N

3000 N

E

1000 N   ↗30°

$\mathbf{F}_{FB}$ $\mathbf{F}_{ED}$

A   30°↑

$\mathbf{F}_{AB}$ *B*   *C*

← 2 m → ← 2 m →← 4 m → $F_{ED}\cos 30°$

4000 N $F_{ED}\sin 30°$

(b)

y

1000 N

E

30°   ↗30°   *x*

$\mathbf{F}_{EF}$ $\mathbf{F}_{ED} = 3000$ N

$\mathbf{F}_{EB}$

(c)

Fig. 6–18

Determine the force in member *EB* of the roof truss shown in Fig. 6–18*a*. Indicate whether the member is in tension or compression.

SOLUTION

Free-Body Diagrams. By the method of sections, any imaginary vertical section that cuts through *EB*, Fig. 6–18*a*, will also have to cut through three other members for which the forces are unknown. For example, section *aa* cuts through *ED*, *EB*, *FB*, and *AB*. If the components of reaction at *A* are calculated first ($A_x = 0$, $A_y = 4000$ N) and a free-body diagram of the left side of this section is considered, Fig. 6–18*b*, it is possible to obtain $\mathbf{F}_{ED}$ by summing moments about *B* to eliminate the other three unknowns; however, $\mathbf{F}_{EB}$ cannot be determined from the remaining two equilibrium equations. One possible way of obtaining $\mathbf{F}_{EB}$ is first to determine $\mathbf{F}_{ED}$ from section *aa*, then use this result on section *bb*, Fig. 6–18*a*, which is shown in Fig. 6–18*c*. Here the force system is concurrent and our sectioned free-body diagram is the same as the free-body diagram for the pin at *E* (method of joints).

Equations of Equilibrium. In order to determine the moment of $\mathbf{F}_{ED}$ about point *B*, Fig. 6–18*b*, we will resolve the force into its rectangular components and, by the principle of transmissibility, extend it to point *C* as shown. The moments of 1000 N, F_{AB}, F_{FB}, F_{EB}, and $F_{ED}\cos 30°$ are all zero about *B*. Therefore,

$$\zeta + \Sigma M_B = 0; \quad 1000\ \text{N}(4\ \text{m}) + 3000\ \text{N}(2\ \text{m}) - 4000\ \text{N}(4\ \text{m}) +$$
$$F_{ED}\sin 30°(4) = 0$$
$$F_{ED} = 3000\ \text{N} \quad (C)$$

Considering now the free-body diagram of section *bb*, Fig. 6–18*c*, we have

$$\xrightarrow{+}\Sigma F_x = 0; \qquad F_{EF}\cos 30° - 3000\cos 30°\ \text{N} = 0$$
$$F_{EF} = 3000\ \text{N} \quad (C)$$
$$+\uparrow\Sigma F_y = 0; \qquad 2(3000\sin 30°\ \text{N}) - 1000\ \text{N} - F_{EB} = 0$$
$$F_{EB} = 2000\ \text{N} \quad (T) \qquad\qquad\qquad Ans.$$

PROBLEMS

6–31. Determine the force in members *BC*, *HC*, and *HG* of the bridge truss and state if these members are in tension or compression.

*6–32. Determine the force in members *GF*, *CF*, and *CD* of the bridge truss and state if these members are in tension or compression.

6–35. The roof truss supports the vertical loading shown. Determine the force in members *BC*, *CK*, and *KJ* and state if these members are in tension or compression.

*6–36. The roof truss supports the vertical loading shown. Determine the force in members *DE* and *DJ* and state if these members are in tension or compression.

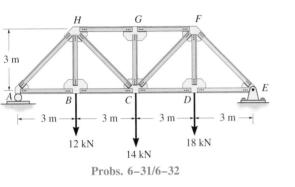

Probs. 6–31/6–32

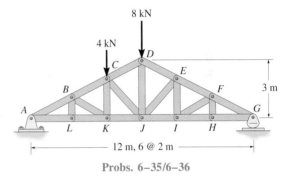

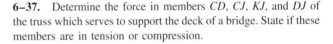

Probs. 6–35/6–36

6–33. The *Howe bridge truss* is subjected to the loading shown. Determine the force in members *HD*, *CD*, and *GD* and state if these members are in tension or compression.

6–34. The *Howe bridge truss* is subjected to the loading shown. Determine the force in members *HI*, *HB*, and *BC* and state if these members are in tension or compression.

6–37. Determine the force in members *CD*, *CJ*, *KJ*, and *DJ* of the truss which serves to support the deck of a bridge. State if these members are in tension or compression.

6–38. Determine the force in members *EI* and *JI* of the truss which serves to support the deck of a bridge. State if these members are in tension or compression.

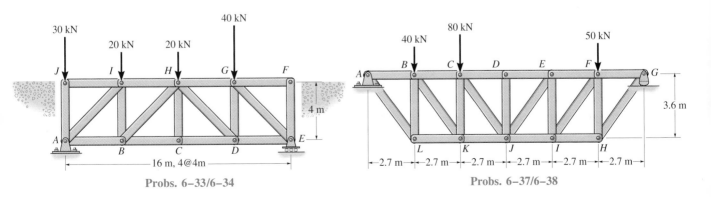

Probs. 6–33/6–34   Probs. 6–37/6–38

6–39. Determine the force in members *CE*, *FE*, and *CD* and state if these members are in tension or compression. *Hint:* The force acting at the pin *G* is directed along bar *GD*. Why?

***6–40.** Determine the force in members *BC*, *FC*, and *FE* and state if these members are in tension or compression. *Hint:* The force acting at the pin *G* is directed along bar *GD*. Why?

6–43. Determine the force in members *BC*, *HC*, and *HG*. After the truss is sectioned use a single equation of equilibrium for the calculation of each force. State if these members are in tension or compression.

***6–44.** Determine the force in members *CD*, *CF*, and *CG* and state if these members are in tension or compression.

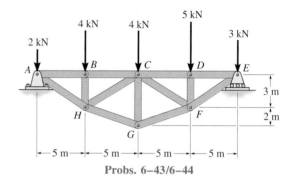

Probs. 6–43/6–44

Probs. 6–39/6–40

6–41. Determine the force developed in members *GB* and *GF* of the bridge truss and state if these members are in tension or compression.

6–42. Determine the force developed in members *FC* and *BC* of the bridge truss and state if these members are in tension or compression.

6–45. Determine the force in member *BC* of the truss and state if this member is in tension or compression.

6–46. Determine the force in member *GJ* of the truss and state if this member is in tension or compression.

6–47. Determine the force in member *GC* of the truss and state if this member is in tension or compression.

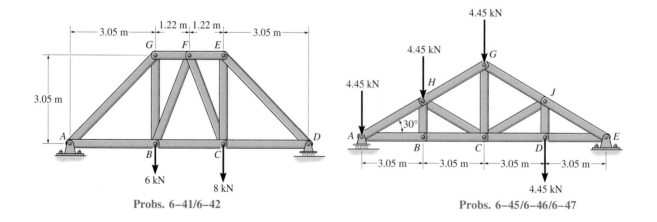

Probs. 6–41/6–42

Probs. 6–45/6–46/6–47

*6–48. The truss is used to support two electrical power lines that exert the forces shown on the structure. Determine the force developed in members *BC, BD,* and *DE* and state if these members are in tension or compression.

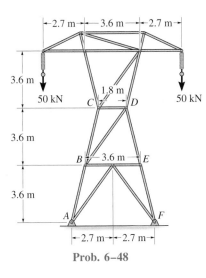

Prob. 6–48

6–49. Determine the force in members *IC* and *CG* of the truss and state if these members are in tension or compression. Also, indicate all zero-force members.

6–50. Determine the force in members *JE* and *GF* of the truss and state if these members are in tension or compression. Also, indicate all zero-force members.

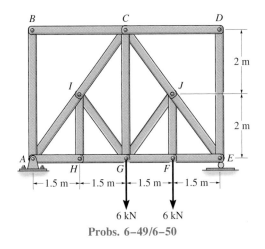

Probs. 6–49/6–50

6–51. The skewed truss carries the load shown. Determine the force in members *CB, BE,* and *EF* and state if these members are in tension or compression. Assume that all joints are pinned.

*6–52. The skewed truss carries the load shown. Determine the force in members *AB, BF,* and *EF* and state if these members are in tension or compression. Assume that all joints are pinned.

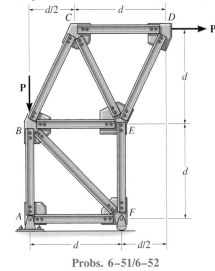

Probs. 6–51/6–52

6–53. The suspension tower consists of an overhead truss which supports the 2-kN cable weights at *A, C, E,* and *G.* If the truss can be assumed pin-supported at *A* and roller-supported at *G,* determine the force in members *EH, EF,* and *IH* and state if these members are in tension or compression.

6–54. The suspension tower consists of an overhead truss which supports the 2-kN cable weights at *A, C, E,* and *G.* If the truss can be assumed pin-supported at *A* and roller-supported at *G,* determine the force in members *KD, CD,* and *KJ* and state if these members are in tension or compression.

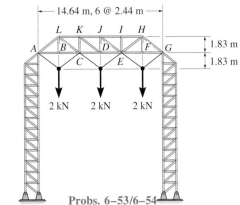

Probs. 6–53/6–54

6–55. Determine the force in members *HG, HC,* and *BC* of the truss and state if these members are in tension or compression. After the truss is sectioned, use a single equation of equilibrium for the calculation of each force.

***6–56.** Determine the force in members *GF, CF,* and *CD* of the truss and state if these members are in tension or compression. After the truss is sectioned, use a single equation of equilibrium for the calculation of each force.

6–58. Determine the force in members *DE, JI,* and *DO* of the *K* truss and state if these members are in tension or compression. *Hint:* Use sections *aa* and *bb*.

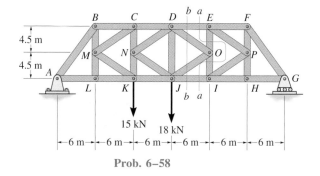

Prob. 6–58

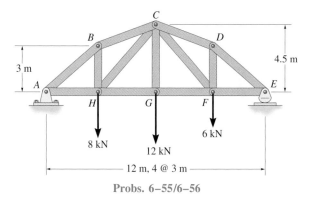

Probs. 6–55/6–56

6–57. Determine the force in members *CB, CG,* and *GF* of the symmetrical truss and state if these members are in tension or compression.

6–59. Determine the force in members *CD* and *KJ* of the *K truss* and state if these members are in tension or compression.

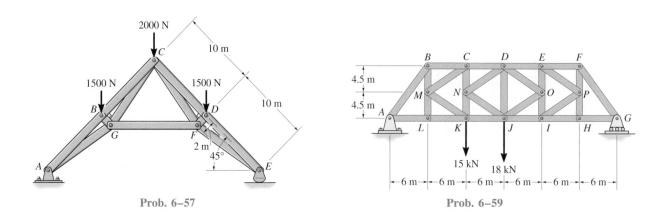

Prob. 6–57

Prob. 6–59

*6.5 Space Trusses

A *space truss* consists of members joined together at their ends to form a stable three-dimensional structure. The simplest element of a space truss is a *tetrahedron,* formed by connecting six members together, as shown in Fig. 6–19. Any additional members added to this basic element would be redundant in supporting the force **P.** A *simple space truss* can be built from this basic tetrahedral element by adding three additional members and a joint forming a system of multiconnected tetrahedrons.

Assumptions for Design. The members of a space truss may be treated as two-force members provided the external loading is applied at the joints and the joints consist of ball-and-socket connections. These assumptions are justified if the welded or bolted connections of the joined members intersect at a common point and the weight of the members can be neglected. In cases where the weight of a member is to be included in the analysis, it is generally satisfactory to apply it as a vertical force, half of its magnitude applied at each end of the member.

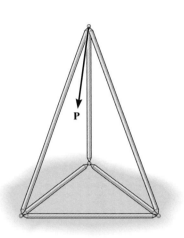

Fig. 6–19

PROCEDURE FOR ANALYSIS

Either the method of joints or the method of sections can be used to determine the forces developed in the members of a simple space truss.

Method of Joints.   Generally, if the forces in *all* the members of the truss must be determined, the method of joints is most suitable for the analysis. When using the method of joints, it is necessary to solve the three scalar equilibrium equations $\Sigma F_x = 0$, $\Sigma F_y = 0$, $\Sigma F_z = 0$ at each joint. The solution of many simultaneous equations can be avoided if the force analysis begins at a joint having at least one known force and at most three unknown forces. If the three-dimensional geometry of the force system at the joint is hard to visualize, it is recommended that a Cartesian vector analysis be used for the solution.

Method of Sections.   If only a *few* member forces are to be determined, the method of sections may be used. When an imaginary section is passed through a truss, and the truss is separated into two parts, the force system acting on one of the parts must satisfy the *six* scalar equilibrium equations: $\Sigma F_x = 0$, $\Sigma F_y = 0$, $\Sigma F_z = 0$, $\Sigma M_x = 0$, $\Sigma M_y = 0$, $\Sigma M_z = 0$ (Eqs. 5–6). By proper choice of the section and axes for summing forces and moments, many of the unknown member forces in a space truss can be computed *directly,* using a single equilibrium equation.

Example 6–8

Determine the forces acting in the members of the space truss shown in Fig. 6–20a. Indicate whether the members are in tension or compression.

SOLUTION

Since there are one known force and three unknown forces acting at joint A, the force analysis of the truss will begin at this joint.

Joint A (Fig. 6–20b).   Expressing each force that acts on the free-body diagram of joint A in vector notation, we have

$$\mathbf{P} = \{-4\mathbf{j}\}\text{ kN}, \qquad \mathbf{F}_{AB} = F_{AB}\mathbf{j}, \qquad \mathbf{F}_{AC} = -F_{AC}\mathbf{k},$$

$$\mathbf{F}_{AE} = F_{AE}\left(\frac{\mathbf{r}_{AE}}{r_{AE}}\right) = F_{AE}(0.577\mathbf{i} + 0.577\mathbf{j} - 0.577\mathbf{k})$$

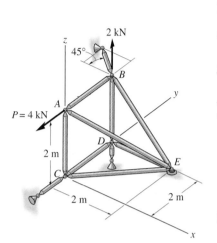

(a)

For equilibrium,

$$\Sigma\mathbf{F} = \mathbf{0}; \qquad\qquad \mathbf{P} + \mathbf{F}_{AB} + \mathbf{F}_{AC} + \mathbf{F}_{AE} = \mathbf{0}$$

$$-4\mathbf{j} + F_{AB}\mathbf{j} - F_{AC}\mathbf{k} + 0.577F_{AE}\mathbf{i} + 0.577F_{AE}\mathbf{j} - 0.577F_{AE}\mathbf{k} = \mathbf{0}$$

$$\Sigma F_x = 0; \qquad\qquad 0.577F_{AE} = 0$$

$$\Sigma F_y = 0; \qquad\qquad -4 + F_{AB} + 0.577F_{AE} = 0$$

$$\Sigma F_z = 0; \qquad\qquad -F_{AC} - 0.577F_{AE} = 0$$

$$F_{AC} = F_{AE} = 0 \qquad\qquad\text{Ans.}$$

$$F_{AB} = 4\text{ kN}\quad\text{(T)} \qquad\qquad\text{Ans.}$$

(b)

Since F_{AB} is known, joint B may be analyzed next.

Joint B (Fig. 6–20c)

$$\Sigma F_x = 0; \qquad\qquad -R_B \cos 45° + 0.707F_{BE} = 0$$

$$\Sigma F_y = 0; \qquad\qquad -4 + R_B \sin 45° = 0$$

$$\Sigma F_z = 0; \qquad\qquad 2 + F_{BD} - 0.707F_{BE} = 0$$

$$R_B = F_{BE} = 5.66\text{ kN}\quad\text{(T)}, \qquad F_{BD} = 2\text{ kN}\quad\text{(C)} \qquad\text{Ans.}$$

(c)

Fig. 6–20

The *scalar* equations of equilibrium may also be applied directly to the force systems on the free-body diagrams of joints D and C, since the force components are easily determined. Show that

$$F_{DE} = F_{DC} = F_{CE} = 0 \qquad\qquad\text{Ans.}$$

PROBLEMS

*6–60. Determine the force in each member of the space truss and state if the members are in tension or compression. The truss is supported by ball-and-socket joints at *D, C,* and *E. Hint:* The support reaction at *E* acts along member *EB*. Why?

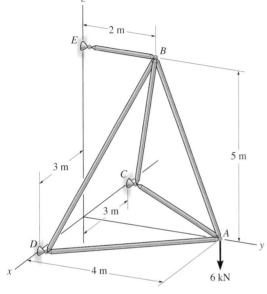

Prob. 6–60

6–61. The tetrahedral truss rests on roller supports at points *A, B,* and *C.* Determine the force in each member and state if the members are in tension or compression.

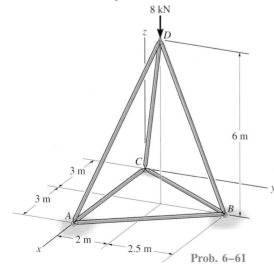

Prob. 6–61

6–62. Determine the force developed in each member of the space truss and state if the members are in tension or compression. The crate has a weight of 150 N.

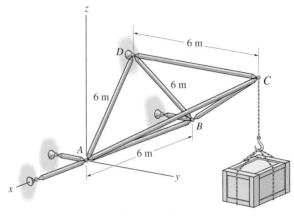

Prob. 6–62

6–63. The space truss is used to support vertical forces at joints *B, C,* and *D.* Determine the force in each member and state if the members are in tension or compression.

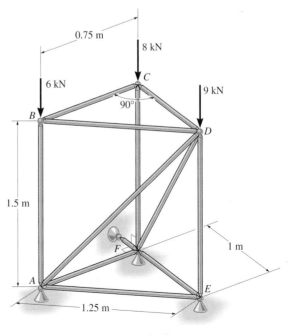

Prob. 6–63

*6–64. The space truss is supported by a ball-and-socket joint at D and short links at C and E. Determine the force in each member and state if the members are in tension or compression. Take $F_1 = \{-500\mathbf{k}\}$ N and $F_2 = \{400\mathbf{j}\}$ N.

6–65. The space truss is supported by a ball-and-socket joint at D and short links at C and E. Determine the force in each member and state if the members are in tension or compression. Take $F_1 = \{200\mathbf{i} + 300\mathbf{j} - 500\mathbf{k}\}$ N and $F_2 = \{400\mathbf{j}\}$ N.

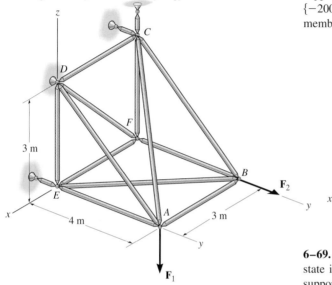

Probs. 6–64/6–65

6–66. Determine the force in each member of the space truss and state if the members are in tension or compression. The truss is supported by a ball-and-socket joint at A and short links at B and C.

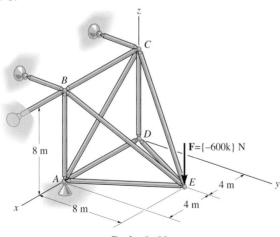

Prob. 6–66

6–67. Determine the force in each member of the space truss and state if the members are in tension or compression. The truss is supported by ball-and-socket joints at A, B, and E. Set $F = \{800\mathbf{j}\}$ N. *Hint:* The support reaction at E acts along member EC. Why?

*6–68. Determine the force in each member of the space truss and state if the members are in tension or compression. The truss is supported by ball-and-socket joints at A, B, and E. Set $F = \{-200\mathbf{i} + 400\mathbf{j}\}$ N. *Hint:* The support reaction at E acts along member EC. Why?

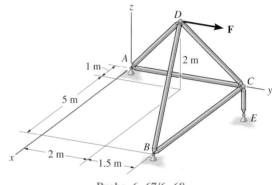

Probs. 6–67/6–68

6–69. Determine the force in each member of the space truss and state if the members are in tension or compression. The truss is supported by ball-and-socket joints at C, D, E, and G.

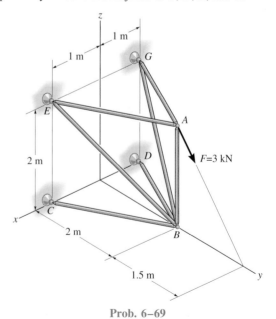

Prob. 6–69

6.6 Frames and Machines

Frames and machines are two common types of structures which are often composed of pin-connected *multiforce members,* i.e., members that are subjected to more than two forces. *Frames* are generally stationary and are used to support loads, whereas *machines* contain moving parts and are designed to transmit and alter the effect of forces. Provided a frame or machine is properly constrained and contains no more supports or members than are necessary to prevent collapse, the forces acting at the joints and supports can be determined by applying the equations of equilibrium to each member. Once the forces at the joints are obtained, it is then possible to *design* the size of the members, connections, and supports using the theory of mechanics of materials and an appropriate engineering design code.

Free-Body Diagrams. In order to determine the forces acting at the joints and supports of a frame or machine, the structure must be disassembled and the free-body diagrams of its parts must be drawn. In this regard, the following important points *must* be observed:

1. Isolate each part by drawing its *outlined shape.* Then show all the forces and/or couple moments that act on the part. Make sure to *label* or *identify* each known and unknown force and couple moment with reference to an established *x, y* coordinate system. Also, indicate any dimensions used for taking moments. Most often the equations of equilibrium are easier to apply if the forces are represented by their rectangular components. As usual, the sense of an unknown force or couple moment can be assumed.
2. Identify all the two-force members in the structure, and represent their free-body diagrams as having two equal but opposite forces acting at their points of application. The line of action of the forces is defined by the line joining the two points where the forces act (see Sec. 5.4). By recognizing the two-force members, we can avoid solving an unnecessary number of equilibrium equations. (See Example 6–14.)
3. Forces common to any two *contacting* members act with equal magnitudes but opposite sense on the respective members. If the two members are treated as a *''system'' of connected members,* then these forces are *''internal''* and are *not shown* on the *free-body diagram of the system;* however, if the free-body diagram of *each member* is drawn, the forces are *''external''* and *must* be shown on each of the free-body diagrams.

The following examples graphically illustrate application of these points in drawing the free-body diagrams of a dismembered frame or machine. In all cases, the weight of the members is neglected.

Example 6–9

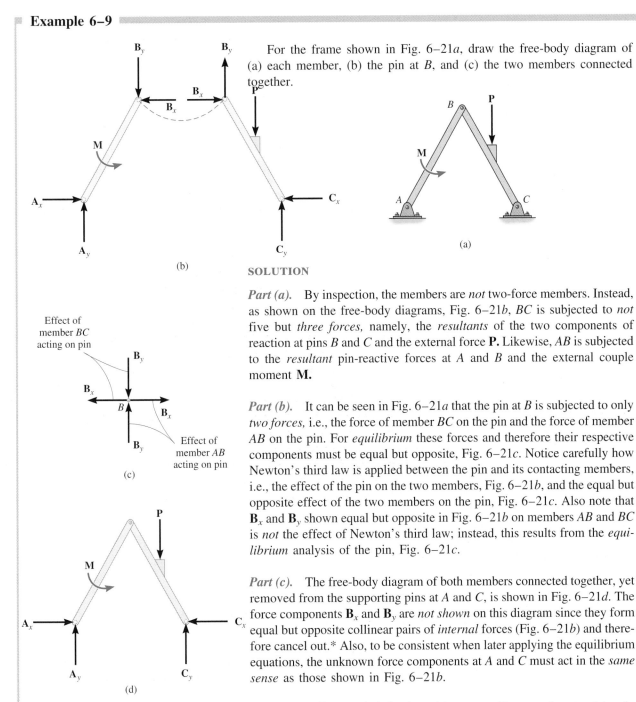

For the frame shown in Fig. 6–21a, draw the free-body diagram of (a) each member, (b) the pin at B, and (c) the two members connected together.

SOLUTION

Part (a).   By inspection, the members are *not* two-force members. Instead, as shown on the free-body diagrams, Fig. 6–21b, BC is subjected to *not* five but *three forces,* namely, the *resultants* of the two components of reaction at pins B and C and the external force **P.** Likewise, AB is subjected to the *resultant* pin-reactive forces at A and B and the external couple moment **M.**

Part (b).   It can be seen in Fig. 6–21a that the pin at B is subjected to only *two forces,* i.e., the force of member BC on the pin and the force of member AB on the pin. For *equilibrium* these forces and therefore their respective components must be equal but opposite, Fig. 6–21c. Notice carefully how Newton's third law is applied between the pin and its contacting members, i.e., the effect of the pin on the two members, Fig. 6–21b, and the equal but opposite effect of the two members on the pin, Fig. 6–21c. Also note that B_x and B_y shown equal but opposite in Fig. 6–21b on members AB and BC is *not* the effect of Newton's third law; instead, this results from the *equilibrium* analysis of the pin, Fig. 6–21c.

Part (c).   The free-body diagram of both members connected together, yet removed from the supporting pins at A and C, is shown in Fig. 6–21d. The force components B_x and B_y are *not shown* on this diagram since they form equal but opposite collinear pairs of *internal* forces (Fig. 6–21b) and therefore cancel out.* Also, to be consistent when later applying the equilibrium equations, the unknown force components at A and C must act in the *same sense* as those shown in Fig. 6–21b.

*This is similar to not including internal forces exerted between adjacent particles of a rigid body when drawing the free-body diagram of the entire rigid body.

Fig. 6–21

Example 6–10

For the frame shown in Fig. 6–22a, draw the free-body diagrams of (a) each of the three members, and (b) members *ABC* and *BD* together.

SOLUTION

Part (a).   By inspection, none of the three members of the frame are two-force members. Instead, each is subjected to *three* forces. The components of these forces are shown on the free-body diagrams in Fig. 6–22b. Notice that equal but opposite force reactions occur at *B*, *C*, and *D*. Draw a free-body diagram of one of the pins at *B*, *C*, or *D* and show why this is so.

Part (b).   The free-body diagram of *ABC* and *BD* together is shown in Fig. 6–22c. Since the entire frame is in equilibrium, the force system on these two members also satisfies the equilibrium equations. Why not show the force components $\mathbf{B}_x$ and $\mathbf{B}_y$ on this diagram?

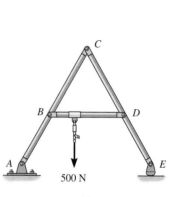

500 N

(a)

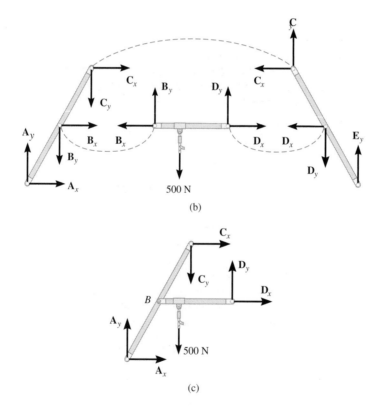

(b)

(c)

Fig. 6–22

Example 6–11

Draw the free-body diagram of each part of the smooth piston and link mechanism used to recycle crushed cans, which is shown in Fig. 6–23*a*.

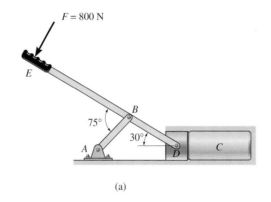

(a)

SOLUTION

By inspection, member *AB* is a two-force member. The free-body diagrams of the parts are shown in Fig. 6–23*b*. Since the pins at *B* and *D* *connect only two parts together,* the forces there are shown as equal but opposite on the separate free-body diagrams of their connected members. In particular, four components of force act on the piston: $\mathbf{D}_x$ and $\mathbf{D}_y$ represent the effect of the pin (or lever *EBD*), $\mathbf{N}_w$ is the *resultant force* of the cylinder's wall, and $\mathbf{P}$ is the resultant compressive force caused by the can *C*.

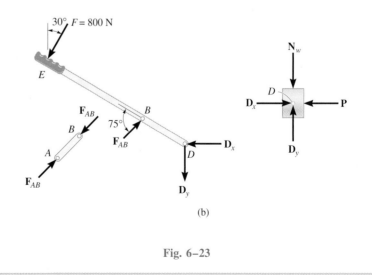

(b)

Fig. 6–23

Example 6–12

For the frame shown in Fig. 6–24*a*, draw the free-body diagrams of (a) the entire frame including the pulleys and cords, (b) the frame without the pulleys and cords, and (c) each of the pulleys.

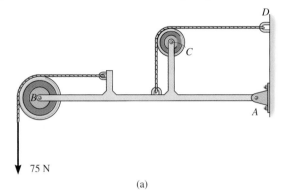

(a)

SOLUTION

Part (a):   When the entire frame including the pulleys and cords is considered, the interactions at the points where the pulleys and cords are connected to the frame become pairs of *internal forces* which cancel each other and therefore are not shown on the free-body diagram, Fig. 6–24*b*.

Part (b):   When the cords and pulleys are removed, their effect *on the frame* must be shown, Fig. 6–24*c*.

Part (c):   The force components B_x, B_y, C_x, C_y of the pins on the pulleys, Fig. 6–24*d*, are equal but opposite to the force components exerted by the pins on the frame, Fig. 6–24*c*. Why?

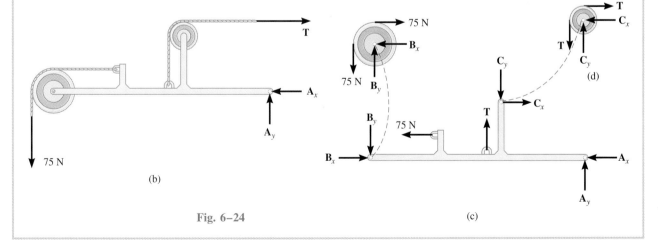

Fig. 6–24

Example 6–13

The hydraulic truck-mounted crane shown in Fig. 6–25a is used to lift a beam that has a mass of 1 Mg. Draw the free-body diagrams of each of its parts, including the pins at A and C.

Fig. 6–25

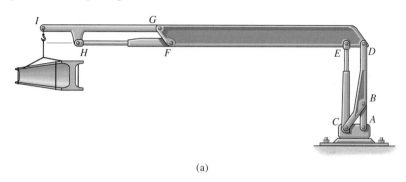

(a)

SOLUTION

By inspection, *HF*, *EC*, and *AB* are all two-force members. The free-body diagrams are shown in Fig. 6–25b. The pin at A is subjected to only *two* forces, namely, the force of the link *AB* and the force of the support. For equilibrium, these forces must be equal in magnitude but opposite in direction. The pin at C, however, is subjected to *three* forces. The force $\mathbf{F}_{EC}$ is caused by the hydraulic cylinder, the force components $\mathbf{C}_x$ and $\mathbf{C}_y$ are caused by member *CBD*, and finally, $\mathbf{C}_x'$ and $\mathbf{C}_y'$ are caused by the support. These components can be related by the equations of force equilibrium applied to the pin.

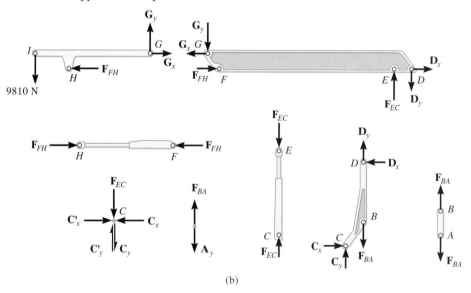

(b)

Before proceeding, it is recommended to cover the solutions to the previous examples and attempt to draw the requested free-body diagrams. When doing so, make sure the work is neat and that all the forces and couple moments are properly labeled.

Equations of Equilibrium. Provided the structure (frame or machine) is properly supported and contains no more supports or members than are necessary to prevent its collapse, then the unknown forces at the supports and connections can be determined from the equations of equilibrium. If the structure lies in the $x-y$ plane, then for *each* free-body diagram drawn the loading must satisfy $\Sigma F_x = 0$, $\Sigma F_y = 0$, and $\Sigma M_O = 0$. The selection of the free-body diagrams used for the analysis is *completely arbitrary*. They may represent each of the members of the structure, a portion of the structure, or its entirety. For example, consider finding the six components of the pin reactions at A, B, and C for the frame shown in Fig. 6–26a. If the frame is dismembered, Fig. 6–26b, these unknowns can be determined by applying the three equations of equilibrium to each of the two members (total of six equations). The free-body diagram of the *entire frame* can also be used for part of the analysis, Fig. 6–26c. Hence, if so desired, all six unknowns can be determined by applying the three equilibrium equations to the entire frame, Fig. 6–26c, and also to either one of its members. Furthermore, the answers can be checked in part by applying the three equations of equilibrium to the remaining "second" member. In general, then, this problem can be solved by writing *at most* six equilibrium equations using free-body diagrams of the members and/or the combination of connected members. Any more than six equations written would *not* be unique from the original six and would serve only to check the results.

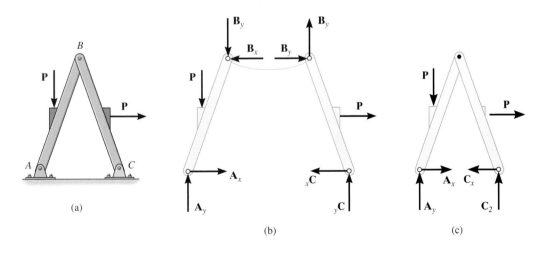

(a)

(b)

(c)

Fig. 6–26

PROCEDURE FOR ANALYSIS

The following procedure provides a method for determining the *joint reactions* of frames and machines (structures) composed of multiforce members.

Free-Body Diagrams.   Draw the free-body diagram of the entire structure, a portion of the structure, or each of its members. The choice should be made so that it leads to the most direct solution to the problem.

Forces common to two members which are in contact act with equal magnitude but opposite sense on the respective free-body diagrams of the members. Recall that all *two-force members,* regardless of their shape, have equal but opposite collinear forces acting at the ends of the member. The unknown forces acting at the joints of multiforce members should be represented by their rectangular components. In many cases it is possible to tell by inspection the proper sense of the unknown forces; however, if this seems difficult, the sense can be assumed.

Equations of Equilibrium.   Count the total number of unknowns to make sure that an equivalent number of equilibrium equations can be written for solution. Recall that in general three equilibrium equations can be written for each rigid body represented in two dimensions. Many times, the solution for the unknowns will be straightforward if moments are summed about a point that lies at the intersection of the lines of action of as many unknown forces as possible. If after obtaining the solution an unknown force magnitude is found to be negative, it means the sense of the force is the reverse of that shown on the free-body diagrams.

The examples that follow illustrate this procedure. All these examples should be *thoroughly understood* before proceeding to solve the problems.

Example 6–14

Determine the horizontal and vertical components of force which the pin at C exerts on member CB of the frame in Fig. 6–27a.

SOLUTION I

Free-Body Diagrams. By inspection it can be seen that AB is a two-force member. The free-body diagrams are shown in Fig. 6–27b.

Equations of Equilibrium. The *three unknowns*, C_x, C_y, and F_{AB}, can be determined by applying the three equations of equilibrium to member CB.

$\zeta+\Sigma M_C = 0;\quad 2000 \text{ N}(2 \text{ m}) - (F_{AB} \sin 60°)(4 \text{ m}) = 0 \quad F_{AB} = 1154.7 \text{ N}$

$\xrightarrow{+}\Sigma F_x = 0;\qquad\qquad 1154.7 \cos 60° \text{ N} - C_x = 0 \quad C_x = 577 \text{ N} \quad Ans.$

$+\uparrow\Sigma F_y = 0;\quad 1154.7 \sin 60° \text{ N} - 2000 \text{ N} + C_y = 0 \quad C_y = 1000 \text{ N} Ans.$

SOLUTION II

Free-Body Diagrams. If one does not recognize that AB is a two-force member, then more work is involved in solving this problem. The free-body diagrams are shown in Fig. 6–27c.

Equations of Equilibrium. The *six unknowns*, A_x, A_y, B_x, B_y, C_x, C_y, are determined by applying the three equations of equilibrium to each member.

Member AB

$\zeta+\Sigma M_A = 0;\qquad B_x(3 \sin 60° \text{ m}) - B_y(3 \cos 60° \text{ m}) = 0 \qquad (1)$

$\xrightarrow{+}\Sigma F_x = 0;\qquad\qquad A_x - B_x = 0 \qquad (2)$

$+\uparrow\Sigma F_y = 0;\qquad\qquad A_y - B_y = 0 \qquad (3)$

Member BC

$\zeta+\Sigma M_C = 0;\qquad 2000 \text{ N}(2 \text{ m}) - B_y(4 \text{ m}) = 0 \qquad (4)$

$\xrightarrow{+}\Sigma F_x = 0;\qquad\qquad B_x - C_x = 0 \qquad (5)$

$+\uparrow\Sigma F_y = 0;\qquad\qquad B_y - 2000 \text{ N} + C_y = 0 \qquad (6)$

The results for C_x and C_y can be determined by solving these equations in the following sequence: 4, 1, 5, then 6. The results are

$$B_y = 1000 \text{ N}$$
$$B_x = 577 \text{ N}$$
$$C_x = 577 \text{ N} \qquad Ans.$$
$$C_y = 1000 \text{ N} \qquad Ans.$$

By comparison, Solution I is simpler since the requirement that $\mathbf{F}_{AB}$ in Fig. 6–27b be equal, opposite, and collinear at the ends of member AB automatically satisfies Eqs. 1, 2, and 3 above and therefore eliminates the need to write these equations. *As a result, always identify the two-force members before starting the analysis!*

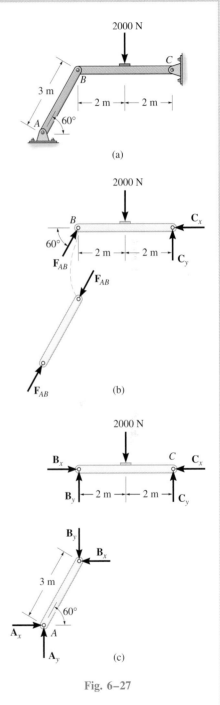

Fig. 6–27

Example 6–15

The compound beam shown in Fig. 6–28a is pin-connected at B. Determine the reactions at its supports. Neglect its weight and thickness.

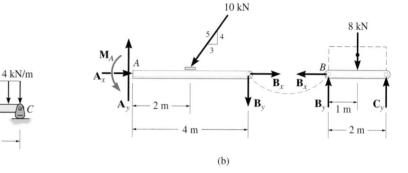

(a)

(b)

Fig. 6–28

SOLUTION

Free-Body Diagrams.   By inspection, if we consider a free-body diagram of the entire beam ABC, there will be three unknown reactions at A and one at C. These four unknowns cannot all be obtained from the three equations of equilibrium, and so it will become necessary to dismember the beam into its two segments as shown in Fig. 6–28b.

Equations of Equilibrium.   The six unknowns are determined as follows:

   Segment BC

$\xrightarrow{+} \Sigma F_x = 0$; $B_x = 0$

$\zeta + \Sigma M_B = 0$; $-8 \text{ kN}(1 \text{ m}) + C_y(2 \text{ m}) = 0$

$+ \uparrow \Sigma F_y = 0$; $B_y - 8 \text{ kN} + C_y = 0$

   Segment AB

$\xrightarrow{+} \Sigma F_x = 0$; $A_x - (10 \text{ kN})(\tfrac{3}{5}) + B_x = 0$

$\zeta + \Sigma M_A = 0$; $M_A - (10 \text{ kN})(\tfrac{4}{5})(2 \text{ m}) - B_y(4 \text{ m}) = 0$

$+ \uparrow \Sigma F_y = 0$; $A_y - (10 \text{ kN})(\tfrac{4}{5}) - B_y = 0$

Solving each of these equations successively, using previously calculated results, we obtain

$$A_x = 6 \text{ kN} \qquad A_y = 12 \text{ kN} \qquad M_A = 32 \text{ kN} \cdot \text{m} \qquad \textit{Ans.}$$

$$B_x = 0 \qquad B_y = 4 \text{ kN}$$

$$C_y = 4 \text{ kN} \qquad\qquad\qquad\qquad\qquad\qquad\qquad \textit{Ans.}$$

Example 6–16

Determine the horizontal and vertical components of force which the pin at C exerts on member $ABCD$ of the frame shown in Fig. 6–29a.

SOLUTION

Free-Body Diagrams.   By inspection, the three components of reaction that the supports exert on $ABCD$ can be determined from a free-body diagram of the entire frame, Fig. 6–29b. Also, the free-body diagram of each frame member is shown in Fig. 6–29c. Notice that member BE is a two-force member. As shown by the colored dashed lines, the forces at B, C, and E have equal magnitudes but opposite directions on the separate free-body diagrams.

Equations of Equilibrium.   The six unknowns A_x, A_y, F_B, C_x, C_y, and D_x will be determined from the equations of equilibrium applied to the entire frame and then to member CEF. We have

Entire Frame

$\zeta + \Sigma M_A = 0;$ $-981 \text{ N}(2 \text{ m}) + D_x(2.8 \text{ m}) = 0$ $D_x = 700.7 \text{ N}$

$\xrightarrow{+} \Sigma F_x = 0;$ $A_x - 700.7 \text{ N} = 0$ $A_x = 700.7 \text{ N}$

$+ \uparrow \Sigma F_y = 0;$ $A_y - 981 \text{ N} = 0$ $A_y = 981 \text{ N}$

Member CEF

$\zeta + \Sigma M_C = 0;$ $-981 \text{ N}(2 \text{ m}) - (F_B \sin 45°)(1.6 \text{ m}) = 0$

$$F_B = -1734.2 \text{ N}$$

$\xrightarrow{+} \Sigma F_x = 0;$ $-C_x - (-1734.2 \cos 45° \text{ N}) = 0$

$$C_x = 1230 \text{ N} \qquad \qquad Ans.$$

$+ \uparrow \Sigma F_y = 0;$ $C_y - (-1734.2 \sin 45° \text{ N}) - 981 \text{ N} = 0$

$$C_y = -245 \text{ N} \qquad \qquad Ans.$$

Since the magnitudes of forces $\mathbf{F}_B$ and $\mathbf{C}_y$ were calculated as negative quantities, they were assumed to be acting in the wrong sense on the free-body diagrams, Fig. 6–29c. The correct sense of these forces might have been determined "by inspection" *before* applying the equations of equilibrium to member CEF. As shown in Fig. 6–29c, moment equilibrium about point E on member CEF indicates that $\mathbf{C}_y$ must actually act *downward* to counteract the moment created by the 981-N force about point E. Similarly, summing moments about point C, it is seen that the vertical component of force $\mathbf{F}_B$ must actually act *upward,* and so $\mathbf{F}_B$ must act upward to the right.

The above calculations can be checked by applying the three equilibrium equations to member $ABCD$, Fig. 6–29c.

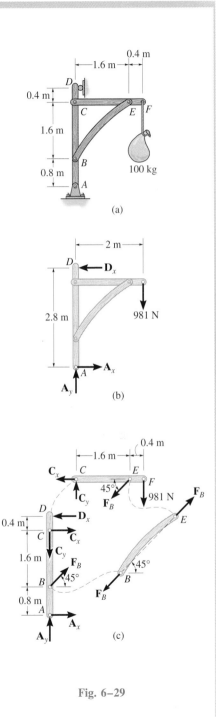

Fig. 6–29

Example 6–17

The smooth disk shown in Fig. 6–30a is pinned at D and has a weight of 20 N. Neglecting the weights of the other members, determine the horizontal and vertical components of reaction at pins B and D.

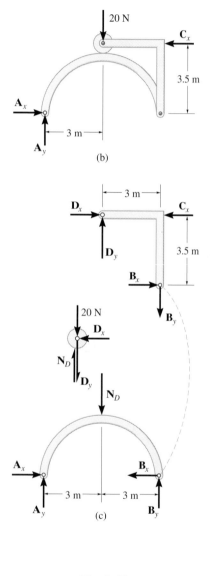

(a)

SOLUTION

Free-Body Diagrams. By inspection, the three components of reaction at the supports can be determined from a free-body diagram of the entire frame, Fig. 6–30b. Also, free-body diagrams of the members are shown in Fig. 6–30c.

Equations of Equilibrium. The eight unknowns can of course be obtained by applying the eight equilibrium equations to each member—three to member AB, three to member BCD, and two to the disk. (Moment equilibrium is automatically satisfied for the disk.) If this is done, however, all the results can be obtained only from a simultaneous solution of some of the equations. (Try it and find out.) To avoid this situation, it is best to first determine the three support reactions on the *entire* frame; then, using these results, the remaining five equilibrium equations can be applied to two other parts in order to solve successively for the other unknowns.

Entire Frame

$$\zeta +\Sigma M_A = 0; \quad -20 \text{ N}(3 \text{ m}) + C_x(3.5 \text{ m}) = 0 \quad C_x = 17.1 \text{ N}$$
$$\xrightarrow{+}\Sigma F_x = 0; \quad\quad A_x - 17.1 \text{ N} = 0 \quad\quad A_x = 17.1 \text{ N}$$
$$+\uparrow \Sigma F_y = 0; \quad\quad A_y - 20 \text{ N} = 0 \quad\quad A_y = 20 \text{ N}$$

Member AB

$$\xrightarrow{+}\Sigma F_x = 0; \quad\quad 17.1 \text{ N} - B_x = 0 \quad\quad B_x = 17.1 \text{ N} \quad\quad\textit{Ans.}$$
$$\zeta +\Sigma M_B = 0; \quad -20 \text{ N}(6 \text{ m}) + N_D(3 \text{ m}) = 0 \quad N_D = 40 \text{ N}$$
$$+\uparrow \Sigma F_y = 0; \quad 20 \text{ N} - 40 \text{ N} + B_y = 0 \quad\quad B_y = 20 \text{ N} \quad\quad\textit{Ans.}$$

Disk

$$\xrightarrow{+}\Sigma F_x = 0; \quad\quad\quad\quad D_x = 0 \quad\quad\quad\quad\textit{Ans.}$$
$$+\uparrow \Sigma F_y = 0; \quad 40 \text{ N} - 20 \text{ N} - D_y = 0 \quad D_y = 20 \text{ N} \quad\quad\textit{Ans.}$$

Fig. 6–30

Example 6–18

Determine the tension in the cables and also the force **P** required to support the 600-N force using the frictionless pulley system shown in Fig. 6–31a.

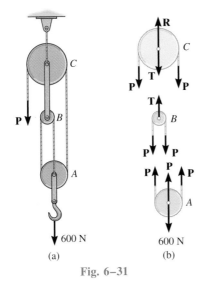

(a)   (b)

Fig. 6–31

SOLUTION

Free-Body Diagrams. A free-body diagram of each pulley *including* its pin and a portion of the contacting cable is shown in Fig. 6–31b. Since the cable is *continuous* and the pulleys are frictionless, the cable has a *constant tension P* acting throughout its length (see Example 5–7). The link connection between pulleys B and C is a two-force member and therefore it has an unknown tension T acting on it. Notice that the *principle of action, equal but opposite reaction* must be carefully observed for forces **P** and **T** when the *separate* free-body diagrams are drawn.

Equations of Equilibrium. The three unknowns are obtained as follows:

Pulley A

$$+\uparrow \Sigma F_y = 0; \qquad 3P - 600 \text{ N} = 0 \qquad P = 200 \text{ N} \qquad\qquad Ans.$$

Pulley B

$$+\uparrow \Sigma F_y = 0; \qquad T - 2P = 0 \qquad T = 400 \text{ N} \qquad\qquad Ans.$$

Pulley C

$$+\uparrow \Sigma F_y = 0; \qquad R - 2P - T = 0 \qquad R = 800 \text{ N} \qquad\qquad Ans.$$

Example 6–19

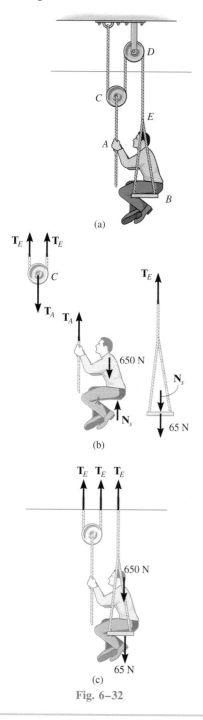

(a)

(b)

(c)

Fig. 6–32

A man having a weight of 650 N supports himself by means of the cable and pulley system shown in Fig. 6–32a. If the seat has a weight of 65 N, determine the equilibrium force that he must exert on the cable at A and the force he exerts on the seat. Neglect the weight of the cables and pulleys.

SOLUTION I

Free-Body Diagrams. The free-body diagrams of the man, seat, and pulley C are shown in Fig. 6–32b. The *two* cables are subjected to tensions $\mathbf{T}_A$ and $\mathbf{T}_E$, respectively. The man is subjected to three forces: his weight, the tension $\mathbf{T}_A$ of cable AC, and the reaction $\mathbf{N}_s$ of the seat.

Equations of Equilibrium. The three unknowns are obtained as follows:

    Man

$$+ \uparrow \Sigma F_y = 0; \qquad T_A + N_s - 650 \text{ N} = 0 \qquad (1)$$

    Seat

$$+ \uparrow \Sigma F_y = 0; \qquad T_E - N_s - 65 \text{ N} = 0 \qquad (2)$$

    Pulley C

$$+ \uparrow \Sigma F_y = 0; \qquad 2T_E - T_A = 0 \qquad (3)$$

The magnitude of force $\mathbf{T}_E$ can be determined by adding Eqs. 1 and 2 to eliminate N_s and then using Eq. 3. The other unknowns are then obtained by resubstitution of T_E.

$$T_A = 476.7 \text{ N} \qquad\qquad Ans.$$
$$T_E = 238.3 \text{ N}$$
$$N_s = 173.3 \text{ N} \qquad\qquad Ans.$$

SOLUTION II

Free-Body Diagrams. By using the blue section shown in Fig. 6–32a, the man, pulley, and seat can be considered as a *single system*, Fig. 6–32c. Here $\mathbf{N}_s$ and $\mathbf{T}_A$ are *internal* forces and hence are not included on the "combined" free-body diagram.

Equations of Equilibrium. Applying $\Sigma F_y = 0$ yields a *direct* solution for T_E.

$$+ \uparrow \Sigma F_y = 0; \qquad 3T_E - 65 \text{ N} - 650 \text{ N} = 0 \qquad T_E = 238.3 \text{ N}$$

The other unknowns can be obtained from Eqs. 2 and 3.

Example 6–20

The hand exerts a force of 35 N on the grip of the spring compressor shown in Fig. 6–33a. Determine the force in the spring needed to maintain equilibrium of the mechanism in the position shown.

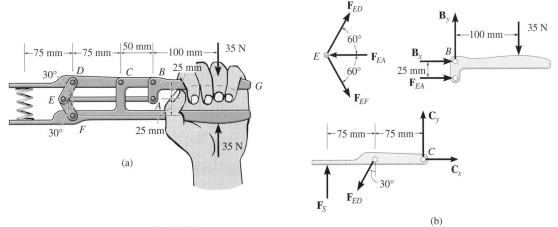

Fig. 6–33

SOLUTION

Free-Body Diagrams. By inspection, members *EA*, *ED*, and *EF* are all two-force members. The free-body diagrams for parts *DC* and *ABG* are shown in Fig. 6–33b. The pin at *E* has also been included here since *three* force interactions occur on this pin. They represent the effects of members *ED*, *EA*, and *EF*. Note carefully how equal and opposite force reactions occur between each of the parts.

Equations of Equilibrium. By studying the free-body diagrams, the most direct way to obtain the spring force is to apply the equations of equilibrium in the following sequence:

Lever *ABG*

$$\downarrow + \Sigma M_B = 0; \quad F_{EA}(25 \text{ mm}) - 35 \text{ N}(100 \text{ mm}) = 0 \quad F_{EA} = 140 \text{ N}$$

Pin *E*

$$+ \uparrow \Sigma F_y = 0; \quad F_{ED} \sin 60° - F_{EF} \sin 60° = 0 \quad F_{ED} = F_{EF} = F$$
$$\xrightarrow{+} \Sigma F_x = 0; \quad 2F \cos 60° - 140 \text{ N} = 0 \quad F = 140 \text{ N}$$

Arm *DC*

$$\downarrow + \Sigma M_C = 0; \quad -F_s(150 \text{ mm}) + 140 \cos 30° \text{ N}(75 \text{ mm}) = 0$$
$$F_s = 60.62 \text{ N} \qquad \qquad \textit{Ans.}$$

Example 6–21

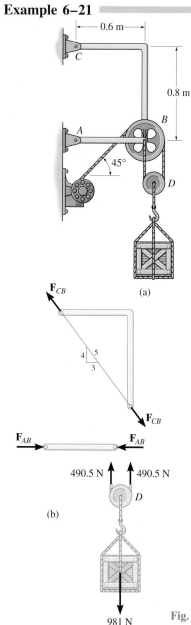

Fig. 6–34

(a)

(b)

The 100-kg block is held in equilibrium by means of the pulley and continuous cable system shown in Fig. 6–34*a*. If the cable is attached to the pin at *B*, compute the forces which this pin exerts on each of its connecting members.

SOLUTION

Free-Body Diagrams.   A free-body diagram of each member of the frame is shown in Fig. 6–34*b*. By inspection, members *AB* and *CB* are two-force members. Furthermore, the cable must be subjected to a force of 490.5 N in order to hold pulley *D* and the block in equilibrium. A free-body diagram of the pin at *B* is needed, since *four interactions* occur at this pin. These are caused by the attached cable (490.5 N), member *AB* ($\mathbf{F}_{AB}$), member *CB* ($\mathbf{F}_{CB}$), and pulley *B* ($\mathbf{B}_x$ and $\mathbf{B}_y$).

Equations of Equilibrium.   Applying the equations of force equilibrium to pulley *B*, we have

$$\xrightarrow{+}\Sigma F_x = 0; \qquad B_x - 490.5 \cos 45° \text{ N} = 0 \qquad B_x = 346.8 \text{ N} \qquad \textit{Ans.}$$
$$+\uparrow\Sigma F_y = 0; \qquad B_y - 490.5 \sin 45° \text{ N} - 490.5 \text{ N} = 0$$
$$B_y = 837.3 \text{ N} \qquad\qquad \textit{Ans.}$$

Using these results, equilibrium of the pin requires that

$$+\uparrow\Sigma F_y = 0; \quad \tfrac{4}{5}F_{CB} - 837.3 \text{ N} - 490.5 \text{ N} = 0 \qquad F_{CB} = 1660 \text{ N} \quad \textit{Ans.}$$
$$\xrightarrow{+}\Sigma F_x = 0; \quad F_{AB} - \tfrac{3}{5}(1660 \text{ N}) - 346.8 \text{ N} = 0 \qquad F_{AB} = 1343 \text{ N} \quad \textit{Ans.}$$

It may be noted that the two-force member *CB* is subjected to bending as caused by the force $\mathbf{F}_{CB}$. From the standpoint of design, it would be better to make this member *straight* (from *C* to *B*) so that the force $\mathbf{F}_{CB}$ would only create tension in the member.

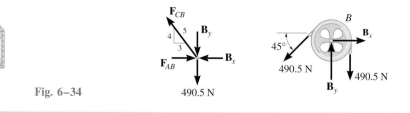

Before solving the following problems, it is suggested that a brief review be made of all the previous examples. This may be done by covering each solution and trying to locate the two-force members, drawing the free-body diagrams, and conceiving ways of applying the equations of equilibrium to obtain the solution.

PROBLEMS

6–70. A force of $P = 80$ N is applied to the handles of the pliers. Determine the force developed on the smooth bolt B and the reaction that pin A exerts on its attached members.

6–71. Determine the force P that must be applied to the handles of the pliers so that it develops a force of 1000 N on the smooth bolt at B. Also, what is the magnitude of the resultant force acting on the pin at A?

6–73. The link is used to hold the rod in place. Determine the required axial force on the screw at E if the largest force to be exerted on the rod at B, C, or D is to be 100 N. Also, find the magnitude of the force reaction at pin A. Assume all surfaces of contact are smooth.

6–74. The link is used to hold the rod in place. Determine the force on the rod at B, C, and D and the magnitude of the reaction at pin A if the axial load on the screw E is 200 N.

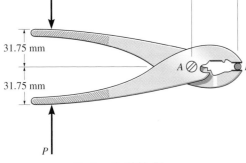

Probs. 6–70/6–71

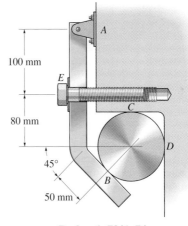

Probs. 6–73/6–74

***6–72.** The eye hook has a positive locking latch when it supports the load because its two parts are pin-connected at A and they bear against one another along the smooth surface at B. Determine the resultant force at the pin and the normal force at B when the eye hook supports a load of 800 N.

6–75. Determine the horizontal and vertical components of force at pins A, B, and C, and the reactions at the fixed support D of the three-member frame.

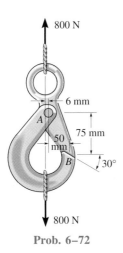

Prob. 6–72

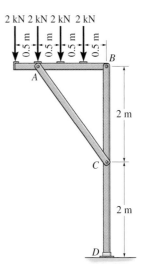

Prob. 6–75

*6–76. The smooth block is held in place using the vice clamp. If the screw exerts a force of 500 N on the block, determine the magnitude of the resultant force on the pin at *A*.

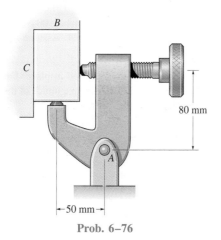

Prob. 6–76

6–77. The two ends of the spanner wrench fit loosely into the smooth slots of the bolt head. Determine the required force *P* on the handle in order to develop a torque of $M = 50$ N · m on the bolt. Also, what is the resultant force on the pin at *B*?

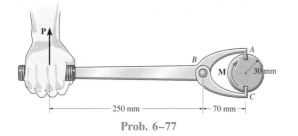

Prob. 6–77

6–78. The two ends of the spanner wrench fit loosely into the smooth slots of the bolt head. Determine the torque *M* on the bolt and the resultant force on the pin at *B* when a force of $P = 80$ N is applied to the handle.

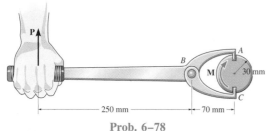

Prob. 6–78

6–79. The three-hinged arch supports the loads $F_1 = 8$ kN and $F_2 = 5$ kN. Determine the horizontal and vertical components of reaction at the pin supports *A* and *B*. Take $h = 2$ m.

*6–80. The three-hinged arch supports the loads $F_1 = 4$ kN and $F_2 = 7$ kN. Determine the horizontal and vertical components of reaction at the pin supports *A* and *B*. Take $h = 0$.

6–81. The three-hinged arch supports the loads $F_1 = 8$ kN and $F_2 = 0$. Determine the horizontal and vertical components of reaction at the pin supports *A* and *B*. Take $h = 3$ m.

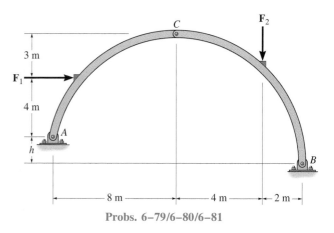

Probs. 6–79/6–80/6–81

6–82. Determine the horizontal and vertical components of force at pin *A*. Take $P = 600$ N.

6–83. Determine the greatest force *P* that can be applied to the frame if the largest force resultant acting at *A* can have a magnitude of 2 kN.

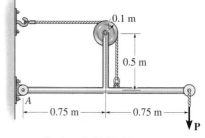

Probs. 6–82/6–83

*6–84. Determine the horizontal and vertical components of force that the pins at *A*, *B*, and *C* exert on their connecting members.

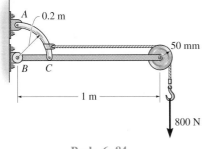

Prob. 6–84

6–85. Determine the force *P* needed to support the 20-kg mass using the *Spanish Burton rig*. Also, what are the reactions at the supporting hooks *A*, *B*, and *C*?

6–86. Determine the force *P* needed to support the 50-kg mass using the *Spanish Burton rig*. The pulleys have a mass of $m_D = 10$ kg, $m_E = m_F = 5$ kg, and $m_G = m_H = 2$ kg. Also, what are the reactions at the supporting hooks *A*, *B*, and *C*?

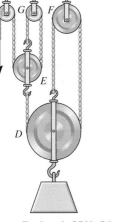

Probs. 6–85/6–86

6–87. Determine the force *P* needed to suspend the 100-N weight. Each pulley has a weight of 10 N. Also, what are the cord reactions at *A* and *B*?

*6–88. If each cord can support a maximum tension of 500 N, determine the largest weight that can be supported by the pulley system. Each pulley has a weight of 10 N.

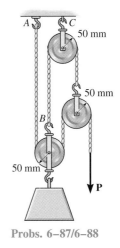

Probs. 6–87/6–88

6–89. Determine the force *P* on the cord, and the angle θ that the pulley-supporting link *AB* makes with the vertical. Neglect the mass of the pulleys and the link. The block has a weight of 200 N and the cord is attached to the pin at *B*. The pulleys have radii of $r_1 = 50$ mm and $r_2 = 25$ mm.

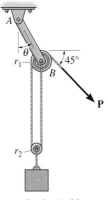

Prob. 6–89

6–90. Determine the horizontal and vertical components of force at *C* which member *ABC* exerts on member *CEF*.

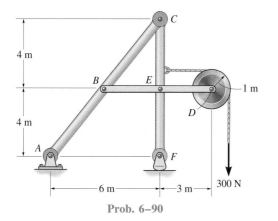

Prob. 6–90

6–91. Determine the horizontal and vertical components of force which the connecting pins at *B*, *E*, and *D* exert on member *BED*.

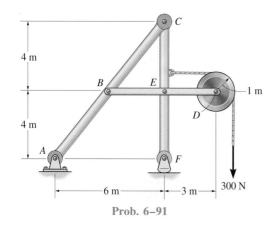

Prob. 6–91

*6–92. Determine the force that the smooth roller *C* exerts on beam *AB*. Also, what are the horizontal and vertical components of reaction at pin *A*? Neglect the weight of the frame and roller.

6–93. Solve Prob. 6–92 if roller *C* has a weight of 20 N.

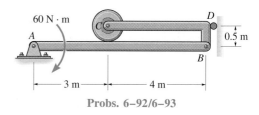

Probs. 6–92/6–93

6–94. Determine the horizontal and vertical components of force which the pins exert on member *ABC*.

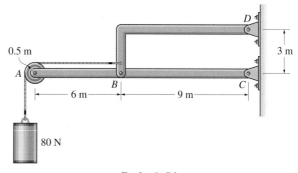

Prob. 6–94

6–95. Determine the horizontal and vertical components of force at pins *B* and *C*.

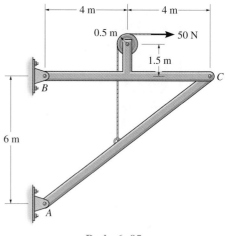

Prob. 6–95

*6–96.   Determine the horizontal and vertical components of force at each pin. The suspended cylinder has a weight of 80 N.

6–97.   Determine the largest weight of the cylinder if the maximum reaction at the rocker A is 150 N.

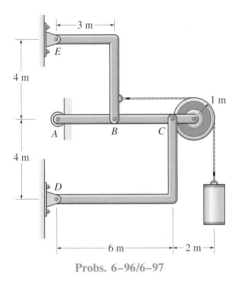

Probs. 6–96/6–97

6–98.   Determine the force P on the cable if the spring is compressed 0.05 m when the mechanism is in the position shown. The spring has a stiffness of k = 800 N/m.

6–99.   Determine the compression of the spring if the cable force is to be P = 80 N. The spring has a stiffness of k = 800 N/m and the mechanism is to be held in the position shown when the spring is compressed.

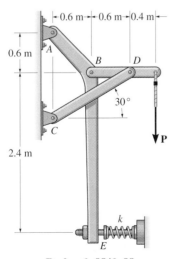

Probs. 6–98/6–99

*6–100.   If a force of P = 60 N is applied perpendicular to the handle of the mechanism, determine the magnitude of force **F** for equilibrium. The members are pin-connected at A, B, C, and D.

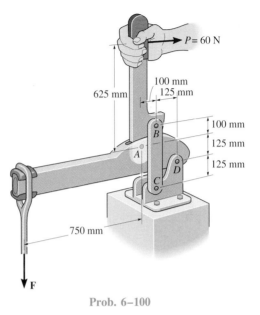

Prob. 6–100

6–101.   The clamp is used to hold the smooth strut S in place. If the tensile force in the bolt GH is 300 N, determine the force exerted at points A and B.

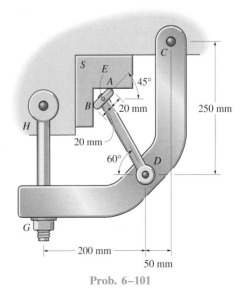

Prob. 6–101

6–102. Determine the required mass of the suspended cylinder if the tension in the chain wrapped around the freely turning gear is to be 2 kN. Also, what is the magnitude of the resultant force on pin A?

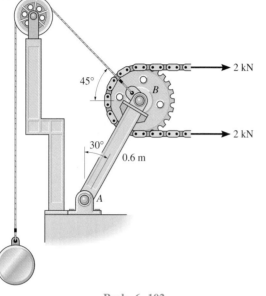

Prob. 6–102

6–103. The derrick is pin-connected to the pivot at A. Determine the largest mass that can be supported by the derrick if the maximum force that can be sustained by the pin at A is 18 kN.

*6–104. The derrick is pin-connected to the pivot at A. Determine the force in the cable at C and in the hoisting cable at D if the suspended crate is 900 kg. Also, what is the resultant force acting on the pin at A?

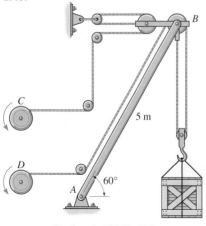

Probs. 6–103/6–104

6–105. Determine the horizontal and vertical components of force which the pins at A, B, and C exert on member ABC of the frame.

6–106. Determine the horizontal and vertical components of force which the pins at D and E exert on member DE of the frame.

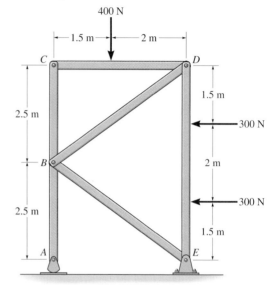

Probs. 6–105/6–106

6–107. By squeezing on the hand brake of the bicycle, the rider subjects the brake cable to a tension of 500 N. If the caliper mechanism is pin-connected to the bicycle frame at B, determine the normal force each brake pad exerts on the rim of the wheel. Is this the force that stops the wheel from turning? Explain.

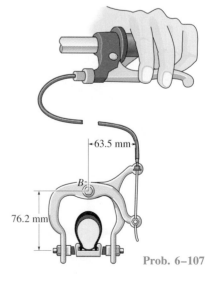

Prob. 6–107

*6–108.   A man having a weight of 778 N attempts to lift himself using one of the two methods shown. Determine the total force he must exert on bar *AB* in each case and the normal reaction he exerts on the platform at *C*. Neglect the weight of the platform.

6–109.   A man having a weight of 778 N attempts to lift himself using one of the two methods shown. Determine the total force he must exert on bar *AB* in each case and the normal reaction he exerts on the platform at *C*. The platform has a weight of 133 N.

6–111.   Determine the horizontal and vertical components of force at pin *B* and the normal force the pin at *C* exerts on the smooth slot. Also, determine the moment and horizontal and vertical reactions of force at *A*. There is a pulley at *E*.

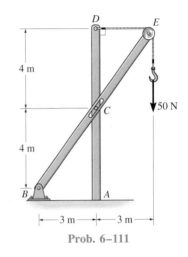

Prob. 6–111

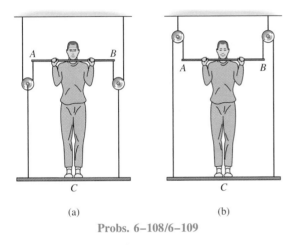

(a)

(b)

Probs. 6–108/6–109

6–110.   Multiple levers can be used in a *compound arrangement* such as shown for the pan scale. If the mass on the pan is 4 kg, determine the reactions at pins *A*, *B*, and *C* and the distance *x* of the 25-g mass to keep the scale in balance.

*6–112.   The flat-bed trailer has a weight of 70 kN and center of gravity at G_T. It is pin-connected to the cab at *D*. The cab has a weight of 60 kN and center of gravity at G_C. Determine the range of values *x* for the position of the 20-kN load *L* so that when it is placed over the rear axle, no axle is subjected to more than 55 N. The load has a center of gravity at G_L.

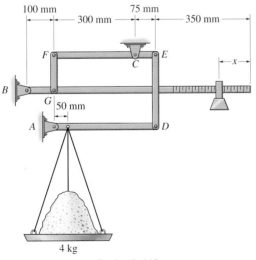

Prob. 6–110

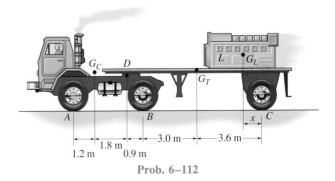

Prob. 6–112

6–113. The flat-bed trailer has a weight of 70 kN and center of gravity at G_T. It is pin-connected to the cab at D. The cab has a weight of 60 kN and center of gravity at G_C. Determine the largest load L that can be placed on the bed if no axle is to be subjected to more than 55 kN. The center of gravity G_L of the load is at $x = 1.2$ m.

6–115. The mechanism is used to hide kitchen appliances under a cabinet by allowing the shelf to rotate downward. If the mixer weighs 100 N, is centered on the shelf, and has a mass center at G, determine the stretch in the spring necessary to hold the shelf in the equilibrium position shown. There is a similar mechanism on each side of the shelf, so that each mechanism supports 50 N of the load. The springs each have a stiffness of $k = 1600$ N/m spring.

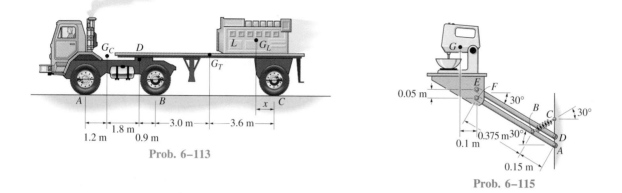

Prob. 6–113

Prob. 6–115

6–114. The aircraft-hangar door opens and closes slowly by means of a motor which draws in the cable AB. If the door is made in two sections (bifold) and each section has a uniform weight W and length L, determine the force in the cable as a function of the door's position θ. The sections are pin-connected at C and D and the bottom is attached to a roller that travels along the vertical track.

***6–116.** The tractor boom supports the uniform mass of 500 kg in the bucket which has a center of mass at G. Determine the force in each hydraulic cylinder AB and CD and the resultant force at pins E and F. The load is supported equally on each side of the tractor by a similar mechanism.

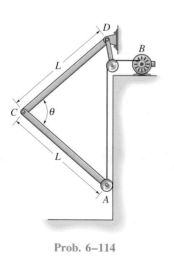

Prob. 6–114

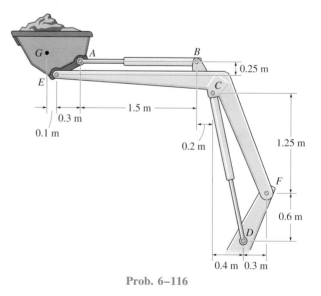

Prob. 6–116

6–117. The pruning shears are subjected to a squeezing force of $P = 80$ N at the grip. Determine the normal force developed on the twig at the blade E.

6–118. Determine the required force P that must be applied at the blade of the pruning shears so that the blade exerts a normal force of 200 N on the twig at E.

*6–120. The linkage for a hydraulic jack is shown. If the load on the jack is 20 kN, determine the pressure acting on the fluid when the jack is in the position shown. All lettered points are pins. The piston at H has a cross-sectional area of $A = 1250$ mm². *Hint:* First find the force F acting along link EH. The pressure in the fluid is $p = F/A$.

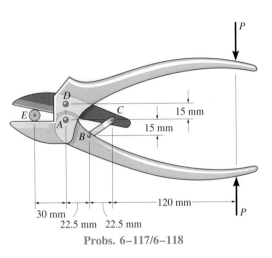

Probs. 6–117/6–118

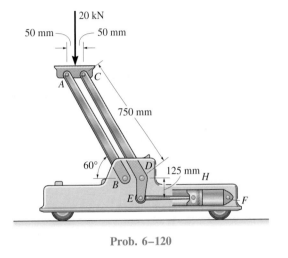

Prob. 6–120

6–119. The three power lines exert the forces shown on the truss joints, which in turn are pin-connected to the poles AH and EG. Determine the force in the guy cable AI and the pin reaction at the support H.

6–121. The hydraulic crane is used to lift the 14-kN load. Determine the force in the hydraulic cylinder AB and the force in links AC and AD when the load is held in the position shown.

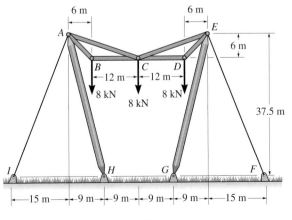

Prob. 6–119

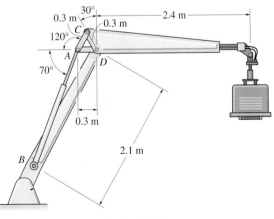

Prob. 6–121

6–122. The kinetic sculpture requires that each of the three pinned beams be in perfect balance at all times during its slow motion. If each member has a uniform weight of 2 N/m and length of 3 m, determine the necessary counterweights W_1, W_2, and W_3 which must be added to the ends of each member to keep the system in balance for any position. Neglect the size of the counterweights.

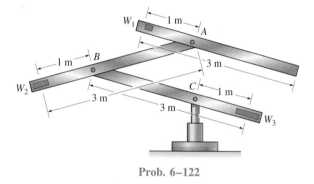

Prob. 6–122

6–123. The spring has an unstretched length of 0.3 m. Determine the angle θ for equilibrium if the uniform links each have a mass of 5 kg.

6–124. The spring has an unstretched length of 0.3 m. Determine the mass m of each uniform link if the angle $\theta = 15°$ for equilibrium.

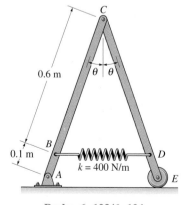

Probs. 6–123/6–124

6–125. Determine the horizontal force F required to maintain equilibrium of the slider mechanism when $\theta = 60°$.

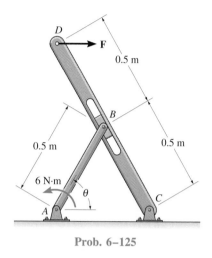

Prob. 6–125

6–126. The two-bar mechanism consists of a lever arm AB and smooth link CD, which has a fixed collar at its end C and a roller at the other end D. Determine the force P needed to hold the lever in the position θ. The spring has a stiffness k and unstretched length of $2L$. The roller contacts either the top or bottom portion of the horizontal guide.

6–127. The two-bar mechanism consists of a lever arm AB and smooth link CD, which has a fixed collar at its end C and a roller at the other end D. If a force P is applied to the lever, determine the required stiffness of the spring so that the lever will reach the equilibrium position when $\theta = 45°$. The unstretched length of the spring is $2L$, and the roller contacts either the top or bottom portion of the horizontal guide. Express k in terms of P and L.

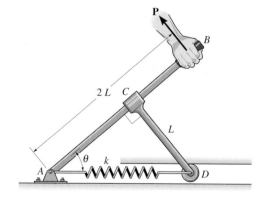

Probs. 6–126/6–127

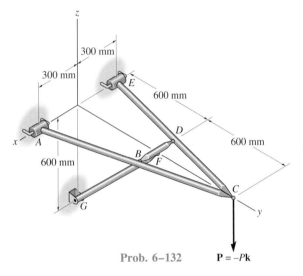

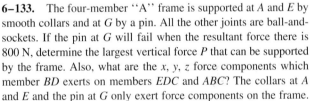

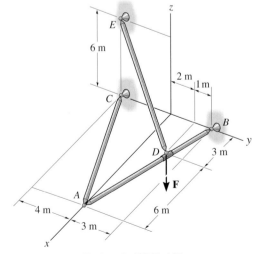

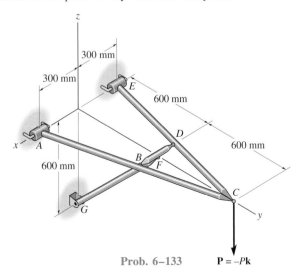

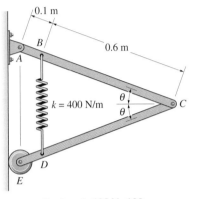

*6–128. The spring has an unstretched length of 0.3 m. Determine the angle θ for equilibrium if the uniform links each have a mass of 5 kg.

6–129. The spring has an unstretched length of 0.3 m. Determine the mass m of each uniform link if the angle $\theta = 20°$ for equilibrium.

Probs. 6–128/6–129

6–130. The three-member frame is connected at its ends using ball-and-socket joints. Determine the x, y, z components of reaction at B and the tension in member ED. The force acting at D is $\mathbf{F} = \{250\mathbf{i} - 350\mathbf{k}\}$ N.

6–131. The three-member frame is connected at its ends using ball-and-socket joints. Determine the x, y, z components of reaction at B and the tension in member ED. The force acting at D is $\mathbf{F} = \{135\mathbf{i} + 200\mathbf{j} - 180\mathbf{k}\}$ N.

Probs. 6–130/6–131

*6–132. The four-member "A" frame supports a vertical force of $P = 600$ N. If it is assumed that the supports at A and E are smooth collars, G is a pin, and all other joints are ball-and-sockets, determine the x, y, z force components which member BD exerts on members EDC and FG. The collars at A and E and the pin at G only exert force components on the frame.

Prob. 6–132 $\mathbf{P} = -P\mathbf{k}$

6–133. The four-member "A" frame is supported at A and E by smooth collars and at G by a pin. All the other joints are ball-and-sockets. If the pin at G will fail when the resultant force there is 800 N, determine the largest vertical force P that can be supported by the frame. Also, what are the x, y, z force components which member BD exerts on members EDC and ABC? The collars at A and E and the pin at G only exert force components on the frame.

Prob. 6–133 $\mathbf{P} = -P\mathbf{k}$

6–134.   The structure is subjected to the loading shown. Member *AD* is supported by a cable *AB* and roller at *C* and fits through a smooth circular hole at *D*. Member *ED* is supported by a roller at *D* and a pole that fits in a smooth snug circular hole at *E*. Determine the *x*, *y*, *z* components of reaction at *E* and the tension in cable *AB*.

6–135.   The structure is subjected to the force of 450 N which lies in a plane parallel to the *y-z* plane. Member *AB* is supported by a ball-and-socket joint at *A* and fits through a snug hole at *B*. Member *CD* is supported by a pin at *C*. Determine the *x*, *y*, *z* components of reaction at *A* and *C*.

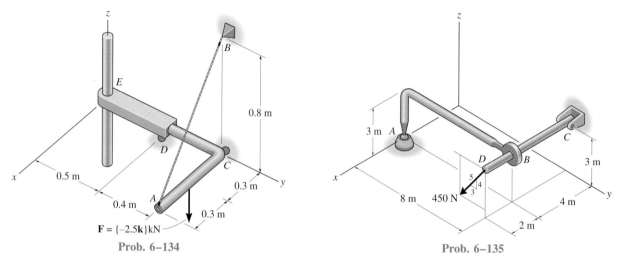

Prob. 6–134

Prob. 6–135

REVIEW PROBLEMS

*6–136.   Determine the horizontal and vertical components of force that the pins at *A* and *C* exert on the two-bar mechanism.

6–137.   Determine the horizontal and vertical components of force at pins *A* and *C* of the two-member frame.

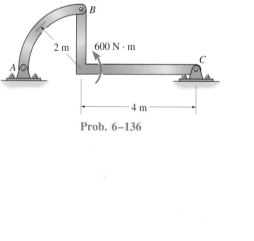

Prob. 6–136

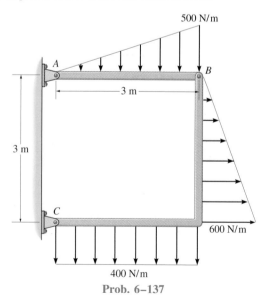

Prob. 6–137

6–138.   Determine the force in each member of the truss and state if the members are in tension or compression.

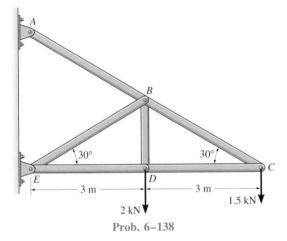

Prob. 6–138

6–139.   The compound beam is fixed at *A* and supported by a rocker at *B* and *C*. There are pins at *D* and *E*. Determine the reactions at the supports.

***6–140.**   Determine the force in each member of the truss and state if the members are in tension or compression.

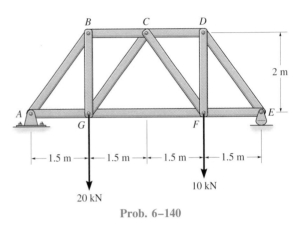

Prob. 6–140

6–141.   The pipe cutter is clamped around the pipe *P*. If the cutting wheel at *A* exerts a normal force of $F_A = 80$ N on the pipe, determine the normal forces of wheels *B* and *C* on the pipe. Also compute the pin reaction on the wheel at *C*. The three wheels each have a radius of 7 mm and the pipe has an outer radius of 10 mm.

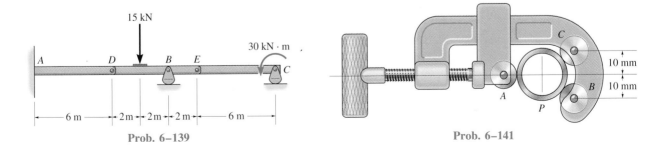

Prob. 6–139

Prob. 6–141

6–142. Determine the resultant forces at pins B and C on member ABC of the four-member frame.

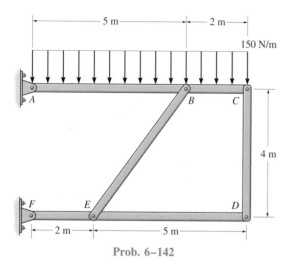

Prob. 6–142

6–143. The mechanism consists of identical meshed gears A and B and arms which are fixed to the gears. The spring attached to the ends of the arms has an unstretched length of 100 mm and a stiffness of $k = 250$ N/m. If a torque of $M = 6$ N · m is applied to gear A, determine the angle θ through which each arm rotates. The gears are each pinned to fixed supports at their centers.

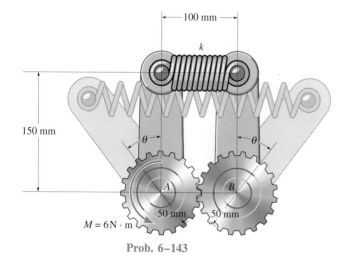

Prob. 6–143

*6–144.** The mechanism consists of identical meshed gears A and B and arms which are fixed to the gears. If a torque of $M = 6$ N · m is applied to gear A as shown, determine the required stiffness k of the spring so that each arm rotates $\theta = 30°$. The gears are each pinned to fixed supports at their centers. The spring has an unstretched length of 100 mm.

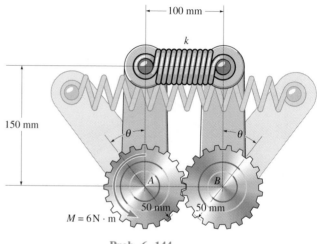

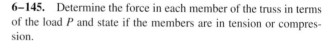

Prob. 6–144

6–145. Determine the force in each member of the truss in terms of the load P and state if the members are in tension or compression.

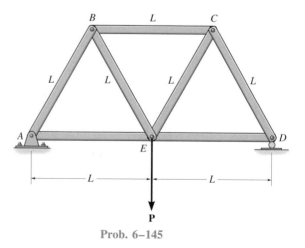

Prob. 6–145

6–146. Each member of the truss is uniform and has a weight W. Remove the external force **P** and determine the approximate force in each member due to the weight of the truss. State if the members are in tension or compression. Solve the problem by *assuming* the weight of each member can be represented as a vertical force, half of which is applied at each end of the member.

*6–148. The man has a weight of 600 N and stands on the uniform plank having a weight of 160 N. Determine the force he exerts on the plank if he pulls with just enough force to lift the plank off the support at B. The plank rests on the smooth surface at A.

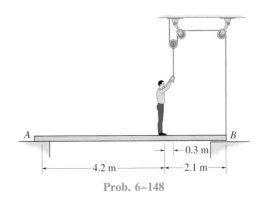

Prob. 6–148

Prob. 6–146

6–147. The spring mechanism is used as a shock absorber for a load applied to the drawbar AB. Determine the equilibrium length of each spring when the 80-N force is applied. Each spring has an unstretched length of 200 mm, and the drawbar slides along the smooth guideposts CG and EF. The bottom springs pass around the guideposts, and the ends of all springs are attached to their respective members.

6–149. The bucket of the backhoe and its contents have a weight of 12 kN and a center of gravity at G. Determine the forces in the hydraulic cylinder AB and in links AC and AD in order to hold the load in the position shown. The bucket is pinned at E. Hint: AB, AC, and AD are all two-force members.

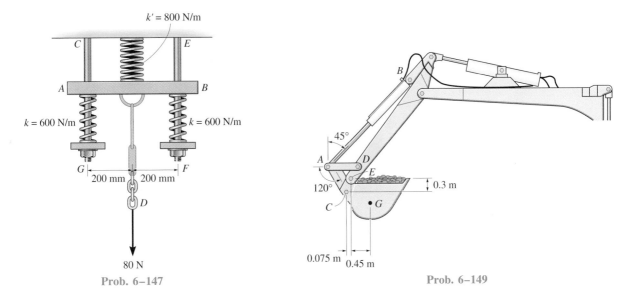

Prob. 6–147

Prob. 6–149

The design and analysis of any structural member requires knowledge of the internal loadings acting within it, not only when it is in place and subjected to service loads, but also when it is being hoisted as shown here. In this chapter we will discuss how engineers determine these loadings.

7

Internal Forces

In this chapter we will develop a technique for finding the internal loading at specific *points* within a structural member, and then we will generalize this method to find the point-to-point variation of loading along the axis of the member. A graph showing this variation of internal load will allow us to find the critical points where the *maximum* internal loading occurs. The analysis of cables will be treated in the last part of the chapter.

7.1 Internal Forces Developed in Structural Members

The design of a structural member requires an investigation of the loadings acting *within* the member which are necessary to balance the loadings acting external to it. The *method of sections* can be used for this purpose. In order to illustrate the procedure, consider the "simply supported" beam shown in Fig. 7–1a, which is subjected to the forces $\mathbf{F}_1$ and $\mathbf{F}_2$. The *support reactions* $\mathbf{A}_x$, $\mathbf{A}_y$, and $\mathbf{B}_y$ can be determined by applying the equations of equilibrium, using the free-body diagram of the *entire beam*, Fig. 7–1b. If the *internal loadings* at point C are to be determined, it is necessary to pass an imaginary section through the beam, cutting it into two segments at that point, Fig. 7–1a. Doing this "exposes" the internal loadings as *external* on the free-body diagram of

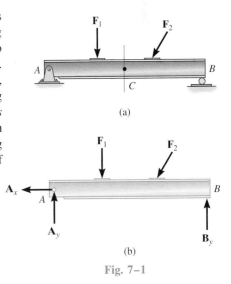

(a)

(b)

Fig. 7–1

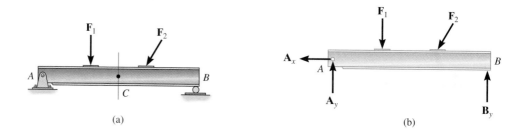

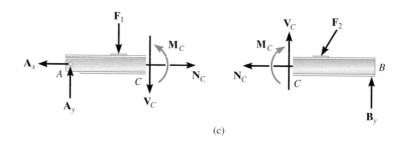

(a)

(b)

(c)

Fig. 7–1

each segment, Fig. 7–1c. Since both segments (*AC* and *CB*) were prevented from translating and rotating *before* the beam was sectioned, equilibrium of each segment is maintained if rectangular force components N_C and V_C and a resultant couple moment M_C are developed at the cut section. Note that these loadings must be equal in magnitude and opposite in direction on each of the segments (Newton's third law). The magnitude of each unknown can now be determined by applying the three equations of equilibrium to either segment *AC* or *CB*. A *direct solution* for N_C is obtained by applying $\Sigma F_x = 0$; V_C is obtained directly from $\Sigma F_y = 0$; and M_C is determined by summing moments about point *C*, $\Sigma M_C = 0$, in order to eliminate the moments of the unknowns N_C and V_C.

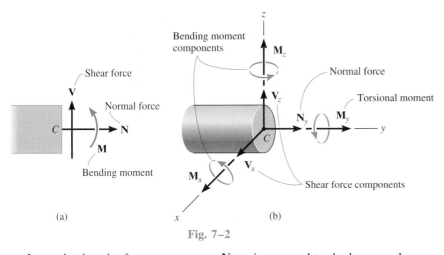

Bending moment components

M_z

V_z

Shear force

Normal force

V

N_y M_y

Normal force

C N

Torsional moment

M

C

y

Bending moment

M_x V_x

Shear force components

(a)

x

(b)

Fig. 7–2

In mechanics, the force components **N,** acting normal to the beam at the cut section, and **V,** acting tangent to the section, are termed the *normal or axial force* and the *shear force,* respectively. The couple moment **M** is referred to as the *bending moment,* Fig. 7–2a. In three dimensions, a general internal force and couple moment resultant will act at the section. The *x, y, z* components of these loadings are shown in Fig. 7–2b. Here N_y is the *normal force,* V_x and V_z are *shear force components,* M_y is a *torsional or twisting moment,* and M_x and M_z are *bending moment components.* For most applications, these *resultant loadings* will act at the geometric center or centroid (*C*) of the section's cross-sectional area. Although the magnitude for each loading generally will be different at various points along the axis of the member, the method of sections can always be used to determine their values.

Free-Body Diagrams. Since frames and machines are composed of *multiforce members,* each of these members will generally be subjected to internal normal, shear, and bending loadings. For example, consider the frame shown in Fig. 7–3a. If the blue section is passed through the frame to determine the internal loadings at points *H, G,* and *F,* the resulting free-body diagram of the top portion of this section is shown in Fig. 7–3b. At each point where a member is sectioned there is an unknown normal force, shear force, and bending moment. As a result, we cannot apply the *three* equations of equilibrium to this section in order to obtain these *nine unknowns.** Instead, to solve this problem we must *first dismember* the frame and determine the reactions at the connections of the members using the techniques of Sec. 6.6. Once this is done, *each member* may then be sectioned at its appropriate point and the three equations of equilibrium can be applied to determine **N, V,** and **M.** For example, the free-body diagram of segment *DG,* Fig. 7–3c, can be used to determine the internal loadings at *G* provided the reactions of the pin, **D**$_x$ and **D**$_y$, are known.

*Recall that this method of analysis worked well for trusses, since truss members are *straight two-force members* which support only an axial or normal load.

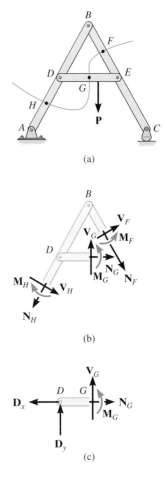

(a)

(b)

(c)

Fig. 7–3

PROCEDURE FOR ANALYSIS

The following procedure provides a means for applying the method of sections to determine the internal loadings at a specific location in a member.

Support Reactions.   Before the member is "cut" or sectioned, it may first be necessary to determine the member's support reactions, so that the equilibrium equations are used only to solve for the internal loadings when the member is sectioned. If the member is part of a frame or machine, the reactions at its connections are determined using the methods of Sec. 6.6.

Free-Body Diagram.   Keep all distributed loadings, couple moments, and forces acting on the member in their *exact location,* then pass an imaginary section through the member, perpendicular to its axis at the point where the internal loading is to be determined. Draw a free-body diagram of one of the "cut" segments on either side of the section and indicate the x, y, z components of the force and couple moment resultants at the section. In particular, if the member is subjected to a *coplanar* system of forces, only **N, V,** and **M** act at the section. In many cases it may be possible to tell by inspection the proper sense of the unknown loadings; however, if this seems difficult, the sense can be assumed.

Equations of Equilibrium.   Apply the equations of equilibrium to obtain the unknown internal loadings. Generally, moments should be summed at the section about axes passing through the *centroid* or geometric center of the member's cross-sectional area, in order to eliminate the unknown normal and shear forces and thereby obtain direct solutions for the moment components. If the solution of the equilibrium equations yields a negative scalar, the assumed sense of the quantity is opposite to that shown on the free-body diagram.

The following examples numerically illustrate this procedure.

Example 7–1

The bar is fixed at its end and is loaded as shown in Fig. 7–4a. Determine the internal normal force at points B and C.

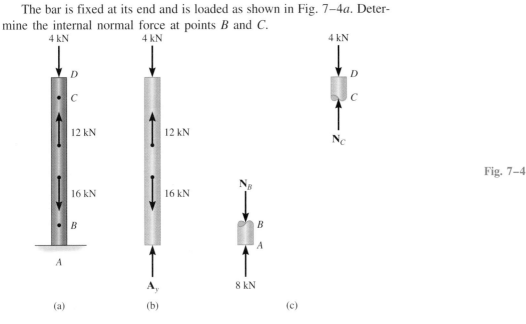

Fig. 7–4

SOLUTION

Support Reactions.   A free-body diagram of the entire bar is shown in Fig. 7–4b. By inspection, only a normal force $\mathbf{A}_y$ acts at the fixed support, since the loads are applied symmetrically along the bar's axis.

$$+\uparrow \Sigma F_y = 0; \qquad A_y - 16 \text{ kN} + 12 \text{ kN} - 4 \text{ kN} = 0 \qquad A_y = 8 \text{ kN}$$

Free-Body Diagrams.   The internal forces at B and C will be found using the free-body diagrams of the sectioned bar shown in Fig. 7–4c. In particular, segments AB and DC will be chosen here, since they contain the *least* number of forces.

Equations of Equilibrium
   Segment AB

$$+\uparrow \Sigma F_y = 0; \qquad 8 \text{ kN} - N_B = 0 \qquad N_B = 8 \text{ kN} \qquad \textit{Ans.}$$

   Segment DC

$$+\uparrow \Sigma F_y = 0; \qquad N_C - 4 \text{ kN} = 0 \qquad N_C = 4 \text{ kN} \qquad \textit{Ans.}$$

Try working this problem in the following manner: Determine N_B from segment BD. (Note that this approach does *not require* solution for the support reaction at A.) Using the result for N_B, isolate segment BC to determine N_C.

Example 7–2

The circular shaft is subjected to three concentrated torques as shown in Fig. 7–5a. Determine the internal torques at points B and C.

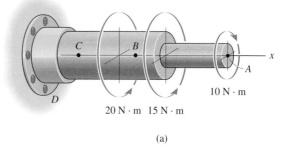

(a)

SOLUTION

Support Reactions.   Since the shaft is subjected only to collinear torques, a torque reaction occurs at the support, Fig. 7–5b. Using the right-hand rule to define the positive directions of the torques, we require

$$\Sigma M_x = 0; \quad -10\ \text{N} \cdot \text{m} + 15\ \text{N} \cdot \text{m} + 20\ \text{N} \cdot \text{m} - T_D = 0$$
$$T_D = 25\ \text{N} \cdot \text{m}$$

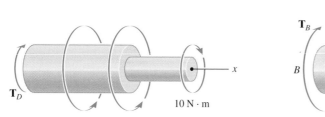

(b)

(c)

Fig. 7–5

Free-Body Diagrams.   The internal torques at B and C will be found using the free-body diagrams of the shaft segments AB and CD shown in Fig. 7–5c.

Equations of Equilibrium.   Applying the equation of moment equilibrium along the shaft's axis, we have

Segment AB

$$\Sigma M_x = 0; \quad -10\ \text{N} \cdot \text{m} + 15\ \text{N} \cdot \text{m} - T_B = 0 \qquad T_B = 5\ \text{N} \cdot \text{m} \qquad \textit{Ans.}$$

Segment CD

$$\Sigma M_x = 0; \qquad T_C - 25\ \text{N} \cdot \text{m} = 0 \qquad T_C = 25\ \text{N} \cdot \text{m} \qquad \textit{Ans.}$$

Try to solve for T_C by using segment CA. Note that this approach does not require a solution for the support reaction at D.

Example 7–3

The beam supports the loading shown in Fig. 7–6a. Determine the internal normal force, shear force, and bending moment acting just to the left, point B, and just to the right, point C, of the 6-kN force.

SOLUTION

Support Reactions.   The free-body diagram of the beam is shown in Fig. 7–6b. When determining the *external reactions*, realize that the 9-kN · m couple moment is a free vector and therefore it can be placed *anywhere* on the free-body diagram of the entire beam. Here we will only determine $\mathbf{A}_y$, since segments AB and AC will be used for the analysis.

$$\zeta+\Sigma M_D = 0; \quad 9 \text{ kN} \cdot \text{m} + (6 \text{ kN})(6 \text{ m}) - A_y(9 \text{ m}) = 0$$
$$A_y = 5 \text{ kN}$$

Free-Body Diagrams.   The free-body diagrams of the left segments AB and AC of the beam are shown in Figs. 7–6c and 7–6d. In this case the 9-kN · m couple moment is *not included* on these diagrams, since it must be kept in its *original position* until *after* the section is made and the appropriate body isolated. In other words, the internal loadings, when determined from the left segments of the beam, are not influenced by the effect of the couple moment since this moment does not actually act on these segments.

Equations of Equilibrium
   Segment AB

$\xrightarrow{+}\Sigma F_x = 0;$	$N_B = 0$	*Ans.*
$+\uparrow\Sigma F_y = 0;$	$5 \text{ kN} - V_B = 0$ $V_B = 5 \text{ kN}$	*Ans.*
$\zeta+\Sigma M_B = 0;$	$-(5 \text{ kN})(3 \text{ m}) + M_B = 0$ $M_B = 15 \text{ kN} \cdot \text{m}$	*Ans.*

   Segment AC

$\xrightarrow{+}\Sigma F_x = 0;$	$N_C = 0$	*Ans.*
$+\uparrow\Sigma F_y = 0;$	$5 \text{ kN} - 6 \text{ kN} + V_C = 0$ $V_C = 1 \text{ kN}$	*Ans.*
$\zeta+\Sigma M_C = 0;$	$-(5 \text{ kN})(3 \text{ m}) + M_C = 0$ $M_C = 15 \text{ kN} \cdot \text{m}$	*Ans.*

Here the moment arm for the 5-kN force in both cases is approximately 3 m, since B and C are "almost" coincident.

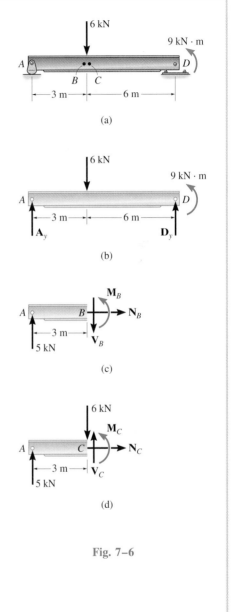

(a)

(b)

(c)

(d)

Fig. 7–6

Example 7–4

Determine the internal normal force, shear force, and bending moment acting at point B of the two-member frame shown in Fig. 7–7a.

SOLUTION

Support Reactions. A free-body diagram of each member is shown in Fig. 7–7b. Since CD is a two-force member, the equations of equilibrium need to be applied only to member AC.

$$\zeta + \Sigma M_A = 0; \quad -400 \text{ N}(4 \text{ m}) + (\tfrac{3}{5})F_{DC}(8 \text{ m}) = 0 \quad F_{DC} = 333.3 \text{ N}$$

$$\xrightarrow{+} \Sigma F_x = 0; \quad -A_x + (\tfrac{4}{5})(333.3 \text{ N}) = 0 \quad A_x = 266.7 \text{ N}$$

$$+\uparrow \Sigma F_y = 0; \quad A_y - 400 \text{ N} + \tfrac{3}{5}(333.3 \text{ N}) = 0 \quad A_y = 200 \text{ N}$$

Free-Body Diagrams. Passing an imaginary section perpendicular to the axis of member AC through point B yields the free-body diagrams of segments AB and BC shown in Fig. 7–7c. When constructing these diagrams it is important to keep the distributed loading exactly as it is until *after* the section is made. Only then can it be replaced by a single resultant force. Why? Also, notice that $\mathbf{N}_B$, $\mathbf{V}_B$, and $\mathbf{M}_B$ act with equal magnitude but opposite direction on each segment—Newton's third law.

Equations of Equilibrium. Applying the equations of equilibrium to segment AB, we have

$$\xrightarrow{+} \Sigma F_x = 0; \quad N_B - 266.7 \text{ N} = 0 \quad N_B = 267 \text{ N} \quad Ans.$$

$$+\uparrow \Sigma F_y = 0; \quad 200 \text{ N} - 200 \text{ N} - V_B = 0 \quad V_B = 0 \quad Ans.$$

$$\zeta + \Sigma M_B = 0; \quad M_B - 200 \text{ N}(4 \text{ m}) + 200 \text{ N}(2 \text{ m}) = 0$$

$$M_B = 400 \text{ N} \cdot \text{m} \quad Ans.$$

As an exercise, try to obtain these same results using segment BC.

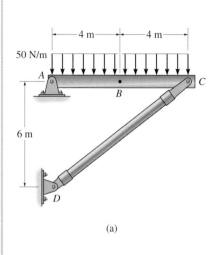

(a)

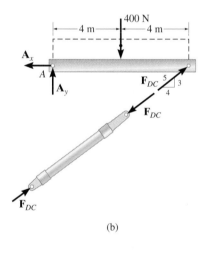

(b)

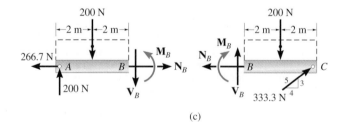

(c)

Fig. 7–7

Example 7–5

Determine the normal force, shear force, and bending moment acting at point E of the frame loaded as shown in Fig. 7–8a.

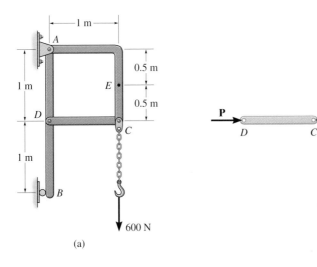

(a)

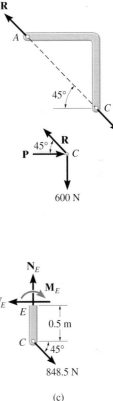

(b)

Fig. 7–8

(c)

SOLUTION

Support Reactions.   By inspection, members AC and CD are two-force members, Fig. 7–8b. In order to determine the internal loadings at E, we must first determine the force $\mathbf{R}$ at the end of member AC. To do this we must analyze the equilibrium of the pin at C. Why?

Summing forces in the vertical direction on the pin, Fig. 7–8b, we have

$$+\uparrow \Sigma F_y = 0; \qquad R \sin 45° - 600 \text{ N} = 0 \qquad R = 848.5 \text{ N}$$

Free-Body Diagram.   The free-body diagram of segment CE is shown in Fig. 7–8c.

Equations of Equilibrium

$$\xrightarrow{+} \Sigma F_x = 0; \quad 848.5 \cos 45° \text{ N} - V_E = 0 \qquad V_E = 600 \text{ N} \qquad Ans.$$
$$+\uparrow \Sigma F_y = 0; \ -848.5 \sin 45° \text{ N} + N_E = 0 \qquad N_E = 600 \text{ N} \qquad Ans.$$
$$\zeta + \Sigma M_E = 0; \ 848.5 \cos 45° \text{ N}(0.5 \text{ m}) - M_E = 0 \ M_E = 300 \text{ N} \cdot \text{m} \qquad Ans.$$

As in Example 6–21, member AC should be *straight* (from A to C) so that bending within the member is eliminated, and the internal force would only create tension in the member.

Example 7–6

A force of $\mathbf{F} = \{-3\mathbf{i} + 7\mathbf{j} - 4\mathbf{k}\}$ kN acts at the corner of a beam extended from a fixed wall as shown in Fig. 7–9a. Determine the internal loadings at A.

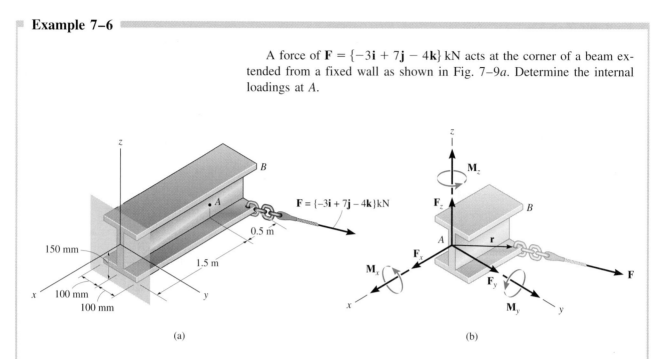

(a)   (b)

SOLUTION   Fig. 7–9

This problem will be solved by considering segment AB of the beam, since it does *not* involve the support reactions.

Free-Body Diagram.   The free-body diagram of segment AB is shown in Fig. 7–9b. The components of the resultant force $\mathbf{F}_A$ and moment $\mathbf{M}_A$ pass through the *centroid* or geometric center of the cross-sectional area at A.

Equations of Equilibrium

$$\Sigma \mathbf{F} = 0; \quad \mathbf{F}_A + \mathbf{F} = 0 \quad \mathbf{F}_A - 3\mathbf{i} + 7\mathbf{j} - 4\mathbf{k} = 0$$

$$\mathbf{F}_A = \{3\mathbf{i} - 7\mathbf{j} + 4\mathbf{k}\} \text{ kN} \qquad \textit{Ans.}$$

$$\Sigma \mathbf{M}_A = 0; \quad \mathbf{M}_A + \mathbf{r} \times \mathbf{F} = \mathbf{M}_A + \begin{vmatrix} \mathbf{i} & \mathbf{j} & \mathbf{k} \\ -0.5 & 0.1 & -0.15 \\ -3 & 7 & -4 \end{vmatrix} = 0$$

$$\mathbf{M}_A = \{-0.650\mathbf{i} + 1.55\mathbf{j} + 3.20\mathbf{k}\} \text{ kN} \cdot \text{m} \qquad \textit{Ans.}$$

Here $\mathbf{F}_x = \{3\mathbf{i}\}$ kN represents the normal force N, whereas $\mathbf{F}_y = \{-7\mathbf{j}\}$ kN and $\mathbf{F}_z = \{4\mathbf{k}\}$ kN are components of the shear force $V = \sqrt{F_y^2 + F_z^2}$. Also, the torsional moment is $\mathbf{M}_x = \{-0.65\mathbf{i}\}$ kN $\cdot$ m, and the bending moment is determined from its components $\mathbf{M}_y = \{1.55\mathbf{j}\}$ kN $\cdot$ m and $\mathbf{M}_z = \{3.20\mathbf{k}\}$ kN $\cdot$ m; i.e., $M_b = \sqrt{M_y^2 + M_z^2}$.

PROBLEMS

7–1.   Three torques act on the shaft. Determine the torques at sections passing through points *A*, *B*, *C*, and *D*.

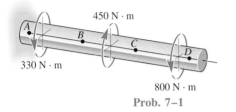

450 N · m

A

B

C

D

330 N · m

800 N · m

Prob. 7–1

7–2.   The shaft is supported by smooth bearings at *A* and *B* and subjected to the torques shown. Determine the torques at sections passing through points *C*, *D*, and *E*.

A   400 N · m

550 N · m

950 N · m

B

C

D

E

Prob. 7–2

7–3.   Three torques act on the shaft as shown. Determine the torques at sections passing through points *A*, *B*, *C*, and *D*.

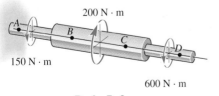

200 N · m

A

B

C

D

150 N · m

600 N · m

Prob. 7–3

*7–4.   The shaft is supported by a journal bearing at *A* and a thrust bearing at *B*. Determine the normal force, shear force, and moment at a section passing through (a) point *C*, which is just to the right of the bearing at *A*, and (b) point *D*, which is just to the left of the 3000-N force.

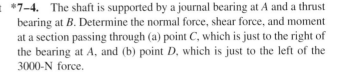

2500 N

3000 N

75 N/m

C

D

B

A

6 m

12 m

2 m

Prob. 7–4

7–5.   Determine the normal force, shear force, and moment at a section passing through point *C*. Assume the support at *A* can be approximated by a pin and *B* as a roller.

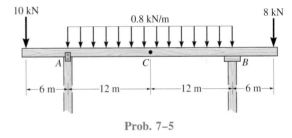

10 kN

0.8 kN/m

8 kN

A

C

B

6 m

12 m

12 m

6 m

Prob. 7–5

7–6.   Determine the normal force, shear force, and moment at a section passing through point *C*.

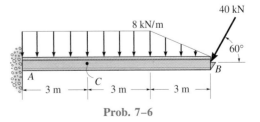

40 kN

8 kN/m

60°

A

C

B

3 m

3 m

3 m

Prob. 7–6

7–7. Determine the normal force, shear force, and moment at a section passing through point D. Take $w = 150$ N/m.

*7–8. The beam AB will fail if the maximum internal moment at D reaches 800 N · m or the normal force in member BC becomes 1500 N. Determine the largest load w it can support.

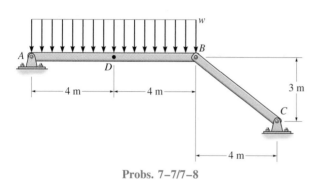

Probs. 7–7/7–8

7–9. Determine the shear force and moment acting at a section passing through point C in the beam.

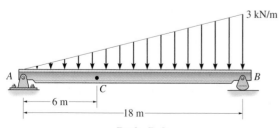

Prob. 7–9

7–10. The cantilevered rack is used to support each end of a smooth pipe that has a total weight of 300 N. Determine the normal force, shear force, and moment that act in the arm at its fixed support A along a vertical section.

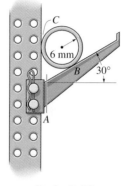

Prob. 7–10

7–11. The A-frame ladder rests on the smooth floor. If a 700-N man stands uniformly on it at D, determine the moment each of the two lock joints at C must resist to hold the ladder stationary. Assume the man exerts only a vertical reaction at D and no other force on the ladder.

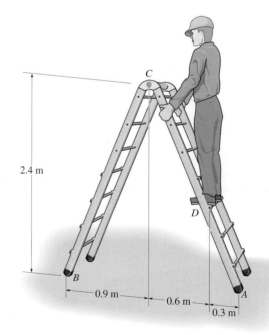

Prob. 7–11

*7–12. Determine the normal force, shear force, and moment at a section passing through point D of the two-member frame.

7–14. Determine the normal force, shear force, and moment at a section passing through point D of the two-member frame.

7–15. Determine the normal force, shear force, and moment at a section passing through point E of the two-member frame.

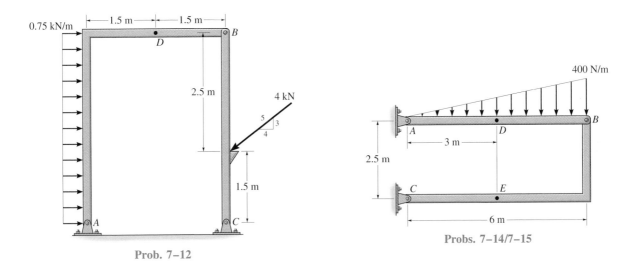

Prob. 7–12

Probs. 7–14/7–15

7–13. The boom DF of the jib crane and the column DE have a uniform weight of 1000 N/m. If the hoist and load weigh 1.8 kN, determine the normal force, shear force, and moment in the crane at sections passing through points A, B, and C.

*7–16. Determine the normal force, shear force, and moment at a section passing through point C. Take $P = 8$ kN.

7–17. The cable will fail when subjected to a tension of 2 kN. Determine the largest vertical load P the frame will support and calculate the internal normal force, shear force, and moment at a section passing through point C for this loading.

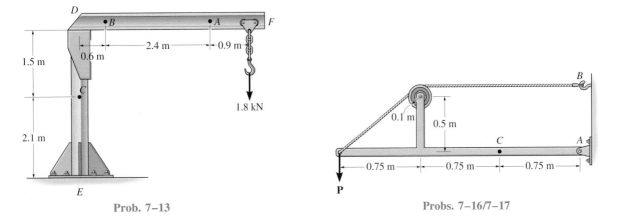

Prob. 7–13

Probs. 7–16/7–17

7–18. The wishbone construction of the power pole supports the three lines, each exerting a force of 3.56 kN on the bracing struts. If the struts are pin-connected at A, B, and C, determine the normal force, shear force, and moment at sections passing through points D, E, and F.

7–21. Determine the normal force, shear force, and moment in the beam at sections passing through points D and E. Point E is just to the right of the 4-kN load.

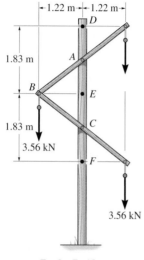

Prob. 7–18

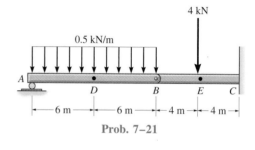

Prob. 7–21

7–19. Determine the normal force, shear force, and moment acting at a section passing through point C.

*__7–20.__ Determine the normal force, shear force, and moment acting at a section passing through point D.

7–22. The 9-kN force is supported by the floor panel DE, which in turn is simply supported at its ends by floor beams. These beams transmit their loads to the girder AB. Determine the shear and moment acting at a section passing through point C in the girder.

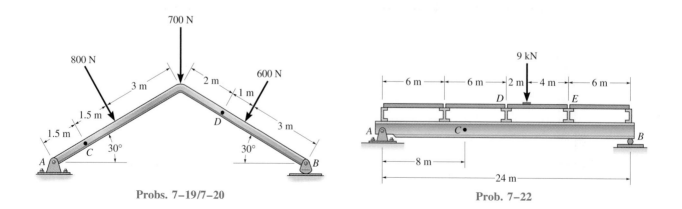

Probs. 7–19/7–20

Prob. 7–22

7–23. Determine the normal force, shear force, and moment in the beam at sections passing through points D and E. Point E is just to the right of the 3-kN load.

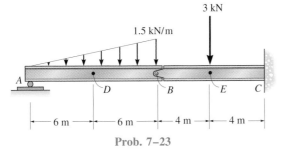

Prob. 7–23

7–26. Determine the normal force, shear force, and moment acting at sections passing through points B and C on the curved rod.

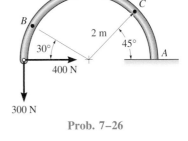

Prob. 7–26

*7–24. Determine the normal force, shear force, and moment at a section passing through point D of the two-member frame.

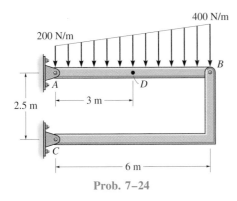

Prob. 7–24

7–27. The semicircular arch is subjected to a uniform distributed load along its axis of w_0 per unit length. Determine the normal force, shear force, and moment in the arch at $\theta = 45°$.

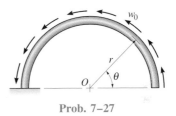

Prob. 7–27

7–25. Determine the normal force, shear force, and moment at sections passing through points E and F. Member BC is pinned at B and there is a smooth slot in it at C. The pin at C is fixed to member CD.

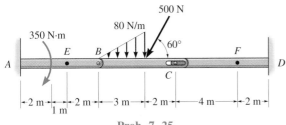

Prob. 7–25

*7–28. The distributed loading $w = w_0 \sin \theta$, measured per unit length, acts on the curved rod. Determine the normal force, shear force, and moment in the rod at $\theta = 45°$.

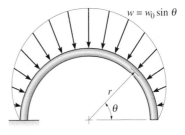

Prob. 7–28

■*7–29. Determine the shear force and moment acting at a section through point C of the beam. For the calculation use Simpson's rule to evaluate the integrals.

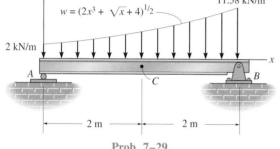

$w = (2x^3 + \sqrt{x} + 4)^{1/2}$

11.58 kN/m

2 kN/m

2 m   2 m

Prob. 7–29

7–30. The man has a weight of 2.5-kN and a center of gravity at G. If the bench upon which he is sitting is fix-connected to the center support, determine the internal x, y, z components of internal loading at sections passing through point A and the base B.

7–31. Determine the x, y, z components of internal loading at a section passing through point C in the pipe assembly. Neglect the weight of the pipe. Take $\mathbf{F}_1 = \{350\mathbf{j} - 400\mathbf{k}\}$ N and $\mathbf{F}_2 = \{150\mathbf{i} - 300\mathbf{k}\}$ N.

*7–32. Determine the x, y, z components of internal loading at a section passing through point C in the pipe assembly. Neglect the weight of the pipe. Take $\mathbf{F}_1 = \{-80\mathbf{i} + 200\mathbf{j} - 300\mathbf{k}\}$ N and $\mathbf{F}_2 = \{250\mathbf{i} - 150\mathbf{j} - 200\mathbf{k}\}$ N.

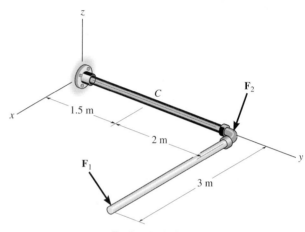

Probs. 7–31/7–32

7–33. Determine the x, y, z components of internal loading at a section passing through point D of the rod. Take $\mathbf{F} = \{-7\mathbf{i} + 12\mathbf{j} - 5\mathbf{k}\}$ kN. The supports at A, B, and C are journal bearings.

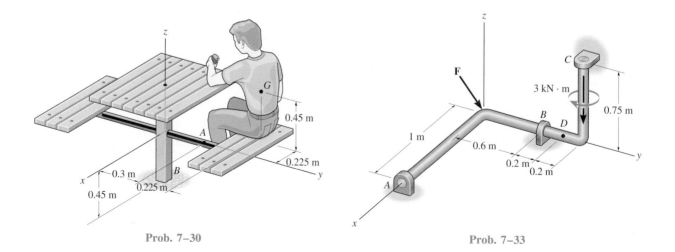

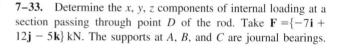

Prob. 7–30

Prob. 7–33

*7.2 Shear and Moment Equations and Diagrams

Beams are structural members which are designed to support loadings applied perpendicular to their axes. In general, beams are long, straight bars having a constant cross-sectional area. Often they are classified as to how they are supported. For example, a *simply supported beam* is pinned at one end and roller-supported at the other, Fig. 7–10, whereas a *cantilevered beam* is fixed at one end and free at the other. The actual design of a beam requires a detailed knowledge of the *variation* of the internal shear force V and bending moment M acting at *each point* along the axis of the beam. After this force and bending-moment analysis is complete, one can then use the theory of mechanics of materials and an appropriate engineering design code to determine the beam's required cross-sectional area.

The *variations* of V and M as a function of the position x along the beam's axis can be obtained by using the method of sections discussed in Sec. 7.1. Here, however, it is necessary to section the beam at an arbitrary distance x from one end rather than at a specified point. If the results are plotted, the graphical variations of V and M as a function of x are termed the *shear diagram* and *bending-moment diagram,* respectively.

In general, the internal shear and bending-moment functions will be discontinuous, or their slope will be discontinuous at points where a distributed load changes or where concentrated forces or couple moments are applied. Because of this, these functions must be determined for *each segment* of the beam located between any two discontinuities of loading. For example, sections located at x_1, x_2, and x_3 will have to be used to describe the variation of V and M throughout the length of the beam in Fig. 7–10. These functions will be valid *only* within regions from O to a for x_1, from a to b for x_2, and from b to L for x_3.

The internal normal force will not be considered in the following discussion for two reasons. In most cases, the loads applied to a beam act perpendicular to the beam's axis and hence produce only an internal shear force and bending moment. For design purposes, the beam's resistance to shear, and particularly to bending, is more important than its ability to resist a normal force.

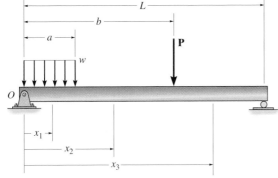

Fig. 7–10

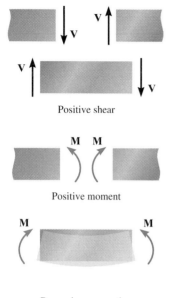

Positive shear

Positive moment

Beam sign convention
Fig. 7–11

Sign Convention. Before presenting a method for determining the shear and bending moment as functions of x and later plotting these functions (shear and bending-moment diagrams), it is first necessary to establish a *sign convention* so as to define a "positive" and "negative" shear force and bending moment acting in the beam. [This is analogous to assigning coordinate directions x positive to the right and y positive upward when plotting a function $y = f(x)$.] Although the choice of a sign convention is arbitrary, here we will choose the one used for the majority of engineering applications. It is illustrated in Fig. 7–11. Here the positive directions are denoted by an internal *shear force* that causes *clockwise rotation* of the member on which it acts, and an internal *moment* that causes *compression, or pushing on the upper part* of the member. Also, positive moment would tend to bend the member if it were elastic, concave upward. Loadings that are opposite to these are considered negative.

PROCEDURE FOR ANALYSIS

The following procedure provides a method for constructing the shear and bending-moment diagrams for a beam.

Support Reactions.   Determine all the reactive forces and couple moments acting on the beam, and resolve all the forces into components acting perpendicular and parallel to the beam's axis.

Shear and Moment Functions.   Specify separate coordinates x having an origin at the beam's *left end* and extending to regions of the beam *between* concentrated forces and/or couple moments, or where there is no discontinuity of distributed loading. Section the beam perpendicular to its axis at each distance x. On the free-body diagram, be sure **V** and **M** are shown acting in their *positive sense,* in accordance with the sign convention given in Fig. 7–11. V is obtained by summing forces perpendicular to the beam's axis, and M is obtained by summing moments about the sectioned end of the segment.

Shear and Moment Diagrams.   Plot the shear diagram (V versus x) and the moment diagram (M versus x). If computed values of the functions describing V and M are *positive,* the values are plotted above the x axis, whereas *negative* values are plotted below the x axis. Generally, it is convenient to plot the shear and bending-moment diagrams directly below the free-body diagram of the beam.

The following examples illustrate this procedure numerically.

Example 7–7

Draw the shear and bending-moment diagrams for the shaft shown in Fig. 7–12a. The support at A is a thrust bearing and at C a journal bearing.

SOLUTION

Support Reactions. The support reactions have been computed, as shown on the shaft's free-body diagram, Fig. 7–12d.

Shear and Moment Functions. The shaft is sectioned at an arbitrary distance x from point A, extending within the region AB, and the free-body diagram of the left segment is shown in Fig. 7–12b. The unknowns **V** and **M** are assumed to act in the *positive sense* on the right-hand face of the segment according to the established sign convention. Why? Applying the equilibrium equations yields

$$+\uparrow \Sigma F_y = 0; \qquad\qquad V = 2.5 \text{ kN} \tag{1}$$

$$\zeta+\Sigma M = 0; \qquad\qquad M = 2.5x \text{ kN} \cdot \text{m} \tag{2}$$

A free-body diagram for a left segment of the shaft extending a distance x within the region BC is shown in Fig. 7–12c. As always, **V** and **M** are shown acting in the positive sense. Hence,

$$+\uparrow \Sigma F_y = 0; \qquad 2.5 \text{ kN} - 5 \text{ kN} - V = 0$$

$$V = -2.5 \text{ kN} \tag{3}$$

$$\zeta+\Sigma M = 0; \qquad M + 5 \text{ kN}(x - 2 \text{ m}) - 2.5 \text{ kN}(x) = 0$$

$$M = (10 - 2.5x) \text{ kN} \cdot \text{m} \tag{4}$$

Shear and Moment Diagrams. When Eqs. 1 through 4 are plotted within the regions in which they are valid, the shear and bending-moment diagrams shown in Fig. 7–12d are obtained. The shear diagram indicates that the internal shear force is always 2.5 kN (positive) within shaft segment AB. Just to the right of point B, the shear force changes sign and remains at a constant value of −2.5 kN for segment BC. The moment diagram starts at zero, increases linearly to point B at x = 2 m, where $M_{max} = 2.5 \text{ kN}(2 \text{ m}) = 5 \text{ kN} \cdot \text{m}$, and thereafter decreases back to zero.

It is seen in Fig. 7–12d that the graph of the shear and moment diagrams is discontinuous at points of concentrated force, i.e., points A, B, and C. For this reason, as stated earlier, it is necessary to express both the shear and bending-moment functions separately for regions between concentrated loads. It should be realized, however, that all loading discontinuities are mathematical, arising from the *idealization of a concentrated force and couple moment*. Physically, loads are always applied over a finite area, and if this load variation could be accounted for, the shear and bending-moment diagrams would actually be continuous over the shaft's entire length.

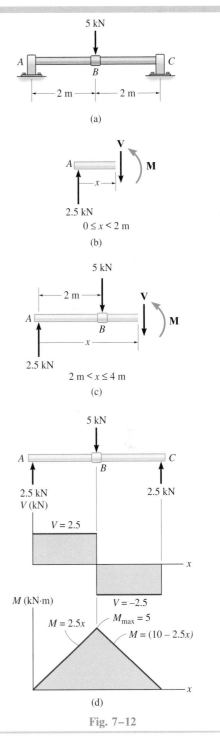

(a)

(b)

(c)

(d)

Fig. 7–12

Example 7–8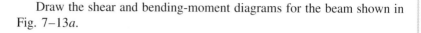

Draw the shear and bending-moment diagrams for the beam shown in Fig. 7–13a.

SOLUTION

Support Reactions. The support reactions have been computed as shown on the beam's free-body diagram, Fig. 7–13c.

Shear and Moment Functions. A free-body diagram for a left segment of the beam having a length x is shown in Fig. 7–13b. The distributed loading acting on this segment has an intensity of $\frac{2}{3}x$ at its end and is replaced by a resultant force *after* the segment is isolated as a free-body diagram. The *magnitude* of the resultant force is equal to $\frac{1}{2}(x)(\frac{2}{3}x) = \frac{1}{3}x^2$. This force *acts through the centroid* of the distributed loading area, a distance $\frac{1}{3}x$ from the right end. Applying the two equations of equilibrium yields

$$+\uparrow \Sigma F_y = 0; \qquad 9 - \frac{1}{3}x^2 - V = 0$$

$$V = \left(9 - \frac{x^2}{3}\right) \text{ kN} \qquad (1)$$

$$\big\downarrow +\Sigma M = 0; \qquad M + \frac{1}{3}x^2\left(\frac{x}{3}\right) - 9x = 0$$

$$M = \left(9x - \frac{x^3}{9}\right) \text{ kN} \cdot \text{m} \qquad (2)$$

Shear and Moment Diagrams. The shear and bending-moment diagrams shown in Fig. 7–13c are obtained by plotting Eqs. 1 and 2.

The point of *zero shear* can be found using Eq. 1:

$$V = 9 - \frac{x^2}{3} = 0$$

$$x = 5.20 \text{ m}$$

This value of x happens to represent the point on the beam where the *maximum moment* occurs (see Sec. 7.3). Using Eq. (2), we have

$$M_{max} = \left(9(5.20) - \frac{(5.20)^3}{9}\right) \text{ kN} \cdot \text{m}$$

$$= 31.2 \text{ kN} \cdot \text{m}$$

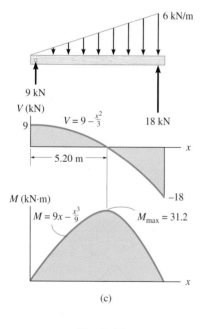

Fig. 7–13

PROBLEMS

For each of the following problems, establish the x axis with the origin at the left side of the beam, and obtain the internal shear and moment as a function of x. Use these results to plot the shear and moment diagrams.

7–34. Draw the shear and moment diagrams for the shaft (a) in terms of the parameters shown; (b) set $P = 9$ kN, $a = 2$ m, $L = 6$ m. There is a thrust bearing at A and a journal bearing at B.

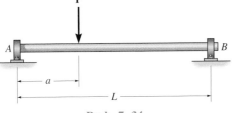

Prob. 7–34

7–35. Draw the shear and moment diagrams for the beam (a) in terms of the parameters shown; (b) set $P = 800$ N, $a = 5$ m, $L = 12$ m.

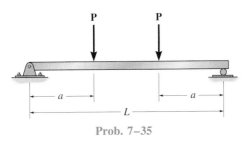

Prob. 7–35

*7–36. Draw the shear and moment diagrams for the beam (a) in terms of the parameters shown; (b) set $M_0 = 500$ N · m, $L = 8$ m.

7–37. If $L = 9$ m, the beam will fail when the maximum shear force is $V_{max} = 5$ kN or the maximum bending moment is $M_{max} = 2$ kN · m. Determine the magnitude M_0 of the largest couple moments it will support.

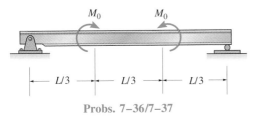

Probs. 7–36/7–37

7–38. The shaft is supported by a thrust bearing at A and a journal bearing at B. Draw the shear and moment diagrams for the shaft (a) in terms of the parameters shown; (b) set $w = 500$ N/m, $L = 10$ m.

7–39. The shaft is supported by a thrust bearing at A and a journal bearing at B. If $L = 10$ m, the shaft will fail when the maximum moment is $M_{max} = 5$ kN · m. Determine the largest uniform distributed load w the shaft will support.

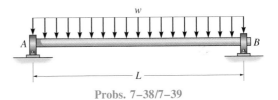

Probs. 7–38/7–39

*7–40. The shaft is supported by a journal bearing at A and a thrust bearing at B. Draw the shear and moment diagrams for the shaft.

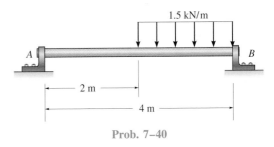

Prob. 7–40

7–41. Draw the shear and moment diagrams for the beam.

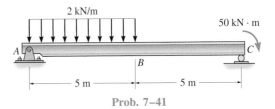

Prob. 7–41

7–42. Draw the shear and moment diagrams for the beam.

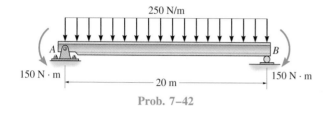

Prob. 7–42

7–43. Draw the shear and moment diagrams for the beam.

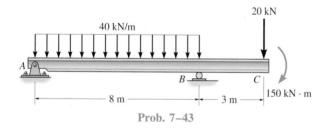

Prob. 7–43

*7–44. Draw the shear and moment diagrams for the beam.

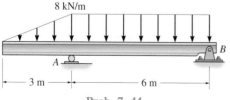

Prob. 7–44

7–45. Draw the shear and moment diagrams for beam *ABC*. Note that there is a pin at *B*. Solve the problem (a) in terms of the parameters shown; (b) set $w = 5$ kN/m, $L = 12$ m.

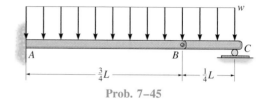

Prob. 7–45

7–46. Draw the shear and moment diagrams for the beam.

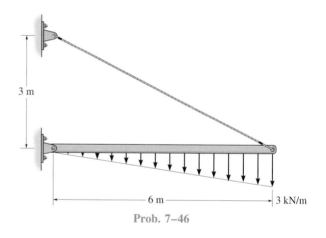

Prob. 7–46

7–47. Determine the distance *a* between the bearings in terms of the shaft's length *L* so that the moment in the *symmetric* shaft is zero at its center.

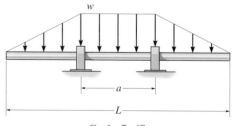

Prob. 7–47

*7–48.   Draw the shear and moment diagrams for the beam (a) in terms of the parameters shown; (b) set w = 250 N/m, L = 12 m.

7–49.   If L = 18 m, the beam will fail when the maximum shear force is V_{max} = 800 N, or the maximum moment is M_{max} = 1200 N · m. Determine the largest intensity w of the distributed loading it will support.

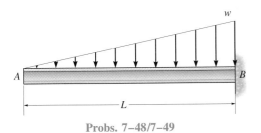

Probs. 7–48/7–49

7–50.   The beam will fail when the maximum internal moment is M_{max}. Determine the position x of the concentrated force **P** and its smallest magnitude that will cause failure.

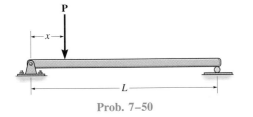

Prob. 7–50

7–51.   Determine the distance a as a fraction of the beam's length L for locating the roller support so that the moment in the beam at B is zero.

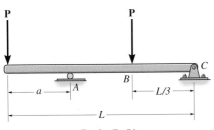

Prob. 7–51

*7–52.   Express the x, y, z components of internal loading in the rod as a function of y, where $0 \le y \le 4$ m.

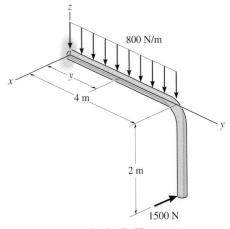

Prob. 7–52

7–53.   Determine the normal force, shear force, and moment in the curved rod as a function of θ.

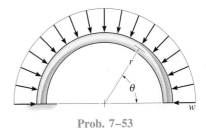

Prob. 7–53

7–54.   Determine the normal force, shear force, and moment in the curved rod as a function of θ.

Prob. 7–54

*7.3 Relations Between Distributed Load, Shear, and Moment

In cases where a beam is subjected to several concentrated forces, couple moments, and distributed loads, the method of constructing the shear and bending-moment diagrams discussed in Sec. 7.2 may become quite tedious. In this section a simpler method for constructing these diagrams is discussed—a method based on differential relations that exist between the load, shear, and bending moment.

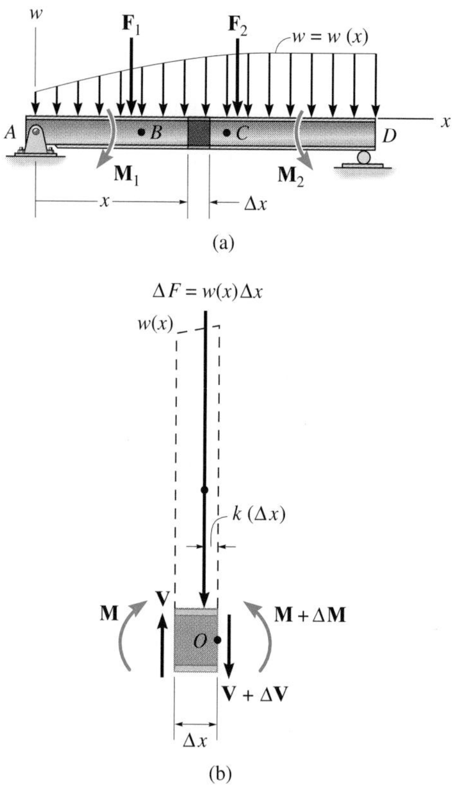

Distributed Load. Consider the beam AD shown in Fig. 7–14a, which is subjected to an arbitrary distributed load $w = w(x)$ and a series of concentrated forces and couple moments. In the following discussion, the *distributed load* will be considered *positive* when the *loading acts downward* as shown. The free-body diagram for a small segment of the beam having a length Δx is shown in Fig. 7–14b. Since this segment has been chosen at a point x along the beam which is *not* subjected to a concentrated force or couple moment, any results obtained will not apply at points of concentrated loading. The internal shear force and bending moment shown on the free-body diagram are assumed to act in the *positive sense* according to the established sign convention, Fig. 7–14b. Note that both the shear force and moment acting on the right-hand face must be increased by a small, finite amount in order to keep the segment in equilibrium. The distributed loading has been replaced by a resultant force $\Delta F = w(x)\,\Delta x$ that acts at a fractional distance $k\,(\Delta x)$ from the right end, where $0 < k < 1$ [for example, if $w(x)$ is *uniform*, $k = \frac{1}{2}$]. Applying the equations of equilibrium, we have

(a)

(b)

Fig. 7–14

$$+\uparrow \Sigma F_y = 0; \qquad V - w(x)\,\Delta x - (V + \Delta V) = 0$$
$$\Delta V = -w(x)\,\Delta x$$

$$\big\backslash +\Sigma M_O = 0; \quad -V\,\Delta x - M + w(x)\,\Delta x[k(\Delta x)] + (M + \Delta M) = 0$$
$$\Delta M = V\,\Delta x - w(x)k(\Delta x)^2$$

Dividing by Δx and taking the limit as $\Delta x \to 0$, these two equations become

$$\frac{dV}{dx} = -w(x)$$

$$\begin{array}{c} \text{Slope of} \\ \text{shear diagram} \end{array} = \begin{array}{c} \text{Negative of distributed} \\ \text{load intensity} \end{array}$$

(7–1)

$$\frac{dM}{dx} = V$$

$$\begin{array}{c} \text{Slope of} \\ \text{moment diagram} \end{array} = \text{Shear}$$

(7–2)

These two equations provide a convenient means for plotting the shear and moment diagrams for a beam. At a specific point in a beam, Eq. 7–1 states that the *slope of the shear diagram is equal to the negative of the intensity of the distributed load,* while Eq. 7–2 states that the *slope of the moment diagram is equal to the shear.* In particular, if the shear is equal to zero, $dM/dx = 0$, and therefore *a point of zero shear corresponds to a point of maximum (or possibly minimum) moment.*

Equations 7–1 and 7–2 may also be rewritten in the form $dV = -w(x)\,dx$ and $dM = V\,dx$. Noting that $w(x)\,dx$ and $V\,dx$ represent differential areas under the distributed-loading and shear diagrams, respectively, we can integrate these areas between two points B and C along the beam, Fig. 7–14a, and write

$$\Delta V_{BC} = -\int w(x)\,dx$$

$$\begin{array}{c} \text{Change} \\ \text{in shear} \end{array} = \begin{array}{c} \text{Negative of area under} \\ \text{loading curve} \end{array}$$

(7–3)

and

$$\Delta M_{BC} = \int V\,dx$$

$$\begin{array}{c} \text{Change} \\ \text{in moment} \end{array} = \begin{array}{c} \text{Area under} \\ \text{shear diagram} \end{array}$$

(7–4)

Equation 7–3 states that the *change in shear between points B and C is equal to the negative of the area under the distributed-loading curve between these points.* Similarly, from Eq. 7–4, the *change in moment between B and C is equal to the area under the shear diagram within region BC.* Because two integrations are involved, first to determine the change in shear, Eq. 7–3, then to determine the change in moment, Eq. 7–4, we can state that if the loading curve $w = w(x)$ is a polynomial of degree n, then $V = V(x)$ will be a curve of degree $n + 1$, and $M = M(x)$ will be a curve of degree $n + 2$.

As stated previously, the above equations do not apply at points where a *concentrated* force or couple moment acts. These two special cases create *discontinuities* in the shear and moment diagrams, and as a result, each deserves separate treatment.

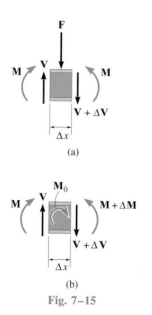

Fig. 7–15

Force. A free-body diagram of a small segment of the beam in Fig. 7–14a, taken from under one of the forces, is shown in Fig. 7–15a. Here it can be seen that force equilibrium requires

$$+\uparrow\Sigma F_y = 0; \qquad\qquad \Delta V = -F \qquad\qquad (7–5)$$

Thus, the *change in shear is negative,* so that on the shear diagram the shear "jumps" *downward when* **F** *acts downward* on the beam. Likewise, the jump in shear (ΔV) is upward when **F** acts upward.

Couple Moment. If we remove a segment of the beam in Fig. 7–14a that is located at the couple moment, the free-body diagram shown in Fig. 7–15b results. In this case letting $\Delta x \rightarrow 0$, moment equilibrium requires

$$\zeta+\Sigma M = 0; \qquad\qquad \Delta M = M_0 \qquad\qquad (7–6)$$

Thus, the *change in moment is positive,* or the moment diagram "jumps" *upward if* **M**$_0$ *is clockwise.* Likewise, the jump ΔM is downward when **M**$_0$ is counterclockwise.

The following examples illustrate application of the above equations for the construction of the shear and moment diagrams. After working through these examples, it is recommended that Examples 7–7 and 7–8 be solved using this method.

Example 7–9

Draw the shear and bending-moment diagrams for the beam shown in Fig. 7–16a.

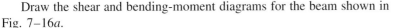

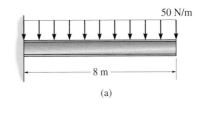

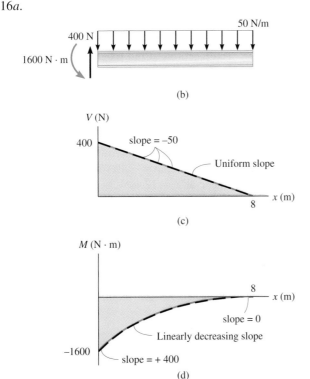

(b)

(c)

(d)

(a)

Fig. 7–16

SOLUTION

Support Reactions.   The reactions at the fixed support have been calculated and are shown on the free-body diagram of the beam, Fig. 7–16b.

Shear Diagram.   The shear at the end points is plotted first, Fig. 7–16c. From the sign convention, Fig. 7–11, $V = +400$ at $x = 0$ and $V = 0$ at $x = 8$. Since $dV/dx = -w = -50$, a straight, *negative* sloping line connects the end points.

Moment Diagram.   From our sign convention, Fig. 7–11, the moments at the beam's end points, $M = -1600$ at $x = 0$ and $M = 0$ at $x = 8$, are plotted first, Fig. 7–16d. Successive values of shear taken from the shear diagram, Fig. 7–16c, indicate that the *slope* $dM/dx = V$ of the moment diagram, Fig. 7–16d, is always positive yet *linearly decreasing* from $dM/dx = 400$ at $x = 0$ to $dM/dx = 0$ at $x = 8$. Thus, due to the integrations, w a constant becomes V a sloping line (first-degree curve) and M a parabola (second-degree curve).

Example 7–10

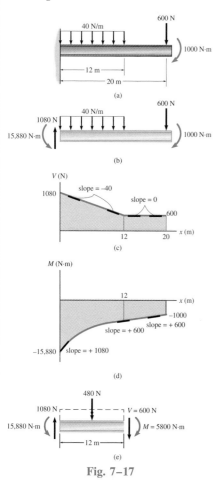

Fig. 7–17

Draw the shear and bending-moment diagrams for the beam shown in Fig. 7–17a.

SOLUTION

Support Reactions. The reactions at the fixed support have been calculated and are shown on the free-body diagram of the beam, Fig. 7–17b.

Shear Diagram. Using the established sign convention, Fig. 7–11, the shear at the ends of the beam is plotted first; i.e., $x = 0$, $V = +1080$; $x = 20$, $V = +600$, Fig. 7–17c.

Since the uniform distributed load is downward and *constant,* the slope of the shear diagram is $dV/dx = -w = -40$ for $0 \leq x < 12$ as indicated.

The magnitude of shear at $x = 12$ is $V = +600$. This can be determined by first finding the area under the load diagram between $x = 0$ and $x = 12$. This represents the change in shear. That is, $\Delta V = -\int w(x)\,dx = -40(12) = -480$. Thus $V|_{x=12} = V|_{x=0} + (-480) = 1080 - 480 = 600$. Also, we can obtain this value by using the method of sections, Fig. 7–17e, where for equilibrium $V = +600$.

Since the load between $12 < x \leq 20$ is $w = 0$, the slope $dV/dx = 0$ as indicated. This brings the shear to the required value of $V = +600$ at $x = 20$.

Moment Diagram. Again, using the established sign convention, Fig. 7–11, the moments at the ends of the beam are plotted first; i.e., $x = 0$, $M = -15,880$; $x = 20$, $M = -1000$, Fig. 7–17d.

Each value of shear gives the slope of the moment diagram since $dM/dx = V$. As indicated, at $x = 0$, $dM/dx = +1080$; and at $x = 12$, $dM/dx = +600$. For $0 \leq x < 12$, specific values of the shear diagram are positive but linearly decreasing. Hence, the moment diagram is parabolic with a linear, decreasing, positive slope.

The magnitude of moment at $x = 12$ is -5800. This can be found by first determining the trapezoidal area under the shear diagram, which represents the change in moment, $\Delta M = \int V\,dx = 600(12) + \frac{1}{2}(1080 - 600)(12) = +10,080$. Thus, $M|_{x=12} = M|_{x=0} + 10,080 = -15,880 + 10,080 = -5800$. The more ''basic'' method of sections can also be used, where equilibrium at $x = 12$ requires $M = -5800$, Fig. 7–17e.

The moment diagram has a constant slope for $12 < x \leq 20$ since, from the shear diagram, $dM/dx = V = +600$. This brings the value of $M = -1000$ at $x = 20$, as required.

Example 7–11

Draw the shear and moment diagrams for the shaft in Fig. 7–18a. The support at A is a thrust bearing and at B a journal bearing.

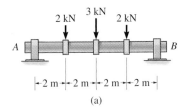

(a)

SOLUTION

Support Reactions.   The reactions at the supports are shown on the free-body diagram in Fig. 7–18b.

Shear Diagram.   The end points $x = 0$, $V = +3.5$ and $x = 8$, $V = -3.5$ are plotted first, as shown in Fig. 7–18c.

Since there is no distributed load on the shaft, the slope of the shear diagram throughout the shaft's length is zero; i.e., $dV/dx = -w = 0$. There is a discontinuity or "jump" of the shear diagram, however, at each concentrated force. From Eq. 7–5, $\Delta V = -F$, the change in shear is negative when the force acts downward and positive when the force acts upward. Stated another way, the "jump" follows the force, i.e., a downward force causes a downward jump, and vice versa. Thus, the 2-kN force at $x = 2$ m changes the shear from 3.5 kN to 1.5 kN; the 3-kN force at $x = 4$ m changes the shear from 1.5 kN to -1.5 kN, etc. We can *also* obtain numerical values for the shear at a specified point in the shaft by using the method of sections, as for example, $x = 2^+$ m, $V = 1.5$ kN in Fig. 7–18e.

Moment Diagram.   The end points $x = 0$, $M = 0$ and $x = 8$, $M = 0$ are plotted first, as shown in Fig. 7–18d.

Since the shear is constant in each region of the shaft, the moment diagram has a corresponding constant positive or negative slope as indicated on the diagram. Numerical values for the change in moment at any point can be computed from the *area* under the shear diagram. For example, at $x = 2$ m, $\Delta M = \int V\, dx = 3.5(2) = 7$. Thus, $M|_{x=2} = M|_{x=0} + 7 = 0 + 7 = 7$. Also, by the method of sections, we can determine the moment at a specified point, as for example, $x = 2^+$ m, $M = 7$ kN $\cdot$ m, Fig. 7–18e.

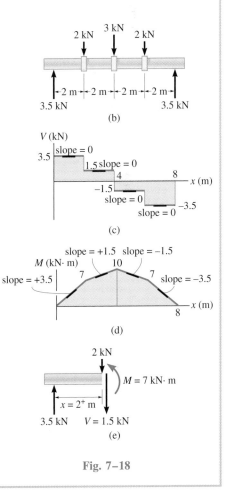

Fig. 7–18

Example 7–12

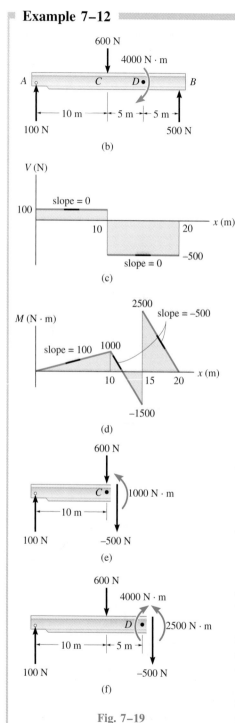

(b)

(c)

(d)

(e)

(f)

Fig. 7–19

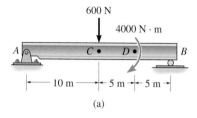

(a)

Sketch the shear and bending-moment diagrams for the beam shown in Fig. 7–19a.

SOLUTION

Support Reactions. The reactions are calculated and indicated on the free-body diagram, Fig. 7–19b.

Shear Diagram. As in Example 7–11, the shear diagram can be constructed by ''following the load'' on the free-body diagram. In this regard, beginning at A, $V_A = +100$ N. No load acts between A and C, so the shear remains constant; i.e., $dV/dx = -w(x) = 0$, Fig. 7–19c. At C the 600-N force acts downward, so the shear jumps down 600 N, from 100 N to -500 N. Again the shear is constant (no load) and ends at -500 N, point B. Notice that no jump or discontinuity in shear occurs at D, the point where the 4000-N · m couple moment is applied, Fig. 7–19b. This is because, for force equilibrium, $\Delta V = 0$ in Fig. 7–15c.

Moment Diagram. The moment at each end of the beam is zero. These two points are plotted first, Fig. 7–19d. The slope of the moment diagram from A to C is constant since $dM/dx = V = +100$. The value of the moment at C can be determined by the method of sections, Fig. 7–19e where $M_C = +1000$ N · m; or by first computing the rectangular area under the shear diagram between A and C to obtain the change in moment $\Delta M_{AC} = (100\ \text{N})(10\ \text{m}) = 1000$ N · m. Since $M_A = 0$, then $M_C = 0 + 1000$ N · m $= 1000$ N · m. From C to D the slope of the moment diagram is $dM/dx = V = -500$, Fig. 7–19c. The area under the shear diagram between points C and D is $\Delta M_{CD} = M_D - M_C = (-500\ \text{N})(5\ \text{m}) = -2500$ N · m, so that $M_D = 1000 - 2500 = -1500$ N · m. A jump in the moment diagram occurs at point D, which is caused by the concentrated couple moment of 4000 N · m. From Eq. 7–6, the jump is *positive* since the couple moment is *clockwise*. Thus, at $x = 15^+$ m, the moment is $M_D = -1500 + 4000 = 2500$ N · m. This value can *also* be determined by the method of sections, Fig. 7–19f. From point D the slope of $dM/dx = -500$ is maintained until the diagram closes to zero at B, Fig. 7–19d.

PROBLEMS

7–55. Draw the shear and moment diagrams for the beam in Prob. 7–35.

*7–56. Draw the shear and moment diagrams for the beam in Prob. 7–40.

7–57. Draw the shear and moment diagrams for the beam in Prob. 7–38.

7–58. Draw the shear and moment diagrams for the beam in Prob. 7–42.

7–59. Draw the shear and moment diagrams for the beam in Prob. 7–46.

*7–60. Draw the shear and moment diagrams for the beam in Prob. 7–48.

7–61. Draw the shear and moment diagrams for the beam.

7–62. Draw the shear and moment diagrams for the beam.

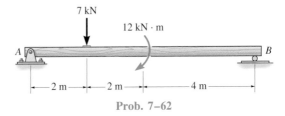

Prob. 7–62

7–63. Draw the shear and moment diagrams for the shaft. The support at A is a thrust bearing and at B it is a journal bearing.

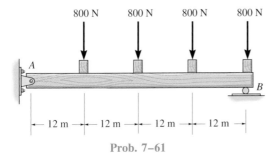

Prob. 7–61

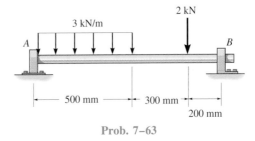

Prob. 7–63

*7–64. Draw the shear and moment diagrams for the beam.

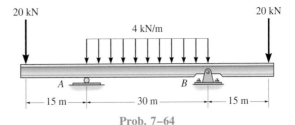

Prob. 7–64

7–65. Draw the shear and moment diagrams for the shaft. The support at A is a journal bearing and at B it is a thrust bearing.

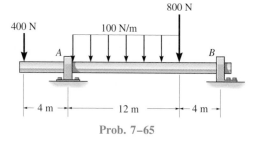

Prob. 7–65

7–66. Draw the shear and moment diagrams for the shaft. The support at A is a journal bearing and at B it is a thrust bearing.

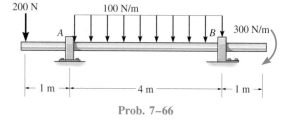

Prob. 7–66

7–67. Draw the shear and moment diagrams for the beam.

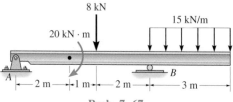

Prob. 7–67

*7–68. Draw the shear and moment diagrams for the beam (a) in terms of the parameters shown; (b) set $w = 500$ N/m, $L = 3$ m.

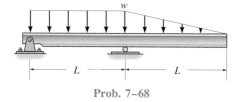

Prob. 7–68

7–69. Draw the shear and moment diagrams for the beam. Take $w = 200$ N/m.

7–70. The beam will fail when the maximum moment is $M_{max} = 30$ kN · m or the maximum shear is $V_{max} = 8$ kN. Determine the largest distributed load w the beam will support.

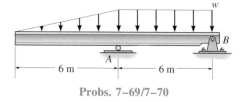

Probs. 7–69/7–70

7–71. Draw the shear and moment diagrams for the beam.

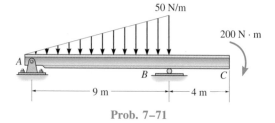

Prob. 7–71

*7–72. Draw the shear and moment diagrams for the beam.

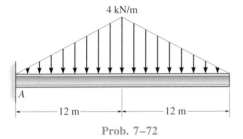

4 kN/m

A

|← 12 m →|← 12 m →|

Prob. 7–72

7–73. Draw the shear and moment diagrams for the beam.

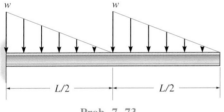

w   *w*

|← L/2 →|← L/2 →|

Prob. 7–73

7–74. Draw the shear and moment diagrams for the beam.

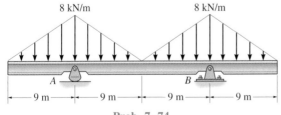

8 kN/m   8 kN/m

A   *B*

|← 9 m →|← 9 m →|← 9 m →|← 9 m →|

Prob. 7–74

7–75. Draw the shear and moment diagrams for the beam.

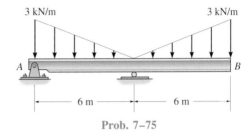

3 kN/m   3 kN/m

A   *B*

|← 6 m →|← 6 m →|

Prob. 7–75

*7–76. Draw the shear and moment diagrams for the beam.

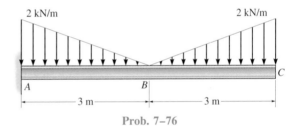

2 kN/m   2 kN/m

A   *B*   *C*

|← 3 m →|← 3 m →|

Prob. 7–76

7–77. The compound beam consists of two segments pin-connected at *B*. Draw the shear and moment diagrams for the beam.

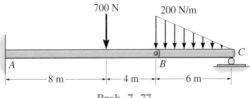

700 N   200 N/m

A   *B*   *C*

|← 8 m →|← 4 m →|← 6 m →|

Prob. 7–77

7–78.   Draw the shear and moment diagrams for the compound beam. The beam is pin-connected at E and F.

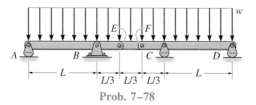

Prob. 7–78

7–79.   Draw the shear and moment diagrams for the beam.

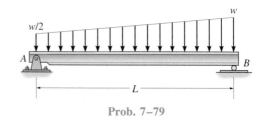

Prob. 7–79

*7.4 Cables

Flexible cables and chains are often used in engineering structures for support and to transmit loads from one member to another. When used to support suspension bridges and trolley wheels, cables form the main load-carrying element of the structure. In the force analysis of such systems, the weight of the cable itself may be neglected because it is often small compared to the load it carries. On the other hand, when cables are used as transmission lines and guys for radio antennas and derricks, the cable weight may become important and must be included in the structural analysis. Three cases will be considered in the analysis that follows: (1) a cable subjected to concentrated loads; (2) a cable subjected to a distributed load; and (3) a cable subjected to its own weight. Regardless of which loading conditions are present, provided the loading is coplanar with the cable, the requirements for equilibrium are formulated in an identical manner.

When deriving the necessary relations between the force in the cable and its slope, we will make the assumption that the cable is *perfectly flexible* and *inextensible*. Due to its flexibility, the cable offers no resistance to bending, and therefore, the tensile force acting in the cable is always tangent to the cable at points along its length. Being inextensible, the cable has a constant length both before and after the load is applied. As a result, once the load is applied, the geometry of the cable remains fixed, and the cable or a segment of it can be treated as a rigid body.

Cable Subjected to Concentrated Loads. When a cable of negligible weight supports several concentrated loads, the cable takes the form of several straight-line segments, each of which is subjected to a constant tensile force. Consider, for example, the cable shown in Fig. 7–20, where the distances h, L_1, L_2, and L_3 and the loads $\mathbf{P}_1$ and $\mathbf{P}_2$ are known. The problem here is to determine the *nine unknowns* consisting of the tension in each of the *three* segments, the *four* components of reaction at A and B, and the sags y_C and y_D at the *two* points C and D. For the solution we can write *two* equations of force equilibrium at each of points A, B, C, and D. This results in a total of *eight equations.** To complete the solution, it will be necessary to know something about the geometry of the cable in order to obtain the necessary ninth equation. For example, if the cable's total *length L* is specified, then the Pythagorean theorem can be used to relate each of the three segmental lengths, written in terms of h, y_C, y_D, L_1, L_2, and L_3, to the total length L. Unfortunately, this type of problem cannot be solved easily by hand. Another possibility, however, is to specify one of the sags, either y_C or y_D, instead of the cable length. By doing this, the equilibrium equations are then sufficient for obtaining the unknown forces and the remaining sag. Once the sag at each point of loading is obtained, the length of the cable can be determined by trigonometry. The following example illustrates a procedure for performing the equilibrium analysis for a problem of this type.

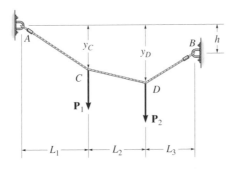

Fig. 7–20

---

*As will be shown in the following example, the eight equilibrium equations can *also* be written for the entire cable, or any part thereof. But *no more* than *eight* equations are available.

Example 7–13

Determine the tension in each segment of the cable shown in Fig. 7–21a.

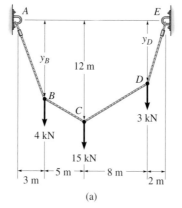

(a)

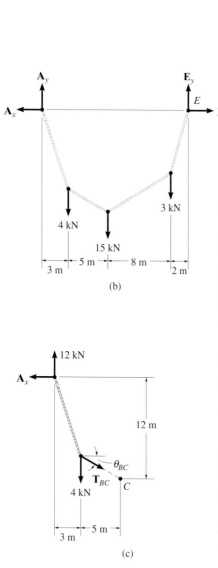

(b)

(c)

Fig. 7–21

SOLUTION

By inspection, there are four unknown external reactions (A_x, A_y, E_x, and E_y) and four unknown cable tensions, one in each cable segment. These eight unknowns along with the two unknown sags y_B and y_D can be determined from *ten* available equilibrium equations. One method is to apply these equations as force equilibrium ($\Sigma F_x = 0$, $\Sigma F_y = 0$) to each of the five points A through E. Here, however, we will take a more direct approach.

Consider the free-body diagram for the entire cable, Fig. 7–21b. Thus,

$$\xrightarrow{+} \Sigma F_x = 0; \qquad -A_x + E_x = 0$$

$$\stackrel{\curvearrowleft}{+} \Sigma M_E = 0; \quad -A_y(18 \text{ m}) + 4 \text{ kN}(15 \text{ m}) + 15 \text{ kN}(10 \text{ m}) + 3 \text{ kN}(2 \text{ m}) = 0$$
$$A_y = 12 \text{ kN}$$

$$+\uparrow \Sigma F_y = 0; \quad 12 \text{ kN} - 4 \text{ kN} - 15 \text{ kN} - 3 \text{ kN} + E_y = 0$$
$$E_y = 10 \text{ kN}$$

Since the sag $y_C = 12$ m is known, we will now consider the leftmost section, which cuts cable BC, Fig. 7–21c.

$$\stackrel{\curvearrowleft}{+} \Sigma M_C = 0; \; A_x(12 \text{ m}) - 12 \text{ kN}(8 \text{ m}) + 4 \text{ kN}(5 \text{ m}) = 0$$
$$A_x = E_x = 6.33 \text{ kN}$$

$$\xrightarrow{+} \Sigma F_x = 0; \qquad T_{BC} \cos \theta_{BC} - 6.33 \text{ kN} = 0$$

$$+\uparrow \Sigma F_y = 0; \qquad 12 \text{ kN} - 4 \text{ kN} - T_{BC} \sin \theta_{BC} = 0$$

Thus,

$$\theta_{BC} = 51.6°$$
$$T_{BC} = 10.2 \text{ kN} \qquad\qquad Ans.$$

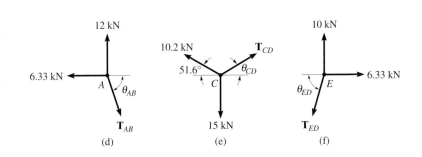

(d)   (e)   (f)

Proceeding now to analyze the equilibrium of points A, C, and E in sequence, we have

Point A   (Fig. 7–21d)

$\xrightarrow{+}\Sigma F_x = 0;$ $T_{AB} \cos \theta_{AB} - 6.33 \text{ kN} = 0$

$+\uparrow \Sigma F_y = 0;$ $-T_{AB} \sin \theta_{AB} + 12 \text{ kN} = 0$

$$\theta_{AB} = 62.2°$$

$$T_{AB} = 13.6 \text{ kN} \qquad\qquad Ans.$$

Point C   (Fig. 7–21e)

$\xrightarrow{+}\Sigma F_x = 0;$ $T_{CD} \cos \theta_{CD} - 10.2 \cos 51.6° \text{ kN} = 0$

$+\uparrow \Sigma F_y = 0;$ $T_{CD} \sin \theta_{CD} + 10.2 \sin 51.6° \text{ kN} - 15 \text{ kN} = 0$

$$\theta_{CD} = 47.9°$$

$$T_{CD} = 9.44 \text{ kN} \qquad\qquad Ans.$$

Point E   (Fig. 7–21f)

$\xrightarrow{+}\Sigma F_x = 0;$ $6.33 \text{ kN} - T_{ED} \cos \theta_{ED} = 0$

$+\uparrow \Sigma F_y = 0;$ $10 \text{ kN} - T_{ED} \sin \theta_{ED} = 0$

$$\theta_{ED} = 57.7°$$

$$T_{ED} = 11.8 \text{ kN} \qquad\qquad Ans.$$

By comparison, the maximum cable tension is in segment AB, since this segment has the greatest slope (θ) and it is required that for any left-hand segment the horizontal component $T \cos \theta = A_x$ (a constant). Also, since the slope angles that the cable segments make with the horizontal have now been determined, it is possible to determine the sags y_B and y_D, Fig. 7–21a, using trigonometry.

Cable Subjected to a Distributed Load. Consider the weightless cable shown in Fig. 7–22a, which is subjected to a loading function $w = w(x)$ *as measured in the x direction.* The free-body diagram of a small segment of the cable having a length Δs is shown in Fig. 7–22b. Since the tensile force in the cable changes continuously in both magnitude and direction along the cable's length, this change is denoted on the free-body diagram by ΔT. The distributed load is represented by its resultant force $w(x)(\Delta x)$, which acts at a fractional distance $k(\Delta x)$ from point O, where $0 < k < 1$. Applying the equations of equilibrium yields

$$\xrightarrow{+}\Sigma F_x = 0; \qquad -T \cos \theta + (T + \Delta T) \cos (\theta + \Delta \theta) = 0$$

$$+\uparrow \Sigma F_y = 0; \quad -T \sin \theta - w(x)(\Delta x) + (T + \Delta T) \sin (\theta + \Delta \theta) = 0$$

$$\underset{\downarrow}{+}\Sigma M_O = 0; \quad w(x)(\Delta x)k(\Delta x) - T \cos \theta \, \Delta y + T \sin \theta \, \Delta x = 0$$

Dividing each of these equations by Δx and taking the limit as $\Delta x \to 0$, and hence $\Delta y \to 0$, $\Delta \theta \to 0$, and $\Delta T \to 0$, we obtain

$$\frac{d(T \cos \theta)}{dx} = 0 \qquad (7\text{–}7)$$

$$\frac{d(T \sin \theta)}{dx} - w(x) = 0 \qquad (7\text{–}8)$$

$$\frac{dy}{dx} = \tan \theta \qquad (7\text{–}9)$$

Integrating Eq. 7–7, we have

$$T \cos \theta = \text{constant} = F_H \qquad (7\text{–}10)$$

Here F_H represents the horizontal component of tensile force at *any point* along the cable.
    Integrating Eq. 7–8 gives

$$T \sin \theta = \int w(x) \, dx \qquad (7\text{–}11)$$

Dividing Eq. 7–11 by Eq. 7–10 eliminates T. Then, using Eq. 7–9, we can obtain the slope

$$\tan \theta = \frac{dy}{dx} = \frac{1}{F_H} \int w(x) \, dx$$

Performing a second integration yields

$$y = \frac{1}{F_H} \int \left(\int w(x) \, dx \right) dx \tag{7-12}$$

This equation is used to determine the curve for the cable, $y = f(x)$. The horizontal force component F_H and the two constants, say C_1 and C_2, resulting from the integration are determined by applying the boundary conditions for the cable.

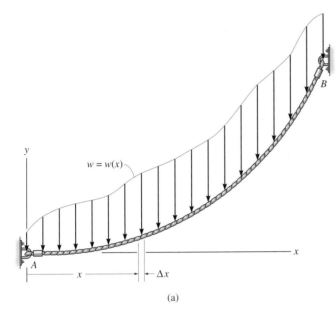

(a)

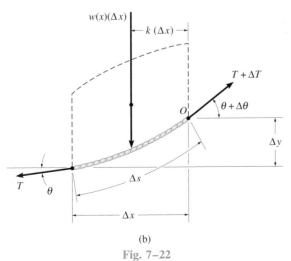

(b)

Fig. 7–22

Example 7–14

The cable of a suspension bridge supports half of the uniform road surface between the two columns at A and B, as shown in Fig. 7–23a. If this distributed loading is w_o, determine the maximum force developed in the cable and the cable's required length. The span length L and sag h are known.

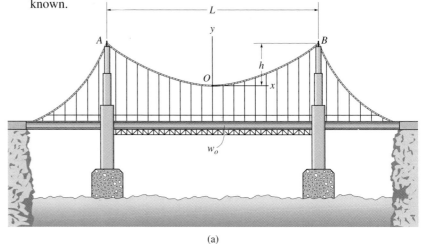

Fig. 7–23

(a)

SOLUTION

We can determine the unknowns in the problem by first finding the curve that defines the shape of the cable by using Eq. 7–12. For reasons of symmetry, the origin of coordinates has been placed at the cable's center. Noting that $w(x) = w_o$, we have

$$y = \frac{1}{F_H} \int \left(\int w_o \, dx \right) dx$$

Integrating this equation twice gives

$$y = \frac{1}{F_H} \left(\frac{w_o x^2}{2} + C_1 x + C_2 \right) \tag{1}$$

The constants of integration may be determined by using the boundary conditions $y = 0$ at $x = 0$ and $dy/dx = 0$ at $x = 0$. Substituting into Eq. 1 yields $C_1 = C_2 = 0$. The curve then becomes

$$y = \frac{w_o}{2F_H} x^2 \tag{2}$$

This is the equation of a *parabola*. The constant F_H may be obtained by using the boundary condition $y = h$ at $x = L/2$. Thus,

$$F_H = \frac{w_o L^2}{8h} \tag{3}$$

Therefore, Eq. 2 becomes

$$y = \frac{4h}{L^2}x^2 \qquad (4)$$

Since F_H is known, the tension in the cable may be determined using Eq. 7–10, written as $T = F_H/\cos \theta$. For $0 \le \theta < \pi/2$, the maximum tension will occur when θ is *maximum*, i.e., at point B, Fig. 7–23a. From Eq. 2, the slope at this point is

$$\frac{dy}{dx}\bigg|_{x=L/2} = \tan \theta_{max} = \frac{w_o}{F_H}x\bigg|_{x=L/2}$$

or

$$\theta_{max} = \tan^{-1}\left(\frac{w_o L}{2F_H}\right) \qquad (5)$$

Therefore,

$$T_{max} = \frac{F_H}{\cos(\theta_{max})} \qquad (6)$$

Using the triangular relationship shown in Fig. 7–23b, which is based on Eq. 5, Eq. 6 may be written as

$$T_{max} = \frac{\sqrt{4F_H^2 + w_o^2 L^2}}{2}$$

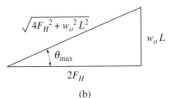

(b)

Substituting Eq. 3 into the above equation yields

$$T_{max} = \frac{w_o L}{2}\sqrt{1 + \left(\frac{L}{4h}\right)^2} \qquad \textit{Ans.}$$

For a differential segment of cable length ds, we can write

$$ds = \sqrt{(dx)^2 + (dy)^2} = \sqrt{1 + \left(\frac{dy}{dx}\right)^2}\, dx$$

Hence, the total length of the cable, $\mathscr{L}$, can be determined by integration. Using Eq. 4, we have

$$\mathscr{L} = \int ds = 2\int_0^{L/2}\sqrt{1 + \left(\frac{8h}{L^2}x\right)^2}\, dx \qquad (7)$$

Integrating and substituting the limits yields

$$\mathscr{L} = \frac{L}{2}\left[\sqrt{1 + \left(\frac{4h}{L}\right)^2} + \frac{L}{4h}\sinh^{-1}\left(\frac{4h}{L}\right)\right] \qquad \textit{Ans.}$$

Cable Subjected to Its Own Weight. When the weight of the cable becomes important in the force analysis, the loading function along the cable becomes a function of the arc length s rather than the projected length x. A generalized loading function $w = w(s)$ acting along the cable is shown in Fig. 7–24a. The free-body diagram for a segment of the cable is shown in Fig. 7–24b. Applying the equilibrium equations to the force system on this diagram, one obtains relationships identical to those given by Eqs. 7–7 through 7–9, but with ds replacing dx. Therefore, it may be shown that

$$T \cos \theta = F_H \tag{7–13}$$

$$T \sin \theta = \int w(s) \, ds$$

$$\frac{dy}{dx} = \frac{1}{F_H} \int w(s) \, ds \tag{7–14}$$

To perform a direct integration of Eq. 7–14, it is necessary to replace dy/dx by ds/dx. Since

$$ds = \sqrt{dx^2 + dy^2}$$

then

$$\frac{dy}{dx} = \sqrt{\left(\frac{ds}{dx}\right)^2 - 1}$$

Therefore,

$$\frac{ds}{dx} = \left\{ 1 + \frac{1}{F_H^2} \left(\int w(s) \, ds \right)^2 \right\}^{1/2}$$

Separating the variables and integrating yields

$$x = \int \frac{ds}{\left\{ 1 + \dfrac{1}{F_H^2} \left(\int w(s) \, ds \right)^2 \right\}^{1/2}} \tag{7–15}$$

The two constants of integration, say C_1 and C_2, are found using the boundary conditions for the cable.

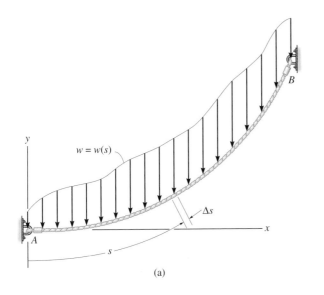

(a)

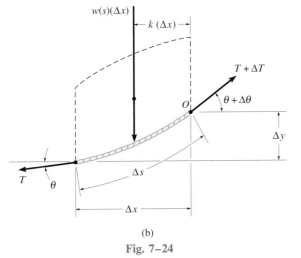

(b)

Fig. 7–24

Example 7–15

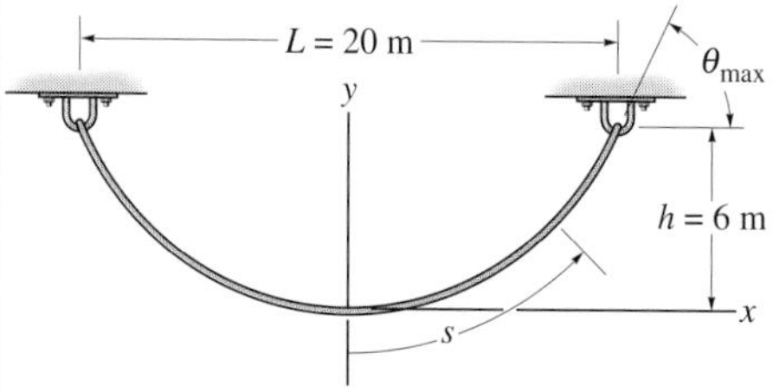

Determine the deflection curve, the length, and the maximum tension in the uniform cable shown in Fig. 7–25. The cable weighs $w_o = 5$ N/m.

SOLUTION

By reasons of symmetry, the origin of coordinates is located at the center of the cable. The deflection curve is expressed as $y = f(x)$. We can determine it by first applying Eq. 7–15, where $w(s) = w_o$.

$$x = \int \frac{ds}{[1 + (1/F_H^2)(\int w_o \, ds)^2]^{1/2}}$$

Integrating the term under the integral sign in the denominator, we have

$$x = \int \frac{ds}{[1 + (1/F_H^2)(w_o s + C_1)^2]^{1/2}}$$

Substituting $u = (1/F_H)(w_o s + C_1)$ so that $du = (w_o/F_H) \, ds$, a second integration yields

$$x = \frac{F_H}{w_o}(\sinh^{-1} u + C_2)$$

or

$$x = \frac{F_H}{w_o}\left\{\sinh^{-1}\left[\frac{1}{F_H}(w_o s + C_1)\right] + C_2\right\} \qquad (1)$$

To evaluate the constants note that, from Eq. 7–14,

$$\frac{dy}{dx} = \frac{1}{F_H}\int w_o \, ds \qquad \text{or} \qquad \frac{dy}{dx} = \frac{1}{F_H}(w_o s + C_1)$$

Since $dy/dx = 0$ at $s = 0$, then $C_1 = 0$. Thus,

$$\frac{dy}{dx} = \frac{w_o s}{F_H} \qquad (2)$$

The constant C_2 may be evaluated by using the condition $s = 0$ at $x = 0$ in Eq. 1, in which case $C_2 = 0$. To obtain the deflection curve, solve for s in Eq. 1, which yields

$$s = \frac{F_H}{w_o}\sinh\left(\frac{w_o}{F_H}x\right) \qquad (3)$$

Now substitute into Eq. 2, in which case

$$\frac{dy}{dx} = \sinh\left(\frac{w_o}{F_H}x\right)$$

Hence

Fig. 7–25

$L = 20$ m

θ_{max}

$h = 6$ m

s

y

x

$$y = \frac{F_H}{w_o} \cosh\left(\frac{w_o}{F_H} x\right) + C_3 \tag{4}$$

If the boundary condition $y = 0$ at $x = 0$ is applied, the constant $C_3 = -F_H/w_o$, and therefore the deflection curve becomes

$$y = \frac{F_H}{w_o}\left[\cosh\left(\frac{w_o}{F_H} x\right) - 1\right]$$

This equation defines the shape of a *catenary curve*. The constant F_H is obtained by using the boundary condition that $y = h$ at $x = L/2$, in which case

$$h = \frac{F_H}{w_o}\left[\cosh\left(\frac{w_o L}{2F_H}\right) - 1\right] \tag{5}$$

Since $w_o = 5$ N/m, $h = 6$ m, and $L = 20$ m, Eqs. 4 and 5 become

$$y = \frac{F_H}{5 \text{ N/m}}\left[\cosh\left(\frac{5 \text{ N/m}}{F_H} x\right) - 1\right] \tag{6}$$

$$6 \text{ m} = \frac{F_H}{5 \text{ N/m}}\left[\cosh\left(\frac{50 \text{ N}}{F_H}\right) - 1\right] \tag{7}$$

Equation 7 can be solved for F_H by using a trial-and-error procedure. The result is

$$F_H = 45.8 \text{ N}$$

and therefore the deflection curve, Eq. 6, becomes

$$y = 9.16[\cosh (0.109x) - 1] \text{ m} \qquad \textit{Ans.}$$

Using Eq. 3, with $x = 10$ m, the half-length of the cable is

$$\frac{\mathcal{L}}{2} = \frac{45.8 \text{ N}}{5 \text{ N/m}} \sinh\left[\frac{5 \text{ N/m}}{45.8 \text{ N}}(10 \text{ m})\right] = 12.1 \text{ m}$$

Hence,

$$\mathcal{L} = 24.2 \text{ m} \qquad \textit{Ans.}$$

Since $T = F_H/\cos \theta$, Eq. 7–13, the maximum tension occurs when θ is maximum, i.e., at $s = \mathcal{L}/2 = 12.1$ m. Using Eq. 2 yields

$$\left.\frac{dy}{dx}\right|_{s=12.1 \text{ m}} = \tan \theta_{\max} = \frac{5 \text{ N/m}(12.1 \text{ m})}{45.8 \text{ N}} = 1.32$$

$$\theta_{\max} = 52.9°$$

Thus,

$$T_{\max} = \frac{F_H}{\cos \theta_{\max}} = \frac{45.8 \text{ N}}{\cos 52.9°} = 75.9 \text{ N} \qquad \textit{Ans.}$$

PROBLEMS

Neglect the weight of the cable in the following problems, *unless* specified.

*7–80.   Determine the tension in each segment of the cable and the cable's total length.

7–82.   The cable supports the three loads shown. Determine the sags y_B and y_D of points B and D. Take $P_1 = 400$ N, $P_2 = 250$ N.

7–83.   The cable supports the three loads shown. Determine the magnitude of $\mathbf{P}_1$ if $P_2 = 300$ N and $y_B = 8$ m. Also find the sag y_D.

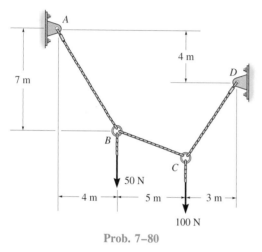

Prob. 7–80

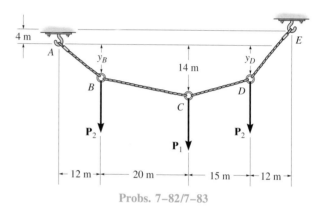

Probs. 7–82/7–83

7–81.   Cable $ABCD$ supports the 4-kg flowerpot E and 6-kg flowerpot F. Determine the maximum tension in the cable and the sag of point B.

*7–84.   The cable supports the loading shown. Determine the distance x_B the force at point B acts from A. Set $P = 40$ N.

7–85.   The cable supports the loading shown. Determine the magnitude of the horizontal force $\mathbf{P}$ so that $x_B = 6$ m.

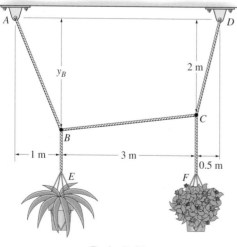

Prob. 7–81

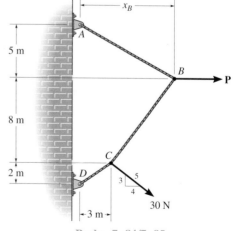

Probs. 7–84/7–85

7–86. Determine the tension in each cable segment and the cable's total length.

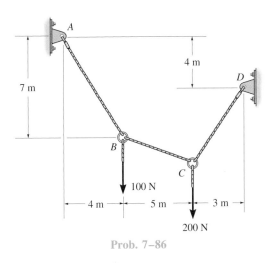

100 N

200 N

Prob. 7–86

7–87. The cable segments support the loading shown. Determine the distance x_B from the force at B to point A. Set $P = 40$ N.

*7–88. The cable segments support the loading shown. Determine the magnitude of the horizontal force **P** so that $x_B = 6$ m.

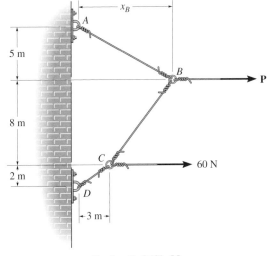

60 N

Probs. 7–87/7–88

7–89. The cable supports a girder which weighs 850 N/m. Determine the tension in the cable at points A, B, and C.

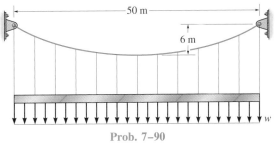

Prob. 7–89

7–90. Determine the maximum uniform loading w N/m that the cable can support if it is capable of sustaining a maximum tension of 3000 N before it will break.

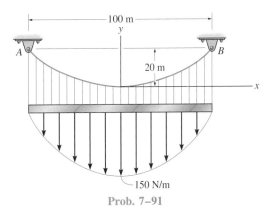

Prob. 7–90

7–91. The cable is subjected to the parabolic loading $w = 150(1 - (x/50)^2)$ N/m, where x is in ft. Determine the equation $y = f(x)$ which defines the cable shape AB and the maximum tension in the cable.

150 N/m

Prob. 7–91

*7–92. The cable is subjected to the triangular loading. If the slope of the cable at point O is zero, determine the equation of the curve $y = f(x)$ which defines the cable shape OB, and the maximum tension developed in the cable.

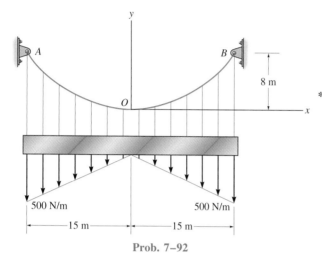

Prob. 7–92

7–93. The cable AB is subjected to a uniform loading of 200 N/m. If the weight of the cable is neglected and the slope angles at points A and B are 30° and 60°, respectively, determine the curve that defines the cable shape and the maximum tension developed in the cable.

7–94. A uniform cord is suspended between two points having the same elevation. Determine the sag-to-span ratio so that the maximum tension in the cord equals the cord's total weight.

7–95. Show that the deflection curve of the cable discussed in Example 7–15 reduces to Eq. (4) in Example 7–14 when the *hyperbolic cosine function* is expanded in terms of a series and only the first two terms are retained. (The answer indicates that the *catenary* may be replaced by a *parabola* in the analysis of problems in which the sag is small. In this case, the cable weight is assumed to be uniformly distributed along the horizontal.)

*■7–96. A cable has a weight of 30 N/m and is supported at points that are 50 m apart and at the same elevation. If it has a length of 60 m, determine the sag.

■7–97. A cable has a weight of 5 N/m. If it can span 300 m and has a sag of 15 m, determine the length of the cable. The ends of the cable are supported at the same elevation.

7–98. A 200-N cable is attached between two points that are 75 m apart and at the same elevation. If the maximum tension developed in the cable is 120 N, determine the length of the cable and the sag.

7–99. The power line is supported at A by the tower. If the cable weighs 0.75 N/m, and the sag $s = 3$ m, determine the resultant horizontal force the cable exerts at A.

*7–100. The power line is supported at A by the tower. If the cable weighs 0.75 N/m, determine the total length of the cable, BAC. Set $s = 3$ m.

7–101. The power line is supported at A by the tower. If the cable weighs 0.75 N/m, determine the required sag s so that the resultant horizontal force the cable exerts at A is zero.

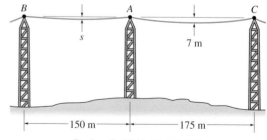

Probs. 7–99/7–100/7–101

Prob. 7–93

7–102. The transmission cable having a weight of 20 N/m is strung across the river as shown. Determine the required force that must be applied to the cable at its points of attachment to the towers at *B* and *C*.

▪7–106. The telephone wire has a mass of 500 g/m. If the cable has a length of 32 m between the poles, determine the maximum tension in the cable and its sag.

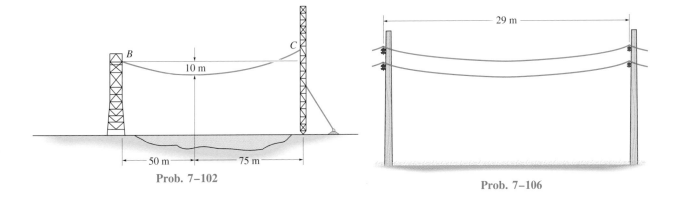

Prob. 7–102

Prob. 7–106

7–103. The uniform beam weighs 800 N/m and is held in the horizontal position by means of the cable *AB*, which has a weight of 10 N/m. If the slope angle of the cable at *A* is 15°, determine the length of the cable.

7–107. A telephone line (cable) stretches between two points which are 150 m apart and at the same elevation. The line sags 5 m and the cable has a weight of 0.3 N/m. Determine the length of the cable and the maximum tension in the cable.

▪*7–108. The cable has a mass of 0.5 kg/m and is 25 m long. Determine the vertical and horizontal components of force it exerts on the top of the tower.

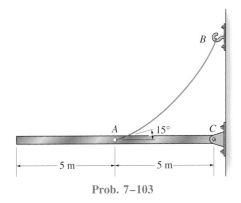

Prob. 7–103

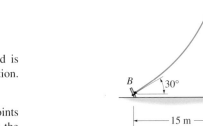

Prob. 7–108

▪*7–104. A 50-m-long chain has a total mass of 100 kg and is suspended between two points 15 m apart at the same elevation. Determine the maximum tension and the sag in the chain.

▪7–105. A chain has a mass of 3 kg/m and is supported at points which are 3 m apart and at the same elevation. If the sag in the chain is 1 m, determine the maximum tension in the chain.

7–109. The buoyant (or vertical) component of force acting on the weather balloon is 600 N. If cable *OB* is 70 m in length and has a mass of 500 g/m, determine the altitude *h* of the balloon.

7–110. The balloon is held in place using a 400-m cord that weighs 0.8 N/m and makes a slope angle of 60°. If the tension in the cord at point *A* is 150 N, determine the length of the cord, *l*, that is lying on the ground and the balloon's height *h*. *Hint:* Establish the coordinate system at *B* as shown.

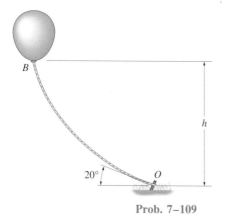

Prob. 7–109

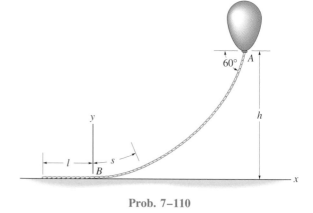

Prob. 7–110

REVIEW PROBLEMS

7–111. Draw the shear and moment diagrams for the beam.

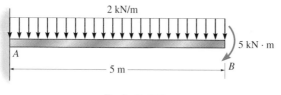

Prob. 7–111

7–113. Draw the shear and moment diagrams for the shaft. The supports at *A* and *B* are journal bearings.

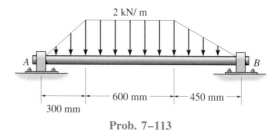

Prob. 7–113

*7–112. The chain is suspended between points *A* and *B*. If it has a weight of 0.5 N/m and the sag is 3 m, determine the maximum tension in the chain.

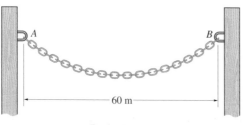

Prob. 7–112

7–114. The 80-m-long chain is fixed at its ends and hoisted at its midpoint B using a crane. If the chain has a weight of 0.5 N/m, determine the minimum height h of the hook in order to lift the chain *completely* off the ground. What is the horizontal force at pin A or C when the chain is in this position? *Hint:* When h is a minimum, the slope at A and C is zero.

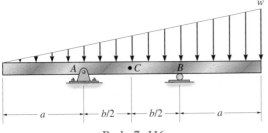

Prob. 7–114

7–115. The two segments of the girder are pin-connected together by a short vertical link BC. Draw the shear and moment diagrams for the girder.

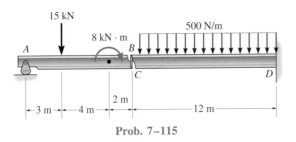

Prob. 7–115

*7–116. Determine the ratio of a/b for which the shear force will be zero at the midpoint C of the beam.

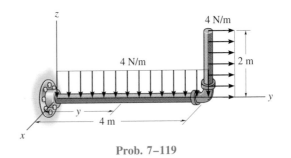

Prob. 7–116

7–117. Draw the shear and moment diagrams for the beam.

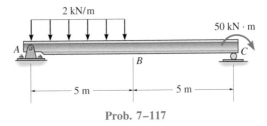

Prob. 7–117

7–118. Determine the normal force, shear force, and moment at sections through points D and E of the frame.

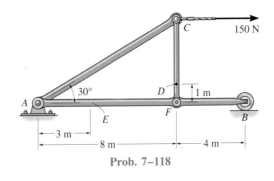

Prob. 7–118

7–119. Express the shear and moment acting in the pipe as a function of y, where $0 \le y \le 4$ m.

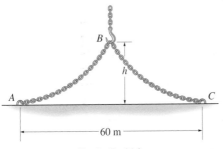

Prob. 7–119

The effective design of a brake system, such as the one for this bicycle, requires efficient capacity for the mechanism to resist frictional forces. In this chapter we will study the nature of friction and show how these forces are considered in engineering analysis.

8

Friction

In the previous chapters the surfaces of contact between two bodies were considered to be perfectly *smooth*. Because of this, the force of interaction between the bodies always acts *normal* to the surface at points of contact. In reality, however, all surfaces are *rough,* and depending on the nature of the problem, the ability of a body to support a *tangential* as well as a *normal* force at its contacting surface must be considered. The tangential force is caused by friction, and in this chapter we will show how to analyze problems involving frictional forces. Specific application will include frictional forces on screws, bearings, disks, and belts. The analysis of rolling resistance is given in the last part of the chapter.

8.1 Characteristics of Dry Friction

Friction may be defined as a force of resistance acting on a body which prevents or retards slipping of the body relative to a second body or surface with which it is in contact. This force always acts *tangent* to the surface at points of contact with other bodies and is directed so as to oppose the possible or existing motion of the body relative to these points.

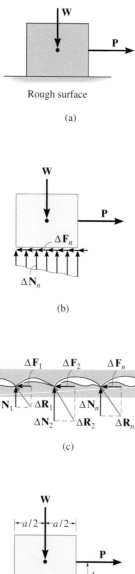

Rough surface

(a)

(b)

(c)

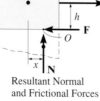

Resultant Normal
and Frictional Forces

(d)

Fig. 8–1a–d

In general, two types of friction can occur between surfaces. *Fluid friction* exists when the contacting surfaces are separated by a film of fluid (gas or liquid). The nature of fluid friction is studied in fluid mechanics, since it depends on knowledge of the velocity of the fluid and the fluid's ability to resist shear force. In this book only the effects of *dry friction* will be presented. This type of friction is often called *Coulomb friction,* since its characteristics were studied extensively by C. A. Coulomb in 1781. Specifically, dry friction occurs between the contacting surfaces of bodies in the absence of a lubricating fluid.

Theory of Dry Friction.

The theory of dry friction can best be explained by considering what effects are caused by pulling horizontally on a block of uniform weight **W** which is resting on a rough horizontal surface, Fig. 8–1a. To properly develop a full understanding of the nature of friction, it is necessary to consider the surfaces of contact to be *nonrigid or deformable.* The other portion of the block, however, will be considered rigid. As shown on the free-body diagram of the block, Fig. 8–1b, the floor exerts a *distribution* of both *normal force* $\Delta \mathbf{N}_n$ and *frictional force* $\Delta \mathbf{F}_n$ along the contacting surface. For equilibrium, the normal forces must act *upward* to balance the block's weight **W,** and the frictional forces act to the left to prevent the applied force **P** from moving the block to the right. Close examination of the contacting surfaces between the floor and block reveals how these frictional and normal forces develop, Fig. 8–1c. It can be seen that many microscopic irregularities exist between the two surfaces and, as a result, reactive forces $\Delta \mathbf{R}_n$ are developed at each of the protuberances.* These forces act at all points of contact and, as shown, each reactive force contributes both a frictional component $\Delta \mathbf{F}_n$ and a normal component $\Delta \mathbf{N}_n$.

Equilibrium.

For simplicity in the following analysis, the effect of the *distributed* normal and frictional loadings will be indicated by their *resultants* **N** and **F**, which are represented on the free-body diagram of the block as shown in Fig. 8–1d. Clearly, the distribution of $\Delta \mathbf{F}_n$ in Fig. 8–1b indicates that **F** always acts *tangent to the contacting surface, opposite* to the direction of **P.** On the other hand, the normal force **N** is determined from the distribution of $\Delta \mathbf{N}_n$ in Fig. 8–1b and is directed upward to balance the block's weight **W.** Notice that **N** acts a distance x to the right of the line of action of **W,** Fig. 8–1d. This location, which coincides with the centroid of the loading diagram in Fig. 8–1b, is necessary in order to balance the "tipping effect" caused by **P.** For example, if **P** is applied at a height h from the surface, Fig. 8–1d, then moment equilibrium about point O is satisfied if $Wx = Ph$ or $x = Ph/W$. In particular, note that the block will be on the verge of *tipping* if **N** acts at the right corner of the block, $x = a/2$.

---

*Besides mechanical interactions as explained here, a detailed treatment of the nature of frictional forces must also include the effects of temperature, density, cleanliness, and atomic or molecular attraction between the contacting surfaces. See D. Tabor, *Journal of Lubrication Technology,* 103, 169, 1981.

Impending Motion. In cases where h is small or the surfaces of contact are rather "slippery," the frictional force **F** may *not* be great enough to balance the magnitude of **P,** and consequently the block will tend to slip *before* it can tip. In other words, as the magnitude of **P** is slowly increased, the magnitude of **F** correspondingly increases until it attains a certain *maximum value F_s,* called the *limiting static frictional force,* Fig. 8–1e. When this value is reached, the block is in *unstable equilibrium,* since any further increase in P will cause deformations and fractures at the points of surface contact and consequently the block will begin to move. Experimentally, it has been determined that the magnitude of the limiting static frictional force **F_s** is *directly proportional* to the magnitude of the resultant normal force **N**. This may be expressed mathematically as

$$F_s = \mu_s N \qquad (8\text{–}1)$$

where the constant of proportionality, μ_s (mu "sub" s), is called the *coefficient of static friction.*

Typical values for μ_s, found in many engineering handbooks, are given in Table 8–1. Although this coefficient is generally less than 1, be aware that in some cases it is possible, as in the case of aluminum on aluminum, for μ_s to be greater than 1. Physically this means, of course, that in this case the frictional force is greater than the corresponding normal force. Furthermore, it should be noted that μ_s is dimensionless and depends only on the characteristics of the two surfaces in contact. A wide range of values is given for each value of μ_s, since experimental testing was done under variable conditions of roughness and cleanliness of the contacting surfaces. For applications, therefore, it is important that both caution and judgment be exercised when selecting a coefficient of friction for a given set of conditions. When an exact calculation of F_s is required, the coefficient of friction should be determined directly by an experiment that involves the two materials to be used.

Motion. If the magnitude of **P** acting on the block is increased so that it becomes greater than F_s, the frictional force at the contacting surfaces drops slightly to a smaller value F_k, called the *kinetic frictional force.* The block will *not* be held in equilibrium ($P > F_k$); instead, it will begin to slide with increasing speed, Fig. 8–1f. The drop made in the frictional force magnitude, from F_s (static) to F_k (kinetic), can be explained by again examining the surfaces of contact, Fig. 8–1g. Here it is seen that when $P > F_s$, then P has the capacity to shear off the peaks at the contact surfaces and cause the block to "lift" somewhat out of its settled position and "ride" on top of the peaks. Once the block begins to slide, high local temperatures at the points of contact cause momentary adhesion (welding) of these points. The continued shearing of these welds is the dominant mechanism creating friction. Since the resultant contact forces $\Delta \mathbf{R}_n$ are aligned slightly more in the vertical direction than before, Fig. 8–1c, they thereby contribute *smaller* frictional components, $\Delta \mathbf{F}_n$, as when the irregularities are meshed.

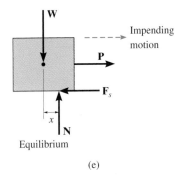

(e)

Table 8–1   Typical Values for μ_s

Contact Materials	Coefficient of Static Friction (μ_s)
Metal on ice	0.03–0.05
Wood on wood	0.30–0.70
Leather on wood	0.20–0.50
Leather on metal	0.30–0.60
Aluminum on aluminum	1.10–1.70

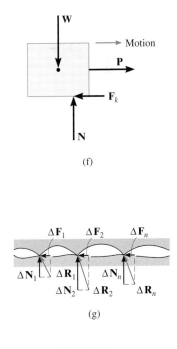

(f)

(g)

Fig. 8–1e–g

Experiments with sliding blocks indicate that the magnitude of the resultant frictional force $\mathbf{F}_k$ is directly proportional to the magnitude of the resultant normal force $\mathbf{N}$. This may be expressed mathematically as

$$F_k = \mu_k N \qquad (8-2)$$

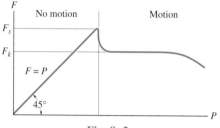

Fig. 8–2

where the constant of proportionality, μ_k, is called the *coefficient of kinetic friction*. Typical values for μ_k are approximately 25 percent *smaller* than those listed in Table 8–1 for μ_s.

the graph in Fig. 8–2, which shows the variation of the frictional force F versus the applied load P. Here the frictional force is categorized in three different ways: namely, F is a *static-frictional force* if equilibrium is maintained; F is a *limiting static-frictional force* $\mathbf{F}_s$ when its magnitude reaches a maximum value needed to maintain equilibrium; and finally, F is termed a *kinetic-frictional force* F_k when sliding occurs at the contacting surface. Notice also from the graph that for very large values of P or for high speeds, because of aerodynamic effects, F_k and likewise μ_k begin to decrease.

Characteristics of Dry Friction.

As a result of *experiments* that pertain to the foregoing discussion, the following rules which apply to bodies subjected to dry friction may be stated.

1. The frictional force acts *tangent* to the contacting surfaces in a direction *opposed* to the *relative motion* or tendency for motion of one surface against another.
2. The magnitude of the maximum static frictional force $\mathbf{F}_s$ that can be developed is independent of the area of contact, provided the normal pressure is not very low nor great enough to severely deform or crush the contacting surfaces of the bodies.
3. The magnitude of the maximum static frictional force is generally greater than the magnitude of the kinetic frictional force for any two surfaces of contact. However, if one of the bodies is moving with a *very low velocity* over the surface of another, F_k becomes approximately equal to F_s, i.e., $\mu_s \approx \mu_k$.
4. When *slipping* at the surface of contact is *about to occur,* the magnitude of the maximum static frictional force is proportional to the magnitude of the normal force, such that $F_s = \mu_s N$, Eq. 8–1.
5. When *slipping* at the surface of contact is *occurring,* the magnitude of the kinetic frictional force is proportional to the magnitude of the normal force, such that $F_k = \mu_k N$, Eq. 8–2.

Angle of Friction. It should be observed that Eqs. 8–1 and 8–2 have a specific yet *limited* use in the solution of friction problems. In particular, the frictional force acting at a contacting surface is determined from $F_k = \mu_k N$ *only* if *relative motion* is occurring between the two surfaces. Furthermore, if two bodies are *stationary*, the magnitude of the frictional force, F, *does not necessarily* equal $\mu_s N$; instead, F must satisfy the inequality $F \le \mu_s N$. Only when *impending motion* occurs does F reach its upper limit, $F = F_s = \mu_s N$. This situation may be better understood by considering the block shown in Fig. 8–3a, which is acted upon by a force **P**. In this case consider $P = F_s$, so that the block is on the *verge of sliding*. For equilibrium, the normal force **N** and frictional force $\mathbf{F}_s$ combine to create a resultant $\mathbf{R}_s$. The angle ϕ_s that $\mathbf{R}_s$ makes with **N** is called the *angle of static friction*. From the figure,

$$\phi_s = \tan^{-1}\left(\frac{F_s}{N}\right) = \tan^{-1}\left(\frac{\mu_s N}{N}\right) = \tan^{-1}\mu_s$$

Provided the block is *not in motion*, any horizontal force $P < F_s$ causes a resultant **R** which has a line of action directed at an angle ϕ from the vertical such that $\phi \le \phi_s$. If **P** creates uniform *motion* of the block, then $P = F_k$. In this case, the resultant $\mathbf{R}_k$ has a line of action defined by ϕ_k, Fig. 8–3b. This angle is referred to as the *angle of kinetic friction*, where

$$\phi_k = \tan^{-1}\left(\frac{F_k}{N}\right) = \tan^{-1}\left(\frac{\mu_k N}{N}\right) = \tan^{-1}\mu_k$$

By comparison, $\phi_s \ge \phi_k$.

Angle of Repose. An experimental method which can be used to measure the coefficient of friction between two contacting surfaces consists of placing a block of one material having a weight W on a plane made of another material, Fig. 8–4a. The plane is inclined to the angle θ_s, at which point the block is on the *verge of sliding* and therefore $F_s = \mu_s N$. The free-body diagram of the block at this instant is shown in Fig. 8–4b. Applying the force equations of equilibrium the normal force $N = W \cos \theta_s$, and the frictional force $F_s = W \sin \theta_s$. Since $F_s = \mu_s N$, then $W \sin \theta_s = \mu_s (W \cos \theta_s)$ or

$$\theta_s = \tan^{-1}\mu_s$$

The angle θ_s is referred to as the *angle of repose*, and by comparison it is equal to the angle of static friction ϕ_s. Once it is measured, the coefficient of static friction is obtained from $\mu_s = \tan \theta_s$. Note that this calculation is independent of the weight of the block, and so for the experiment W does not have to be known.

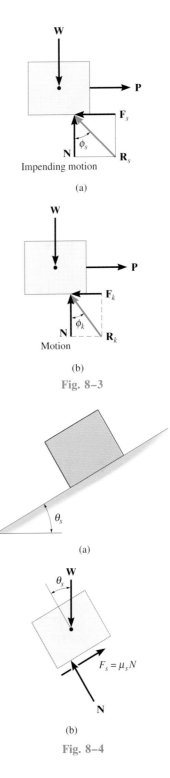

Impending motion

(a)

Motion

(b)

Fig. 8–3

(a)

(b)

Fig. 8–4

8.2 Problems Involving Dry Friction

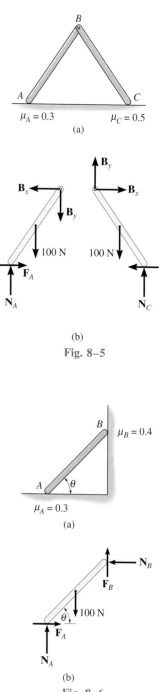

$\mu_A = 0.3$ $\mu_C = 0.5$

(a)

$\mathbf{B}_y$

$\mathbf{B}_x$ $\mathbf{B}_x$

$\mathbf{B}_y$

100 N   100 N

$\mathbf{F}_A$ $\mathbf{F}_C$

N_A N_C

(b)

Fig. 8–5

B $\mu_B = 0.4$

A θ

$\mu_A = 0.3$

(a)

N_B

F_B

100 N

θ

F_A

N_A

(b)

Fig. 8–6

If a rigid body is in equilibrium when it is subjected to a system of forces that includes the effect of friction, the force system must satisfy not only the equations of equilibrium but *also* the laws that govern the frictional forces.

Types of Friction Problems. In general, there are three types of mechanics problems involving dry friction. They can easily be classified once the free-body diagrams are drawn and the total number of unknowns are identified and compared with the total number of available equilibrium equations. Each type of problem will now be explained and illustrated graphically by examples. In all these cases the geometry and dimensions for the problem are assumed to be known.

Equilibrium. Problems in this category are strictly equilibrium problems which require *the total number of unknowns to be equal to the total number of available equilibrium equations*. Once the frictional forces are determined from the solution, however, their numerical values must be checked to be sure they satisfy the inequality $F \leq \mu_s N$; otherwise, slipping will occur and the body will not remain in equilibrium. A problem of this type is shown in Fig. 8–5a. Here we must determine the frictional forces at A and C to check if the equilibrium position of the bars can be maintained. If the bars are uniform and have known weights of 100 N each, then the free-body diagrams are as shown in Fig. 8–5b. There are six unknown force components which can be determined *strictly* from the six equilibrium equations (three for each member). Once F_A, N_A, F_C, and N_C are determined, then the bars will remain in equilibrium provided $F_A \leq 0.3N_A$ and $F_C \leq 0.5N_C$ are satisfied.

Impending Motion at All Points. In this case *the total number of unknowns will equal the total number of available equilibrium equations plus the total number of available frictional equations, $F = \mu N$*. In particular, if *motion is impending* at the points of contact, then $F_s = \mu_s N$; whereas if the body is *slipping*, then $F_k = \mu_k N$. For example, consider the problem of finding the smallest angle θ at which the 100-N bar in Fig. 8–6a can be placed against the wall without slipping. The free-body diagram is shown in Fig. 8–6b. Here there are *five* unknowns: F_A, N_A, F_B, N_B, θ. For the solution there are *three* equilibrium equations and *two* static frictional equations which apply at *both* points of contact, so that $F_A = 0.3N_A$ and $F_B = 0.4N_B$. (It should also be noted that the bar will not be in a state where motion impends *unless* the bar slips at *both* points A and B simultaneously.)

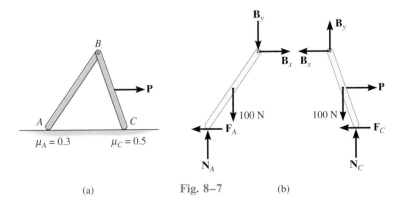

Fig. 8–7

(a)   (b)

Tipping or Impending Motion at Some Points. Here *the total number of unknowns will be less than the number of available equilibrium equations plus the total number of frictional equations or conditional equations for tipping.* As a result, several possibilities for motion or impending motion will exist and the problem will involve a determination of the kind of motion which actually occurs. For example, consider the two-member frame shown in Fig. 8–7a. In this problem we wish to determine the horizontal force **P** needed to cause movement of the frame. If each member has a weight of 100 N, then the free-body diagrams are as shown in Fig. 8–7b. There are *seven* unknowns: N_A, F_A, N_C, F_C, B_x, B_y, P. For a unique solution we must satisfy the *six* equilibrium equations (three for each member) and only *one* of two possible static frictional equations. This means that as **P** increases its magnitude it will either cause slipping at A and no slipping at C, so that $F_A = 0.3N_A$ and $F_C \leq 0.5N_C$; or slipping occurs at C and no slipping at A, in which case $F_C = 0.5N_C$ and $F_A \leq 0.3N_A$. The actual situation can be determined by calculating P for each case and then choosing the case for which P is *smallest.* If in both cases the *same value* for P is calculated, which in practice would be highly improbable, then slipping at both points occurs simultaneously; i.e., the *seven unknowns* will satisfy *eight equations.* As a second example, consider a block having a width b, height h, and weight W which is resting on a rough surface, Fig. 8–8a. The force **P** needed to cause motion is to be determined. Inspection of the free-body diagram, Fig. 8–8b, indicates that there are *four unknowns,* namely, P, F, N, and x. For a unique solution, however, we must satisfy the *three* equilibrium equations and either *one* static friction equation or *one* conditional equation which requires the block not to tip. Hence two possibilities of motion exist. Either the block will *slip,* Fig. 8–8b, in which case $F = \mu_s N$ and the value obtained for x must satisfy $0 \leq x \leq b/2$; or the block will *tip,* Fig. 8–8c, in which case $x = b/2$ and the frictional force will satisfy the inequality $F \leq \mu_s N$. The solution yielding the *smallest* value of P will define the type of motion the block undergoes. If it happens that the same value of P is calculated for both cases, although this would be very improbable, then slipping and tipping will occur simultaneously; i.e., the *four unknowns* will satisfy *five equations.*

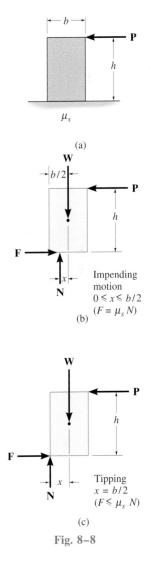

Fig. 8–8

Equilibrium Versus Frictional Equations. It was stated earlier that the frictional force *always* acts so as to either oppose the relative motion or impede the motion of a body over its contacting surface. Realize, however, that we can *assume* the sense of the frictional force in problems which require F to be an ''equilibrium force'' and satisfy the inequality $F < \mu_s N$. The correct sense is made known *after* solving the equations of equilibrium for F. For example, if F is a negative scalar, the sense of $\mathbf{F}$ is the reverse of that which was assumed. This convenience of *assuming* the sense of $\mathbf{F}$ is possible because the equilibrium equations equate to zero the *components of vectors* acting in the *same direction.* In cases where the frictional equation $F = \mu N$ is used in the solution of a problem, however, the convenience of *assuming* the sense of $\mathbf{F}$ is *lost,* since the frictional equation relates only the *magnitudes* of two perpendicular vectors. Consequently, $\mathbf{F}$ *must always* be shown acting with its *correct sense* on the free-body diagram whenever the frictional equation is used for the solution of a problem.

PROCEDURE FOR ANALYSIS

The following procedure provides a method for solving equilibrium problems involving dry friction.

Free-Body Diagrams.   Draw the necessary free-body diagrams and determine the number of unknowns or equations required for a complete solution. Unless stated in the problem, *always* show the frictional forces as *unknowns;* i.e., *do not assume that $F = \mu N$.* Recall that only three equations of coplanar equilibrium can be written for each body. Consequently, if there are more unknowns than equations of equilibrium, it will be necessary to apply the frictional equation at some, if not all, points of contact to obtain the extra equations needed for a complete solution.

Equations of Friction and Equilibrium.   Apply the equations of equilibrium and the necessary frictional equations (or conditional equations if tipping is involved) and solve for the unknowns. If the problem involves a three-dimensional force system such that it becomes difficult to obtain the force components or the necessary moment arms, apply the equations of equilibrium using Cartesian vectors.

The following example problems illustrate this procedure numerically.

Example 8–1

The uniform crate shown in Fig. 8–9a has a mass of 20 kg. If a force $P = 80$ N is applied to the crate, determine if it remains in equilibrium. The coefficient of static friction is $\mu_s = 0.3$.

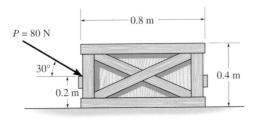

(a)

Fig. 8–9

SOLUTION

Free-Body Diagram.   As shown in Fig. 8–9b, the *resultant* normal force N_C must act a distance x from the crate's center line in order to counteract the tipping effect caused by **P**. There are *three unknowns: F*, N_C, and x, which can be determined strictly from the *three* equations of equilibrium.

Equations of Equilibrium

$$\xrightarrow{+}\Sigma F_x = 0; \qquad 80 \cos 30° \text{ N} - F = 0$$
$$+\uparrow\Sigma F_y = 0; \qquad -80 \sin 30° \text{ N} + N_C - 196.2 \text{ N} = 0$$
$$\zeta+\Sigma M_O = 0; \quad 80 \sin 30° \text{ N}(0.4 \text{ m}) - 80 \cos 30° \text{ N}(0.2 \text{ m}) + N_C(x) = 0$$

Solving,

$$F = 69.3 \text{ N}$$
$$N_C = 236 \text{ N}$$
$$x = -0.00908 \text{ m} = -9.08 \text{ mm}$$

Since x is negative it indicates the *resultant* normal force acts (slightly) to the *left* of the crate's center line. No tipping will occur since $x \leq 0.4$ m. Also, the *maximum* frictional force which can be developed at the surface of contact is $F_{max} = \mu_s N_C = 0.3(236 \text{ N}) = 70.8$ N. Since $F = 69.3$ N < 70.8 N, the crate will *not slip,* although it is very close to doing so.

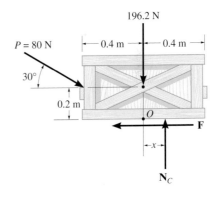

(b)

Example 8–2

The pipe shown in Fig. 8–10a is gripped between two levers that are pinned together at C. If the coefficient of static friction between the levers and the pipe is $\mu = 0.3$, determine the maximum angle θ at which the pipe can be gripped without slipping. Neglect the weight of the pipe.

SOLUTION

Free-Body Diagram. As shown in Fig. 8–10b, there are five unknowns: N_A, F_A, N_B, F_B, and θ. The *three* equations of equilibrium and *two* frictional equations at A and B apply. The frictional forces act toward C to prevent upward motion of the pipe.

Equations of Friction and Equilibrium. The frictional equations are

$$F_s = \mu_s N; \qquad\qquad F_A = \mu N_A$$
$$F_B = \mu N_B$$

Using these results, and applying the equations of equilibrium, yields

$$\xrightarrow{+} \Sigma F_x = 0;$$

$$N_A \cos\left(\frac{\theta}{2}\right) + \mu N_A \sin\left(\frac{\theta}{2}\right) - N_B \cos\left(\frac{\theta}{2}\right) - \mu N_B \sin\left(\frac{\theta}{2}\right) = 0 \quad (1)$$

$$\zeta + \Sigma M_O = 0;$$

$$-\mu N_B(r) + \mu N_A(r) = 0 \qquad\qquad (2)$$

$$+\uparrow \Sigma F_y = 0;$$

$$N_A \sin\left(\frac{\theta}{2}\right) - \mu N_A \cos\left(\frac{\theta}{2}\right) + N_B \sin\left(\frac{\theta}{2}\right) - \mu N_B \cos\left(\frac{\theta}{2}\right) = 0 \quad (3)$$

From either Eq. 1 or 2 it is seen that $N_A = N_B$. This could also have been determined directly from the symmetry of *both* geometry and loading. Substituting the result into Eq. 3, we obtain

$$\sin\left(\frac{\theta}{2}\right) - \mu \cos\left(\frac{\theta}{2}\right) = 0$$

so that

$$\tan\left(\frac{\theta}{2}\right) = \frac{\sin(\theta/2)}{\cos(\theta/2)} = \mu = 0.3$$
$$\theta = 2\tan^{-1} 0.3 = 33.4° \qquad\qquad Ans.$$

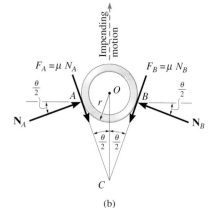

(a)

(b)

Fig. 8–10

Example 8–3

The uniform rod having a weight W and length l is supported at its ends A and B, where the coefficient of static friction is μ, Fig. 8–11a. Determine the greatest angle θ so the rod does not slip. Neglect the thickness of the rod for the calculation.

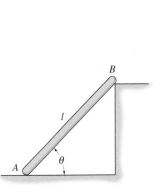

(a)

SOLUTION

Free-Body Diagram. As shown in Fig. 8–11b, there are *five* unknowns: F_A, N_A, F_B, N_B, and θ. These can be determined from the *three* equilibrium equations and *two* frictional equations applied at points A and B. The frictional forces must be drawn with their correct sense so that they oppose the tendency for motion of the rod. Why?

Equations of Friction and Equilibrium. Writing the frictional equations,

$$F = \mu_s N; \qquad\qquad F_A = \mu N_A$$
$$F_B = \mu N_B$$

Using these results and applying the equations of equilibrium yields

$$\xrightarrow{+}\Sigma F_x = 0; \qquad \mu N_A + \mu N_B \cos\theta - N_B \sin\theta = 0 \qquad (1)$$
$$+\uparrow\Sigma F_y = 0; \qquad N_A - W + N_B \cos\theta + \mu N_B \sin\theta = 0 \qquad (2)$$
$$\big\downarrow+\Sigma M_G = 0; \quad -N_A\left(\frac{l}{2}\cos\theta\right) + \mu N_A\left(\frac{l}{2}\sin\theta\right) + N_B\left(\frac{l}{2}\right) = 0 \qquad (3)$$

Moments were summed about the center of the rod G in order to eliminate W. We can solve Eqs. 1 and 3, which reduce to

$$\mu N_A = N_B(\sin\theta - \mu\cos\theta)$$
$$N_B = N_A(\cos\theta - \mu\sin\theta)$$

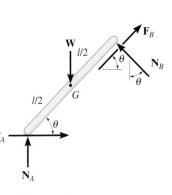

(b)

Fig. 8–11

Thus

$$\mu N_A = N_A(\cos\theta - \mu\sin\theta)(\sin\theta - \mu\cos\theta)$$
$$\mu = \sin\theta\cos\theta - \mu\cos^2\theta - \mu\sin^2\theta + \mu^2\sin\theta\cos\theta$$
$$\mu = (1 + \mu^2)\sin\theta\cos\theta - \mu(\sin^2\theta + \cos^2\theta)$$

Since $\sin^2\theta + \cos^2\theta = 1$ and $\sin 2\theta = 2\sin\theta\cos\theta$, then

$$2\mu = \left(\frac{1 + \mu^2}{2}\right)\sin 2\theta$$

Solving for θ, we have

$$\theta = \frac{1}{2}\sin^{-1}\left(\frac{4\mu}{1 + \mu^2}\right) \qquad\qquad Ans.$$

Example 8–4

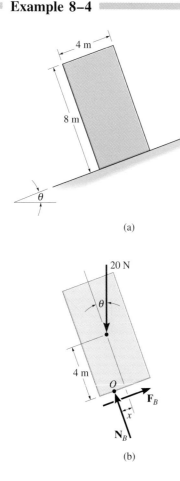

(a)

(b)

Fig. 8–12

The homogeneous block shown in Fig. 8–12a has a weight of 20 N and rests on the incline for which $\mu_s = 0.55$. Determine the largest angle of tilt, θ, of the plane before the block moves.

SOLUTION

Free-Body Diagram.   As shown in Fig. 8–12b, the dimension x is used to locate the position of the resultant normal force $\mathbf{N}_B$ under the block. There are *four* unknowns, θ, N_B, F_B, and x. *Three* equations of equilibrium are available. The *fourth* equation is obtained by investigating the conditions for tipping or sliding of the block.

Equations of Equilibrium.   Applying the equations of equilibrium yields

$$+\swarrow\Sigma F_x = 0; \qquad\qquad 20 \sin\theta\,\text{N} - F_B = 0 \qquad\qquad (1)$$

$$+\nwarrow\Sigma F_y = 0; \qquad\qquad N_B - 20 \cos\theta\,\text{N} = 0 \qquad\qquad (2)$$

$$\curvearrowright+\Sigma M_O = 0; \qquad 20 \sin\theta\,\text{N}(4\text{ m}) - 20 \cos\theta\,\text{N}(x) = 0 \qquad (3)$$

(Impending Motion of Block.)   This requires use of the frictional equation

$$F_s = \mu_s N; \qquad\qquad\qquad F_B = 0.55 N_B \qquad\qquad\qquad (4)$$

Solving Eqs. 1 through 4 yields

$$N_B = 17.5\text{ N} \qquad F_B = 9.64\text{ N} \qquad \theta = 28.8° \qquad x = 2.2\text{ m}$$

Since $x = 2.2$ m > 2 m, the block will tip *before* sliding.

(Tipping of Block.)   This requires

$$x = 2\text{ m} \qquad\qquad\qquad\qquad (5)$$

Solving Eqs. 1 through 3 using Eq. 5 yields

$$N_B = 17.9\text{ N} \qquad F_B = 8.94\text{ N}$$

$$\theta = 26.6° \qquad\qquad\qquad\qquad\qquad Ans.$$

Note: If we *first* assumed that the block tips, then the results for F_B would have to be checked with the maximum *possible* static frictional force; i.e.,

$$F_B = 8.94\text{ N} \overset{?}{<} (0.55)(17.9\text{ N}) = 9.84\text{ N}$$

Since the inequality holds, indeed the block will tip before it slips.

Example 8–5

Beam *AB* is subjected to a uniform load of 200 N/m and is supported at *B* by a post *BC*, Fig. 8–13*a*. If the coefficients of static friction at *B* and *C* are $\mu_B = 0.2$ and $\mu_C = 0.5$, determine the force **P** needed to pull the post out from under the beam. Neglect the weight of the members and the thickness of the post.

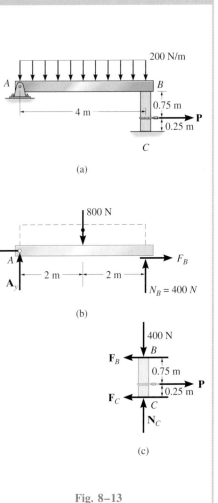

(a)

(b)

(c)

Fig. 8–13

SOLUTION

Free-Body Diagrams. The free-body diagram of beam *AB* is shown in Fig. 8–13*b*. Applying $\Sigma M_A = 0$, we obtain $N_B = 400$ N. This result is shown on the free-body diagram of the post, Fig. 8–13*c*. Referring to this member, the *four* unknowns F_B, P, F_C, and N_C are determined from the *three* equations of equilibrium and *one* frictional equation applied either at *B* or *C*.

Equations of Equilibrium and Friction.

$$\stackrel{+}{\rightarrow}\Sigma F_x = 0; \qquad\qquad P - F_B - F_C = 0 \qquad (1)$$
$$+\uparrow\Sigma F_y = 0; \qquad\qquad N_C - 400\text{ N} = 0 \qquad (2)$$
$$\zeta+\Sigma M_C = 0; \qquad -P(0.25\text{ m}) + F_B(1\text{ m}) = 0 \qquad (3)$$

(Post Slips Only at B.)   This requires $F_C \le \mu N_C$ and

$$F_B = \mu_B N_B; \qquad\qquad F_B = 0.2(400\text{ N}) = 80\text{ N}$$

Using this result and solving Eqs. 1 through 3, we obtain

$$P = 320\text{ N}$$
$$F_C = 240\text{ N}$$
$$N_C = 400\text{ N}$$

Since $F_C = 240$ N $> \mu_C N_C = 0.5(400$ N$) = 200$ N, the other case of movement must be investigated.

(Post Slips Only at C.)   Here $F_B \le \mu_B N_B$ and

$$F_C = \mu_C N_C; \qquad\qquad F_C = 0.5 N_C \qquad (4)$$

Solving Eqs. 1 through 4 yields

$$P = 267\text{ N} \qquad\qquad\qquad Ans.$$
$$N_C = 400\text{ N}$$
$$F_C = 200\text{ N}$$
$$F_B = 66.7\text{ N}$$

Obviously, this case occurs first, since it requires a *smaller* value for *P*.

Example 8-6

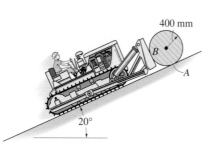

(a)

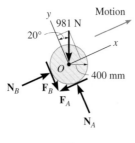

(b)

Fig. 8-14

Determine the normal force that must be exerted on the 100-kg spool shown in Fig. 8–14a to push it up the 20° incline at constant velocity. The coefficients of static and kinetic friction at the points of contact are $(\mu_s)_A = 0.18$, $(\mu_k)_A = 0.15$ and $(\mu_s)_B = 0.45$, $(\mu_k)_B = 0.4$.

SOLUTION

Free-Body Diagram. As shown in Fig. 8–14b, there are four unknowns N_A, F_A, N_B, and F_B acting on the spool. These can be determined from the *three* equations of equilibrium and *one* frictional equation, which applies either at A or B. If slipping only occurs at B, the spool *rolls* up the incline; whereas if slipping only occurs at A, the spool will *slide* up the incline. Here we must calculate N_B.

Equations of Equilibrium and Friction

$$+\nearrow\Sigma F_x = 0; \qquad -F_A + N_B - 981 \sin 20° \text{ N} = 0 \qquad (1)$$

$$+\nwarrow\Sigma F_y = 0; \qquad N_A - F_B - 981 \cos 20° \text{ N} = 0 \qquad (2)$$

$$\downarrow+\Sigma M_O = 0; \qquad F_B(400 \text{ mm}) - F_A(400 \text{ mm}) = 0 \qquad (3)$$

(Spool Rolls up Incline.) In this case $F_A \le 0.18N_A$ and

$$(F_k)_B = (\mu_k)_B N_B; \qquad F_B = 0.40N_B \qquad (4)$$

The direction of the frictional force at B must be specified correctly. Why? Since the spool is being forced up the plane, $\mathbf{F}_B$ acts downward to prevent the clockwise rolling motion of the spool, Fig. 8–14b. Solving Eqs. 1 through 4, we have

$$N_A = 1146 \text{ N} \qquad F_A = 224 \text{ N} \qquad N_B = 559 \text{ N} \qquad F_B = 224 \text{ N}$$

The assumption regarding no slipping at A should be checked.

$$F_A \le (\mu_s)_A N_A; \qquad 224 \text{ N} \overset{?}{\le} 0.18(1146 \text{ N}) = 206 \text{ N}$$

The inequality does *not apply,* and therefore slipping occurs at A and not at B. Hence, the other case of motion must be investigated.

(Spool Slides up Incline.) In this case, $F_B \le 0.45N_B$ and

$$(F_k)_A = (\mu_k)_A N_A; \qquad F_A = 0.15N_A \qquad (5)$$

Solving Eqs. 1 through 3 and 5 yields

$$N_A = 1084 \text{ N} \qquad F_A = 163 \text{ N} \qquad N_B = 498 \text{ N} \qquad F_B = 163 \text{ N}$$

The validity of the solution ($N_B = 498$ N) can be checked by testing the assumption that indeed no slipping occurs at B.

$$F_B \le (\mu_s)_B N_B; \qquad 163 \text{ N} < 0.45(498 \text{ N}) = 224 \text{ N} \qquad \text{(check)}$$

PROBLEMS

8–1. Determine the horizontal force P needed to just start moving the 300-N crate up the plane. Take $\mu_s = 0.3$.

8–2. Determine the range of values for which the horizontal force **P** will prevent the 300-lb crate from slipping down or up the inclined plane. Take $\mu_s = 0.1$.

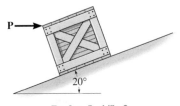

Probs. 8–1/8–2

8–3. If the horizontal force $P = 80$ N, determine the normal and frictional forces acting on the 300-N crate. Take $\mu_s = 0.3$, $\mu_k = 0.2$.

*8–4. If the horizontal force $P = 140$ N, determine the normal and frictional forces acting on the 300-N crate. Take $\mu_s = 0.3$, $\mu_k = 0.2$.

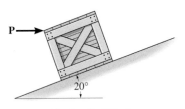

Probs. 8–3/8–4

8–5. Determine the magnitude of force **P** needed to start towing the 40-kg crate. Also determine the location of the resultant normal force acting on the crate, measured from point A. Take $\mu_s = 0.3$.

8–6. Determine the friction force on the 40-kg crate, and the resultant normal force if the force $P = 300$ N. Take $\mu_s = 0.5$ and $\mu_k = 0.2$.

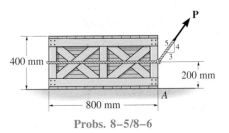

Probs. 8–5/8–6

8–7. The loose-fitting collar is supported by the pipe for which the coefficient of static friction at the points of contact A and B is $\mu_s = 0.2$. Determine the smallest dimension d so the rod will not slip when the load **P** is applied.

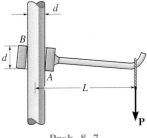

Prob. 8–7

*8–8. An axial force of $T = 800$ N is applied to the bar. If the coefficient of static friction at the jaws C and D is $\mu_s = 0.5$, determine the smallest normal force that the screw at A must exert on the smooth surface of the links at B and C in order to hold the bar stationary. The links are pin-connected at F and G.

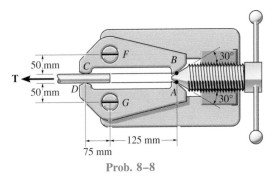

Prob. 8–8

8–9. The block brake consists of a pin-connected lever and friction block at B. The coefficient of static friction between the wheel and the lever is $\mu_s = 0.3$, and a torque of 5 N · m is applied to the wheel. Determine if the brake can hold the wheel stationary when the force applied to the lever is (a) $P = 30$ N, (b) $P = 70$ N.

8–10. Solve Prob. 8–9 if the 5-N · m torque is applied counterclockwise.

8–13. The winch on the truck is used to hoist the garbage bin onto the bed of the truck. If the loaded bin has a weight of 85 kN and center of gravity at G, determine the force in the cable needed to begin the lift. The coefficients of static friction at A and B are $\mu_A = 0.3$ and $\mu_B = 0.2$, respectively. Neglect the height of the support at A.

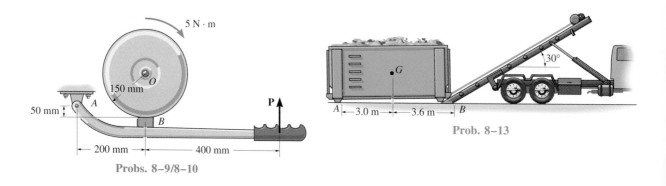

Prob. 8–13

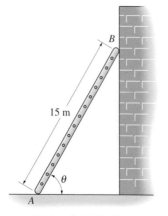

Probs. 8–9/8–10

8–11. The 15-m ladder has a uniform weight of 80 N and rests against the smooth wall at B. If the coefficient of static friction at A is $\mu_A = 0.4$, determine if the ladder will slip. Take $\theta = 60°$.

*8–12. Determine the smallest angle θ at which the ladder in Prob. 8–11 can be placed against the side of the smooth wall without having it slip.

8–14. The car has a mass of 1.6 Mg and center of mass at G. If the coefficient of static friction between the shoulder of the road and the tires is $\mu_s = 0.4$, determine the greatest slope θ the shoulder can have without causing the car to slip or tip over if the car travels along the shoulder at constant velocity.

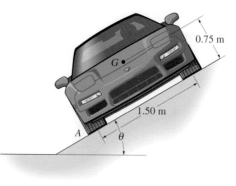

Prob. 8–14

Probs. 8–11/8–12

8–15. The man pushes against the stack of four uniform boxes each weighing 200 N. If the coefficient of static friction between each box is $\mu_s = 0.6$ and between the floor and the bottom box $\mu_s' = 0.4$, determine the greatest horizontal force P that can be applied without causing any slipping or tipping.

*8–16. The man pushes against the stack of four uniform boxes each weighing 200 N. If $P = 100$ N, determine if the stack will slip or tip. The coefficient of static friction between each box is $\mu_s = 0.6$ and between the floor and the bottom box $\mu_s' = 0.4$.

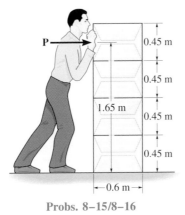

P→

0.45 m

0.45 m

1.65 m

0.45 m

0.45 m

←0.6 m→

Probs. 8–15/8–16

8–17. The uniform hoop of weight W is suspended from the peg at A and a horizontal force $\mathbf{P}$ is slowly applied at B. If the hoop begins to slip at A when $\theta = 30°$, determine the coefficient of static friction between the hoop and the peg.

8–18. The uniform hoop of weight W is suspended from the peg at A and a horizontal force $\mathbf{P}$ is slowly applied at B. If the coefficient of static friction between the hoop and peg is $\mu_s = 0.2$, determine if it is possible for the angle $\theta = 30°$ before the hoop begins to slip.

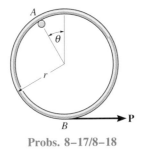

A

θ

r

B   → P

Probs. 8–17/8–18

8–19. The tractor has a weight of 45 kN with center of gravity at G. The driving traction is developed at the rear wheels B, while the front wheels at A are free to roll. If the coefficient of static friction between the wheels at B and the ground is $\mu_s = 0.5$, deter-

mine the largest force P at which it can pull without causing the wheels at B to slip or the front wheels at A to lift off the ground.

*8–20. The tractor has a weight of 45 kN with center of gravity at G. The driving traction is developed at the rear wheels B, while the front wheels at A are free to roll. If the coefficient of static friction between the wheels at B and the ground is $\mu_s = 0.5$, determine if it is possible to pull at $P = 12$ kN without causing the wheels at B to slip or the front wheels at A to lift off the ground.

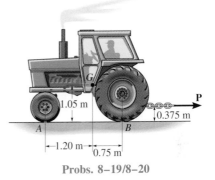

G

1.05 m

P

0.375 m

A   B

←1.20 m→

0.75 m

Probs. 8–19/8–20

8–21. The friction tongs are used to drag the 100-kg pallet and load along the floor. Determine the tension in the chain and the minimum coefficient of static friction at the shoes A and B of the tongs so slipping of the tongs does not occur. The coefficient of static friction between the pallet and the floor is $\mu_s = 0.4$.

8–22. The coefficient of static friction between the shoes at A and B of the tongs and the pallet is $\mu_s' = 0.5$, and between the pallet and the floor $\mu_s = 0.4$. If a horizontal towing force of $P = 300$ N is applied to the tongs, determine the largest mass that can be towed.

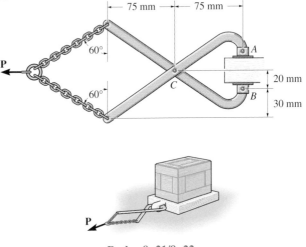

← 75 mm →|← 75 mm →

60°

P

A

20 mm

60°

C

B

30 mm

P

Probs. 8–21/8–22

8–23. Determine the maximum weight W the man can lift with constant velocity using the pulley system, without and then with the "leading block" or pulley at A. The man has a weight of 800 N and the coefficient of static friction between his feet and the ground is $\mu_s = 0.6$.

*8–24. If the weight of the load is $W = 320$ N, determine the normal and frictional forces acting on the 800-N man needed to support the load in each case. The coefficient of static friction between his feet and the ground is $\mu_s = 0.6$.

8–26. The uniform dresser has a weight of 360 N and rests on a tile floor for which $\mu_s = 0.25$. If the man pushes on it in the horizontal direction $\theta = 0°$, determine the smallest magnitude of force **F** needed to move the dresser. Also, if the man has a weight of 600 N, determine the smallest coefficient of static friction between his shoes and the floor so that he does not slip.

8–27. The uniform dresser has a weight of 360 N and rests on a tile floor for which $\mu_s = 0.25$. If the man pushes on it in the direction $\theta = 30°$, determine the smallest magnitude of force **F** needed to move the dresser. Also, if the man has a weight of 600 N, determine the smallest coefficient of static friction between his shoes and the floor so that he does not slip.

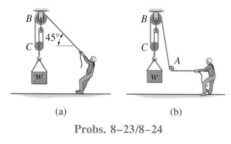

(a)   (b)

Probs. 8–23/8–24

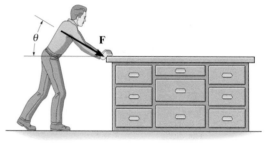

Probs. 8–26/8–27

8–25. The pipe is hoisted using the tongs. If the coefficient of static friction at A and B is μ_s, determine the smallest dimension b so that any pipe of inner diameter d can be lifted.

*8–28. The spool has a mass of 200 kg and rests against the wall and on the beam. If the coefficient of static friction at A and B is $\mu_A = 0.4$ and $\mu_B = 0.5$, respectively, determine the smallest vertical force P that must be applied to the cable that will cause the spool to turn.

8–29. The spool has a mass of 200 kg and rests against the wall and on the beam. If the coefficient of static friction at B is $\mu_B = 0.3$, and the wall is smooth, determine the friction force developed at B when the vertical force applied to the cable is $P = 800$ N.

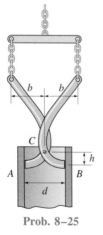

Prob. 8–25

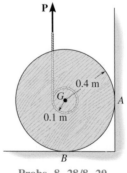

Probs. 8–28/8–29

8–30. Car *A* has a mass of 1.4 Mg and mass center at *G*. If car *B* exerts a horizontal force on *A* of 2 kN, determine if this force is great enough to move car *A*. The coefficients of static and kinetic friction between the tires and the road are $\mu_s = 0.5$ and $\mu_k = 0.35$. Assume *B*'s bumper is smooth and *B* does not slip.

*8–32. Determine how far *d* the man can walk slowly up the plank without causing the plank to slip. The coefficient of static friction at *A* and *B* is $\mu_s = 0.3$. The man has a weight of 800 N and a center of gravity at *G*. Neglect the thickness and weight of the plank.

8–33. Determine how far *d* the man can walk slowly up the 100-N plank without causing it to slip. The coefficient of static friction at *A* is $\mu_A = 0.5$, and the surface at *B* is smooth. The man has a weight of 800 N and a center of gravity at *G*. Neglect the thickness of the plank.

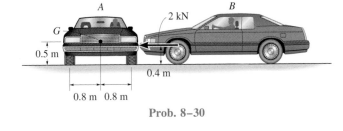

Prob. 8–30

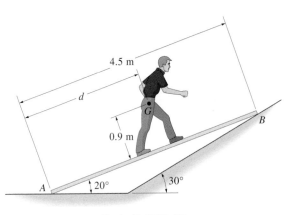

Prob. 8–32/8–33

8–31. The board can be adjusted vertically by tilting it up and sliding the smooth pin *A* along the vertical guide *G*. When placed horizontally, the bottom *C* then bears along the edge of the guide, where $\mu_s = 0.4$. Determine the largest dimension *d* which will support any applied force **F** without causing the board to slip downward.

8–34. The homogeneous semicylinder has a mass *m* and mass center at *G*. Determine the largest angle θ of the inclined plane upon which it rests so that it does not slip down the plane. The coefficient of static friction between the plane and the cylinder is $\mu_s = 0.3$. Also, what is the angle ϕ for this case?

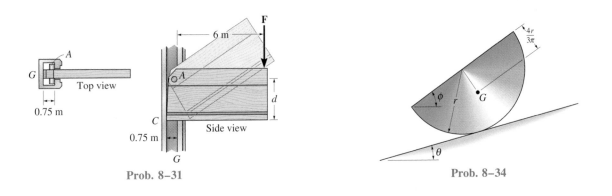

Prob. 8–31

Prob. 8–34

8–35. The 8-kN concrete pipe is being lowered from the truck bed when it is in the position shown. If the coefficient of static friction at the points of support A and B is $\mu_s = 0.4$, determine where it begins to slip first; at A or B, or both at A and B.

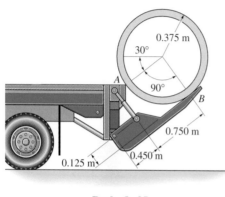

Prob. 8–35

*8–36.** The uniform crate resting on the dolly has a mass of 500 kg and mass center at G. If the front casters contact a high step, and the coefficient of static friction between the crate and the dolly is $\mu_s = 0.45$, determine the greatest force P that can be applied without causing motion of the crate. The dolly does not move.

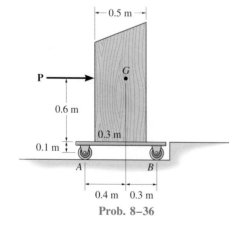

Prob. 8–36

8–37. The door brace AB is to be designed to prevent opening the door. If the brace forms a pin connection under the doorknob and the coefficient of static friction with the floor is $\mu_s = 0.5$, determine the largest length L the brace can have to prevent the door from being opened. Neglect the weight of the brace.

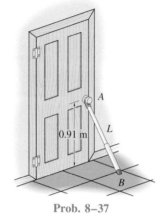

Prob. 8–37

8–38. The double-block brake mechanism is used to prevent the wheel from turning when the wheel is subjected to the torque of $M = 5\,\text{N} \cdot \text{m}$. If the coefficient of static friction between the blocks and the wheel is $\mu_s = 0.8$, determine the smallest vertical force P applied to the handle needed to stop the wheel.

8–39. Solve Prob. 8–38 if the torque is $M = 5\,\text{N} \cdot \text{m}$ clockwise.

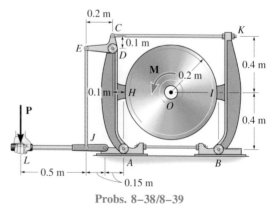

Probs. 8–38/8–39

*8–40. The carpenter slowly pushes the uniform board horizontally over the top of the saw horse. The board has a uniform weight of 100 N/m, and the saw horse has a weight of 150 N and a center of gravity at *G*. Determine if the saw horse will stay in position, slip, or tip if the board is pushed forward when *d* = 3 m. The coefficients of static friction are shown in the figure.

8–41. The carpenter slowly pushes the uniform board horizontally over the top of the saw horse. The board has a uniform weight of 100 N/m, and the saw horse has a weight of 150 N and a center of gravity at *G*. Determine if the saw horse will stay in position, slip, or tip if the board is pushed forward when *d* = 4.2 m. The coefficients of static friction are shown in the figure.

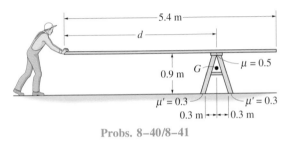

Probs. 8–40/8–41

8–42. The man has a weight of 800 N, and the coefficient of static friction between his shoes and the floor is $\mu_s = 0.5$. Determine where he should position his center of gravity *G* at *d* in order to exert the maximum horizontal force on the door. What is this force?

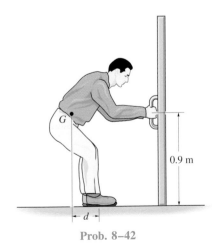

Prob. 8–42

8–43. The smooth barrel has a weight *W* and is to be held on the incline using the chock at *A*. If the coefficient of static friction between the chock and the incline is $\mu_s = 0.5$, determine the design angle θ of the chock so the barrel will not move. Neglect the weight of the chock.

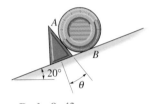

Prob. 8–43

*8–44. Determine the smallest force the man must exert on the rope in order to move the 80-kg crate. Also, what is the angle θ at this moment? The coefficient of static friction between the crate and the floor is $\mu_s = 0.3$.

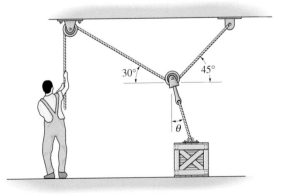

Prob. 8–44

8–45. The friction hook is made from a fixed frame which is shown colored and a cylinder of negligible weight. A piece of paper is placed between the wall and the cylinder. If $\mu_s = 0.3$ at all points of contact, determine the design angle θ so that any weight W of paper p can be held.

8–46. The friction hook is made from a fixed frame which is shown colored and a cylinder of negligible weight. A piece of paper is placed between the smooth wall and the cylinder. If $\theta = 20°$, determine the smallest coefficient of static friction μ at all points of contact so that any weight W of paper p can be held.

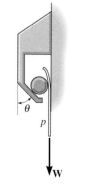

Probs. 8–45/8–46

8–47. The beam AB has a negligible mass and thickness and is subjected to a triangular distributed loading. It is supported at one end by a pin and at the other end by a post having a mass of 50 kg and negligible thickness. Determine the minimum force P needed to move the post. The coefficients of static friction at B and C are $\mu_B = 0.4$ and $\mu_C = 0.2$, respectively.

*8–48. The beam AB has a negligible mass and thickness and is subjected to a triangular distributed loading. It is supported at one end by a pin and at the other end by a post having a mass of 50 kg and negligible thickness. Determine the two coefficients of static friction at B and at C so that when the magnitude of the applied force is increased to $P = 150$ N, the post slips at both B and C simultaneously.

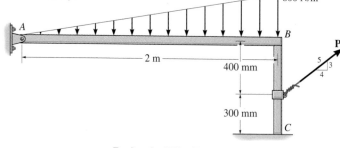

Probs. 8–47/8–48

8–49. The 45-kg disk rests on the surface for which the coefficient of static friction is $\mu_A = 0.2$. Determine the largest couple moment M that can be applied to the bar without causing motion.

8–50. The 45-kg disk rests on the surface for which the coefficient of static friction is $\mu_A = 0.15$. If $M = 50$ N $\cdot$ m, determine the friction force at A.

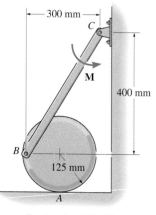

Probs. 8–49/8–50

8–51. The wheel weighs 20 N and rests on a surface for which $\mu_B = 0.2$. A cord wrapped around it is attached to the top of the 30-N homogeneous block. If the coefficient of static friction at D is $\mu_D = 0.3$, determine the smallest vertical force that can be applied tangentially to the wheel which will cause motion to impend.

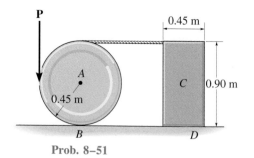

Prob. 8–51

*8–52.   The ring has a mass of 0.5 kg and is resting on the surface of the table. In an effort to move the ring a normal force **P** from the finger is exerted on it. If this force is directed towards the ring's center O as shown, determine its magnitude when the ring is on the verge of slipping at A. The coefficient of static friction at A is $\mu_A = 0.2$ and at B, $\mu_B = 0.3$.

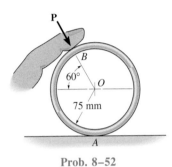

Prob. 8–52

8–54.   The end C of the two-bar linkage rests on the top center of the 50-kg cylinder. If the coefficients of static friction at C and E are $\mu_C = 0.6$ and $\mu_E = 0.3$, determine the largest vertical force P which can be applied at B without causing motion. Neglect the mass of the bars.

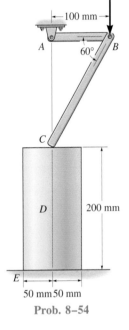

Prob. 8–54

8–53.   Two blocks A and B, each having a mass of 6 kg, are connected by the linkage shown. If the coefficients of static friction at the contacting surfaces are $\mu_B = 0.8$ and $\mu_A = 0.2$, determine the largest vertical force P that may be applied to pin C without causing the blocks to slip. Neglect the weight of the links.

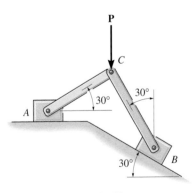

Prob. 8–53

8–55.   Block C has a mass of 50 kg and is confined between two walls by smooth rollers. If the block rests on top of the 40-kg spool, determine the minimum cable force P needed to move the spool. The cable is wrapped around the spool's inner core. The coefficients of static friction at A and B are $\mu_A = 0.3$ and $\mu_B = 0.6$.

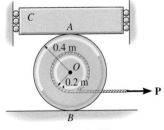

Prob. 8–55

*8–56.   Block C has a mass of 50 kg and is confined between two walls by smooth rollers. If the block rests on top of the 40-kg spool, determine the required coefficients of static friction at A and B so that the spool slips at A and B when the magnitude of the applied force is increased to $P = 300$ N.

8–59.   The uniform rod has a length l, a mass m, and is supported by a ball-and-socket joint at A. If it is located a distance a from the wall, determine the smallest height h for placement against the wall which will not allow the rod to slip. The coefficient of static friction at B is μ.

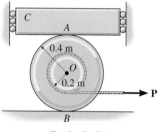

Prob. 8–56

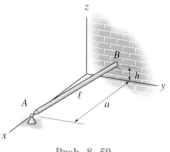

Prob. 8–59

8–57.   The block of weight W is being pulled up the inclined plane of slope α using a force $\mathbf{P}$. If $\mathbf{P}$ acts at the angle ϕ as shown, show that for slipping to occur, $P = W \sin(\alpha + \theta)/\cos(\phi - \theta)$, where θ is the angle of friction; $\theta = \tan^{-1}\mu$.

8–58.   Determine the angle ϕ at which $\mathbf{P}$ should act on the block so that the magnitude of $\mathbf{P}$ is as small as possible. What is the corresponding value of P? The block weighs W and the slope α is known.

*8–60.   The disk has a weight W and lies on the plane which has a coefficient of static friction μ. Determine the maximum height h to which the plane can be lifted without causing the disk to slip.

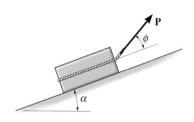

Probs. 8–57/8–58

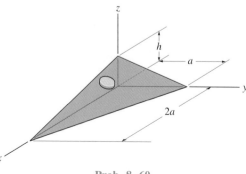

Prob. 8–60

8.3 Wedges

A *wedge* is a simple machine which is often used to transform an applied force into much larger forces, directed at approximately right angles to the applied force. Also, wedges can be used to give small displacements or adjustments to heavy loads.

Consider, for example, the wedge shown in Fig. 8–15a, which is used to *lift* a block of weight W by applying a force P to the wedge. Free-body diagrams of the block and wedge are shown in Fig. 8–15b. Here we have excluded the weight of the wedge since it is usually *small* compared to the weight of the block. Also, note that the frictional forces F_1 and F_2 must oppose the motion of the wedge. Likewise, the frictional force F_3 of the wall on the block must act downward so as to oppose the block's upward motion. The locations of the resultant normal forces are not important in the force analysis, since neither the block nor wedge will "tip." Hence the moment equilibrium equations will not be considered. There are seven unknowns consisting of the applied force P, needed to cause motion of the wedge, and six normal and frictional forces. The seven available equations consist of two force equilibrium equations ($\Sigma F_x = 0$, $\Sigma F_y = 0$) applied to the wedge and block (four equations total) and the frictional equation $F = \mu N$ applied at each surface of contact (three equations total).

If the block is to be *lowered*, the frictional forces will all act in a sense opposite to that shown in Fig. 8–15b. The applied force P will act to the right as shown if the coefficient of friction is very *small* or the wedge angle θ is *large*. Otherwise, P may have the reverse sense of direction in order to *pull* on the wedge to remove it. If P is *removed*, or $P = 0$, and friction forces hold the block in place, then the wedge is referred to as *self-locking*.

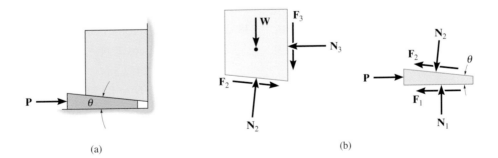

(a)   (b)

Fig. 8–15

Example 8–7

The uniform stone has a mass of 500 kg and is held in the horizontal position using a wedge at B as shown in Fig. 8–16a. If the coefficient of static friction is $\mu_s = 0.3$ at the surface in contact with the wedge, determine the force P needed to remove the wedge. Is the wedge self-locking? Assume that the stone does not slip at A.

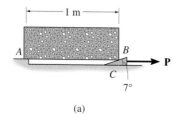

(a)

Fig. 8–16

(b)

SOLUTION

Since the wedge is to be removed, slipping is about to occur at the surfaces of contact. Thus, $F = \mu_s N$ on the wedge, and the free-body diagrams are shown in Fig. 8–16b. Note that on the wedge the friction force opposes the motion and on the stone at A, $F_A \leq \mu_s N_A$, since slipping does not occur there. There are five unknowns F_A, N_A, N_B, N_C, and P. Three equilibrium equations for the stone and two for the wedge are available for solution. From the free-body diagram of the stone,

$$\zeta + \Sigma M_A = 0; \quad -4905 \text{ N}(0.5 \text{ m}) + (N_B \cos 7° \text{ N})(1 \text{ m}) +$$
$$(0.3 N_B \sin 7° \text{ N})(1 \text{ m}) = 0$$
$$N_B = 2383.1 \text{ N}$$

Using this result for the wedge, we have

$$\xrightarrow{+} \Sigma F_x = 0; \quad 2383.1 \sin 7° \text{ N} - 0.3(2383.1 \cos 7° \text{ N}) + P - 0.3 N_C = 0$$
$$+\uparrow \Sigma F_y = 0; \quad N_C - 2383.1 \cos 7° \text{ N} - 0.3(2383.1 \sin 7° \text{ N}) = 0$$
$$N_C = 2452.5 \text{ N}$$
$$P = 1154.9 \text{ N} = 1.15 \text{ kN} \qquad \qquad Ans.$$

Since P is positive, indeed the wedge must be pulled out. Obviously, if P is zero, the wedge will remain in place (self-locking) and the frictional forces $\mathbf{F}_C$ and $\mathbf{F}_B$ developed at the points of contact will satisfy $F_B < \mu_s N_B$ and $F_C < \mu_s N_C$.

*8.4 Frictional Forces on Screws

In most cases screws are used as fasteners; however, in many types of machines they are incorporated to transmit power or motion from one part of the machine to another. A *square-threaded screw* is commonly used for the latter purpose, especially when large forces are applied along its axis. In this section we will analyze the forces acting on square-threaded screws. The analysis of other types of screws, such as the V-thread, is based on the same principles.

A *screw* may be thought of simply as an inclined plane or wedge wrapped around a cylinder. A nut initially at position A on the screw shown in Fig. 8–17a will move up to B when rotated 360° around the screw. This rotation is equivalent to translating the nut up an inclined plane of height l and length $2\pi r$, where r is the mean radius of the thread, Fig. 8–17b. The rise l for a single revolution is referred to as the *lead* of the screw, where the *lead angle* is given by $\theta = \tan^{-1}(l/2\pi r)$.

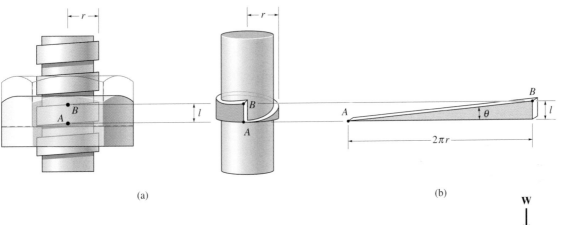

(a)

(b)

Fig. 8–17

Frictional Analysis. When a screw is subjected to large axial loads, the frictional forces developed on the thread become important if we are to determine the moment **M*** needed to turn the screw. Consider, for example, the square-threaded jack screw shown in Fig. 8–18, which supports the vertical load **W**. The reactive forces of the jack to this load are actually distributed over the circumference of the screw thread in contact with the screw hole in the jack, that is, within region h shown in Fig. 8–18. For simplicity, this

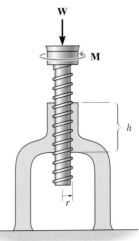

Fig. 8–18

---

*For applications, **M** is developed by applying a horizontal force **P** at a right angle to the end of a lever that would be fixed to the screw.

portion of thread can be imagined as being unwound from the screw and represented as a simple block resting on an inclined plane having the screw's lead angle θ, Fig. 8–19a. Here the inclined plane represents the inside *supporting thread* of the jack base. Three forces act on the block or screw. The force **W** is the total axial load applied to the screw. The horizontal force **S** is caused by the applied moment **M**, such that the magnitudes of these loads can be related by summing moments about the axis of the screw. We require $M = Sr$, where r is the screw's mean radius. As a result of **W** and **S**, the inclined plane exerts a resultant force **R** on the block, which is shown to have components acting normal, **N**, and tangent, **F**, to the contacting surfaces.

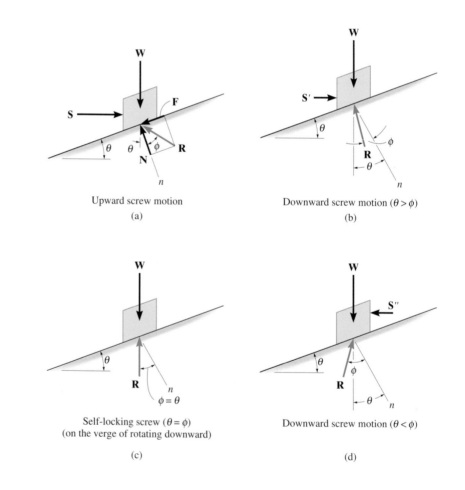

Upward screw motion
(a)

Downward screw motion $(\theta > \phi)$
(b)

Self-locking screw $(\theta = \phi)$
(on the verge of rotating downward)
(c)

Downward screw motion $(\theta < \phi)$
(d)

Fig. 8–19

Upward Screw Motion. Provided M is great enough, the screw (and hence the block) can either be brought to the verge of upward impending motion or motion can be occurring. Under these conditions, **R** acts at an angle $(\theta + \phi)$ from the vertical as shown in Fig. 8–19a, where $\phi = \tan^{-1}(F/N) = \tan^{-1}(\mu N/N) = \tan^{-1}\mu$. Applying the two force equations of equilibrium to the block, we obtain

$$\xrightarrow{+}\Sigma F_x = 0; \qquad\qquad S - R \sin(\theta + \phi) = 0$$
$$+\uparrow\Sigma F_y = 0; \qquad\qquad R \cos(\theta + \phi) - W = 0$$

Eliminating R and solving for S, then substituting this value into the equation $M = Sr$, yields

$$M = Wr \tan(\theta + \phi) \qquad\qquad (8\text{–}3)$$

As indicated, M is the moment necessary to cause upward impending motion of the screw, provided $\phi = \phi_s = \tan^{-1}\mu_s$ (the angle of static friction). If ϕ is replaced by $\phi_k = \tan^{-1}\mu_k$ (the angle of kinetic friction), Eq. 8–3 will give a smaller value M necessary to maintain uniform upward motion of the screw.

Downward Screw Motion $(\theta > \phi)$. If the surface of the screw is very *slippery*, it may be possible for the screw to rotate downward if the magnitude, *not the direction*, of the moment is reduced to, say, $M' < M$. As shown in Fig. 8–19b, this causes the effect of **M'** to become **S'**, and it requires the angle ϕ (ϕ_s or ϕ_k) to lie on the opposite side of the normal n to the plane supporting the block, such that $\theta > \phi$. For this case, Eq. 8–3 becomes

$$M' = Wr \tan(\theta - \phi) \qquad\qquad (8\text{–}4)$$

Self-Locking Screw. If the moment **M** (or its effect **S**) is *removed*, the screw will remain *self-locking*; i.e., it will support the load **W** by *friction forces alone* provided $\phi \geq \theta$. To show this, consider the necessary limiting case when $\phi = \theta$, Fig. 8–19c. Here vertical equilibrium is maintained since **R** is vertical and thus balances **W.**

Downward Screw Motion $(\theta < \phi)$. When the surface of the screw is *very rough*, the screw will not rotate downward as stated above. Instead, the direction of the applied moment must be *reversed* in order to cause the motion. The free-body diagram shown in Fig. 8–19d is representative of this case. Here **S''** is caused by the applied (reverse) moment **M''**. Hence Eq. 8–3 becomes

$$M'' = Wr \tan(\phi - \theta) \qquad\qquad (8\text{–}5)$$

Each of the above cases should be thoroughly understood before proceeding to solve problems.

Example 8–8

The turnbuckle shown in Fig. 8–20 has a square thread with a mean radius of 5 mm and a lead of 2 mm. If the coefficient of static friction between the screw and the turnbuckle is $\mu_s = 0.25$, determine the moment **M** that must be applied to draw the end screws closer together. Is the turnbuckle self-locking?

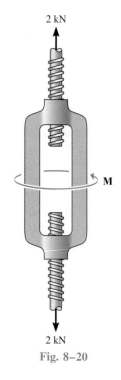

2 kN

M

2 kN

Fig. 8–20

SOLUTION

The moment may be obtained by using Eq. 8–3. Why? Since friction at *two screws* must be overcome, this requires

$$M = 2[Wr \tan(\theta + \phi)] \qquad (1)$$

Here $W = 2000$ N, $r = 5$ mm, $\phi_s = \tan^{-1}\mu_s = \tan^{-1}(0.25) = 14.04°$, and $\theta = \tan^{-1}(l/2\pi r) = \tan^{-1}(2 \text{ mm}/[2\pi(5 \text{ mm})]) = 3.64°$. Substituting these values into Eq. 1 and solving gives

$$M = 2[(2000 \text{ N})(5 \text{ mm}) \tan(14.04° + 3.64°)]$$

$$= 6375.1 \text{ N} \cdot \text{mm} = 6.38 \text{ N} \cdot \text{m} \qquad \textit{Ans.}$$

When the moment is *removed*, the turnbuckle will be self-locking; i.e., it will not unscrew, since $\phi_s > \theta$.

PROBLEMS

8–61. The blocks each have a weight of 50 N. If the coefficient of static friction at A is $\mu_s = 0.2$ and between each block $\mu_s' = 0.4$, determine how many blocks can be stacked as shown before they begin to topple.

20°

20°

A

Prob. 8–61

8–62. Each block has a weight of 400 N. Determine how far the force **P** can compress the spring until block B slips on block A. What is the magnitude of **P** for this to occur?

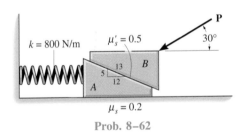

$k = 800$ N/m $\mu_s' = 0.5$ **P**   30°

13   B
5   12
A

$\mu_s = 0.2$

Prob. 8–62

8–63. Column D is subjected to a vertical load of 8000 N. It is supported on two identical wedges A and B for which the coefficient of static friction at the contacting surfaces between A and B and B and C is $\mu_s = 0.4$. Determine the force P needed to raise wedge B and the equilibrium force P′ needed to hold wedge A stationary. The contacting surface between A and D is smooth.

***8–64.** Column D is subjected to a vertical load of 8000 N. It is supported on two identical wedges A and B for which the coefficient of static friction at the contacting surfaces between A and B and B and C is $\mu_s = 0.4$. If the forces **P** and **P′** are removed are the wedges self-locking? The contacting surface between A and D is smooth.

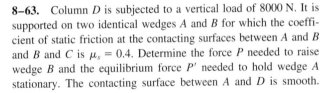

8000 N

D

P → *B* 10° ↑ 10° *A* ← **P′**

C

Probs. 8–63/8–64

8–65. If the spring is compressed 60 mm and the coefficient of static friction between the tapered stub S and the slider A is $\mu_{SA} = 0.5$, determine the horizontal force P needed to move the slider forward. The stub is free to move without friction within the fixed collar C. The coefficient of static friction between A and surface B is $\mu_{AB} = 0.4$. Neglect the weights of the slider and stub.

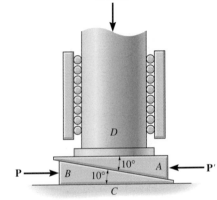

$k = 300$ N/m

C

S

P →   30°
A

B

Prob. 8–65

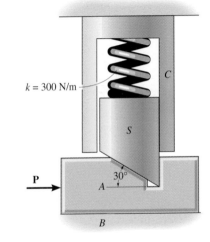

8–66. If the beam AD is loaded as shown, determine the horizontal force P which must be applied to the wedge in order to remove it from under the beam. The coefficients of static friction at the wedge's top and bottom surfaces are $\mu_{CA} = 0.25$ and $\mu_{CB} = 0.35$, respectively. If $P = 0$, is the wedge self-locking?

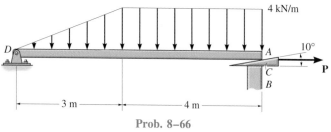

Prob. 8–66

8–69. The coefficient of static friction between wedges B and C is $\mu_s = 0.6$ and between the surfaces of contact B and A and C and D, $\mu_s' = 0.4$. If the spring is compressed 200 mm when in the position shown, determine the smallest force P needed to move wedge C to the left. Neglect the weight of the wedges.

8–70. The coefficient of static friction between the wedges B and C is $\mu_s = 0.6$ and between the surfaces of contact B and A and C and D, $\mu_s' = 0.4$. If $P = 50$ N, determine the largest allowable compression of the spring without causing wedge C to move to the left. Neglect the weight of the wedges.

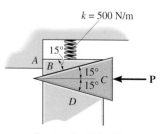

Probs. 8–69/8–70

8–67. The wedge is used to level the member. For the loading shown, determine the horizontal force P that must be applied to move the wedge to the right. The coefficient of static friction between the wedge and its two surfaces of contact is $\mu_s = 0.25$. Neglect the size and weight of the wedge.

*8–68. The wedge is used to level the member. For the loading shown, determine the reversed horizontal force $-\mathbf{P}$ that must be applied to pull the wedge out to the left. The coefficient of static friction between the wedge and its two surfaces of contact is $\mu_s = 0.15$. Neglect the weight of the wedge.

8–71. The wedge blocks are used to hold the specimen in a tension testing machine. Determine the design angle θ of the wedges so that the specimen will not slip regardless of the applied load. The coefficients of static friction are $\mu_A = 0.1$ at A and $\mu_B = 0.6$ at B. Neglect the weight of the blocks.

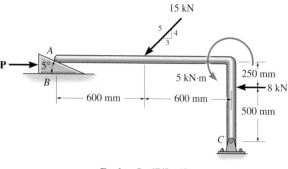

Probs. 8–67/8–68

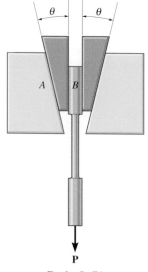

Prob. 8–71

*8–72. The two blocks have weights of $W_A = 600$ N and $W_B = 500$ N. Determine the smallest horizontal force **P** that must be applied to block A in order to move it. The coefficient of static friction between the blocks is $\mu_s = 0.3$ and between the floor and each block $\mu_s' = 0.5$.

8–74. The column is used to support the upper floor. If a force $F = 80$ N is applied perpendicular to the handle to tighten the screw, determine the compressive force in the column. The square-threaded screw on the jack has a coefficient of static friction of $\mu_s = 0.4$, mean diameter of 25 mm, and a lead of 3 mm.

8–75. If the force **F** is removed from the handle of the jack in Prob. 8–74, determine if the screw is self-locking.

Prob. 8–72

8–73. The device is used to pull the battery-cable terminal C from the post of a battery. If the required pulling force is 850 N, determine the torque M that must be applied to the handle on the screw to tighten it. The screw has square threads, a mean diameter of 5 mm, a lead of 2 mm, and a coefficient of static friction of $\mu_s = 0.5$.

Probs. 8–74/8–75

*8–76. The clamp provides pressure from several directions on the edges of the board. If the square-threaded screw has a lead of 3 mm, radius of 10 mm, and the coefficient of static friction is $\mu_s = 0.4$, determine the horizontal force developed on the board at A and the vertical forces developed at B and C if a torque of $M = 1.5$ N · m is applied to the handle to tighten it further. The blocks at B and C are pin-connected to the board.

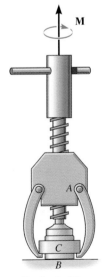

Prob. 8–73

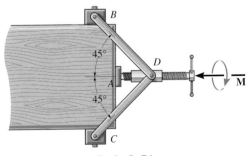

Prob. 8–76

8–77. Determine the clamping force on the board A if the screw of the "C" clamp is tightened with a twist of $M = 8 \, \text{N} \cdot \text{m}$. The single square-threaded screw has a mean radius of 10 mm, a lead of 3 mm, and the coefficient of static friction is $\mu_s = 0.35$.

8–78. If the required clamping force at the board A is to be 50 N, determine the torque M that must be applied to the handle of the "C" clamp to tighten it down. The single square-threaded screw has a mean radius of 10 mm, a lead of 3 mm, and the coefficient of static friction is $\mu_s = 0.35$.

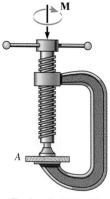

Probs. 8–77/8–78

8–79. The square-threaded screw has a mean diameter of 20 mm and a lead of 4 mm. If the weight of the plate A is 50 N, determine the smallest coefficient of static friction between the screw and the plate so that the plate does not travel down the screw when the plate is suspended as shown.

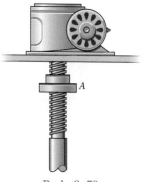

Prob. 8–79

*8–80. Determine the horizontal force P applied perpendicular to the handle of the jack screw necessary to start lifting the 3-kN load. The square-threaded screw has a lead of 5 mm and a mean diameter of 60 mm. The coefficient of static friction for the screw is $\mu_s = 0.2$.

Prob. 8–80

8–81. The shaft has a square-threaded screw with a lead of 9 mm and a mean radius of 15 mm. If it is in contact with a plate gear having a mean radius of 20 mm, determine the resisting torque M on the plate gear which can be overcome if a torque of 7 N · m is applied to the shaft. The coefficient of static friction between the gear and the screw is $\mu_s = 0.2$. Neglect friction of the bearings located at A and B.

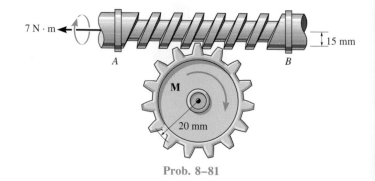

Prob. 8–81

8–82. The fixture clamp consists of a square-threaded screw having a coefficient of static friction of $\mu_s = 0.3$, mean diameter of 3 mm, and a lead of 1 mm. The five points indicated are pin connections. Determine the clamping force at the smooth blocks D and E when a torque of $M = 0.08$ N · m is applied to the handle of the screw.

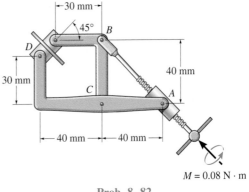

Prob. 8–82

*8.5 Frictional Forces on Flat Belts

Whenever belt drives or band brakes are designed, it is necessary to determine the frictional forces developed between the belt and its contacting surface. In this section we will analyze the frictional forces acting on a flat belt, although the analysis of other types of belts, such as the V-belt, is based on similar principles.

Here we will consider the flat belt shown in Fig. 8–21*a*, which passes over a fixed curved surface, such that the total angle of belt to surface contact in radians is β and the coefficient of friction between the two surfaces is μ. We will determine the tension T_2 in the belt which is needed to pull the belt counterclockwise over the surface and thereby overcome both the frictional forces at the surface of contact and the known tension T_1. Obviously, $T_2 > T_1$.

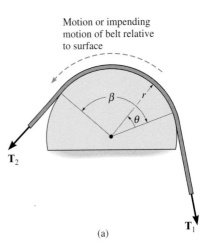

(a)

Frictional Analysis. A free-body diagram of the belt segment in contact with the surface is shown in Fig. 8–21b. Here the normal force **N** and the frictional force **F,** acting at different points along the belt, will vary both in magnitude and direction. Due to this *unknown* force distribution, the analysis of the problem will proceed on the basis of initially studying the forces acting on a differential element of the belt.

A free-body diagram of an element having a length *ds* is shown in Fig. 8–21c. Assuming either impending motion or motion of the belt, the magnitude of the frictional force $dF = \mu \, dN$. This force opposes the sliding motion of the belt and thereby increases the magnitude of the tensile force acting in the belt by *dT*. Applying the two force equations of equilibrium, we have

$$\xrightarrow{+}\Sigma F_x = 0; \quad T\cos\left(\frac{d\theta}{2}\right) + \mu \, dN - (T + dT)\cos\left(\frac{d\theta}{2}\right) = 0$$

$$+\uparrow\Sigma F_y = 0; \quad dN - (T + dT)\sin\left(\frac{d\theta}{2}\right) - T\sin\left(\frac{d\theta}{2}\right) = 0$$

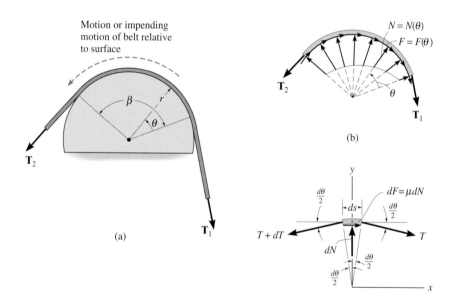

(a)

(b)

(c)

Fig. 8–21

Since $d\theta$ is of *infinitesimal size*, $\sin(d\theta/2)$ and $\cos(d\theta/2)$ can be replaced by $d\theta/2$ and 1, respectively. Also, the *product* of the two infinitesimals dT and $d\theta/2$ may be neglected when compared to infinitesimals of the first order. The above two equations therefore reduce to

$$\mu \, dN = dT$$

and

$$dN = T \, d\theta$$

Eliminating dN yields

$$\frac{dT}{T} = \mu \, d\theta$$

Integrating this equation between all the points of contact that the belt makes with the drum, and noting that $T = T_1$ at $\theta = 0$ and $T = T_2$ at $\theta = \beta$, yields

$$\int_{T_1}^{T_2} \frac{dT}{T} = \mu \int_0^{\beta} d\theta$$

$$\ln \frac{T_2}{T_1} = \mu\beta$$

Solving for T_2, we obtain

$$T_2 = T_1 e^{\mu\beta} \tag{8-6}$$

where $T_2, T_1 =$ belt tensions; $\mathbf{T}_1$ opposes the direction of motion (or impending motion) of the belt measured relative to the surface, while $\mathbf{T}_2$ acts in the direction of the relative belt motion (or impending motion); because of friction, $T_2 > T_1$

$\mu =$ coefficient of static or kinetic friction between the belt and the surface of contact

$\beta =$ angle of belt to surface contact, measured in radians

$e = 2.718. \ldots$, base of the natural logarithm

Note that Eq. 8–6 is *independent* of the *radius* of the drum and instead depends on the angle of belt to surface contact, β. Furthermore, as indicated by the integration, this equation is valid for flat belts placed on *any shape* of contacting surface. For application, however, keep in mind that Eq. 8–6 is valid only when *impending motion* occurs.

Example 8–9

The maximum tension that can be developed in the belt shown in Fig. 8–22a is 500 N. If the pulley at A is free to rotate and the coefficient of static friction at the fixed drums B and C is $\mu_s = 0.25$, determine the largest mass of the cylinder that can be lifted by the belt. Assume that the force **T** applied at the end of the belt is directed vertically downward, as shown.

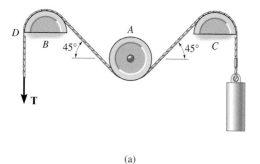

(a)

SOLUTION

Lifting the cylinder, which has a weight $W = mg$, causes the belt to move counterclockwise over the drums at B and C; hence, the maximum tension T_2 in the belt occurs at D. Thus, $T_2 = 500$ N. A section of the belt passing over the drum at B is shown in Fig. 8–22b. Since $180° = \pi$ rad, the angle of contact between the drum and the belt is $\beta = (135°/180°)\pi = 3\pi/4$ rad. Using Eq. 8–6, we have

$$T_2 = T_1 e^{\mu_s \beta}; \qquad 500 \text{ N} = T_1 e^{0.25[(3/4)\pi]}$$

Hence,

$$T_1 = \frac{500 \text{ N}}{e^{0.25[(3/4)\pi]}} = \frac{500 \text{ N}}{1.80} = 277.4 \text{ N}$$

Since the pulley at A is free to rotate, equilibrium requires that the tension in the belt remains the *same* on both sides of the pulley.

The section of the belt passing over the drum at C is shown in Fig. 8–22c. The weight $W < 277.4$ N. Why? Applying Eq. 8–6, we obtain

$$T_2 = T_1 e^{\mu_s \beta}; \qquad 277.4 \text{ N} = W e^{0.25[(3/4)\pi]}$$

$$W = 153.9 \text{ N}$$

so that

$$m = \frac{W}{g} = \frac{153.9 \text{ N}}{9.81 \text{ m/s}^2}$$

$$= 15.7 \text{ kg} \qquad \qquad Ans.$$

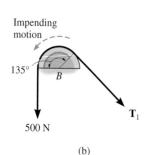

Impending motion

135°

B

500 N

T$_1$

(b)

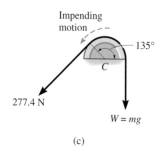

Impending motion

135°

C

277.4 N

$W = mg$

(c)

Fig. 8–22

PROBLEMS

8–83. Determine the *minimum* tension in the rope at points A and B that is necessary to maintain equilibrium. Take $\mu_s = 0.3$ between the rope and the fixed post D.

*8–84. Determine the *maximum* tension in the rope at points A and B that is necessary to maintain equilibrium. Take $\mu_s = 0.3$ between the rope and the fixed post D.

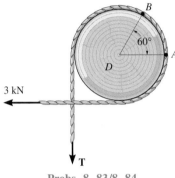

Probs. 8–83/8–84

8–85. A cylinder having a mass of 250 kg is to be supported by the cord which wraps over the pipe. Determine the *smallest* vertical force F needed to support the load if the cord passes (a) once over the pipe, $\beta = 180°$, and (b) two times over the pipe, $\beta = 540°$. Take $\mu_s = 0.2$.

8–86. A cylinder having a mass of 250 kg is to be supported by the cord which wraps over the pipe. Determine the *largest* vertical force F that can be applied to the cord without moving the cylinder. The cord passes (a) once over the pipe, $\beta = 180°$, and (b) two times over the pipe, $\beta = 540°$. Take $\mu_s = 0.2$.

Probs. 8–85/8–86

8–87. The truck, which has a mass of 3.4 Mg, is to be lowered down the slope by a rope that is wrapped around a tree. If the wheels are free to roll and the man at A can resist a pull of 300 N, determine the minimum number of turns the rope should be wrapped around the tree to lower the truck at a constant speed. The coefficient of kinetic friction between the tree and rope is $\mu_k = 0.3$.

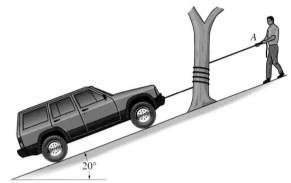

Prob. 8–87

*8–88. The boat has a weight of 500 N and is held in position off the side of a ship by the spars at A and B. A sailor having a weight of 130 N gets in the boat, wraps a rope around an overhead boom at C, and ties it to the ends of the boat as shown. If the boat is disconnected from the spars, determine the *minimum number* of *half turns* the rope must make around the boom so that the boat can be safely lowered into the water. The coefficient of kinetic friction between the rope and the boom is $\mu_k = 0.15$. *Hint:* The problem requires that the normal force between the man's feet and the boat be as small as possible.

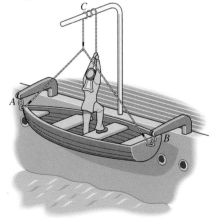

Prob. 8–88

8–89. A cord having a weight of 0.5 N/m and a total length of 10 m is suspended over a peg P as shown. If the coefficient of static friction between the peg and cord is $\mu_s = 0.5$, determine the longest length h which one side of the suspended cord can have without causing motion. Neglect the size of the peg and the length of cord draped over it.

Prob. 8–89

8–90. The uniform concrete pipe has a weight of 800 N and is unloaded slowly from the truck bed using the rope and skids shown. If the coefficient of kinetic friction between the rope and pipe is $\mu_k = 0.3$, determine the force the worker must exert on the rope to lower the pipe at constant speed. There is a pulley at B, and the pipe does not slip on the skids. The lower portion of the rope is parallel to the skids.

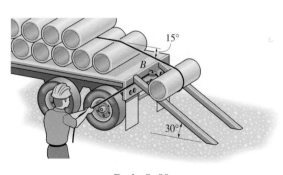

Prob. 8–90

8–91. The choker sling is used to lift the smooth pipe that has a mass of 600 kg. If the coefficient of static friction between the loop at the end A of the sling and the rope is $\mu_s = 0.3$, determine the angle θ at the connection.

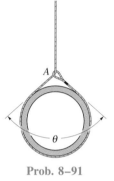

Prob. 8–91

8–92. The belt on the portable dryer wraps around the drum D, idler pulley A, and motor pulley B. If the motor can develop a maximum torque of $M = 0.80$ N · m, determine the smallest spring tension required to hold the belt from slipping. The coefficient of static friction between the belt and the drum and motor pulley is $\mu_s = 0.3$.

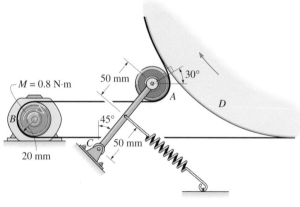

Prob. 8–92

8–93. Blocks A and B weigh 50 N and 30 N, respectively. Using the coefficients of static friction indicated, determine the greatest weight of block D without causing motion.

8–94. Blocks A, B, and D weigh 50 N, 30 N, and 12 N, respectively. Using the coefficients of static friction indicated, determine the frictional force between blocks A and B and between block A and the floor C.

8–97. The simple band brake is constructed so that the ends of the friction strap are connected to the pin at A and the lever arm at B. If the wheel is subjected to a torque of $M = 80$ N · m, determine the smallest force P applied to the lever that is required to hold the wheel stationary. The coefficient of static friction between the strap and wheel is $\mu_s = 0.5$.

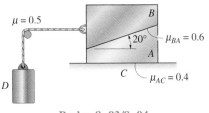

Probs. 8–93/8–94

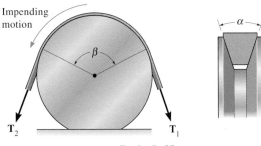

Prob. 8–97

8–95. Determine the smallest lever force P needed to prevent the wheel from rotating if it is subjected to a torque of $M = 250$ N · m. The coefficient of static friction between the belt and the wheel is $\mu_s = 0.3$. The wheel is pin-connected at its center, B.

*8–96.** Determine the torque M that can be resisted by the band brake if a force of $P = 30$ N is applied to the handle of the lever. The coefficient of static friction between the belt and the wheel is $\mu_s = 0.3$. The wheel is pin-connected at its center, B.

8–98. Show that the frictional relationship between the belt tensions, the coefficient of friction μ, and the angular contacts α and β for the V-belt is $T_2 = T_1 e^{\mu\beta/\sin(\alpha/2)}$.

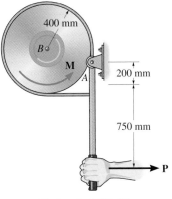

Probs. 8–95/8–96

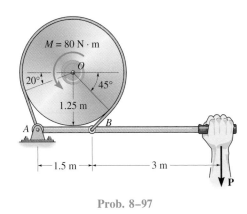

Prob. 8–98

8–99. Block *A* has a mass of 50 kg and rests on surface *B* for which $\mu_s = 0.25$. If the coefficient of static friction between the cord and the fixed peg at *C* is $\mu_s' = 0.3$, determine the greatest mass of the suspended cylinder *D* without causing motion.

*8–100. Block *A* has a mass of 50 kg and rests on surface *B* for which $\mu_s = 0.25$. If the mass of the suspended cylinder *D* is 4 kg, determine the frictional force of the surface on *A*. The coefficient of static friction between the cord and the fixed peg at *C* is $\mu_s' = 0.3$.

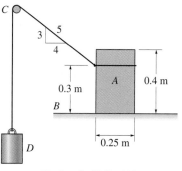

Probs. 8–99/8–100

*8.6 Frictional Forces on Collar Bearings, Pivot Bearings, and Disks

Pivot and *collar bearings* are commonly used in machines to support an *axial load* on a rotating shaft. These two types of support are shown in Fig. 8–23. Provided the bearings are not lubricated, or are only partially lubricated, the laws of dry friction may be applied to determine the moment **M** needed to turn the shaft when it supports an axial force **P**.

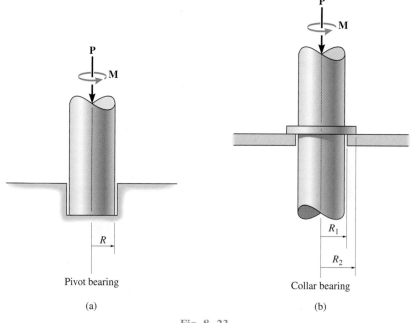

Pivot bearing

(a)

Collar bearing

(b)

Fig. 8–23

Frictional Analysis. The collar bearing on the shaft shown in Fig. 8–24 is subjected to an axial force **P** and has a total bearing or contact area $\pi(R_2^2 - R_1^2)$. In the following analysis, the normal pressure p is considered to be *uniformly distributed* over this area—a reasonable assumption provided the bearing is new and evenly supported. Since $\Sigma F_z = 0$, p, measured as a force per unit area, is $p = P/\pi(R_2^2 - R_1^2)$.

The moment needed to cause impending rotation of the shaft can be determined from moment equilibrium of the frictional forces dF developed at the bearing surface by applying $\Sigma M_z = 0$. A small area element $dA = (r\,d\theta)(dr)$, shown in Fig. 8–24, is subjected to both a normal force $dN = p\,dA$ and an associated frictional force,

$$dF = \mu_s\,dN = \mu_s\,p\,dA = \frac{\mu_s P}{\pi(R_2^2 - R_1^2)}\,dA$$

The normal force does not create a moment about the z axis of the shaft; however, the frictional force does: namely, $dM = r\,dF$. Integration is needed to compute the total moment created by all the frictional forces acting on differential areas dA. Therefore, for impending rotational motion,

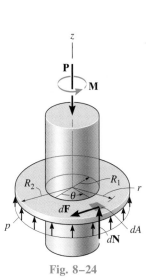

Fig. 8–24

$$\Sigma M_z = 0; \qquad M - \int_A r\,dF = 0$$

Substituting for dF and dA and integrating over the entire bearing area yields

$$M = \int_{R_1}^{R_2}\int_0^{2\pi} r\left[\frac{\mu_s P}{\pi(R_2^2 - R_1^2)}\right](r\,d\theta\,dr) = \frac{\mu_s P}{\pi(R_2^2 - R_1^2)}\int_{R_1}^{R_2} r^2\,dr \int_0^{2\pi} d\theta$$

or

$$M = \tfrac{2}{3}\mu_s P\left(\frac{R_2^3 - R_1^3}{R_2^2 - R_1^2}\right) \qquad (8\text{–}7)$$

This equation gives the magnitude of moment required for impending rotation of the shaft. The frictional moment developed at the end of the shaft, when it is *rotating* at constant speed, can be found by substituting μ_k for μ_s in Eq. 8–7.

When $R_2 = R$ and $R_1 = 0$, as in the case of a pivot bearing, Fig. 8–23a, Eq. 8–7 reduces to

$$M = \tfrac{2}{3}\mu_s PR \qquad (8\text{–}8)$$

Recall from the initial assumption that both Eqs. 8–7 and 8–8 apply only for bearing surfaces subjected to *constant pressure*. If the pressure is not uniform, a variation of the pressure as a function of the bearing area must be determined before integrating to obtain the moment. The following example illustrates this concept.

Example 8–10

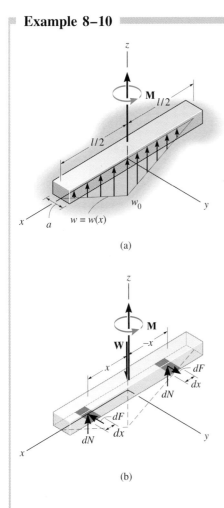

(a)

(b)

Fig. 8–25

The uniform bar shown in Fig. 8–25a has a total mass m. If it is assumed that the normal pressure acting at the contacting surface varies linearly along the length of the bar as shown, determine the couple moment **M** required to rotate the bar. Assume that the bar's width a is negligible in comparison to its length l. The coefficient of static friction is equal to μ.

SOLUTION

A free-body diagram of the bar is shown in Fig. 8–25b. Since the bar has a total weight of $W = mg$, the intensity w_0 of the distributed load at the center $(x = 0)$ is determined from vertical force equilibrium, Fig. 8–25a.

$$+ \uparrow \Sigma F_z = 0; \quad -mg + 2\left[\frac{1}{2}\left(\frac{l}{2}\right)w_0\right] = 0 \qquad w_0 = \frac{2mg}{l}$$

Since $w = 0$ at $x = l/2$, the distributed load expressed as a function of x is

$$w = w_0\left(1 - \frac{2x}{l}\right) = \frac{2mg}{l}\left(1 - \frac{2x}{l}\right)$$

The magnitude of the normal force acting on a segment of area having a length dx is therefore

$$dN = w\, dx = \frac{2mg}{l}\left(1 - \frac{2x}{l}\right)dx$$

The magnitude of the frictional force acting on the same element of area is

$$dF = \mu\, dN = \frac{2\mu mg}{l}\left(1 - \frac{2x}{l}\right)dx$$

Hence, the moment created by this force about the z axis is

$$dM = x\, dF = \frac{2\mu mg}{l}x\left(1 - \frac{2x}{l}\right)dx$$

The summation of moments about the z axis of the bar is determined by integration, which yields

$$\Sigma M_z = 0; \qquad M - 2\int_0^{l/2}\frac{2\mu mg}{l}x\left(1 - \frac{2x}{l}\right)dx = 0$$

$$M = \frac{4\mu mg}{l}\left(\frac{x^2}{2} - \frac{2x^3}{3l}\right)\Bigg|_0^{l/2}$$

$$M = \frac{\mu mgl}{6} \qquad\qquad Ans.$$

*8.7 Frictional Forces on Journal Bearings

When a shaft or axle is subjected to lateral loads, a *journal bearing* is commonly used for support. Well-lubricated journal bearings are subjected to the laws of fluid mechanics, in which the viscosity of the lubricant, the speed of rotation, and the amount of clearance between the shaft and bearing are needed to determine the frictional resistance of the bearing. When the bearing is not lubricated or is only partially lubricated, however, a reasonable analysis of the frictional resistance can be based on the laws of dry friction.

Frictional Analysis. A typical journal-bearing support is shown in Fig. 8–26a. As the shaft rotates in the direction shown in the figure, it rolls up against the wall of the bearing to some point A where slipping occurs. If the lateral load acting at the end of the shaft is **W,** it is necessary that the bearing reactive force **R** acting at A be equal and opposite to **W,** Fig. 8–26b. The moment needed to maintain constant rotation of the shaft can be found by summing moments about the z axis of the shaft; i.e.,

$$\Sigma M_z = 0; \qquad -M + (R \sin \phi_k)r = 0$$

or

$$M = Rr \sin \phi_k \qquad (8\text{–}9)$$

where ϕ_k is the angle of kinetic friction defined by $\tan \phi_k = F/N = \mu_k N/N = \mu_k$. In Fig. 8–26c, it is seen that $r \sin \phi_k = r_f$. The dashed circle with radius r_f is called the *friction circle*, and as the shaft rotates, the reaction **R** will always be tangent to it. If the bearing is partially lubricated, μ_k is small, and therefore $\mu_k = \tan \phi_k \approx \sin \phi_k \approx \phi_k$. Under these conditions, a reasonable *approximation* to the moment needed to overcome the frictional resistance becomes

$$M \approx Rr\mu_k \qquad (8\text{–}10)$$

The following example illustrates a common application of this analysis.

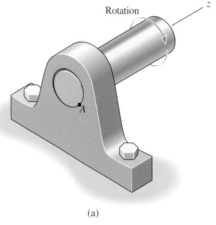

Rotation

(a)

Fig. 8–26

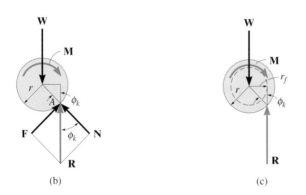

(b)   (c)

Example 8–11

The 100-mm-diameter pulley shown in Fig. 8–27a fits loosely on a 10-mm-diameter shaft for which the coefficient of static friction is μ_s = 0.4. Determine the minimum tension T in the belt needed to (a) raise the 100-kg block and (b) lower the block. Assume that no slipping occurs between the belt and pulley and neglect the weight of the pulley.

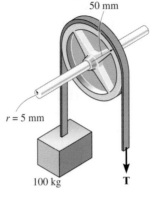

50 mm

$r = 5$ mm

100 kg   **T**

(a)

SOLUTION

Part (a). A free-body diagram of the pulley is shown in Fig. 8–27b. When the pulley is subjected to belt tensions of 981 N each, it makes contact with the shaft at point P_1. As the tension T is *increased*, the pulley will roll around the shaft to point P_2 before motion impends. From the figure, the friction circle has a radius $r_f = r \sin \phi$. Using the simplification $\sin \phi \approx \phi$, $r_f \approx r\mu_s$ = (5 mm)(0.4) = 2 mm, so that summing moments about P_2 gives

$$\zeta + \Sigma M_{P_2} = 0; \qquad 981 \text{ N}(52 \text{ mm}) - T(48 \text{ mm}) = 0$$
$$T = 1063 \text{ N} = 1.06 \text{ kN} \qquad \qquad \textit{Ans.}$$

If a more exact analysis is used, then $\phi = \tan^{-1} 0.4 = 21.8°$. Thus, the radius of the friction circle would be $r_f = r \sin \phi = 5 \sin 21.8°$ = 1.86 mm. Therefore,

$$\zeta + \Sigma M_{P_2} = 0;$$
$$981 \text{ N}(50 \text{ mm} + 1.86 \text{ mm}) - T(50 \text{ mm} - 1.86 \text{ mm}) = 0$$
$$T = 1057 \text{ N} = 1.06 \text{ kN} \qquad \qquad \textit{Ans.}$$

Part (b). When the block is lowered, the resultant force **R** acting on the shaft passes through point P_3, as shown in Fig. 8–27c. Summing moments about this point yields

$$\zeta + \Sigma M_{P_3} = 0; \qquad 981 \text{ N}(48 \text{ mm}) - T(52 \text{ mm}) = 0$$
$$T = 906 \text{ N} \qquad \qquad \textit{Ans.}$$

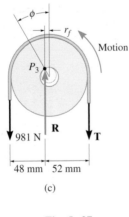

ϕ

r_f

Motion

P_1 P_2

R

981 N   **T**

52 mm   48 mm

(b)

ϕ

r_f

Motion

P_3

R

981 N   **T**

48 mm   52 mm

(c)

Fig. 8–27

*8.8 Rolling Resistance

If a *rigid* cylinder of weight **W** rolls at constant velocity along a *rigid* surface, the normal force exerted by the surface on the cylinder acts at the tangent point of contact, as shown in Fig. 8–28a. Under these conditions, provided the cylinder does not encounter frictional resistance from the air, motion will continue indefinitely. Actually, however, no materials are perfectly rigid, and therefore the reaction of the surface on the cylinder consists of a distribution of normal pressure. For example, consider the cylinder to be made of a very hard material, and the surface on which it rolls to be soft. Due to its weight, the cylinder compresses the surface underneath it, Fig. 8–28b. As the cylinder rolls, the surface material in front of the cylinder *retards* the motion since it is being *deformed*, whereas the material in the rear is *restored* from the deformed state and therefore tends to *push* the cylinder forward. The normal pressures acting on the cylinder in this manner are represented in Fig. 8–28b by their resultant forces $\mathbf{N}_d$ and $\mathbf{N}_r$. Unfortunately, the magnitude of the force of *deformation* and its horizontal component is *always greater* than that of *restoration,* and consequently a horizontal driving force **P** must be applied to the cylinder to maintain the motion, Fig. 8–28b.*

Rolling resistance is caused primarily by this effect, although it is also, to a smaller degree, the result of surface adhesion and relative micro-sliding between the surfaces of contact. Because the actual force **P** needed to overcome these effects is difficult to determine, a simplified method will be developed here to explain one way engineers have analyzed this phenomenon. To do this, we will consider the resultant of the *entire* normal pressure, $\mathbf{N} = \mathbf{N}_d + \mathbf{N}_r$, acting on the cylinder, Fig. 8–28c. As shown in Fig. 8–28d, this force acts at an angle θ with the vertical. To keep the cylinder in equilibrium, i.e., rolling at a constant rate, it is necessary that **N** be *concurrent* with the driving force **P** and the weight **W**. Summing moments about point A gives $Wa = P(r \cos \theta)$. Since the deformations are generally very small in relation to the cylinder's radius, $\cos \theta \approx 1$; hence,

$$Wa \approx Pr$$

or

$$P \approx \frac{Wa}{r} \qquad (8\text{–}11)$$

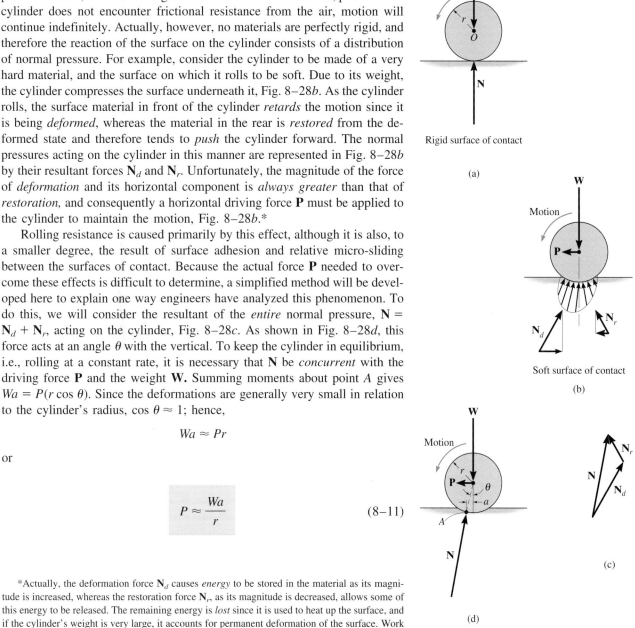

Rigid surface of contact

(a)

Soft surface of contact

(b)

(c)

(d)

Fig. 8–28

*Actually, the deformation force $\mathbf{N}_d$ causes *energy* to be stored in the material as its magnitude is increased, whereas the restoration force $\mathbf{N}_r$, as its magnitude is decreased, allows some of this energy to be released. The remaining energy is *lost* since it is used to heat up the surface, and if the cylinder's weight is very large, it accounts for permanent deformation of the surface. Work must be done by the horizontal force **P** to make up for this loss.

The distance a is termed the *coefficient of rolling resistance*, which has the dimension of length. For instance, $a \approx 0.5$ mm for a wheel rolling on a rail, both of which are made of mild steel. For hardened steel ball bearings on steel, $a \approx 0.1$ mm. Experimentally, though, this factor is difficult to measure, since it depends on such parameters as the rate of rotation of the cylinder, the elastic properties of the contacting surfaces, and the surface finish. For this reason, little reliance is placed on the data for determining a. The analysis presented here does, however, indicate why a heavy load (W) offers greater resistance to motion (P) than a light load under the same conditions. Furthermore, since the force needed to *roll* the cylinder over the surface will be much less than that needed to *slide* the cylinder across the surface, the analysis indicates why roller or ball bearings are often used to minimize the frictional resistance between moving parts.

Example 8–12

A 10-kg steel wheel shown in Fig. 8–29a has a radius of 100 mm and rests on an inclined plane made of wood. If θ is increased so that the wheel begins to roll down the incline with constant velocity when $\theta = 1.2°$, determine the coefficient of rolling resistance.

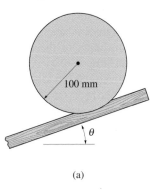

(a)

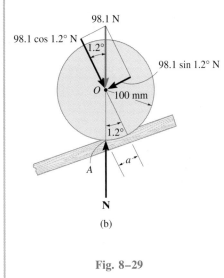

(b)

Fig. 8–29

SOLUTION

As shown on the free-body diagram, Fig. 8–29b, when the wheel has impending motion, the normal reaction **N** acts at point A defined by the dimension a. Resolving the weight into components parallel and perpendicular to the incline, and summing moments about point A, yields (approximately)

$$\zeta + \Sigma M_A = 0; \quad 98.1 \cos 1.2°(a) - 98.1 \sin 1.2°(100) = 0$$

Solving, we obtain

$$a = 2.1 \text{ mm} \qquad \qquad \textit{Ans.}$$

PROBLEMS

8–101. The collar bearing uniformly supports an axial force of $P = 800$ N. If the coefficient of static friction is $\mu_s = 0.3$, determine the torque M required to overcome friction.

8–102. The collar bearing uniformly supports an axial force of $P = 500$ N. If a torque of $M = 3$ N · m is applied to the shaft and causes it to rotate at constant angular velocity, determine the coefficient of kinetic friction at the surface of contact.

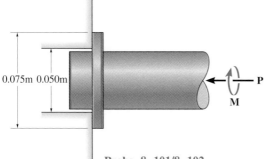

0.075m 0.050m

P

M

Probs. 8–101/8–102

8–103. The annular ring bearing is subjected to a thrust of 800 N. If $\mu_s = 0.35$, determine the torque M that must be applied to overcome friction.

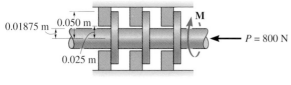

0.01875 m   0.050 m

0.025 m

M

P = 800 N

Prob. 8–103

***8–104.** Assuming the pressure of the sanding disk D on the surface to be uniform, determine the vertical couple forces developed at the handles which are necessary to hold it in equilibrium. The disk is pushed against the surface with a total horizontal force of 75 N. Neglect the weight of the sander and take $\mu_k = 0.60$.

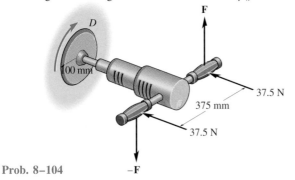

D

100 mm

F

37.5 N

375 mm

37.5 N

Prob. 8–104   –F

8–105. The floor-polishing machine rotates at a constant angular velocity. If it has a weight of 400 N, determine the couple forces F the operator must apply to the handles to hold the machine stationary. The coefficient of kinetic friction between the floor and brush is $\mu_k = 0.3$. Assume the brush exerts a uniform pressure on the floor.

0.45 m

0.60 m

Prob. 8–105

8–106. The plate clutch consists of a flat plate A that slides over the rotating shaft S. The shaft is fixed to the driving plate gear B. If the gear C, which is in mesh with B, is subjected to a torque of $M = 0.8$ N · m, determine the smallest force P, that must be applied via the control arm, to stop the rotation. The coefficient of static friction between the plates A and D is $\mu_s = 0.4$. Assume the bearing pressure between A and D to be uniform.

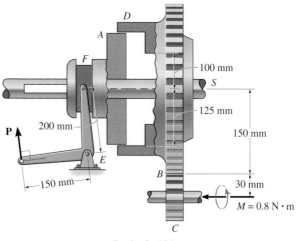

D

A

F

100 mm

S

125 mm

150 mm

200 mm

P

E

150 mm

B

30 mm

M = 0.8 N · m

C

Prob. 8–106

8–107. Because of wearing at the edges, the pivot bearing is subjected to a conical pressure distribution at its surface of contact. Determine the torque M required to overcome friction and turn the shaft, which supports an axial force **P**. The coefficient of static friction is μ. For the solution, it is necessary to determine the peak pressure p_0 in terms of P and the bearing radius R.

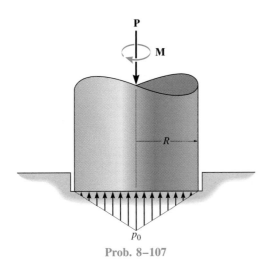

Prob. 8–107

8–109. The pivot bearing is subjected to a pressure distribution at its surface of contact which varies as shown. If the coefficient of static friction is μ, determine the torque M required to overcome friction if the shaft supports an axial force **P**.

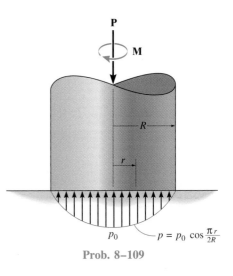

$p = p_0 \cos \dfrac{\pi r}{2R}$

Prob. 8–109

*8–108. The pivot bearing is subjected to a parabolic pressure distribution at its surface of contact. If the coefficient of static friction is μ, determine the torque M required to overcome friction and turn the shaft if it supports an axial force **P**.

$p = p_0 \left(1 - \dfrac{r^2}{R^2}\right)$

Prob. 8–108

8–110. The tractor is used to push the 6600-N pipe. To do this it must overcome the frictional forces at the ground, caused by sand. Assuming that the sand exerts a pressure on the bottom of the pipe as shown, and the coefficient of static friction between the pipe and the sand is $\mu_s = 0.3$, determine the force required to push the pipe forward. Also, determine the peak pressure p_0.

0.38 m

θ

$p = p_0 \cos\theta$

p_0

3.8 m

Prob. 8–110

8–111. The corkscrew is used to remove the 15-mm-diameter cork from the bottle. Determine the smallest vertical force P that must be applied to the handle if the gauge pressure in the bottle is $p = 175$ kPa and the cork pushes against the sides of the bottle's neck with a uniform pressure of 90 kPa. The coefficient of static friction between the bottle and the cork is $\mu_s = 0.4$. *Hint:* The force exerted on the bottom of the cork is $F = pA$, where A is the surface area of the cork's bottom and p is the gauge pressure.

8–113. The 5-kg pulley has a diameter of 240 mm and the axle has a diameter of 40 mm. If the coefficient of kinetic friction between the axle and the pulley is $\mu_k = 0.15$, determine the vertical force P on the rope required to lift the 80-kg block at constant velocity.

8–114. Solve Prob. 8–113 if the force **P** is applied horizontally to the right.

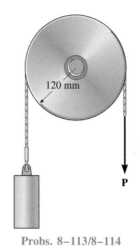

Probs. 8–113/8–114

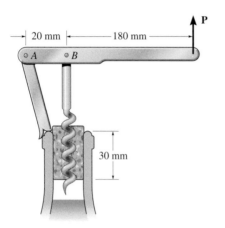

Prob. 8–111

8–115. The pulley has a radius of 300 mm and fits loosely on the 50-mm-diameter shaft. If the loadings acting on the belt cause the pulley to rotate with constant angular velocity, determine the frictional force between the shaft and the pulley and compute the coefficient of kinetic friction. The pulley weighs 180 N.

*8–112. If the smallest tension force $T_A = 500$ N is required to pull the belt downward at A over the shaft S, determine the coefficient of kinetic friction between the loosely fitting collar bushing B and the shaft. Assume that the belt does not slip on the collar; rather, the collar slips on the shaft.

*8–116. The pulley has a radius of 300 mm and fits loosely on the 50-mm-diameter shaft. If the loadings acting on the belt cause the pulley to rotate with constant angular velocity, determine the frictional force between the shaft and the pulley and compute the coefficient of kinetic friction. Neglect the weight of the pulley.

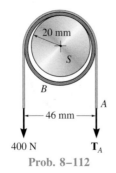

Prob. 8–112

Probs. 8–115/8–116

8–117. The collar fits *loosely* around a fixed shaft that has a radius of 20 mm. If the coefficient of kinetic friction between the shaft and the collar is $\mu_k = 0.3$, determine the force P on the horizontal segment of the belt so that the collar rotates counterclockwise with a constant angular velocity. Assume that the belt does not slip on the collar; rather, the collar slips on the shaft. Neglect the weight and thickness of the belt and collar. The radius, measured from the center of the collar to the mean thickness of the belt, is 22.5 mm.

8–118. The collar fits *loosely* around a fixed shaft that has a radius of 2 mm. If the coefficient of kinetic friction between the shaft and the collar is $\mu_k = 0.3$, determine the force P on the horizontal segment of the belt so that the collar rotates clockwise with a constant angular velocity. Assume that the belt does not slip on the collar; rather, the collar slips on the shaft. Neglect the weight and thickness of the belt and collar. The radius, measured from the center of the collar to the mean thickness of the belt, is 22.5 mm.

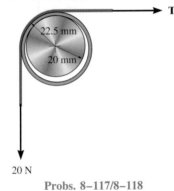

Probs. 8–117/8–118

8–119. A pulley having a diameter of 80 mm and a mass of 1.25 kg is supported loosely on a shaft having a diameter of 20 mm. Determine the torque M that must be applied to the pulley to cause it to rotate with constant angular velocity. The coefficient of kinetic friction between the shaft and pulley is $\mu_k = 0.4$. Also calculate the angle θ which the normal force at the point of contact makes with the horizontal. The shaft itself cannot rotate.

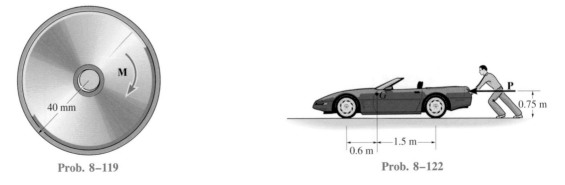

Prob. 8–119

*8–120. The weight of the body on the tibiotalar joint J is 500 N. If the radius of curvature of the talus surface of the ankle is 35 mm, and the coefficient of static friction between the bones is $\mu_s = 0.1$, determine the force T developed in the Achilles tendon necessary to rotate the joint.

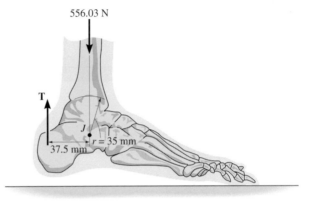

Prob. 8–120

8–121. A car has a mass of 1.5 Mg. Determine the horizontal force P that must be applied to overcome the rolling resistance of the tires if the coefficient of rolling resistance is 1.25 mm. The tires have a diameter of 825 mm.

8–122. The car has a weight of 13 kN and center of gravity at G. Determine the horizontal force P that must be applied to overcome the rolling resistance of the wheels. The coefficient of rolling resistance is 12.5 mm. The tires have a diameter of 825 mm.

Prob. 8–122

8–123. The hand cart has wheels with a diameter of 80 mm. If a crate having a mass of 500 kg is placed on the cart so that each wheel carries an equal load, determine the horizontal force P that must be applied to the handle to overcome the rolling resistance. The coefficient of rolling resistance is 2 mm. Neglect the mass of the cart.

8–126. A large stone having a mass of 500 kg is moved along the incline using a series of 150-mm-diameter rollers for which the coefficient of rolling resistance is 3 mm at the ground and 4 mm at the bottom surface of the stone. Determine the magnitude of force T needed to allow the stone to descend the plane at a constant speed. *Hint:* Use the result of Prob. 8–125.

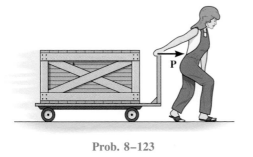

Prob. 8–123

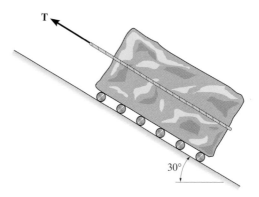

Prob. 8–126

*8–124. Experimentally it is found that a disk having a diameter of 120 mm rolls with a constant speed down an inclined plane having a slope of 12 mm/m. Determine the coefficient of rolling resistance for the disk.

8–125. The cylinder is subjected to a load that has a weight W. If the coefficients of rolling resistance for the cylinder's top and bottom surfaces are a_A and a_B, respectively, show that a force having a magnitude of $P = [W(a_A + a_B)]/2r$ is required to move the load and thereby roll the cylinder forward. Neglect the weight of the cylinder.

8–127. The carriage is used to support a crane load of $F = 5.0$ MN. The nine axles shown in the arrangement serve to equalize the load on each of the eight wheels. (By comparison note that a single supporting beam would *deform* and the wheels would share an unequal amount of load.) Determine the horizontal force that must be applied to the carriage in order to overcome the rolling resistance. The coefficient of rolling resistance is 0.175 mm and each wheel has a diameter of 900 mm.

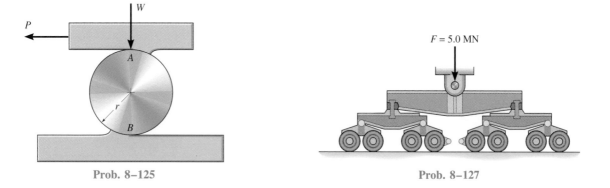

Prob. 8–125

Prob. 8–127

REVIEW PROBLEMS

*8–128.   The man attempts to pull open the door, which requires a force of 855 N. If he has a weight of 900 N and the coefficient of static friction between his shoes and the floor is $\mu_s = 0.5$, determine if he can do it.

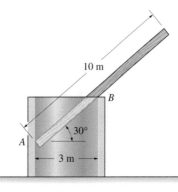

Prob. 8–128

8–129.   The uniform board having a length of 10 m and weight W is placed within the opening of the concrete pipe having an inner diameter of 3 m. Determine the coefficient of static friction μ at A and B if the board is on the verge of slipping when $\theta = 30°$.

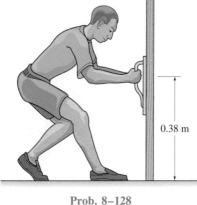

Prob. 8–129

8–130.   The cam or short link is pinned at A and is used to hold mops or brooms against a wall. If the coefficient of static friction between the broomstick and the cam is $\mu_s = 0.2$, determine if it is possible to support the broom having a weight W. The surface at B is smooth. Neglect the weight of the cam.

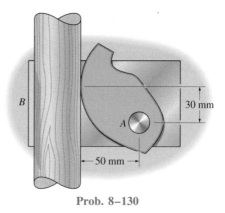

Prob. 8–130

8–131.   The 1.4-Mg machine is to be moved over a level surface using a series of rollers for which the coefficient of rolling resistance is 0.5 mm at the ground and 0.2 mm at the bottom surface of the machine. Determine the appropriate diameter of the rollers so that the machine can be pushed forward with a horizontal force of $P = 250$ N. *Hint:* Use the result of Prob. 8–125.

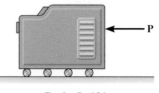

Prob. 8–131

*8-132. The carton clamp on the forklift has a coefficient of static friction of $\mu_s = 0.5$ with any cardboard carton, whereas a cardboard carton has a coefficient of static friction of $\mu_s' = 0.4$ with any other cardboard carton. Compute the smallest horizontal force P the clamp must exert on the sides of a carton so that two cartons A and B each weighing 300 N can be lifted. What smallest clamping force P' is required to lift three 300-N cartons? The third carton C is placed between A and B.

Prob. 8-132

8-133. The differential band brake is constructed so that the ends of the friction strap are connected to the pin at A and the lever at B. If the wheel is subjected to a torque of $M = 30 \text{ N} \cdot \text{m}$, determine the vertical force P applied to the lever required to hold the wheel stationary. The coefficient of static friction between the strap and wheel is $\mu_s = 0.5$.

8-134. A turnbuckle is used to tension member BC of the truss. The coefficient of static friction between the square-threaded screws and the turnbuckle is $\mu_s = 0.5$. The screws have a mean radius of 6 mm and a lead of 3 mm. If a torque of $M = 10 \text{ N} \cdot \text{m}$ is applied to the turnbuckle to draw the screws closer together, determine the force in each member of the truss. No external forces act on the truss.

8-135. A turnbuckle is used to tension member BC of the truss. The coefficient of static friction between the square-threaded screws and the turnbuckle is $\mu_s = 0.5$. The screws have a mean radius of 6 mm and a lead of 3 mm. Determine the torque M that must be applied to the turnbuckle to draw the screws closer together, so that a tension force of 500 N is developed in member BC.

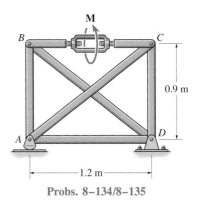

Probs. 8-134/8-135

*8-136. A cable is attached to a 60-N plate B, passes over a fixed disk at C, and is attached to the block at A. Using the coefficients of static friction shown in the figure, determine the smallest weight of block A that will prevent sliding motion of B down the plane.

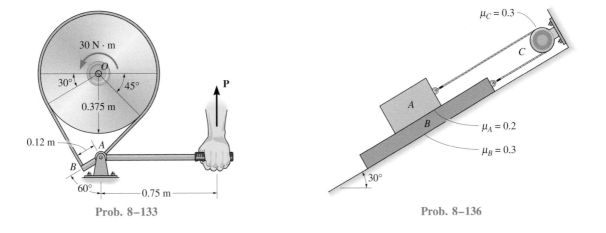

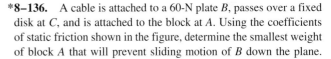

Prob. 8-133

Prob. 8-136

When a pressure vessel is designed it is important to be able to determine the center of gravity of its component parts, calculate its volume and surface area, and reduce three-dimensional distributed loadings to their resultants. These topics are discussed in this chapter.

9

Center of Gravity and Centroid

In this chapter we will discuss the method used to determine the location of the center of gravity and center of mass for a system of discrete particles, and then we will expand its application to include a body of arbitrary shape. The same method of analysis will also be used to determine the geometric center, or centroid, of lines, areas, and volumes. Once the centroid has been located, we will then show how to obtain the area and volume of a surface of revolution and determine the resultants of various types of distributed loadings.

9.1 Center of Gravity and Center of Mass for a System of Particles

Center of Gravity. Consider the system of n particles fixed within a region of space as shown in Fig. 9–1a. The weights of the particles comprise a system of parallel forces* which can be replaced by a single (equivalent) resultant weight and a defined point of application. This point is called the *center of gravity G.* To find its $\bar{x}, \bar{y}, \bar{z}$ coordinates, we must use the principles outlined in Sec. 4.9. This requires that the resultant weight be equal to the total weight of all n particles; that is,

$$W_R = \Sigma W$$

The sum of the moments of the weights of all the particles about the x, y, and z axes is then equal to the moment of the resultant weight about these axes.

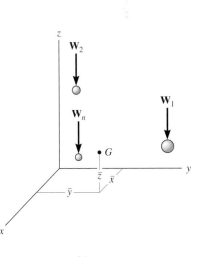

(a)

*This is not true in the exact sense, since the weights are not parallel to each other; rather they are all *concurrent* at the earth's center. Furthermore, the acceleration of gravity g is actually different for each particle, since it depends on the distance from the earth's center to the particle. For all practical purposes, however, both of these effects can be neglected.

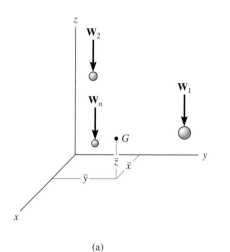

(a)

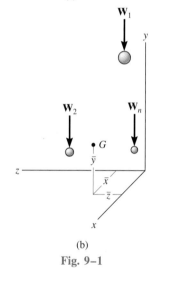

(b)

Fig. 9–1

Thus, to determine the $\bar{x}$ coordinate of G, we can sum moments about the y axis. This yields

$$\bar{x} W_R = \tilde{x}_1 W_1 + \tilde{x}_2 W_2 + \cdots + \tilde{x}_n W_n$$

Likewise, summing moments about the x axis, we can obtain the $\bar{y}$ coordinate; i.e.,

$$\bar{y} W_R = \tilde{y}_1 W_1 + \tilde{y}_2 W_2 + \cdots + \tilde{y}_n W_n$$

Although the weights do not produce a moment about the z axis, we can obtain the z coordinate of G by imagining the coordinate system, with the particles fixed in it, as being rotated 90° about the x (or y) axis, Fig. 9–1b. Summing moments about the x axis, we have

$$\bar{z} W_R = \tilde{z}_1 W_1 + \tilde{z}_2 W_2 + \cdots + \tilde{z}_n W_n$$

We can generalize these formulas, and write them symbolically in the form

$$\bar{x} = \frac{\Sigma \tilde{x} W}{\Sigma W} \qquad \bar{y} = \frac{\Sigma \tilde{y} W}{\Sigma W} \qquad \bar{z} = \frac{\Sigma \tilde{z} W}{\Sigma W}$$

$$(9\text{–}1)$$

Here

$\bar{x}, \bar{y}, \bar{z}$ represent the coordinates of the center of gravity G of the system of particles, regardless of the orientation of the x, y, z axes.

$\tilde{x}, \tilde{y}, \tilde{z}$ represent the coordinates of each particle in the system.

$W_R = \Sigma W$ is the total sum of the weights of all the particles in the system.

Formulas having this same form will be presented throughout this chapter to represent other "quantities" for a system. In all cases, however, keep in mind that they simply represent a balance between the sum of the moments of the "quantity" for *each part* of the system and the moment of the *resultant* "quantity" for the system.

Center of Mass. To study problems concerning the motion of *matter* under the influence of force, i.e., dynamics, it is necessary to locate a point called the *center of mass*. Provided the acceleration of gravity g for every particle is constant, then $W = mg$. Substituting into Eqs. 9–1 and canceling g from both the numerator and denominator yields

$$\bar{x} = \frac{\Sigma \tilde{x} m}{\Sigma m} \qquad \bar{y} = \frac{\Sigma \tilde{y} m}{\Sigma m} \qquad \bar{z} = \frac{\Sigma \tilde{z} m}{\Sigma m}$$

$$(9\text{–}2)$$

By comparison, then, the location of the center of gravity *coincides* with that of the center of mass. Recall, however, that particles have "weight" only when under the influence of a gravitational attraction, whereas the center of mass is independent of gravity. For example, it would be meaningless to define the center of gravity of a system of particles representing the planets of our solar system, while the center of mass of this system is important.

9.2 Center of Gravity, Center of Mass, and Centroid for a Body

Center of Gravity. When the principles used to determine Eqs. 9–1 are applied to a system of particles composing a rigid body having a total weight W, one obtains the same form as these equations except that each particle located at $(\widetilde{x}, \widetilde{y}, \widetilde{z})$ is thought to have a *differential weight dW*, Fig. 9–2. As a result, *integration* is required rather than a discrete summation of the terms. The resulting equations are

$$\bar{x} = \frac{\int \widetilde{x}\, dW}{\int dW} \qquad \bar{y} = \frac{\int \widetilde{y}\, dW}{\int dW} \qquad \bar{z} = \frac{\int \widetilde{z}\, dW}{\int dW} \qquad (9\text{–}3)$$

In order to use these equations properly, the differential weight dW must be expressed in terms of its associated volume dV. If γ represents the *specific weight* of the body, measured as a weight per unit volume, then $dW = \gamma\, dV$ and therefore

$$\bar{x} = \frac{\int_V \widetilde{x}\gamma\, dV}{\int_V \gamma\, dV} \qquad \bar{y} = \frac{\int_V \widetilde{y}\gamma\, dV}{\int_V \gamma\, dV} \qquad \bar{z} = \frac{\int_V \widetilde{z}\gamma\, dV}{\int_V \gamma\, dV} \qquad (9\text{–}4)$$

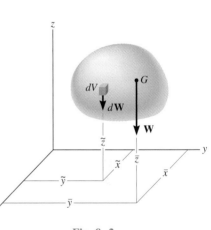

Fig. 9–2

Here integration must be performed throughout the entire volume of the body.

Center of Mass. The *density ρ*, or mass per unit volume, is related to γ by the equation $\gamma = \rho g$, where g is the acceleration of gravity. Substituting this relationship into Eqs. 9–4 and canceling g from both the numerators and denominators yields similar equations (with ρ replacing γ) that can be used to determine the body's *center of mass*.

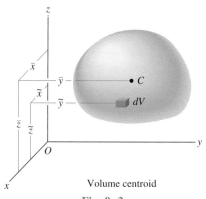

Volume centroid

Fig. 9–3

Centroid. The *centroid* is a point which defines the *geometric center* of an object. Its location can be determined from formulas similar to those used to determine the body's center of gravity or center of mass. In particular, if the material composing a body is uniform or *homogeneous*, the *density or specific weight* will be *constant* throughout the body, and therefore this term will factor out of the integrals and *cancel* from both the numerators and denominators of Eqs. 9–4. The resulting formulas define the centroid of the body since they are independent of the body's weight and instead depend only on the body's geometry. Three specific cases will be considered.

Volume. If an object is subdivided into volume elements dV, Fig. 9–3, the location of the centroid C $(\bar{x}, \bar{y}, \bar{z})$ for the volume of the object can be determined by computing the "moments" of the elements about the coordinate axes. The resulting formulas are

$$\bar{x} = \frac{\displaystyle\int_V \widetilde{x}\, dV}{\displaystyle\int_V dV} \qquad \bar{y} = \frac{\displaystyle\int_V \widetilde{y}\, dV}{\displaystyle\int_V dV} \qquad \bar{z} = \frac{\displaystyle\int_V \widetilde{z}\, dV}{\displaystyle\int_V dV} \qquad (9\text{–}5)$$

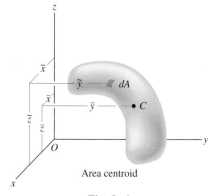

Area centroid

Fig. 9–4

Area. In a similar manner, the centroid for the surface area of an object, such as a plate or shell, Fig. 9–4, can be found by subdividing the area into differential elements dA and computing the "moments" of these area elements about the coordinate axes, namely,

$$\bar{x} = \frac{\displaystyle\int_A \widetilde{x}\, dA}{\displaystyle\int_A dA} \qquad \bar{y} = \frac{\displaystyle\int_A \widetilde{y}\, dA}{\displaystyle\int_A dA} \qquad \bar{z} = \frac{\displaystyle\int_A \widetilde{z}\, dA}{\displaystyle\int_A dA} \qquad (9\text{–}6)$$

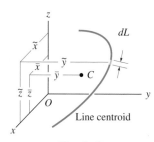

Line centroid

Fig. 9–5

Line. If the geometry of the object, such as a thin rod or wire, takes the form of a line, Fig. 9–5, the manner of finding its centroid is identical to the procedure outlined above. The results are

$$\bar{x} = \frac{\displaystyle\int_L \widetilde{x}\, dL}{\displaystyle\int_L dL} \qquad \bar{y} = \frac{\displaystyle\int_L \widetilde{y}\, dL}{\displaystyle\int_L dL} \qquad \bar{z} = \frac{\displaystyle\int_L \widetilde{z}\, dL}{\displaystyle\int_L dL} \qquad (9\text{–}7)$$

It is important to remember that when applying Eqs. 9–4 through 9–7 it is best to choose a coordinate system that simplifies as much as possible the equation used to describe the object's boundary. For example, polar coordinates are generally appropriate for objects having circular boundaries. Also, if a rectangular coordinate system is used, the terms $\tilde{x}$, $\tilde{y}$, $\tilde{z}$ in the equations refer to the "moment arms" or coordinates of the *center of gravity or centroid for the differential element* used. If possible, this differential element should be chosen such that it has a differential size or thickness in only *one direction*. When this is done, only a single integration is required to cover the entire region.

Symmetry. Notice that in all of the above cases the location of C does not necessarily have to be within the object; rather, it can be located off the object in space. Also, the *centroids* of some shapes may be partially or completely specified by using conditions of *symmetry*. In cases where the shape has an axis of symmetry, the centroid of the shape will lie along that axis. For example, the centroid C for the line shown in Fig. 9–6 must lie along the y axis, since for every elemental length dL at a distance $+\tilde{x}$ to the right of the y axis, there is an identical element at a distance $-\tilde{x}$ to the left. The total moment for all the elements about the axis of symmetry will therefore cancel; i.e., $\int \tilde{x}\, dL = 0$ (Eq. 9–7), so that $\bar{x} = 0$. In cases where a shape has two or three axes of symmetry, it follows that the centroid lies at the intersection of these axes, Fig. 9–7 and Fig. 9–8.

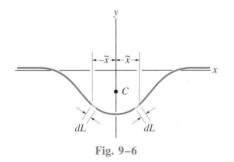

Fig. 9–6

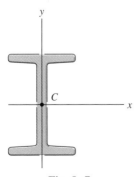

Fig. 9–7

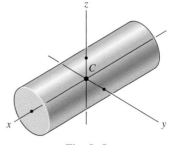

Fig. 9–8

PROCEDURE FOR ANALYSIS

The following procedure provides a method for determining the center of gravity or centroid of an object or shape using a single integration.

Differential Element.   Select an appropriate coordinate system and specify the coordinate axes. Then choose an appropriate differential element for integration. If x, y, z axes are used, then for lines this element dL is represented as a differential line segment; for areas the element dA is generally a rectangle having a finite length and differential width; and for volumes the element dV is either a circular disk having a finite radius and differential thickness, or a shell having a finite length and radius and differential thickness. Locate the element so that it intersects the *boundary* of the shape at an *arbitrary point* (x, y, z).

Size and Moment Arms.   Express the length dL, area dA, or volume dV of the element in terms of the coordinates used to define the boundary of the shape. Determine the coordinates or moment arms $\tilde{x}$, $\tilde{y}$, $\tilde{z}$ for the centroid or center of gravity of the element.

Integrations.   Substitute the data computed above into the appropriate equations (Eqs. 9–4 through 9–7) and perform the integrations.* Note that integration can be accomplished only when the function in the integrand is expressed in terms of the *same variable as the differential thickness of the element.* The limits of the integral are then defined from the two extreme locations of the element's differential thickness, so that when the elements are "summed" or the integration performed, the entire region is covered.

  *Formulas for integration are given in Appendix A.

The following examples illustrate this procedure numerically.

Example 9–1

Locate the centroid of the rod bent into the shape of a parabolic arc, shown in Fig. 9–9.

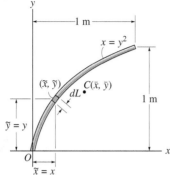

Fig. 9–9

SOLUTION

Differential Element.   The differential element is shown in Fig. 9–9. It is located on the curve at the *arbitrary point* (x, y).

Length and Moment Arms.   The differential length of the element dL can be expressed in terms of the differentials dx and dy by using the Pythagorean theorem.

$$dL = \sqrt{(dx)^2 + (dy)^2} = \sqrt{\left(\frac{dx}{dy}\right)^2 + 1}\ dy$$

Since $x = y^2$, then $dx/dy = 2y$. Therefore, expressing dL in terms of y and dy, we have

$$dL = \sqrt{(2y)^2 + 1}\ dy$$

The centroid is located at $\widetilde{x} = x$, $\widetilde{y} = y$.

Integrations.   Applying Eqs. 9–7 and integrating with respect to y using the formulas in Appendix A, we have

$$\bar{x} = \frac{\displaystyle\int_L \widetilde{x}\ dL}{\displaystyle\int_L dL} = \frac{\displaystyle\int_0^1 x\sqrt{4y^2 + 1}\ dy}{\displaystyle\int_0^1 \sqrt{4y^2 + 1}\ dy} = \frac{\displaystyle\int_0^1 y^2\sqrt{4y^2 + 1}\ dy}{\displaystyle\int_0^1 \sqrt{4y^2 + 1}\ dy}$$

$$= \frac{0.746}{1.479} = 0.504\ \text{m} \qquad\qquad\qquad\qquad \textit{Ans.}$$

$$\bar{y} = \frac{\displaystyle\int_L \widetilde{y}\ dL}{\displaystyle\int_L dL} = \frac{\displaystyle\int_0^1 y\sqrt{4y^2 + 1}\ dy}{\displaystyle\int_0^1 \sqrt{4y^2 + 1}\ dy} = \frac{0.848}{1.479} = 0.573\ \text{m} \qquad \textit{Ans.}$$

Example 9–2

Locate the centroid of the circular wire segment shown in Fig. 9–10.

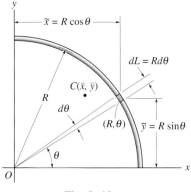

Fig. 9–10

SOLUTION

Polar coordinates will be used to solve this problem since the arc is circular.

Differential Element.   A differential circular arc is selected as shown in the figure. This element intersects the curve at (R, θ).

Length and Moment Arm.   The differential length of the element is $dL = R\,d\theta$, and its centroid is located at $\widetilde{x} = R\cos\theta$ and $\widetilde{y} = R\sin\theta$.

Integrations.   Applying Eqs. 9–7 and integrating with respect to θ, we obtain

$$\bar{x} = \frac{\displaystyle\int_L \widetilde{x}\,dL}{\displaystyle\int_L dL} = \frac{\displaystyle\int_0^{\pi/2}(R\cos\theta)R\,d\theta}{\displaystyle\int_0^{\pi/2}R\,d\theta} = \frac{R^2\displaystyle\int_0^{\pi/2}\cos\theta\,d\theta}{R\displaystyle\int_0^{\pi/2}d\theta} = \frac{2R}{\pi} \quad Ans.$$

$$\bar{y} = \frac{\displaystyle\int_L \widetilde{y}\,dL}{\displaystyle\int_L dL} = \frac{\displaystyle\int_0^{\pi/2}(R\sin\theta)R\,d\theta}{\displaystyle\int_0^{\pi/2}R\,d\theta} = \frac{R^2\displaystyle\int_0^{\pi/2}\sin\theta\,d\theta}{R\displaystyle\int_0^{\pi/2}d\theta} = \frac{2R}{\pi} \quad Ans.$$

Example 9–3

Determine the distance $\bar{y}$ to the centroid of the area of the triangle shown in Fig. 9–11.

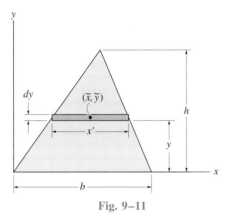

Fig. 9–11

SOLUTION

Differential Element.   Consider a rectangular element having thickness dy and *variable length x′*, Fig. 9–11. By similar triangles, $b/h = x'/(h - y)$ or $x' = \dfrac{b}{h}(h - y)$. The element intersects the sides of the triangle at a height y above the x axis.

Area and Moment Arms.   The area of the element is $dA = x'\, dy = \dfrac{b}{h}(h - y)\, dy$, and its centroid is located a distance $\tilde{y} = y$ from the x axis.

Integrations.   Applying the second of Eqs. 9–6, and integrating with respect to y, yields

$$\bar{y} = \frac{\displaystyle\int_A \tilde{y}\, dA}{\displaystyle\int_A dA} = \frac{\displaystyle\int_0^h y\frac{b}{h}(h - y)\, dy}{\displaystyle\int_0^h \frac{b}{h}(h - y)\, dy} = \frac{\frac{1}{6}bh^2}{\frac{1}{2}bh}$$

$$= \frac{h}{3} \hspace{4cm} \textit{Ans.}$$

Example 9–4

Locate the centroid for the area of a quarter circle shown in Fig. 9–12a.

SOLUTION I

Differential Element.   Polar coordinates will be used since the boundary is circular. We choose the element in the shape of a *triangle*, Fig. 9–12a. (Actually the shape is a circular sector; however, neglecting higher-order differentials, the element becomes triangular.) The element intersects the curve at point (R, θ).

Area and Moment Arms.   The area of the element is

$$dA = \tfrac{1}{2}(R)(R\,d\theta) = \frac{R^2}{2}d\theta$$

and using the results of Example 9–3, the centroid of the (triangular) element is located at $\widetilde{x} = \tfrac{2}{3}R\cos\theta$, $\widetilde{y} = \tfrac{2}{3}R\sin\theta$.

Integrations.   Applying Eqs. 9–6, and integrating with respect to θ, we obtain

$$\bar{x} = \frac{\displaystyle\int_A \widetilde{x}\,dA}{\displaystyle\int_A dA} = \frac{\displaystyle\int_0^{\pi/2}\left(\frac{2}{3}R\cos\theta\right)\frac{R^2}{2}\,d\theta}{\displaystyle\int_0^{\pi/2}\frac{R^2}{2}\,d\theta}$$

$$= \frac{\left(\dfrac{2}{3}R\right)\displaystyle\int_0^{\pi/2}\cos\theta\,d\theta}{\displaystyle\int_0^{\pi/2}d\theta} = \frac{4R}{3\pi} \qquad Ans.$$

$$\bar{y} = \frac{\displaystyle\int_A \widetilde{y}\,dA}{\displaystyle\int_A dA} = \frac{\displaystyle\int_0^{\pi/2}\left(\frac{2}{3}R\sin\theta\right)\frac{R^2}{2}\,d\theta}{\displaystyle\int_0^{\pi/2}\frac{R^2}{2}\,d\theta}$$

$$= \frac{\left(\dfrac{2}{3}R\right)\displaystyle\int_0^{\pi/2}\sin\theta\,d\theta}{\displaystyle\int_0^{\pi/2}d\theta} = \frac{4R}{3\pi} \qquad Ans.$$

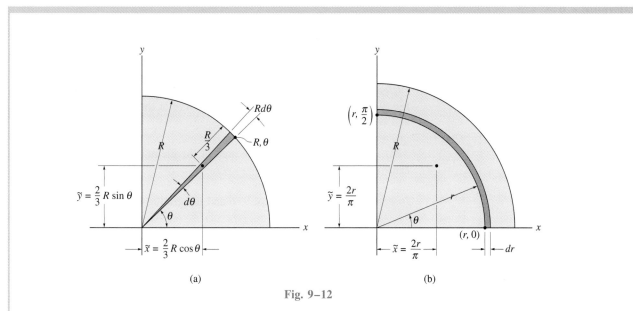

Fig. 9–12

SOLUTION II

Differential Element.   The differential element may be chosen in the form of a *circular arc* having a thickness dr as shown in Fig. 9–12b. The element intersects the axes at points $(r, 0)$ and $(r, \pi/2)$.

Area and Moment Arms.   The area of the element is $dA = (2\pi r/4)\, dr$. Since the centroid of a 90° circular arc was determined in Example 9–2, then for the element $\widetilde{x} = 2r/\pi$, $\widetilde{y} = 2r/\pi$.

Integrations.   Using Eqs. 9–6, and integrating with respect to r, we obtain

$$\bar{x} = \frac{\displaystyle\int_A \widetilde{x}\, dA}{\displaystyle\int_A dA} = \frac{\displaystyle\int_0 \frac{2r}{\pi}\left(\frac{2\pi r}{4}\right) dr}{\displaystyle\int_0^R \frac{2\pi r}{4}\, dr} = \frac{\displaystyle\int_0 r^2\, dr}{\dfrac{\pi}{2}\displaystyle\int_0^R r\, dr} = \frac{4}{3}\frac{R}{\pi} \qquad Ans.$$

$$\bar{y} = \frac{\displaystyle\int_A \widetilde{y}\, dA}{\displaystyle\int_A dA} = \frac{\displaystyle\int_0^R \frac{2r}{\pi}\left(\frac{2\pi r}{4}\right) dr}{\displaystyle\int_0^R \frac{2\pi r}{4}\, dr} = \frac{\displaystyle\int_0^R r^2\, dr}{\dfrac{\pi}{2}\displaystyle\int_0^R r\, dr} = \frac{4}{3}\frac{R}{\pi} \qquad Ans.$$

Example 9–5

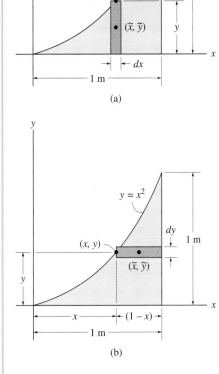

(a)

(b)

Fig. 9–13

Locate the centroid of the area shown in Fig. 9–13a.

SOLUTION I

Differential Element. A differential element of thickness dx is shown in Fig. 9–13a. The element intersects the curve at the *arbitrary point* (x, y), and so it has a height y.

Area and Moment Arms. The area of the element is $dA = y\,dx$, and its centroid is located at $\widetilde{x} = x$, $\widetilde{y} = y/2$.

Integrations. Applying Eqs. 9–6 and integrating with respect to x yields

$$\bar{x} = \frac{\displaystyle\int_A \widetilde{x}\,dA}{\displaystyle\int_A dA} = \frac{\displaystyle\int_0^1 xy\,dx}{\displaystyle\int_0^1 y\,dx} = \frac{\displaystyle\int_0^1 x^3\,dx}{\displaystyle\int_0^1 x^2\,dx} = \frac{0.250}{0.333} = 0.75 \text{ m} \qquad Ans.$$

$$\bar{y} = \frac{\displaystyle\int_A \widetilde{y}\,dA}{\displaystyle\int_A dA} = \frac{\displaystyle\int_0^1 (y/2)y\,dx}{\displaystyle\int_0^1 y\,dx} = \frac{\displaystyle\int_0^1 (x^2/2)x^2\,dx}{\displaystyle\int_0^1 x^2\,dx} = \frac{0.100}{0.333} = 0.3 \text{ m} \quad Ans.$$

SOLUTION II

Differential Element. The differential element of thickness dy is shown in Fig. 9–13b. The element intersects the curve at the *arbitrary point* (x, y), and so it has a length $(1 - x)$.

Area and Moment Arms. The area of the element is $dA = (1 - x)\,dy$, and its centroid is located at

$$\widetilde{x} = x + \left(\frac{1 - x}{2}\right) = \frac{1 + x}{2}, \quad \widetilde{y} = y$$

Integrations. Applying Eqs. 9–6 and integrating with respect to y, we obtain

$$\bar{x} = \frac{\displaystyle\int_A \widetilde{x}\,dA}{\displaystyle\int_A dA} = \frac{\displaystyle\int_0^1 [(1 + x)/2](1 - x)\,dy}{\displaystyle\int_0^1 (1 - x)\,dy} = \frac{\dfrac{1}{2}\displaystyle\int_0^1 (1 - y)\,dy}{\displaystyle\int_0^1 (1 - \sqrt{y})\,dy} = \frac{0.250}{0.333} = 0.75 \text{ m} \quad Ans.$$

$$\bar{y} = \frac{\displaystyle\int_A \widetilde{y}\,dA}{\displaystyle\int_A dA} = \frac{\displaystyle\int_0^1 y(1 - x)\,dy}{\displaystyle\int_0^1 (1 - x)\,dy} = \frac{\displaystyle\int_0^1 (y - y^{3/2})\,dy}{\displaystyle\int_0^1 (1 - \sqrt{y})\,dy} = \frac{0.100}{0.333} = 0.3 \text{ m} \quad Ans.$$

Example 9–6

Locate the $\bar{x}$ centroid of the shaded area bounded by the two curves $y = x$ and $y = x^2$, Fig. 9–14.

SOLUTION I

Differential Element.   A differential element of thickness dx is shown in Fig. 9–14a. The element intersects the curves at *arbitrary points* (x, y_1) and (x, y_2), and so it has a height $(y_2 - y_1)$.

Area and Moment Arm.   The area of the element is $dA = (y_2 - y_1)\,dx$, and its centroid is located at $\tilde{x} = x$.

Integrations.   Applying Eq. 9–6, we have

$$\bar{x} = \frac{\displaystyle\int_A \tilde{x}\,dA}{\displaystyle\int_A dA} = \frac{\displaystyle\int_0^1 x(y_2 - y_1)\,dx}{\displaystyle\int_0^1 (y_2 - y_1)\,dx} = \frac{\displaystyle\int_0^1 x(x - x^2)\,dx}{\displaystyle\int_0^1 (x - x^2)\,dx} = \frac{\frac{1}{12}}{\frac{1}{6}} = 0.5 \text{ m} \quad Ans.$$

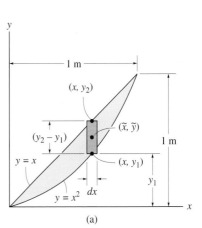

(a)

SOLUTION II

Differential Element.   A differential element having a thickness dy is shown in Fig. 9–14b. The element intersects the curves at the *arbitrary points* (x_2, y) and (x_1, y), and so it has a length $(x_1 - x_2)$.

Area and Moment Arm.   The area of the element is $dA = (x_1 - x_2)\,dy$, and its centroid is located at

$$\tilde{x} = x_2 + \frac{x_1 - x_2}{2} = \frac{x_1 + x_2}{2}$$

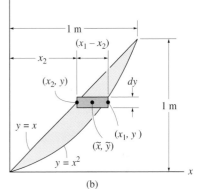

(b)

Fig. 9–14

Integrations.   Applying Eq. 9–6, we have

$$\bar{x} = \frac{\displaystyle\int_A \tilde{x}\,dA}{\displaystyle\int_A dA} = \frac{\displaystyle\int_0^1 [(x_1 + x_2)/2](x_1 - x_2)\,dy}{\displaystyle\int_0^1 (x_1 - x_2)\,dy} = \frac{\displaystyle\int_0^1 [(\sqrt{y} + y)/2](\sqrt{y} - y)\,dy}{\displaystyle\int_0^1 (\sqrt{y} - y)\,dy}$$

$$= \frac{\frac{1}{2}\displaystyle\int_0^1 (y - y^2)\,dy}{\displaystyle\int_0^1 (\sqrt{y} - y)\,dy} = \frac{\frac{1}{12}}{\frac{1}{6}} = 0.5 \text{ m} \qquad\qquad Ans.$$

Example 9–7

Locate the $\bar{y}$ centroid for the paraboloid of revolution, which is generated by revolving the shaded area shown in Fig. 9–15a about the y axis.

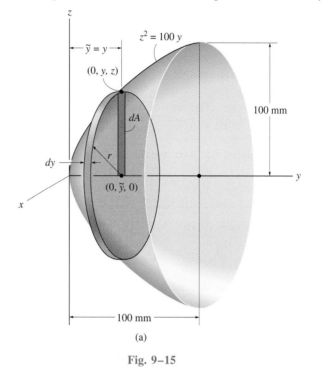

(a)

Fig. 9–15

SOLUTION I

Differential Element. An element having the shape of a *thin disk* is chosen, Fig. 9–15a. This element has a thickness *dy*. In this "disk" method of analysis, the element of planar area, *dA*, is always taken *perpendicular* to the axis of revolution. Here the element intersects the generating curve at the *arbitrary point* (0, y, z), and so its radius is *r = z.*

Volume and Moment Arm. The volume of the element is $dV = (\pi z^2)\,dy$, and its centroid is located at $\widetilde{y} = y$.

Integrations. Applying the second of Eqs. 9–5 and integrating with respect to y yields

$$\bar{y} = \frac{\int_V \widetilde{y}\,dV}{\int_V dV} = \frac{\int_0^{100} y(\pi z^2)\,dy}{\int_0^{100} (\pi z^2)\,dy} = \frac{100\pi \int_0^{100} y^2\,dy}{100\pi \int_0^{100} y\,dy} = 66.7 \text{ mm} \qquad \textit{Ans.}$$

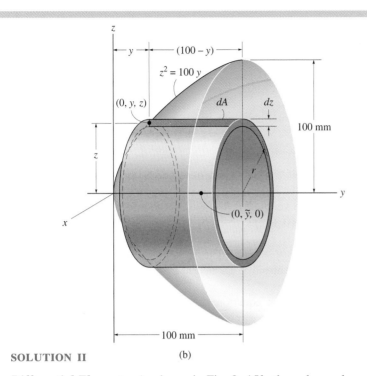

(b)

SOLUTION II

Differential Element.   As shown in Fig. 9–15b, the volume element can be chosen in the form of a *thin cylindrical shell,* where the shell's thickness is dz. In this "shell" method of analysis, the element of planar area, dA, is always taken *parallel* to the axis of revolution. Here the element intersects the generating curve at point $(0, y, z)$, and so the radius of the shell is $r = z$.

Volume and Moment Arm.   The volume of the element is $dV = 2\pi r\, dA = 2\pi z(100 - y)\, dz$, and its centroid is located at $\tilde{y} = y + (100 - y)/2 = (100 + y)/2$.

Integrations.   Applying the second of Eqs. 9–5 and integrating with respect to z yields

$$\bar{y} = \frac{\displaystyle\int_V \tilde{y}\, dV}{\displaystyle\int_V dV} = \frac{\displaystyle\int_0^{100} [(100 + y)/2]\, 2\pi z(100 - y)\, dz}{\displaystyle\int_0^{100} 2\pi z(100 - y)\, dz}$$

$$= \frac{\pi \displaystyle\int_0^{100} z(10^4 - 10^{-4} z^4)\, dz}{2\pi \displaystyle\int_0^{100} z(100 - 10^{-2} z^2)\, dz} = 66.7 \text{ mm} \qquad \textit{Ans.}$$

Example 9–8

Determine the location of the center of mass of the cylinder shown in Fig. 9–16a if its density varies directly with its distance from the base, such that $\rho = 200z$ kg/m³.

SOLUTION

For reasons of material symmetry,

$$\bar{x} = \bar{y} = 0 \qquad \textit{Ans.}$$

Differential Element.   A disk element of radius 0.5 m and thickness dz is chosen for integration, Fig. 9–16a, since the *density of the entire element is constant* for a given value of z. The element is located along the z axis at the *arbitrary point* $(0, 0, z)$.

Volume and Moment Arm.   The volume of the element is $dV = \pi(0.5)^2\, dz$, and its centroid is located at $\tilde{z} = z$.

Integrations.   Using an equation similar to the third of Eqs. 9–4 and integrating with respect to z, noting that $\rho = 200z$, we have

$$\bar{z} = \frac{\displaystyle\int_V \tilde{z}\rho\, dV}{\displaystyle\int_V \rho\, dV} = \frac{\displaystyle\int_0^1 z(200z)\pi(0.5)^2\, dz}{\displaystyle\int_0^1 (200z)\pi(0.5)^2\, dz}$$

$$= \frac{\displaystyle\int_0^1 z^2\, dz}{\displaystyle\int_0^1 z\, dz} = 0.667 \text{ m} \qquad \textit{Ans.}$$

Note: It is not possible to use a shell element for integration such as shown in Fig. 9–16b, since the density of the material composing the shell would *vary* along the shell's height, and hence the location of $\tilde{z}$ for the element cannot be specified.

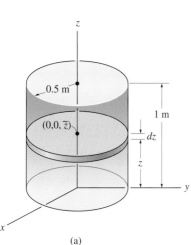

(a)

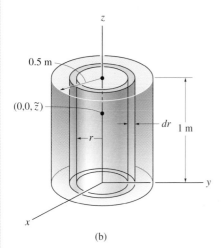

(b)

Fig. 9–16

PROBLEMS

9–1. Locate the center of gravity $\bar{x}$ of the homogeneous rod bent in the form of a semicircular arc. The rod has a weight per unit length of 0.5 N/m. Also, determine the horizontal reaction at the smooth support B and the x and y components of reaction at the pin A.

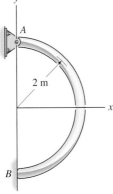

Prob. 9–1

9–2. Locate the center of gravity of the rod having a constant cross-sectional area if its density varies according to $\rho = kx^2$, where k is a constant.

Prob. 9–2

9–3. Locate the centroid $\bar{x}$ of the circular rod. Express the answer in terms of the radius r and semiarc angle α.

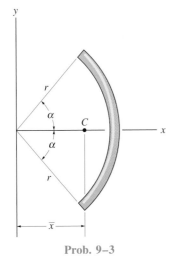

Prob. 9–3

*9–4. Locate the center of mass of the homogeneous rod bent into the shape of a circular arc.

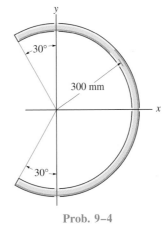

Prob. 9–4

9–5. Determine the distance $\bar{x}$ to the center of gravity of the homogeneous rod bent into the parabolic shape. If the rod has a weight per unit length of 0.5 N/m, determine the reactions at the fixed support O.

9–6. Determine the distance $\bar{y}$ to the center of gravity of the homogeneous rod bent into the parabolic shape.

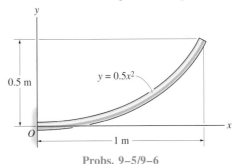

Probs. 9–5/9–6

9–7. Locate the centroid of the shaded area.

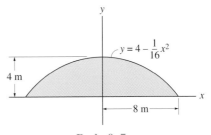

Prob. 9–7

*9–8. Locate the centroid of the shaded area.

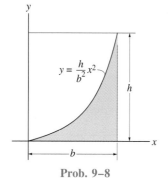

Prob. 9–8

9–9. Locate the centroid of the parabolic area.

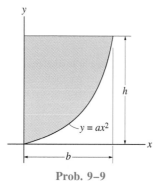

Prob. 9–9

9–10. Locate the centroid of the shaded area.

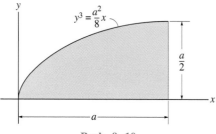

Prob. 9–10

9–11. Locate the centroid of the shaded area.

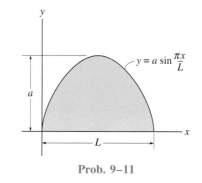

Prob. 9–11

*9–12. Locate the centroid of the shaded area.

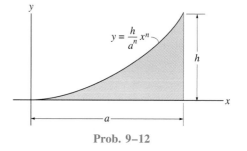

Prob. 9–12

9–13. Locate the centroid of the shaded area.

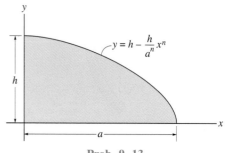

Prob. 9–13

9–14. Locate the centroid of the shaded area bounded by the parabola and the line $y = a$.

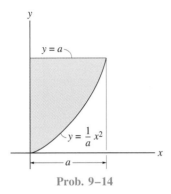

Prob. 9–14

9–15. Locate the centroid of the quarter elliptical area.

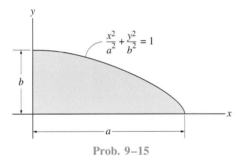

Prob. 9–15

*9–16. Locate the centroid of the shaded area.

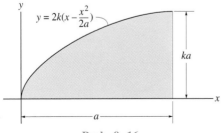

Prob. 9–16

9–17. Locate the centroid of the quarter circle.

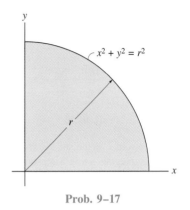

Prob. 9–17

9–18. Locate the centroid of the spandrel area.

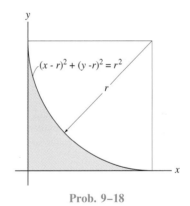

Prob. 9–18

9–19. Locate the centroid of the semicircular area.

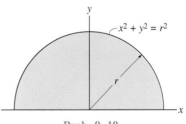

Prob. 9–19

*9–20.   Locate the centroid of the shaded area.

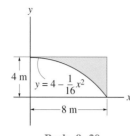

$y = 4 - \dfrac{1}{16}x^2$

4 m

8 m

Prob. 9–20

■9–21.   Locate the centroid $\bar{x}$ of the shaded area. Solve the problem by evaluating the integrals using Simpson's rule.

■9–22.   Locate the centroid $\bar{y}$ of the shaded area. Solve the problem by evaluating the integrals using Simpson's rule.

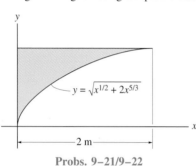

$y = \sqrt{x^{1/2} + 2x^{5/3}}$

2 m

Probs. 9–21/9–22

9–23.   The steel plate is 0.3 m thick and has a density of 7850 kg/m³. Determine the location of its center of mass. Also compute the reactions at the pin and roller support.

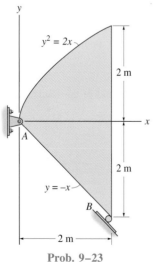

$y^2 = 2x$

2 m

2 m

$y = -x$

A

B

2 m

Prob. 9–23

*■9–24.   Locate the centroid $\bar{x}$ of the shaded area. Solve the problem by evaluating the integrals using Simpson's rule.

■9–25.   Locate the centroid $\bar{y}$ of the shaded area. Solve the problem by evaluating the integrals using Simpson's rule.

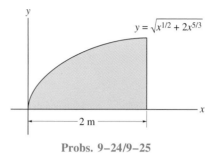

$y = \sqrt{x^{1/2} + 2x^{5/3}}$

2 m

Probs. 9–24/9–25

■9–26.   Locate the centroid $\bar{x}$ of the shaded area. Solve the problem by evaluating the integrals using Simpson's rule.

■9–27.   Locate the centroid $\bar{y}$ of the shaded area. Solve the problem by evaluating the integrals using Simpson's rule.

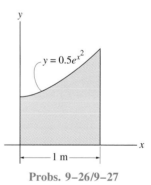

$y = 0.5e^{x^2}$

1 m

Probs. 9–26/9–27

*■9–28.   Determine the location $\bar{r}$ of the centroid C for the loop of the lemniscate. $r^2 = 2a^2\cos 2\,\theta$, $(-45° \le \theta \le 45°)$.

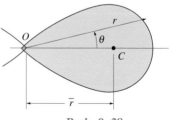

O

θ

r

C

$\bar{r}$

Prob. 9–28

9–29. Determine the location $\bar{r}$ of the centroid C of the upper portion of the cardioid, $r = a(1 - \cos\theta)$.

9–31. Locate the centroid of the solid.

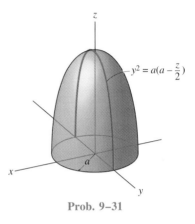

$y^2 = a\left(a - \dfrac{z}{2}\right)$

Prob. 9–31

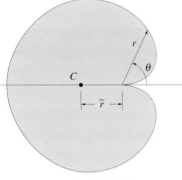

Prob. 9–29

9–30. Locate the centroid of the ellipsoid of revolution.

*9–32. Locate the centroid of the solid.

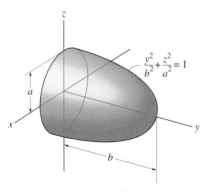

$\dfrac{y^2}{b^2} + \dfrac{z^2}{a^2} = 1$

Prob. 9–30

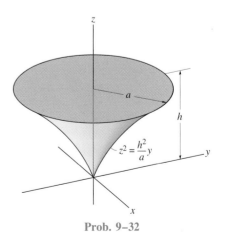

$z^2 = \dfrac{h^2}{a}y$

Prob. 9–32

9–33. Locate the centroid $\bar{x}$ of the thin homogeneous hemispherical shell. *Suggestion:* Choose a ring element having a center at $(x, 0, 0)$, radius z, and thickness $dL = \sqrt{(dx)^2 + (dz)^2}$.

9–35. The king's chamber of the Great Pyramid of Gîza is located at its centroid. Assuming the pyramid to be a solid, prove that this point is at $\bar{z} = \frac{1}{4}h$. *Suggestion:* Use a rectangular differential plate element having a thickness dz and area $(2x)(2y)$.

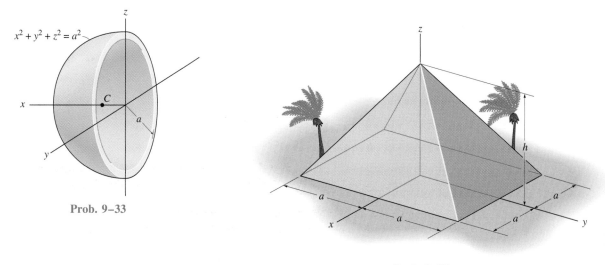

Prob. 9–33

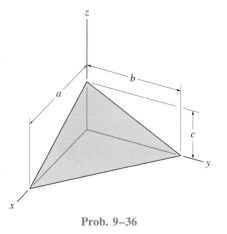

Prob. 9–35

Prob. 9–34 Locate the centroid $\bar{z}$ of the thin conical shell. *Suggestion:* Use ring elements having a center at $(0, 0, z)$, radius y, and thickness $dL = \sqrt{(dy)^2 + (dz)^2}$.

*9–36. Determine the location $\bar{z}$ of the centroid for the tetrahedron. *Suggestion:* Use a triangular "plate" element parallel to the x-y plane and of thickness dz.

Prob. 9–34

Prob. 9–36

9–37. Locate the centroid of the quarter-cone.

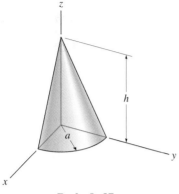

Prob. 9–37

9–38. Locate the center of gravity $\bar{x}$ of the hemisphere. The density of the material varies linearly from zero at the origin O to ρ_0 at the surface. *Suggestion:* Choose a hemispherical shell element for integration and use the result of Prob. 9–33.

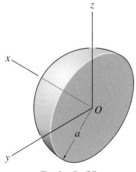

Prob. 9–38

9–39. Locate the centroid $\bar{z}$ of the frustum of the right-circular cone.

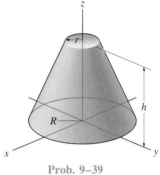

Prob. 9–39

*9–40. The hemisphere of radius r is made from a stack of very thin plates such that the density varies with height $\rho = kz$, where k is a constant. Determine its mass and the distance $\bar{z}$ to the location of its center of mass G.

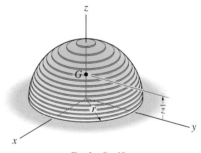

Prob. 9–40

9.3 Composite Bodies

A *composite body* consists of a series of connected "simpler" shaped bodies, which may be rectangular, triangular, semicircular, etc. Such a body can often be sectioned or divided into its composite parts, and provided the *weight* and location of the center of gravity of each of these parts are known, we can eliminate the need for integration to determine the center of gravity for the entire body. The method for doing this requires treating each composite part as a particle and following the procedure outlined in Sec. 9.1. Formulas analogous to Eqs. 9–1 result, since we must account for a finite number of weights. Rewriting these formulas, we have

$$\bar{x} = \frac{\Sigma \tilde{x} W}{\Sigma W} \qquad \bar{y} = \frac{\Sigma \tilde{y} W}{\Sigma W} \qquad \bar{z} = \frac{\Sigma \tilde{z} W}{\Sigma W}$$

(9–8)

Here

$\bar{x}, \bar{y}, \bar{z}$ represent the coordinates of the center of gravity G of the composite body.

$\tilde{x}, \tilde{y}, \tilde{z}$ represent the coordinates of the center of gravity of each composite part of the body.

ΣW is the sum of the weights of all the composite parts of the body, or simply the total weight of the body.

When the body has a *constant density or specific weight,* the center of gravity *coincides* with the centroid of the body. The centroid for composite lines, areas, and volumes can be found using relations analogous to Eqs. 9–8; however, the *W*'s are replaced by *L*'s, *A*'s, and *V*'s, respectively. Centroids for common shapes of lines, areas, shells, and volumes are given in the table on the inside back cover.

PROCEDURE FOR ANALYSIS

The following procedure provides a method for determining the center of gravity of a body or the centroid of a composite geometrical object represented by a line, area, or volume.

Composite Parts.   Using a sketch, divide the body or object into a finite number of composite parts that have simpler shapes. If a composite part has a *hole,* or geometric region having no material, then consider the composite part without the hole, and the hole as an *additional* composite part having *negative* weight or size.

Moment Arms.   Establish the coordinate axes on the sketch and determine the coordinates $\tilde{x}, \tilde{y}, \tilde{z}$ of the center of gravity or centroid of each part.

Summations.   Determine $\bar{x}, \bar{y}, \bar{z}$ by applying the center of gravity equations, Eqs. 9–8, or the analogous centroid equations. If an object is *symmetrical* about an axis, recall that the centroid of the object lies on this axis.

If desired, the calculations can be arranged in tabular form, as indicated in the following three examples.

Example 9–9

Locate the centroid of the wire shown in Fig. 9–17a.

SOLUTION

Composite Parts.   The wire is divided into three segments as shown in Fig. 9–17b.

Moment Arms.   The location of the centroid for each piece is determined and indicated in the figure. In particular, the centroid of segment ① is determined either by integration or using the table on the inside back cover.

Summations.   The calculations are tabulated as follows:

Segment	L (mm)	$\widetilde{x}$ (mm)	$\widetilde{y}$ (mm)	$\widetilde{z}$ (mm)	$\widetilde{x}L$ (mm²)	$\widetilde{y}L$ (mm²)	$\widetilde{z}L$ (mm²)
1	$\pi(60) = 188.5$	60	−38.2	0	11 310	−7200	0
2	40	0	20	0	0	800	0
3	20	0	40	−10	0	800	−200
	$\Sigma L = 248.5$				$\Sigma\widetilde{x}L = 11\ 310$	$\Sigma\widetilde{y}L = -5600$	$\Sigma\widetilde{z}L = -200$

Thus,

$$\bar{x} = \frac{\Sigma\widetilde{x}L}{\Sigma L} = \frac{11\ 310}{248.5} = 45.5 \text{ mm} \qquad\qquad Ans.$$

$$\bar{y} = \frac{\Sigma\widetilde{y}L}{\Sigma L} = \frac{-5600}{248.5} = -22.5 \text{ mm} \qquad\qquad Ans.$$

$$\bar{z} = \frac{\Sigma\widetilde{z}L}{\Sigma L} = \frac{-200}{248.5} = -0.805 \text{ mm} \qquad\qquad Ans.$$

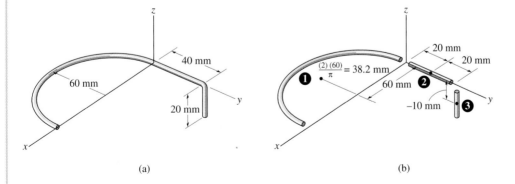

(a)

(b)

Fig. 9–17

Example 9–10

Locate the centroid of the plate area shown in Fig. 9–18a.

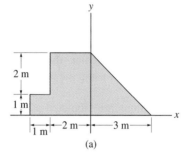

(a)

SOLUTION

Composite Parts.   The plate is divided into three segments as shown in Fig. 9–18b. Here the area of the small rectangle ③ is considered "negative" since it must be subtracted from the larger one ②.

Moment Arms.   The centroid of each segment is located as indicated in the figure. Note that the $\widetilde{x}$ coordinates of ② and ③ are *negative*.

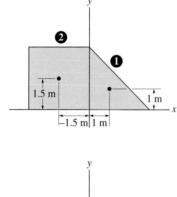

Summations.   Taking the data from Fig. 9–18b, the calculations are tabulated as follows:

Segment	A (m^2)	$\widetilde{x}$ (m)	$\widetilde{y}$ (m)	$\widetilde{x}A$ (m^3)	$\widetilde{y}A$ (m^3)
1	$\frac{1}{2}(3)(3) = 4.5$	1	1	4.5	4.5
2	$(3)(3) = 9$	−1.5	1.5	−13.5	13.5
3	$-(2)(1) = -2$	−2.5	2	5	−4
	$\Sigma A = 11.5$			$\Sigma\widetilde{x}A = -4$	$\Sigma\widetilde{y}A = 14$

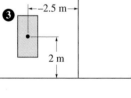

(b)

Fig. 9–18

Thus,

$$\bar{x} = \frac{\Sigma\widetilde{x}A}{\Sigma A} = \frac{-4}{11.5} = -0.348 \text{ m} \qquad Ans.$$

$$\bar{y} = \frac{\Sigma\widetilde{y}A}{\Sigma A} = \frac{14}{11.5} = 1.22 \text{ m} \qquad Ans.$$

Example 9–11

Locate the center of mass of the composite assembly shown in Fig. 9–19a. The conical frustum has a density of $\rho_c = 8$ Mg/m³, and the hemisphere has a density of $\rho_h = 4$ Mg/m³.

SOLUTION

Composite Parts. The assembly can be thought of as consisting of four segments as shown in Fig. 9–19b. For the calculations, ③ and ④ must be considered as "negative" volumes in order that the four segments, when added together, yield the total composite shape shown in Fig. 9–19a.

Moment Arm. Using the table on the inside back cover, the computations for the centroid $\widetilde{z}$ of each piece are shown in the figure.

Summations. Because of *symmetry*, note that

$$\bar{x} = \bar{y} = 0 \qquad \textit{Ans.}$$

Since $W = mg$ and g is constant, the third of Eqs. 9–8 becomes $\bar{z} = \Sigma\widetilde{z}m/\Sigma m$. The mass of each piece can be computed from $m = \rho V$ and used for the calculations. Also, 1 Mg/m³ = 10^{-6} kg/mm³, so that

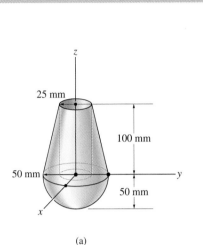

(a)

Fig. 9–19

Segment	m (kg)	$\widetilde{z}$ (mm)	$\widetilde{z}m$ (kg · mm)
1	$8(10^{-6})(\frac{1}{3})\pi(50)^2(200) = 4.189$	50	209.440
2	$4(10^{-6})(\frac{2}{3})\pi(50)^3 = 1.047$	-18.75	-19.635
3	$-8(10^{-6})(\frac{1}{3})\pi(25)^2(100) = -0.524$	$100 + 25 = 125$	-65.450
4	$-8(10^{-6})\pi(25)^2(100) = -1.571$	50	-78.540
	$\Sigma m = 3.141$		$\Sigma\widetilde{z}m = 45.815$

Thus,

$$\bar{z} = \frac{\Sigma\widetilde{z}m}{\Sigma m} = \frac{45.815}{3.141} = 14.6 \text{ mm} \qquad \textit{Ans.}$$

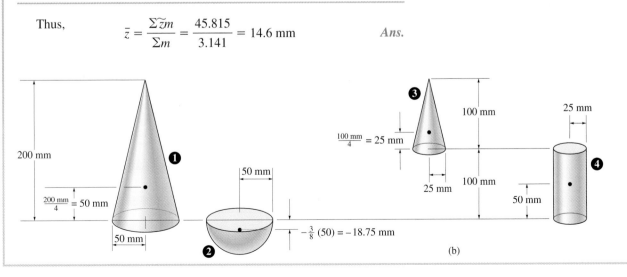

(b)

PROBLEMS

9–41. If the four particles can be replaced by a single 10-kg particle acting at a distance of 2 m to the left of the origin, determine the position $\tilde{x}_p$ and the mass m_p of particle P.

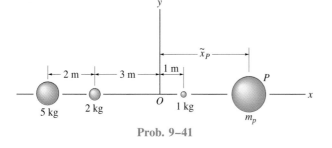

Prob. 9–41

9–42. Determine the location of the center of mass $(\bar{x}, \bar{y}, \bar{z})$ of the three particles.

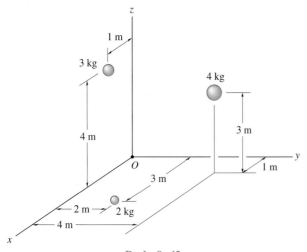

Prob. 9–42

9–43. A "roll-formed" member has the cross section shown. Determine the location $\bar{y}$ of the centroid C. Neglect the thickness of the material and any slight bends at the corners.

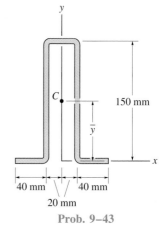

Prob. 9–43

*9–44. The steel and aluminum plate assembly is bolted together and fastened to the wall. Each plate has a constant width in the z direction of 200 mm and thickness of 20 mm. If the density of A and B is $\rho_s = 7.85$ Mg/m³, and for C, $\rho_{al} = 2.71$ Mg/m³, determine the location $\bar{x}$ of the center of mass. Neglect the size of the bolts.

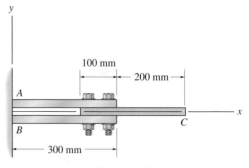

Prob. 9–44

9–45. The truss is made from five members, each having a length of 4 m and a mass of 7 kg/m. If the mass of the gusset plates at the joints and the thickness of the members can be neglected, determine the distance d to where the hoisting cable must be attached, so that the truss does not tip (rotate) when it is lifted.

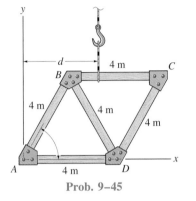

Prob. 9–45

9–46. The three members of the frame each have a weight per unit length of 4 N/m. Locate the position $(\bar{x}, \bar{y})$ of the center of gravity. Neglect the size of the pins at the joints and the thickness of the members. Also, calculate the reactions at the fixed support A.

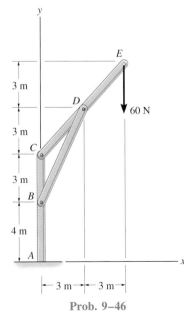

Prob. 9–46

9–47. Determine the location $(\bar{x}, \bar{y})$ of the centroid of the area.

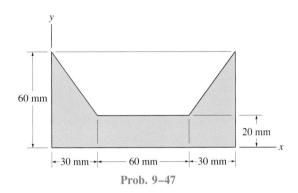

Prob. 9–47

*9–48. Determine the location $\bar{y}$ of the centroid of the area.

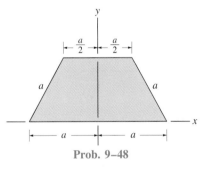

Prob. 9–48

9–49. Determine the weight and location $(\bar{x}, \bar{y})$ of the center of gravity G of the concrete retaining wall. The wall has a length of 10 m, and concrete has a specific gravity of $\gamma = 24$ kN/m³.

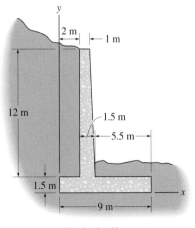

Prob. 9–49

9–50. Locate the centroid $\bar{y}$ of the concrete beam having the tapered cross section shown.

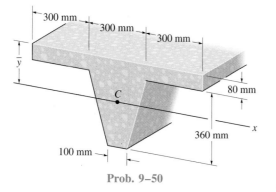

Prob. 9–50

9–51. Determine the distance $\bar{y}$ to the centroid of the trapezoidal area in terms of the dimensions shown.

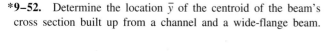

Prob. 9–51

*9–52. Determine the location $\bar{y}$ of the centroid of the beam's cross section built up from a channel and a wide-flange beam.

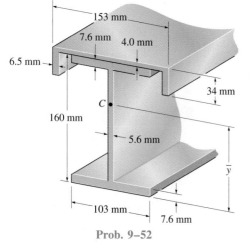

Prob. 9–52

9–53. Determine the location ($\bar{x}$, $\bar{y}$) of the centroid C for the angle's cross-sectional area.

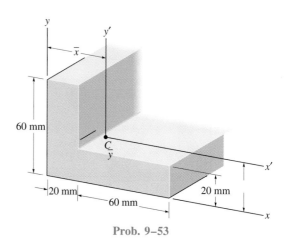

Prob. 9–53

9–54. Determine the location $\bar{y}$ of the centroid of the beam's cross-sectional area.

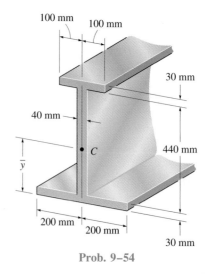

Prob. 9–54

9–55. Determine the location $\bar{y}$ of the centroidal axis $\bar{x}\bar{x}$ of the beam's cross-sectional area. Neglect the size of the corner welds at A and B for the calculation.

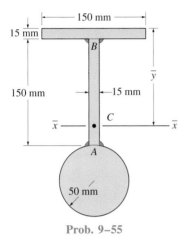

150 mm

15 mm

B

150 mm   15 mm

$\bar{y}$

$\bar{x}$ *C* $\bar{x}$

A

50 mm

Prob. 9–55

*9–56. Determine the location $(\bar{x}, \bar{y})$ of the centroid C of the cross-sectional area for the structural member constructed from two equal-sized channels welded together as shown. Assume all corners are square. Neglect the size of the welds.

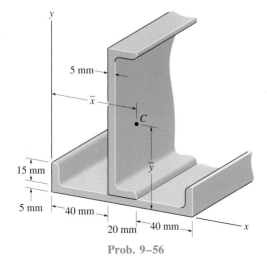

5 mm

$\bar{x}$

C

15 mm

5 mm   40 mm

20 mm   40 mm

$\bar{y}$

Prob. 9–56

9–57. Determine the location $(\bar{x}, \bar{y})$ of the centroid $\dot{C}$ of the area.

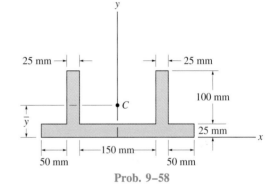

6 mm   3 mm

6 mm

x

6 mm

Prob. 9–57

9–58. Determine the location $\bar{y}$ of the centroid for the cross-sectional area.

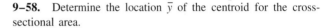

25 mm

25 mm

100 mm

C

$\bar{y}$

25 mm

150 mm

50 mm   50 mm

x

Prob. 9–58

9–59. Determine the location $(\bar{x}, \bar{y})$ of the centroid of the cross-sectional area.

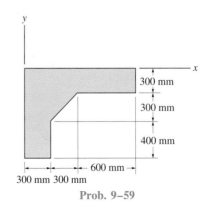

x

300 mm

300 mm

400 mm

600 mm

300 mm 300 mm

Prob. 9–59

*9–60. Determine the location $\bar{y}$ of the centroid of the cross-sectional area.

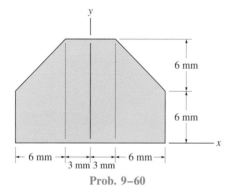

6 mm

6 mm

6 mm   3 mm 3 mm   6 mm

Prob. 9–60

9–61. The tank and compressor have a mass of 15 kg and mass center at G_T, and the motor has a mass of 70 kg and a mass center at G_M. Determine the angle of tilt, θ, of the tank so that the unit will be on the verge of tipping over.

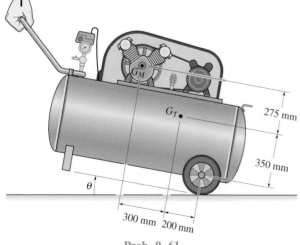

275 mm

350 mm

300 mm   200 mm

θ

Prob. 9–61

9–62. The wooden table is made from a square board having a weight of 60 N. Each of the legs weighs 8 N and is 0.9 m long. Determine how high its center of gravity is from the floor. Also, what is the angle, measured from the horizontal, through which its top surface can be tilted on two of its legs before it begins to overturn? Neglect the thickness of each leg.

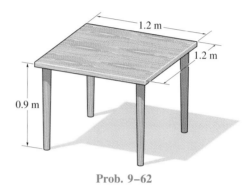

1.2 m

1.2 m

0.9 m

Prob. 9–62

9–63. Determine the location $\bar{x}$ of the centroid C of the shaded area which is part of a circle having a radius r.

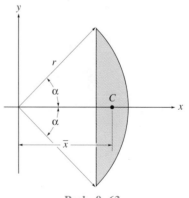

y

r

α

α

C

x

$\bar{x}$

Prob. 9–63

*9–64.   A triangular plate made of homogeneous material has a constant thickness which is very small. If it is folded over as shown, determine the location $\bar{y}$ of the plate's center of gravity G.

9–65.   A triangular plate made of homogeneous material has a constant thickness which is very small. If it is folded over as shown, determine the location $\bar{z}$ of the plate's center of gravity G.

9–67.   Locate the center of mass of the two-block assembly. The densities of materials A and B are $\rho_A = 15$ kN/m^3 and $\rho_B = 40$ kN/m^3, respectively.

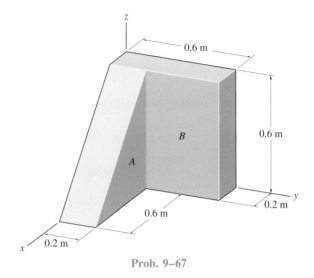

Prob. 9–67

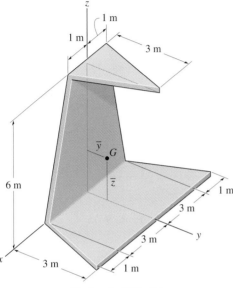

Probs. 9–64/9–65

9–66.   Determine the location $(\bar{x}, \bar{y})$ of the center of gravity of the three-wheeler. The location of the center of gravity of each component and its weight are tabulated in the figure. If the three-wheeler is symmetrical with respect to the x-y plane, determine the normal reactions each of its wheels exerts on the ground.

*9–68.   The buoy is made from two homogeneous cones each having a radius of 1.5 m. If $h = 1.2$ m, find the distance $\bar{z}$ to the buoy's center of gravity G.

9–69.   The buoy is made from two homogeneous cones each having a radius of 1.5 m. If it is required that the buoy's center of gravity G be located at $\bar{z} = 0.5$ m, determine the height h of the top cone.

1. Rear wheels	90 N
2. Mechanical components	425 N
3. Frame	600 N
4. Front wheel	40 N

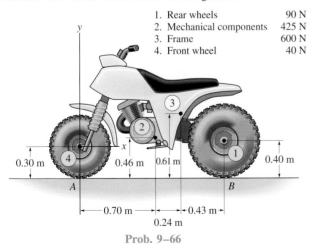

Prob. 9–66

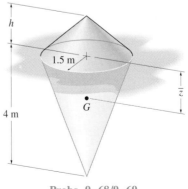

Probs. 9–68/9–69

9–70. The assembly consists of a 400-mm wooden dowel rod and a tight-fitting steel collar. Determine the distance $\bar{x}$ to its center of gravity if the specific weights of the materials are $\gamma_w = 25$ kN/m³ and $\gamma_{st} = 78$ kN/m³.

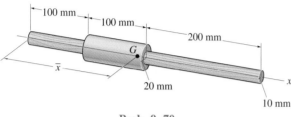

Prob. 9–70

9–71. Determine the distance h to which a 100-mm-diameter hole must be bored into the base of the cone so that the center of mass of the resulting shape is located at $\bar{z} = 115$ mm. The material has a density of 8 Mg/m³.

*9–72. Determine the distance $\bar{z}$ to the centroid of the shape which consists of a cone with a hole of height $h = 50$ mm bored into its base.

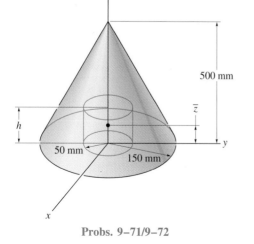

Probs. 9–71/9–72

9–73. Determine the location $\bar{z}$ of the centroid of the top made from a hemisphere and a cone.

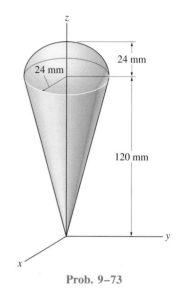

Prob. 9–73

9–74. The assembly is made from a steel hemisphere, $\rho_{st} = 7.80$ Mg/m³, and an aluminum cylinder, $\rho_{al} = 2.70$ Mg/m³. Determine the height h of the cylinder so that the mass center of the assembly is located at $\bar{z} = 160$ mm.

9–75. The assembly is made from a steel hemisphere, $\rho_{st} = 7.80$ Mg/m³, and an aluminum cylinder, $\rho_{al} = 2.70$ Mg/m³. Determine the mass center of the assembly if the height of the cylinder is $h = 200$ mm.

Probs. 9–74/9–75

*9.4 Theorems of Pappus and Guldinus

The two *theorems of Pappus and Guldinus,* which were first developed by Pappus of Alexandria during the third century A.D. and then restated at a later time by the Swiss mathematician Paul Guldin or Guldinus (1577–1643), are used to find the surface area and volume of any object of revolution.

A *surface area of revolution* is generated by revolving a *plane curve* about a nonintersecting fixed axis in the plane of the curve; whereas a *volume of revolution* is generated by revolving a *plane area* about a nonintersecting fixed axis in the plane of the area. For example, if the *line AB* shown in Fig. 9–20 is rotated about a fixed axis, it generates the *surface area* of a cone; if the triangular *area ABC* shown in Fig. 9–21 is rotated about the axis, it generates the *volume* of a cone.

The statements and proofs of the theorems of Pappus and Guldinus follow. The proofs require that the generating curves and areas do *not* cross the axis about which they are rotated; otherwise, two sections on either side of the axis would generate areas or volumes having opposite signs and hence cancel each other.

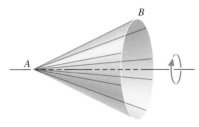

Fig. 9–20

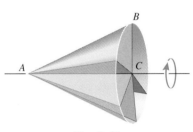

Fig. 9–21

Surface Area

The area of a surface of revolution equals the product of the length of the generating curve and the distance traveled by the centroid of the curve in generating the surface area.

Proof. When a differential length dL of the curve shown in Fig. 9–22 is revolved about an axis through a distance $2\pi r$, it generates a ring having a surface area $dA = 2\pi r\, dL$. The entire surface area, generated by revolving the entire curve about the axis, is therefore $A = 2\pi \int_L r\, dL$. This equation may be simplified, however, by noting that the location $\bar{r}$ of the centroid for the line of total length L can be determined from an equation having the form of Eqs. 9–7, namely, $\bar{r} = \int r\, dL/L$. Thus the total surface area becomes $A = 2\pi\bar{r}L$. In general, though, if the line does not undergo a complete revolution, then

$$A = \theta\bar{r}L \qquad (9\text{–}9)$$

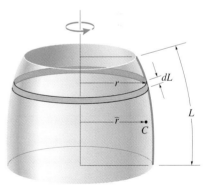

Fig. 9–22

where A = surface area of revolution
   θ = angle of revolution measured in radians, $\theta \leq 2\pi$
   $\bar{r}$ = perpendicular distance from the axis of revolution to the centroid of the generating curve
   L = length of the generating curve

Fig. 9–23

Volume

The volume of a body of revolution equals the product of the generating area and the distance traveled by the centroid of the area in generating the volume.

Proof. When the differential area dA shown in Fig. 9–23 is revolved about an axis through a distance $2\pi r$, it generates a ring having a volume $dV = 2\pi r\, dA$. The entire volume, generated by revolving A about the axis, is therefore $V = 2\pi\int_A r\, dA$. Here the integral can be eliminated by using an equation analogous to Eqs. 9–6, $\int_A r\, dA = \bar{r}A$, where $\bar{r}$ locates the centroid C of the generating area A, and the volume becomes $V = 2\pi\bar{r}A$. In general, though,

$$V = \theta\bar{r}A \tag{9-10}$$

where V = volume of revolution
θ = angle of revolution measured in radians, $\theta \leqslant 2\pi$
$\bar{r}$ = perpendicular distance from the axis of revolution to the centroid of the generating area
A = generating area

Composite Shapes. We may also apply the above two theorems to lines or areas that may be composed of a series of composite parts. In this case the total surface area or volume generated is the addition of the surface areas or volumes generated by each of the composite parts. Since each part undergoes the *same* angle of revolution, θ, and the distance from the axis of revolution to the centroid of each composite part is $\tilde{r}$, then

$$A = \theta\Sigma\tilde{r}L \tag{9-11}$$

and

$$V = \theta\Sigma\tilde{r}A \tag{9-12}$$

Application of the above theorems is illustrated numerically in the following example.

Example 9–12

Show that the surface area of a sphere is $A = 4\pi R^2$ and its volume is $V = \frac{4}{3}\pi R^3$.

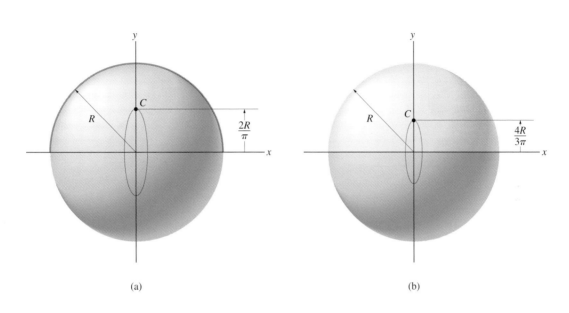

(a)   (b)

Fig. 9–24

SOLUTION

Surface Area.   The surface area of the sphere in Fig. 9–24*a* is generated by rotating a semicircular *arc* about the *x* axis. Using the table on the inside back cover, it is seen that the centroid of this arc is located at a distance $\bar{r} = 2R/\pi$ from the *x* axis of rotation. Since the centroid moves through an angle of $\theta = 2\pi$ rad in generating the sphere, by applying Eq. 9–9 we have

$$A = \theta \bar{r} L; \qquad A = 2\pi \left(\frac{2R}{\pi}\right) \pi R = 4\pi R^2 \qquad \textit{Ans.}$$

Volume.   The volume of the sphere is generated by rotating the semicircular *area* in Fig. 9–24*b* about the *x* axis. Using the table on the inside back cover to locate the centroid and applying Eq. 9–10, we have

$$V = \theta \bar{r} A; \qquad V = 2\pi \left(\frac{4R}{3\pi}\right) \left(\frac{1}{2}\pi R^2\right) = \frac{4}{3}\pi R^3 \qquad \textit{Ans.}$$

PROBLEMS

*9–76. Using integration, determine both the area and the centroidal distance $\bar{x}$ of the shaded area. Then, using the second theorem of Pappus–Guldinus, determine the volume of the solid generated by revolving the area about the y axis.

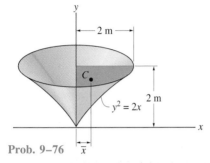

Prob. 9–76

9–77. Using integration, determine the area and the centroidal distance $\bar{y}$ of the shaded area. Then, using the second theorem of Pappus–Guldinus, determine the volume of a paraboloid formed by revolving the area about the x axis.

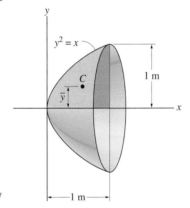

Prob. 9–77

9–78. Using integration, determine the area and the centroidal distance $\bar{x}$ of the shaded area. Then, using the second theorem of Pappus-Guldinus, determine the volume of a solid formed by revolving the area about the y axis.

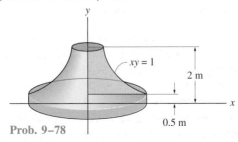

Prob. 9–78

9–79. Determine the surface area of the casting.

*9–80. Determine the volume of material needed to make the casting.

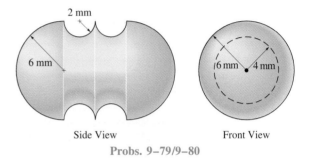

Side View   Front View

Probs. 9–79/9–80

9–81. Using integration, determine both the area and the distance $\bar{y}$ to the centroid of the shaded area. Then using the second theorem of Pappus—Guldinus, determine the volume of the solid generated by revolving the shaded area about the x axis.

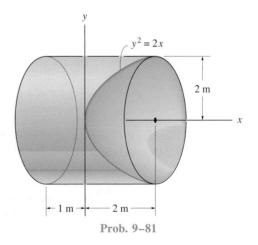

Prob. 9–81

9–82. Using integration, determine both the area and the centroidal distance $\bar{x}$ of the shaded area. Then, using the second theorem of Pappus—Guldinus, determine the volume of the solid generated by revolving the shaded area about the y axis.

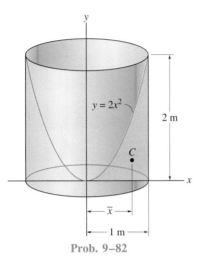

Prob. 9–82

*9–84. Determine the volume of concrete needed to construct the curb.

9–85. Determine the surface area of the curb. Do not include the area of the ends in the calculation.

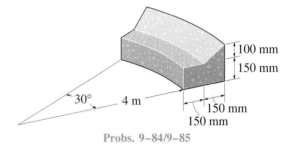

Probs. 9–84/9–85

9–83. Sand is piled between two walls as shown. Assume the pile to be a quarter section of a cone and that 26 percent of this volume is voids (air space). Use the second theorem of Pappus–Guldinus to determine the volume of sand.

9–86. The *rim* of a flywheel has the cross section A-A shown. Determine the volume of material needed for its construction.

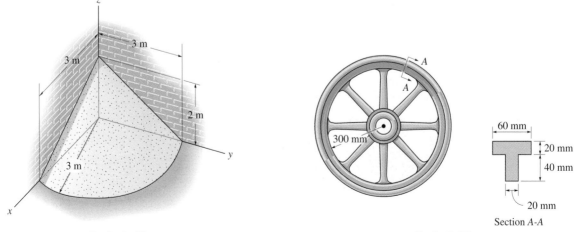

Prob. 9–83

Prob. 9–86

9–87. The anchor ring is made of steel having a specific weight of $\gamma_{st} = 78$ kN/m^3. Determine its weight. The cross section is circular as shown.

*9–88. The anchor ring is made of steel having a specific weight of $\gamma_{st} = 78$ kN/m^3. Determine the surface area of the ring. The cross section is circular as shown.

9–90. The water-supply tank has a hemispherical bottom and cylindrical sides. Determine the weight of water in the tank when it is filled to the top at C. Take $\gamma_w = 9.80$ kN/m^3.

9–91. Determine the volume of paint needed to paint the outside surface of the water-supply tank, which consists of a hemispherical bottom, cylindrical sides, and conical top. Each liter of paint can cover 6 m^2.

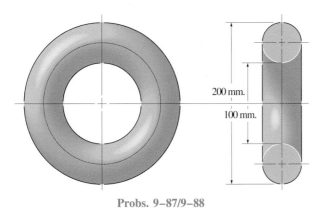

200 mm.

100 mm.

Probs. 9–87/9–88

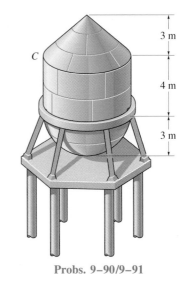

3 m

C

4 m

3 m

Probs. 9–90/9–91

*9–92. The hopper is filled to its top with coal. Determine the volume of coal if the voids (air space) are 35 percent of the volume of the hopper.

9–93. If the hopper is made of steel plate having a thickness of 10 mm, determine its weight when empty. Steel has a density of $\rho_{st} = 7.85$ Mg/m^3.

9–89. A circular sea wall is made of concrete. Determine the total weight of the wall if the concrete has a specific weight of $\gamma_c = 25$ kN/m^3.

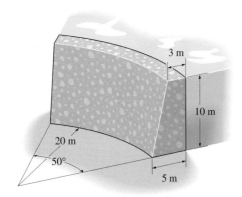

3 m

10 m

20 m

50°

5 m

Prob. 9–89

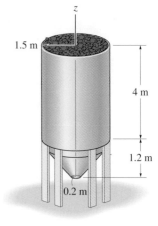

z

1.5 m

4 m

1.2 m

0.2 m

Probs. 9–92/9–93

9–94. The full circular aluminum housing is used in an automotive brake system. Half the cross section is shown in the figure. Determine its weight if aluminum has a specific weight of $\gamma_{al} = 25$ kN/m³.

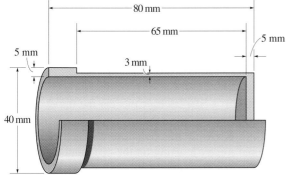

Prob. 9–94

9–95. Determine the approximate amount of aluminum necessary to make the funnel. It consists of a full circular part having a thickness of 2 mm. Its cross section is shown in the figure.

*9–96. Determine the approximate outer surface area of the funnel. It consists of a full circular part of negligible thickness.

Probs. 9–95/9–96

9–97. Determine the height h to which liquid should be poured into the cup so that it contacts half the surface area on the inside of the cup. Neglect the cup's thickness for the calculation.

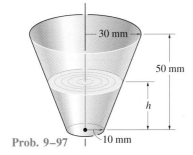

Prob. 9–97

9–98. The process tank is used to store liquids during manufacturing. Estimate both the volume of the tank and its surface area. The tank has a flat top and the plates from which the tank is made have negligible thickness.

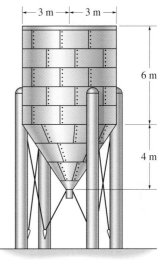

Prob. 9–98

9–99. Determine the volume of steel needed to make the brake piston. It consists of a full circular part. Its cross section is shown in the figure.

*9–100. Determine the volume of oil that can be contained in the brake piston. It consists of a full circular part. Its cross section is shown in the figure.

9–101. Determine the interior surface area of the brake piston. It consists of a full circular part. Its cross section is shown in the figure.

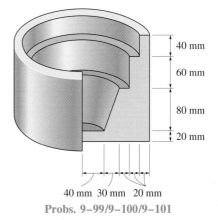

Probs. 9–99/9–100/9–101

*9.5 Resultant of a General Distributed Force System

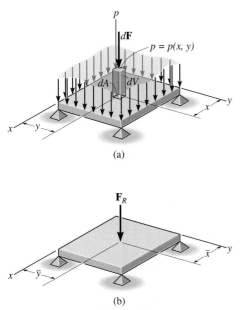

(a)

$\mathbf{F}_R$

(b)

Fig. 9–25

In Sec. 4.10 we discussed the method used to simplify a distributed loading that is uniform along an axis of a rectangular surface. In this section we will generalize this method to include surfaces that have an arbitrary shape and are subjected to a variable load distribution. As a specific application, in Sec. 9.6 we will find the resultant loading acting on the surface of a body that is submerged in a fluid.

Pressure Distribution over a Surface. Consider the flat plate shown in Fig. 9–25a, which is subjected to the loading function $p = p(x, y)$ Pa, where 1 pascal, Pa = 1 N/m². Knowing this function, we can determine the *magnitude* of the infinitesimal force $d\mathbf{F}$ acting on the differential area dA m² of the plate, located at the arbitrary point (x, y). This force magnitude is simply $dF = [p(x, y)$ N/m²]$(dA$ m²$) = [p(x, y)\,dA]$ N. The entire loading on the plate is therefore represented as a system of *parallel forces* infinite in number and each acting on separate differential areas dA. This system of parallel forces will now be simplified to a single resultant force $\mathbf{F}_R$ acting through a unique point $(\bar{x}, \bar{y})$ on the plate, Fig. 9–25b.

Magnitude of Resultant Force. To determine the *magnitude* of $\mathbf{F}_R$, it is necessary to sum each of the differential forces $d\mathbf{F}$ acting over the plate's *entire surface area A*. This sum may be expressed mathematically as an integral:

$$F_R = \Sigma F; \qquad F_R = \int_A p(x, y)\,dA = \int_V dV \qquad (9\text{–}13)$$

Note that $p(x, y)\,dA = dV$, the colored differential *volume element* shown in Fig. 9–25a. Therefore, the result indicates that the *magnitude of the resultant force is equal to the total volume under the distributed-loading diagram.*

Location of Resultant Force. The location $(\bar{x}, \bar{y})$ of $\mathbf{F}_R$ is determined by setting the moments of $\mathbf{F}_R$ equal to the moments of all the forces $d\mathbf{F}$ about the respective y and x axes. From Fig. 9–25a and 9–25b, using Eq. 9–13, we have

$$\bar{x} = \frac{\int_A x p(x, y)\,dA}{\int_A p(x, y)\,dA} = \frac{\int_V x\,dV}{\int_V dV} \qquad \bar{y} = \frac{\int_A y p(x, y)\,dA}{\int_A p(x, y)\,dA} = \frac{\int_V y\,dV}{\int_V dV} \qquad (9\text{–}14)$$

Hence, it can be seen that the *line of action of the resultant force passes through the geometric center or centroid of the volume under the distributed loading diagram.*

*9.6 Fluid Pressure

According to Pascal's law, a fluid at rest creates a pressure p at a point that is the *same* in *all* directions. The magnitude of p, measured as a force per unit area, depends on the specific weight γ or mass density ρ of the fluid and the depth z of the point from the fluid surface.* The relationship can be expressed mathematically as

$$p = \gamma z = \rho g z \qquad (9\text{--}15)$$

where g is the acceleration of gravity. Equation 9–15 is valid only for fluids that are assumed *incompressible,* as in the case of most liquids. Gases are compressible fluids, and since their density changes significantly with both pressure and temperature, Eq. 9–15 cannot be used.

To illustrate how Eq. 9–15 is applied, consider the submerged plate shown in Fig. 9–26. Three points on the plate have been specified. Since points A and B are both at depth z_2 from the liquid surface, the *pressure* at these points has a magnitude $p_2 = \gamma z_2$. Likewise, point C is at depth z_1; hence, $p_1 = \gamma z_1$. In all cases, the pressure acts *normal* to the surface area dA located at the specified point, Fig. 9–26. Using Eq. 9–15 and the results of Sec. 9.5, it is possible to determine the resultant force caused by a liquid pressure distribution, and specify its location on the surface of a submerged plate. Three different shapes of plates will now be considered.

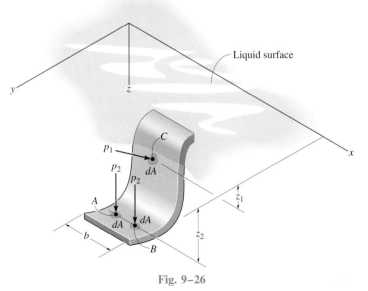

Fig. 9–26

*In particular, for water $\gamma = 62.4$ lb/ft³, or $\gamma = 9810$ N/m³, since $\rho = 1000$ kg/m³ and $g = 9.81$ m/s².

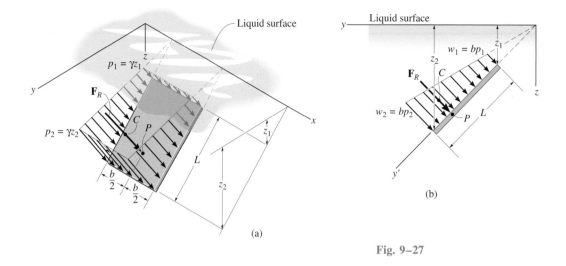

Fig. 9–27

Flat Plate of Constant Width.

A flat rectangular plate of constant width, which is submerged in a liquid having a specific weight γ, is shown in Fig. 9–27a. The plane of the plate makes an angle with the horizontal, such that its top edge is located at a depth z_1 from the liquid surface and its bottom edge is located at a depth z_2. Since pressure varies linearly with depth, Eq. 9–15, the distribution of pressure over the plate's surface is represented by a trapezoidal volume having an intensity of $p_1 = \gamma z_1$ at depth z_1 and $p_2 = \gamma z_2$ at depth z_2. As noted in Sec. 9.5, the magnitude of the *resultant force* $\mathbf{F}_R$ is equal to the *volume* of this loading diagram and $\mathbf{F}_R$ has a *line of action* that passes through the volume's centroid C. Hence $\mathbf{F}_R$ does *not* act at the centroid of the plate; rather, it acts at point P, called the *center of pressure*.

Since the plate has a *constant width,* the loading distribution may also be viewed in two dimensions, Fig. 9–27b. Here the loading intensity is measured as force/length and varies linearly from $w_1 = bp_1 = b\gamma z_1$ to $w_2 = bp_2 = b\gamma z_2$. The magnitude of $\mathbf{F}_R$ in this case equals the trapezoidal *area*, and $\mathbf{F}_R$ has a *line of action* that passes through the area's *centroid C.* For numerical applications, the area and location of the centroid for a trapezoid are tabulated on the inside back cover.

Curved Plate of Constant Width.

When the submerged plate is curved, the pressure acting normal to the plate continually changes direction, and therefore calculation of the magnitude of $\mathbf{F}_R$ and its location P is more difficult than for a flat plate. Three- and two-dimensional views of the loading distribution are shown in Figs. 9–28a and 9–28b, respectively. Here integration can be used to determine both F_R and the location of the centroid C or center of pressure P.

A simpler method exists, however, for calculating the magnitude of $\mathbf{F}_R$ and its location along a curved (or flat) plate having a *constant width.* This

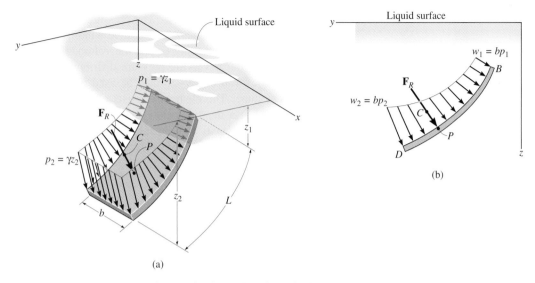

$p_1 = \gamma z_1$

F_R

C P

$p_2 = \gamma z_2$

b

L

z_1

z_2

x

Liquid surface

y

(a)

Liquid surface

y

$w_1 = bp_1$

B

F_R

$w_2 = bp_2$

C

P

D

z

(b)

method requires separate calculations for the horizontal and vertical *compo-nents* of $\mathbf{F}_R$. For example, the distributed loading acting on the curved plate *DB* in Fig. 9–28b can be represented by the *equivalent loading* shown in Fig. 9–29. Here the plate supports the weight of liquid W_f contained within the block *BDA*. This force has a magnitude $W_f = (\gamma b)(\text{area}_{BDA})$ and acts through the centroid of *BDA*. In addition, there are the pressure distributions caused by the liquid acting along the vertical and horizontal sides of the block. Along the vertical side *AD*, the force $\mathbf{F}_{AD}$ has a magnitude that equals the area under the trapezoid and acts through the centroid C_{AD} of this area. The distributed load-ing along the horizontal side *AB* is constant, since all points lying in this plane are at the same depth from the surface of the liquid. The magnitude of $\mathbf{F}_{AB}$ is simply the area of the rectangle. This force acts through the area's centroid C_{AB} or the midpoint of *AB*. Summing the three forces in Fig. 9–29 yields $\mathbf{F}_R = \Sigma\mathbf{F} = \mathbf{F}_{AD} + \mathbf{F}_{AB} + \mathbf{W}_f$, which is shown in Fig. 9–28. Finally, the loca-tion of the center of pressure *P* on the plate is determined by applying the equation $M_{R_o} = \Sigma M_O$, which states that the moment of the resultant force about a convenient reference point, Fig. 9–28, is equal to the sum of the moments of the three forces in Fig. 9–29 about the same point.

Fig. 9–28

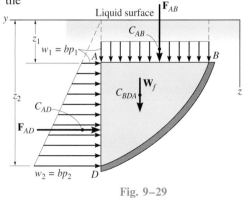

Liquid surface

y

z_1

$w_1 = bp_1$

A

z_2

C_{AD}

$\mathbf{F}_{AD}$

$w_2 = bp_2$ D

$\mathbf{F}_{AB}$

C_{AB}

B

C_{BDA} $\mathbf{W}_f$

z

Fig. 9–29

Flat Plate of Variable Width. The pressure distribution acting on the surface of a submerged plate having a variable width is shown in Fig. 9–30. The resultant force of this loading equals the volume described by the plate area as its base and linear varying pressure distribution as its altitude. The shaded element shown in Fig. 9–30 may be used if integration is chosen to determine this volume. The element consists of a rectangular strip of area $dA = x\,dy'$ located at a depth z below the liquid surface. Since a uniform pressure $p = \gamma z$ (force/area) acts on dA, the magnitude of the differential force $d\mathbf{F}$ is equal to $dF = dV = p\,dA = \gamma z(x\,dy')$. Integrating over the entire volume yields Eq. 9–13; i.e.,

$$F_R = \int_A p\,dA = \int_V dV = V$$

From Eq. 9–14, the centroid of V defines the point through which $\mathbf{F}_R$ acts. The center of pressure, which lies on the surface of the plate just below C, has coordinates $P(\bar{x}, \bar{y}')$ defined by the equations

$$\bar{x} = \frac{\displaystyle\int_V \tilde{x}\,dV}{\displaystyle\int_V dV} \qquad \bar{y}' = \frac{\displaystyle\int_V \tilde{y}'\,dV}{\displaystyle\int_V dV}$$

This point should *not* be mistaken for the centroid of the plate's *area*.

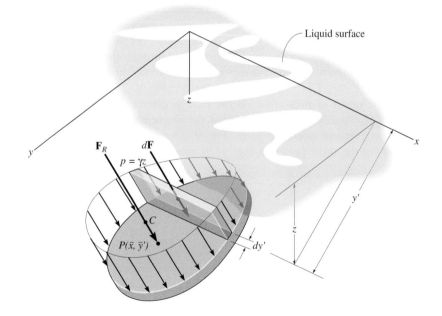

Fig. 9–30

Example 9–13

Determine the magnitude and location of the resultant hydrostatic force acting on the submerged rectangular plate AB shown in Fig. 9–31a. The plate has a width of 1.5 m; $\rho_w = 1000$ kg/m³.

SOLUTION

The water pressures at depths A and B are

$$p_A = \rho_w g z_A = (1000 \text{ kg/m}^3)(9.81 \text{ m/s}^2)(2 \text{ m}) = 19.62 \text{ kPa}$$
$$p_B = \rho_w g z_B = (1000 \text{ kg/m}^3)(9.81 \text{ m/s}^2)(5 \text{ m}) = 49.05 \text{ kPa}$$

Since the plate has a constant width, the distributed loading can be viewed in two dimensions as shown in Fig. 9–31b. The intensity of the load at A and B is

$$w_A = bp_A = (1.5 \text{ m})(19.62 \text{ kPa}) = 29.4 \text{ kN/m}$$
$$w_B = bp_B = (1.5 \text{ m})(49.05 \text{ kPa}) = 73.6 \text{ kN/m}$$

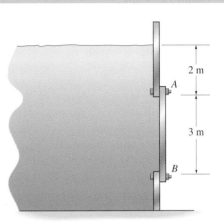

(a)

From the table on the inside back cover, the magnitude of the resultant force $\mathbf{F}_R$ created by the distributed load is

$$F_R = \text{area of trapezoid}$$
$$= \tfrac{1}{2}(3)(29.4 + 73.6) = 154.5 \text{ kN} \qquad \textit{Ans.}$$

This force acts through the centroid of the area,

$$h = \frac{1}{3}\left(\frac{2(29.4) + 73.6}{29.4 + 73.6}\right)(3) = 1.29 \text{ m} \qquad \textit{Ans.}$$

measured upward from B, Fig. 9–31b.

The same results can be obtained by considering two components of $\mathbf{F}_R$ defined by the triangle and rectangle shown in Fig. 9–31c. Each force acts through its associated centroid and has a magnitude of

$$F_{Re} = (29.4 \text{ kN/m})(3 \text{ m}) = 88.2 \text{ kN}$$
$$F_t = \tfrac{1}{2}(44.2 \text{ kN/m})(3 \text{ m}) = 66.3 \text{ kN}$$

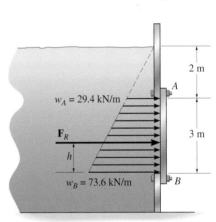

(b)

Hence,

$$F_R = F_{Re} + F_t = 88.2 + 66.3 = 154.5 \text{ kN} \qquad \textit{Ans.}$$

The location of $\mathbf{F}_R$ is determined by summing moments about B, Fig. 9–31b and c, i.e.,

$$\uparrow+(M_R)_B = \Sigma M_B; \quad (154.5)h = 88.2(1.5) + 66.3(1)$$
$$h = 1.29 \text{ m} \qquad \textit{Ans.}$$

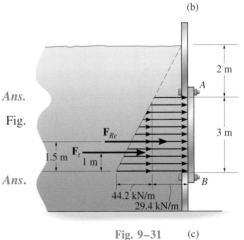

Fig. 9–31 (c)

Example 9–14

Determine the magnitude of the resultant hydrostatic force acting on the surface of a seawall shaped in the form of a parabola as shown in Fig. 9–32a. The wall is 5 m long; $\rho_w = 1020 \text{ kg/m}^3$.

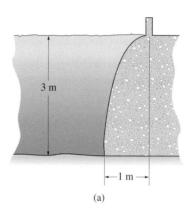

(a)

SOLUTION

The horizontal and vertical components of the resultant force will be calculated, Fig. 9–32b. Since

$$p_B = \rho_w g z_B = (1020 \text{ kg/m}^3)(9.81 \text{ m/s}^2)(3 \text{ m}) = 30.02 \text{ kPa}$$

then

$$w_B = b p_B = 5 \text{ m}(30.02 \text{ kPa}) = 150.1 \text{ kN/m}$$

Thus,

$$F_x = \tfrac{1}{2}(3 \text{ m})(150.1 \text{ kN/m}) = 225.2 \text{ kN}$$

The area of the parabolic sector ABC can be determined using the table on the inside back cover. Hence, the weight of water within this region is

$$F_y = (\rho_w g b)(\text{area}_{ABC})$$
$$= (1020 \text{ kg/m}^3)(9.81 \text{ m/s}^2)(5 \text{ m})[\tfrac{1}{3}(1 \text{ m})(3 \text{ m})] = 50.0 \text{ kN}$$

The resultant force is therefore

$$F_R = \sqrt{F_x^2 + F_y^2} = \sqrt{(225.2)^2 + (50.0)^2}$$
$$= 231 \text{ kN} \qquad \qquad \textit{Ans.}$$

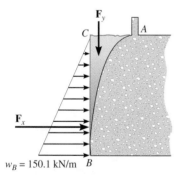

(b)

Fig. 9–32

Example 9–15

Determine the magnitude and location of the resultant force acting on the triangular end plates of the water trough shown in Fig. 9–33a; $\rho_w = 1000 \text{ kg/m}^3$.

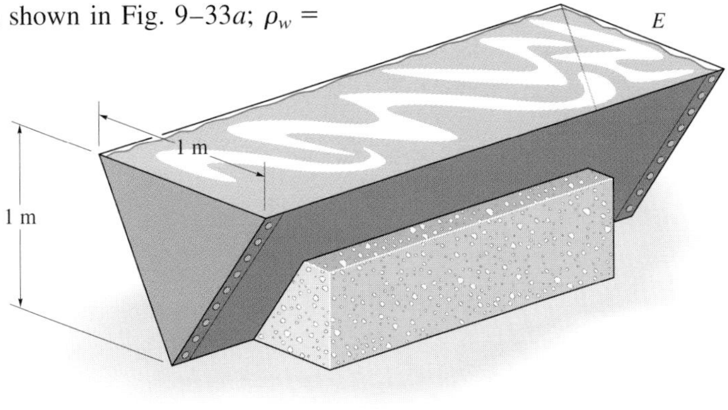

(a)

SOLUTION

The pressure distribution acting on the end plate E is shown in Fig. 9–33b. The magnitude of the resultant force **F** is equal to the volume of this loading distribution. We will solve the problem by integration. Choosing the differential volume element shown in the figure, we have

$$dF = dV = p \, dA = \rho_w gz(2x \, dz) = 19\,620zx \, dz$$

The equation of line AB is

$$x = 0.5(1 - z)$$

Hence, substituting and integrating with respect to z from $z = 0$ to $z = 1$ m yields

$$F = V = \int_V dV = \int_0^1 (19\,620)z[0.5(1 - z)] \, dz$$

$$= 9810 \int_0^1 (z - z^2) \, dz = 1635 \text{ N} = 1.64 \text{ kN} \qquad \textbf{\textit{Ans.}}$$

This resultant passes through the *centroid of the volume*. Because of symmetry,

$$\bar{x} = 0 \qquad \textbf{\textit{Ans.}}$$

Since $\tilde{z} = z$ for the volume element in Fig. 9–33b, then

$$\bar{z} = \frac{\int_V \tilde{z} \, dV}{\int_V dV} = \frac{\int_0^1 z(19\,620)z[0.5(1 - z)] \, dz}{1635} = \frac{9810 \int_0^1 (z^2 - z^3) \, dz}{1635}$$

$$= 0.5 \text{ m} \qquad \textbf{\textit{Ans.}}$$

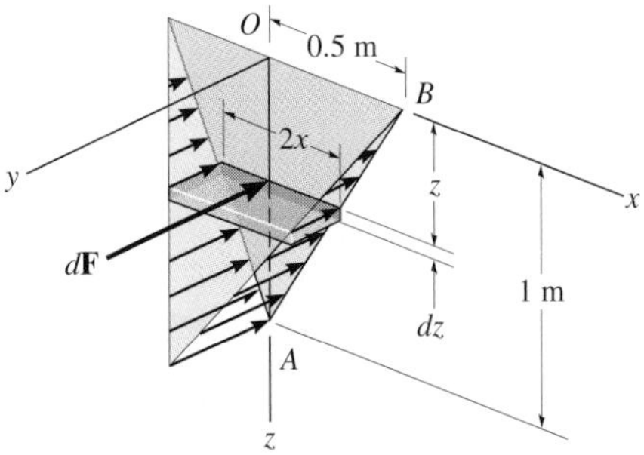

(b)

Fig. 9–33

PROBLEMS

9–102. Determine the magnitude of the resultant hydrostatic force acting on the dam and its location, measured from the top surface of the water. The width of the dam is 8 m; $\rho_w = 1.0\ \text{Mg/m}^3$.

*9–104. The concrete dam is designed so that its face AB has a gradual slope into the water as shown. Because of this, the frictional force at the base BD of the dam is increased due to the hydrostatic force of the water acting on the dam. Calculate the hydrostatic force acting on the face AB of the dam. The dam is 18 m wide. $\gamma_w = 9.80\ \text{kN/m}^3$.

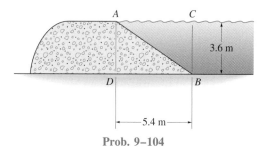

Prob. 9–104

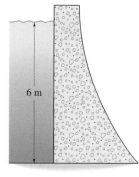

Prob. 9–102

9–105. The storage tank contains oil having a specific weight of $\gamma_o = 8.80\ \text{kN/m}^3$. If the tank is 1.8 m wide, calculate the resultant force acting on the inclined side BC of the tank, caused by the oil, and specify its location along BC, measured from B. Also compute the total resultant force acting on the bottom of the tank.

9–103. The tank is filled with water to a depth of $d = 4$ m. Determine the resultant force the water exerts on side A and side B of the tank. If oil instead of water is placed in the tank, to what depth d should it reach so that it creates the same resultant forces? $\rho_o = 900\ \text{kg/m}^3$ and $\rho_w = 1000\ \text{kg/m}^3$.

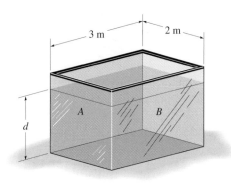

Prob. 9–103

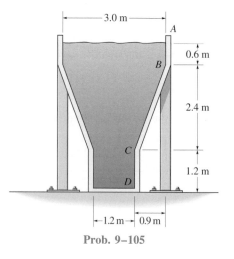

Prob. 9–105

9–106. The semicircular tunnel passes under a river which is 9 m deep. Determine the vertical resultant hydrostatic force acting per meter of length along the length of the tunnel. The tunnel is 6 m wide; $\rho_w = 1.0 \text{ Mg/m}^3$.

Prob. 9–106

9–107. The storage tank contains oil having a density of $\rho_o = 0.90 \text{ Mg/m}^3$. If the tank is 1.5 m wide, calculate the resultant force acting on the inclined side AB of the tank, caused by the oil, and specify its location along AB, measured from A.

*9–108. The storage tank contains oil having a density of $\rho_o = 0.90 \text{ Mg/m}^3$. If the tank is 1.5 m wide, calculate the resultant force acting on side BC of the tank, caused by the oil, and specify its location along BC, measured from C.

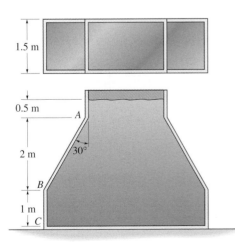

Probs. 9–107/9–108

9–109. The gate AB is 8 m wide. Determine the horizontal and vertical components of force acting on the pin at B and the vertical reaction at the smooth support A. $\rho_w = 1.0 \text{ Mg/m}^3$.

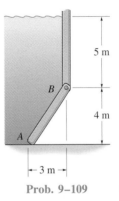

Prob. 9–109

9–110. The structure shown is used for temporary storage of oil at sea for later loading into ships. When it is empty the water level is at A (sea level). As oil is loaded, the water is displaced through exit ports at D. If the riser EC is filled with oil, i.e., to a depth of C, determine the height h to B of the oil level above sea level. $\rho_o = 900 \text{ kg/m}^3$ and $\rho_w = 1020 \text{ kg/m}^3$.

9–111. If the structure in Prob. 9–110 is totally filled with oil, i.e., until it reaches a depth of 58 m below sea level, how high h will the oil level extend above sea level?

Probs. 9–110/9–111

*9–112. The arched surface AB is shaped in the form of a quarter circle. If it is 8 m long, determine the horizontal and vertical components of the resultant force caused by the water acting on the surface. $\rho_w = 1.0 \text{ Mg/m}^3$.

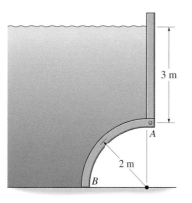

Prob. 9–112

9–114. Determine the magnitude of the resultant hydrostatic force acting per meter of length on the sea wall. $\rho_w = 1.0 \text{ Mg/m}^3$.

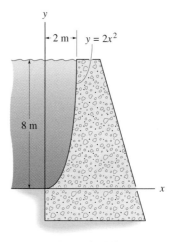

Prob. 9–114

9–113. Determine the magnitude and location of the resultant hydrostatic force acting on each of the cover plates A and B. $\rho_w = 1.0 \text{ Mg/m}^3$.

9–115. Determine the magnitude of the resultant hydrostatic force acting per meter of length on the sea wall; $\gamma_w = 9.80 \text{ kN/m}^3$.

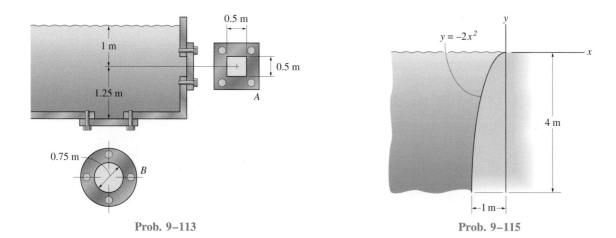

Prob. 9–113

Prob. 9–115

*9–116. The rear end of the boat has a trapezoidal form with the dimensions shown. If the water on the outside of the boat is 75 mm from the top, determine the resultant force and its location, measured from the top of the boat. $\gamma_w = 10$ kN/m^3.

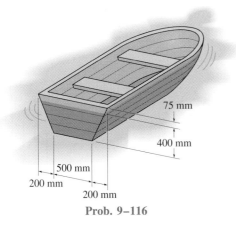

Prob. 9–116

9–117. The tank is filled to the top ($y = 0.5$ m) with water having a density of $\rho_w = 1.0$ Mg/m^3. Determine the resultant force of the water pressure acting on the flat end plate C of the tank, and its location, measured from the top of the tank.

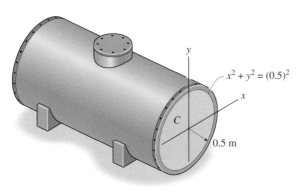

Prob. 9–117

9–118. The end plates of the trough are in the form of a semiellipse. If it is filled to the top with water, determine the resultant force the water exerts on these plates and the location of the center of pressure, measured from the top of the trough. $\gamma_w = 10$ kN/m^3.

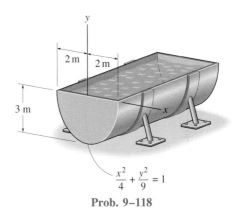

$$\frac{x^2}{4} + \frac{y^2}{9} = 1$$

Prob. 9–118

9–119. The wind blows uniformly on the front surface of the metal building with a pressure of 30 kN/m^2. Determine the resultant force it exerts on the surface and the position of this resultant.

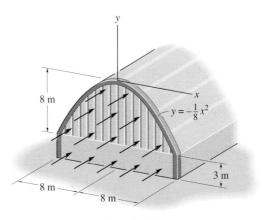

Prob. 9–119

*9–120. The pressure loading on the plate is described by the function $p = [-240/(x + 1) + 340]$ Pa. Determine the magnitude of the resultant force and the coordinates $(\bar{x}, \bar{y})$ of the point where the line of action of the force intersects the plate.

9–122. The loading acting on a square plate is represented by a parabolic pressure distribution. Determine the magnitude of the resultant force and the coordinates $(\bar{x}, \bar{y})$ of the point where the line of action of the force intersects the plate. Also, what are the reactions at the rollers B and C and the ball-and-socket joint A? Neglect the weight of the plate.

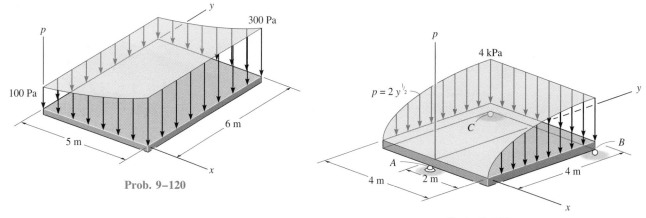

Prob. 9–120

Prob. 9–122

9-121. The pressure loading on the plate varies uniformly along each of its edges. Determine the magnitude of the resultant force and the coordinates $(\bar{x}, \bar{y})$ of the point where the line of action of the force intersects the plate. *Hint:* The equation defining the boundary of the load has the form $p = ax + by + c$, where the constants a, b, and c have to be determined.

9–123. The load over the plate varies linearly along the sides of the plate such that $p = \frac{2}{3}[x(4 - y)]$ kPa. Determine the magnitude of the resultant force and the coordinates $(\bar{x}, \bar{y})$ of the point where the line of action of the force intersects the plate.

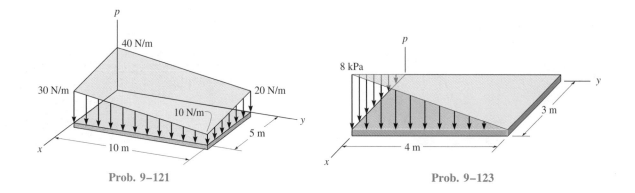

Prob. 9–121

Prob. 9–123

REVIEW PROBLEMS

*9–124. A circular V-belt has an inner radius of 600 mm and a cross-sectional area as shown. Determine the volume of material required to make the belt.

9–125. A circular V-belt has an inner radius of 600 mm and a cross-sectional area as shown. Determine the surface area of the belt.

9–127. Determine the distance $\bar{y}$ of the centroid of the cross-sectional area.

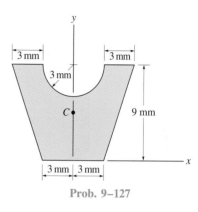

Prob. 9–127

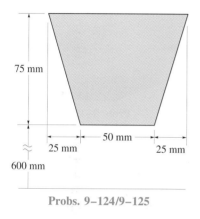

Probs. 9–124/9–125

9–126. The starter for an electric motor has the cross-sectional area shown. If copper wiring has a density of $\rho_{cu} = 8.90 \text{ Mg/m}^3$ and the steel frame has a density of $\rho_{st} = 7.80 \text{ Mg/m}^3$, estimate the total mass of the starter. Neglect the voids within the copper wiring.

*9–128. Locate the centroid for the cold-formed metal strut having the cross section shown. Neglect the thickness of the material and slight bends at the corners.

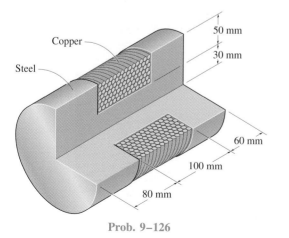

Prob. 9–126

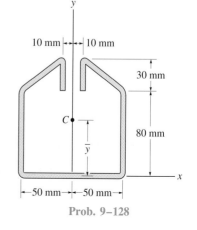

Prob. 9–128

9–129. Locate the center of gravity of the homogeneous rod. The rod has a weight of 2 kN/m. Also, compute the x, y, z components of reaction at the fixed support A.

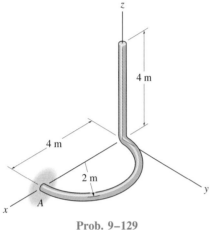

4 m

2 m

4 m

z

y

x

A

Prob. 9–129

9–130. The thin-walled channel and stiffener have the cross section shown. If the material has a constant thickness, determine the location $\bar{y}$ of its centroid. The dimensions are indicated to the center of each segment.

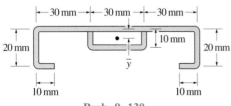

30 mm   30 mm   30 mm

20 mm   10 mm   20 mm

$\bar{y}$

10 mm   10 mm

Prob. 9–130

9–131. Determine the volume of steel needed to produce the tapered part. The cross section is shown, although the part is 360° around. Also, compute the outside surface area of the part, excluding its ends.

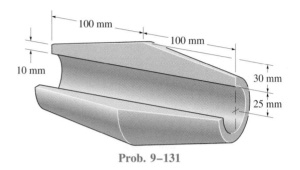

100 mm

100 mm

10 mm

30 mm

25 mm

Prob. 9–131

*9–132. The rectangular bin is filled with coal, which creates a pressure distribution along wall A that varies as shown, i.e., $p = 4z^3$ kN/m², where z is in feet. Compute the resultant force created by the coal, and its location, measured from the top surface of the coal.

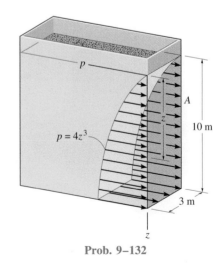

p

A

$p = 4z^3$

10 m

3 m

z

Prob. 9–132

9–133.   Determine the location $(\bar{x}, \bar{y})$ of the center of mass for the compressor assembly. The locations of the centers of mass of the various components and their masses are indicated and tabulated in the figure. What are the vertical reactions at blocks A and B needed to support the platform?

9–134.   Locate the centroid of the channel's cross-sectional area.

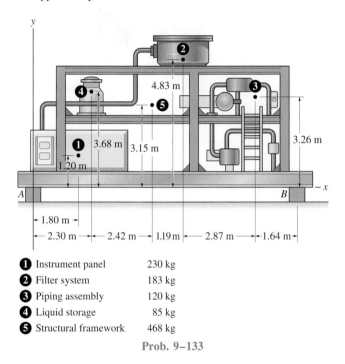

	Instrument panel	230 kg
❷	Filter system	183 kg
❸	Piping assembly	120 kg
❹	Liquid storage	85 kg
❺	Structural framework	468 kg

Prob. 9–133

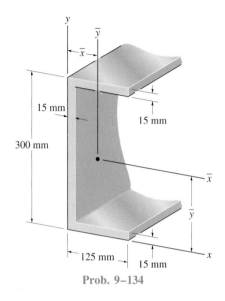

Prob. 9–134

The design of a structural member, such as a beam or column, requires calculation of its cross-sectional moment of inertia. In this chapter we will discuss how this is done.

10

Moments of Inertia

In this chapter we will develop a method for determining the moment of inertia for both an area and a body having a specified mass. The moment of inertia for an area is an important property in engineering, since it must be determined or specified if one is to analyze or design a structural member or a mechanical part. On the other hand, a body's mass moment of inertia must be known if one is studying the motion of the body.

10.1 Definition of Moments of Inertia for Areas

In the last chapter we determined the centroid for an area by considering the first moment of the area about an axis; that is, for the computation we had to evaluate an integral of the form $\int x \, dA$. Integrals of the second moment of an area, such as $\int x^2 \, dA$, are referred to as the *moment of inertia* for the area. The terminology "moment of inertia" as used here is actually a misnomer; however, it has been adopted because of the similarity with integrals of the same form related to mass.

The moment of inertia of an area originates whenever one has to compute the moment of a distributed load that varies linearly from the moment axis. A typical example of this kind of loading occurs due to the pressure of a liquid acting on the surface of a submerged plate. It was pointed out in Sec. 9.6 that the pressure, or force per unit area, exerted at a point located a distance z below the surface of a liquid is $p = \gamma z$, Eq. 9–15, where γ is the specific weight of the liquid. Thus, the magnitude of force exerted by a liquid on the

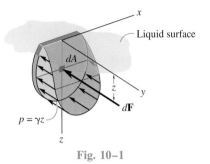

Fig. 10–1

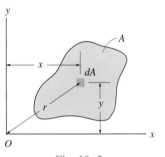

Fig. 10–2

area dA of the submerged plate shown in Fig. 10–1 is $dF = p\,dA = \gamma z\,dA$. The moment of this force about the x axis of the plate is $dM = z\,dF = \gamma z^2\,dA$, and therefore the moment created by the *entire* pressure distribution is $M = \gamma \int z^2\,dA$. Here the integral represents the moment of inertia of the area of the plate about the x axis. Since integrals of this form often arise in formulas used in fluid mechanics, mechanics of materials, structural mechanics, and machine design, the engineer should become familiar with the methods used for their computation.

Moment of Inertia. Consider the area A, shown in Fig. 10–2, which lies in the x–y plane. By definition, the moments of inertia of the differential planar area dA about the x and y axes are $dI_x = y^2\,dA$ and $dI_y = x^2\,dA$, respectively. For the entire area the *moments of inertia* are determined by integration; i.e.,

$$I_x = \int_A y^2\,dA$$
$$I_y = \int_A x^2\,dA$$

(10–1)

We can also formulate the second moment of the differential area dA about the pole O or z axis, Fig. 10–2. This is referred to as the polar moment of inertia, $dJ_O = r^2\,dA$. Here r is the perpendicular distance from the pole (z axis) to the element dA. For the entire area the *polar moment of inertia* is

$$J_O = \int_A r^2\,dA = I_x + I_y$$

(10–2)

The relationship between J_O and I_x, I_y is possible since $r^2 = x^2 + y^2$, Fig. 10–2.

From the above formulations it is seen that I_x, I_y, and J_O will *always* be *positive*, since they involve the product of distance squared and area. Furthermore, the units for moment of inertia involve length raised to the fourth power, e.g., m^4, mm^4.

10.2 Parallel-Axis Theorem for an Area

If the moment of inertia for an area is known about an axis passing through its centroid, it is convenient to determine the moment of inertia of the area about a corresponding parallel axis using the *parallel-axis theorem*. To derive this theorem, consider finding the moment of inertia of the shaded area shown in Fig. 10–3 about the x axis. In this case, a differential element dA is located at an arbitrary distance y' from the *centroidal x' axis*, whereas the *fixed distance* between the parallel x and x' axes is defined as d_y. Since the moment of inertia of dA about the x axis is $dI_x = (y' + d_y)^2 \, dA$, then for the entire area,

$$I_x = \int_A (y' + d_y)^2 \, dA$$

$$= \int_A y'^2 \, dA + 2d_y \int_A y' \, dA + d_y^2 \int_A dA$$

The first integral represents the moment of inertia of the area about the centroidal axis, $\bar{I}_{x'}$. The second integral is zero since the x' axis passes through the area's centroid C; i.e., $\int y' \, dA = \bar{y} \int dA = 0$ since $\bar{y} = 0$. Realizing that the third integral represents the total area A, the final result is therefore

$$\boxed{I_x = \bar{I}_{x'} + Ad_y^2} \qquad (10\text{--}3)$$

A similar expression can be written for I_y; i.e.,

$$\boxed{I_y = \bar{I}_{y'} + Ad_x^2} \qquad (10\text{--}4)$$

And finally, for the polar moment of inertia about an axis perpendicular to the x–y plane and passing through the pole O (z axis), Fig. 10–3, we have

$$\boxed{J_O = \bar{J}_C + Ad^2} \qquad (10\text{--}5)$$

The form of each of these equations states that *the moment of inertia of an area about an axis is equal to the moment of inertia of the area about a parallel axis passing through the area's centroid plus the product of the area and the square of the perpendicular distance between the axes.*

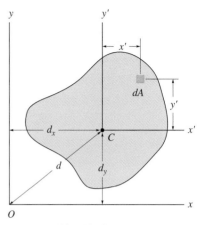

Fig. 10–3

10.3 Radius of Gyration of an Area

The *radius of gyration* of a planar area has units of length and is a quantity that is often used for the design of columns in structural mechanics. Provided the areas and moments of inertia are *known,* the radii of gyration are determined from the formulas

$$\boxed{k_x = \sqrt{\dfrac{I_x}{A}} \quad k_y = \sqrt{\dfrac{I_y}{A}} \quad k_O = \sqrt{\dfrac{J_O}{A}}} \qquad (10\text{--}6)$$

The form of these equations is easily remembered, since it is similar to that for finding the moment of inertia of a differential area about an axis. For example, $I_x = k_x^2 A$; whereas for a differential area, $dI_x = y^2 \, dA$.

10.4 Moments of Inertia for an Area by Integration

When the boundaries for a planar area are expressed by mathematical functions, Eqs. 10–1 may be integrated to determine the moments of inertia for the area. If the element of area chosen for integration has a differential size in two directions as shown in Fig. 10–2, a double integration must be performed to evaluate the moment of inertia. Most often, however, it is easier to perform only a single integration by choosing an element having a differential size or thickness in only one direction.

PROCEDURE FOR ANALYSIS

If a single integration is performed to determine the moment of inertia of an area about an axis, it will first be necessary to specify the differential element dA. Most often this element will be rectangular, such that it will have a finite length and differential width. The element should be located so that it intersects the boundary of the area at the *arbitrary point (x, y).* There are two possible ways to orient the element with respect to the axis about which the moment of inertia is to be determined.

Case 1.   The *length* of the element can be oriented *parallel* to the axis. This situation occurs when the rectangular element shown in Fig. 10–4 is used to determine I_y for the area. Direct application of Eq. 10–1, i.e., $I_y = \int x^2 \, dA$, can be made in this case, since the element has an infinitesimal thickness dx and therefore *all parts* of the element lie at the *same* moment-arm distance x from the y axis.*

Case 2.   The *length* of the element can be oriented *perpendicular* to the axis. Here Eq. 10–1 *does not apply,* since all parts of the element will *not* lie at the same moment-arm distance from the axis. For example, if the rectangular element in Fig. 10–4 is used for determining I_x for the area, it will first be necessary to calculate the moment of inertia of the *element* about a horizontal axis passing through the element's centroid and then determine the moment of inertia of the *element* about the x axis by using the parallel-axis theorem. Integration of this result will yield I_x.

*In the case of the element $dA = dx \, dy$, Fig. 10–2, the moment arms y and x are appropriate for the formulation of I_x and I_y (Eq. 10–1) since the *entire* element, because of its infinitesimal size, lies at the specified y and x perpendicular distances from the x and y axes.

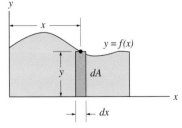

Fig. 10–4

Application of these cases is illustrated in the following examples.

Example 10–1

Determine the moment of inertia for the rectangular area shown in Fig. 10–5 with respect to (a) the centroidal x' axis, (b) the axis x_b passing through the base of the rectangle, and (c) the pole or z' axis perpendicular to the x'–y' plane and passing through the centroid C.

SOLUTION (*CASE 1*)

Part (a). The differential element shown in Fig. 10–5 is chosen for integration. Because of its location and orientation, the *entire element* is at a distance y' from the x' axis. Here it is necessary to integrate from $y' = -h/2$ to $y' = h/2$. Since $dA = b\,dy'$, then

$$\bar{I}_{x'} = \int_A y'^2\,dA = \int_{-h/2}^{h/2} y'^2(b\,dy') = b\int_{-h/2}^{h/2} y'^2\,dy'$$

$$= \frac{1}{12}bh^3 \qquad\qquad Ans.$$

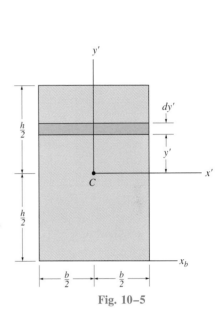

Fig. 10–5

Part (b). The moment of inertia about an axis passing through the base of the rectangle can be obtained by using the result of part (a) and applying the parallel-axis theorem, Eq. 10–3.

$$I_{x_b} = \bar{I}_{x'} + Ad_y^2$$

$$= \frac{1}{12}bh^3 + bh\left(\frac{h}{2}\right)^2 = \frac{1}{3}bh^3 \qquad\qquad Ans.$$

Part (c). To obtain the polar moment of inertia about point C, we must first obtain $\bar{I}_{y'}$, which may be found by interchanging the dimensions b and h in the result of part (a), i.e.,

$$\bar{I}_{y'} = \frac{1}{12}hb^3$$

Using Eq. 10–2, the polar moment of inertia about C is therefore

$$\bar{J}_C = \bar{I}_{x'} + \bar{I}_{y'} = \frac{1}{12}bh(h^2 + b^2) \qquad\qquad Ans.$$

Example 10–2

Determine the moment of inertia of the shaded area shown in Fig. 10–6a about the x axis.

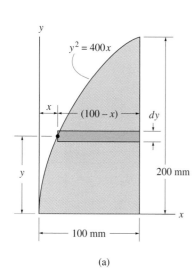

(a)

SOLUTION I (CASE 1)

A differential element of area that is *parallel* to the x axis, as shown in Fig. 10–6a, is chosen for integration. Since the element has a thickness dy and intersects the curve at the *arbitrary point* (x, y), the area is $dA = (100 - x)\, dy$. Furthermore, all parts of the element lie at the same distance y from the x axis. Hence, integrating with respect to y, from $y = 0$ to $y = 200$ mm, yields

$$I_x = \int_A y^2\, dA = \int_A y^2(100 - x)\, dy$$

$$= \int_0^{200} y^2\left(100 - \frac{y^2}{400}\right) dy = 100\int_0^{200} y^2\, dy - \frac{1}{400}\int_0^{200} y^4\, dy$$

$$= 107(10^6)\ \text{mm}^4 \qquad\qquad \textit{Ans.}$$

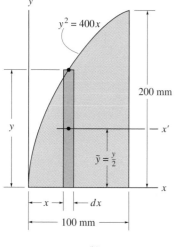

(b)

Fig. 10–6

SOLUTION II (CASE 2)

A differential element *parallel* to the y axis, as shown in Fig. 10–6b, is chosen for integration. It intersects the curve at the *arbitrary point* (x, y). In this case, all parts of the element do *not* lie at the same distance from the x axis, and therefore the parallel-axis theorem must be used to determine the *moment of inertia of the element* with respect to this axis. For a rectangle having a base b and height h, the moment of inertia about its centroidal axis has been determined in part (a) of Example 10–1. There it was found that $\bar{I}_{x'} = \frac{1}{12}bh^3$. For the differential element shown in Fig. 10–6b, $b = dx$ and $h = y$, and thus $d\bar{I}_{x'} = \frac{1}{12}dx\, y^3$. Since the centroid of the element is at $\tilde{y} = y/2$ from the x axis, the moment of inertia of the element about this axis is

$$dI_x = d\bar{I}_{x'} + dA\,\tilde{y}^2 = \frac{1}{12}dx\, y^3 + y\, dx\left(\frac{y}{2}\right)^2 = \frac{1}{3}y^3\, dx$$

[This result can also be concluded from part (b) of Example 10–1.] Integrating with respect to x, from $x = 0$ to $x = 100$ mm, yields

$$I_x = \int dI_x = \int_A \frac{1}{3}y^3\, dx = \int_0^{100} \frac{1}{3}(400x)^{3/2}\, dx$$

$$= 107(10^6)\ \text{mm}^4 \qquad\qquad \textit{Ans.}$$

Example 10–3

Determine the moment of inertia with respect to the x axis of the circular area shown in Fig. 10–7a.

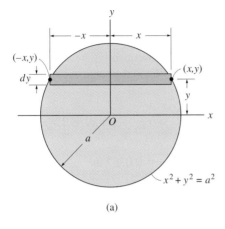

(a)

SOLUTION I (*CASE 1*)

Using the differential element shown in Fig. 10–7a, since $dA = 2x\, dy$, we have

$$I_x = \int_A y^2\, dA = \int_A y^2(2x)\, dy$$

$$= \int_{-a}^{a} y^2(2\sqrt{a^2 - y^2})\, dy = \frac{\pi a^4}{4} \qquad Ans.$$

SOLUTION II (*CASE 2*)

When the differential element is chosen as shown in Fig. 10–7b, the centroid for the element happens to lie on the x axis, and so, applying Eq. 10–3, noting that $d_y = 0$, we have

$$dI_x = \frac{1}{12}\, dx\, (2y)^3$$

$$= \frac{2}{3} y^3\, dx$$

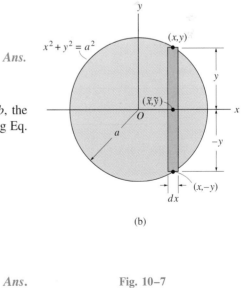

(b)

Integrating with respect to x yields

$$I_x = \int_{-a}^{a} \frac{2}{3}(a^2 - x^2)^{3/2}\, dx = \frac{\pi a^4}{4} \qquad Ans.$$

Fig. 10–7

Example 10–4

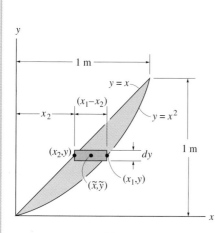

(a)

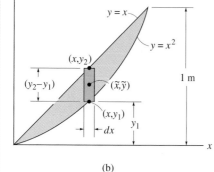

(b)

Fig. 10–8

Determine the moment of inertia of the shaded area shown in Fig. 10–8a about the x axis.

SOLUTION I *(CASE 1)*

The differential element of area parallel to the x axis is chosen for integration, Fig. 10–8a. The element intersects the curve at the *arbitrary points* (x_2, y) and (x_1, y). Consequently, its area is $dA = (x_1 - x_2)\, dy$. Since all parts of the element lie at the same distance y from the x axis, we have

$$I_x = \int_A y^2\, dA = \int_0^1 y^2(x_1 - x_2)\, dy = \int_0^1 y^2(\sqrt{y} - y)\, dy$$

$$I_x = \frac{2}{7}y^{7/2} - \frac{1}{4}y^4 \Big|_0^1 = 0.0357 \text{ m}^4 \qquad\qquad Ans.$$

SOLUTION II *(CASE 2)*

The differential element of area parallel to the y axis is shown in Fig. 10–8b. It intersects the curves at the *arbitrary points* (x, y_2) and (x, y_1). Since all parts of its entirety do *not* lie at the same distance from the x axis, we must first use the parallel-axis theorem to find the element's moment of inertia about the x axis, then integrate this result to determine I_x. Thus,

$$dI_x = d\bar{I}_{x'} + dA\, \tilde{y}^2 = \frac{1}{12}\, dx\, (y_2 - y_1)^3 +$$

$$(y_2 - y_1)\, dx \left(y_1 + \frac{y_2 - y_1}{2} \right)^2$$

$$= \frac{1}{3}(y_2^3 - y_1^3)\, dx = \frac{1}{3}(x^3 - x^6)\, dx$$

$$I_x = \frac{1}{3}\int_0^1 (x^3 - x^6)\, dx = \frac{1}{12}x^4 - \frac{1}{21}x^7 \Big|_0^1 = 0.0357 \text{ m}^4 \quad Ans.$$

By comparison, Solution I requires much less computation. If an integral using a particular element appears difficult to evaluate, try solving the problem using an element oriented in the other direction.

PROBLEMS

10–1.   The 28-mm^2 area has a moment of inertia about the xx axis of 3325 mm^4. Determine its moment of inertia about the $x'x'$ axis. The $\overline{xx}$ axis passes through the centroid C of the area.

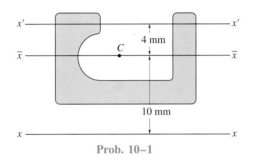

Prob. 10–1

10–3.   Determine the moment of inertia of the shaded area about the x axis.

*10–4.   Determine the moment of inertia of the shaded area about the y axis.

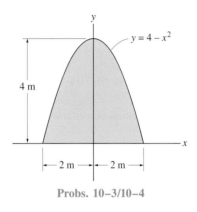

Probs. 10–3/10–4

10–2.   The $15(10^3)$-mm^2 area has a moment of inertia about the yy axis of $25(10^6)$ mm^4. Determine its moment of inertia about the $y'y'$ axis. The $\overline{yy}$ axis passes through the centroid C of the area.

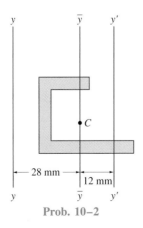

Prob. 10–2

10–5.   Determine the moment of inertia of the shaded area about the x axis.

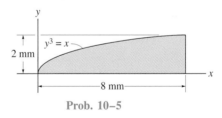

Prob. 10–5

10–6.   Determine the moment of inertia of the shaded area about the y axis.

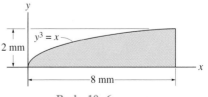

Prob. 10–6

10–7. Determine the moment of inertia of the shaded area about the x axis.

*10–8. Determine the moment of inertia of the shaded area about the y axis.

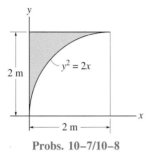

Probs. 10–7/10–8

10–9. Determine the moment of inertia of the shaded area about the x axis.

10–10. Determine the moment of inertia of the shaded area about the y axis.

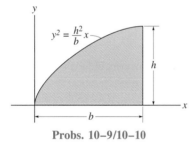

Probs. 10–9/10–10

10–11. Determine the moment of inertia of the area about the x axis. Solve the problem in two ways, using rectangular differential elements: (a) having a thickness dx, and (b) having a thickness dy.

*10–12. Determine the moment of inertia of the area about the y axis. Solve the problem in two ways, using rectangular differential elements: (a) having a thickness dx, and (b) having a thickness dy.

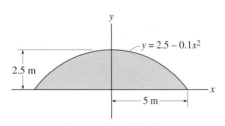

Probs. 10–11/10–12

10–13. Determine the moment of inertia of the shaded area about the y axis.

10–14. Determine the moment of inertia of the shaded area about the x axis.

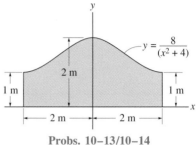

Probs. 10–13/10–14

10–15. Determine the moment of inertia of the shaded area about the x axis.

*10–16. Determine the moment of inertia of the shaded area about the y axis.

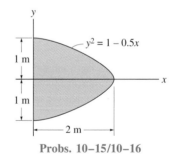

Probs. 10–15/10–16

10–17. Determine the moment of inertia of the shaded area about the x axis.

10–18. Determine the moment of inertia of the shaded area about the y axis.

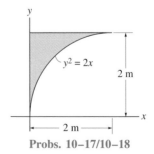

Probs. 10–17/10–18

10–19. Determine the moment of inertia of the equilateral triangle about the x' axis passing through its centroid.

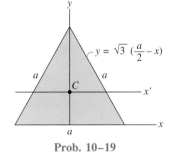

$$y = \sqrt{3}\left(\frac{a}{2} - x\right)$$

Prob. 10–19

*10–20. Determine the moment of inertia of the area enclosed by the arch of the curve $y = \sin x$ about the x axis.

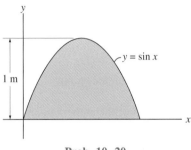

$y = \sin x$

1 m

Prob. 10–20

■**10–21.** Determine the moment of inertia of the area enclosed by the arch of the curve $y = \sin x$ about a vertical axis passing through its centroid.

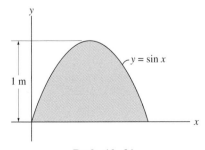

$y = \sin x$

1 m

Prob. 10–21

10–22. Determine the moment of inertia of the shaded area about the x axis.

10–23. Determine the moment of inertia of the shaded area about the y axis.

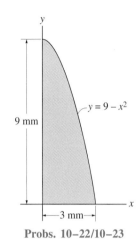

$y = 9 - x^2$

9 mm

3 mm

Probs. 10–22/10–23

*■10–24. Determine the moment of inertia of the area about the y axis. Use Simpson's rule to evaluate the integral.

*■10–25. Determine the moment of inertia of the area about the x axis. Use Simpson's rule to evaluate the integral.

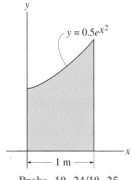

$y = 0.5e^{x^2}$

1 m

Probs. 10–24/10–25

10.5 Moments of Inertia for Composite Areas

A composite area consists of a series of connected ''simpler'' parts or shapes, such as semicircles, rectangles, and triangles. Provided the moment of inertia of each of these parts is known or can be determined about a common axis, then the moment of inertia of the composite area equals the *algebraic sum* of the moments of inertia of all its parts.

PROCEDURE FOR ANALYSIS

The following procedure provides a method for determining the moment of inertia of a composite area about a reference axis.

Composite Parts.   Using a sketch, divide the area into its composite parts and indicate the perpendicular distance from the *centroid* of each part to the reference axis.

Parallel-Axis Theorem.   The moment of inertia of each part should be determined about its centroidal axis, which is parallel to the reference axis. For the calculation use the table given on the inside back cover. If the centroidal axis does not coincide with the reference axis, the parallel-axis theorem, $I = \bar{I} + Ad^2$, should be used to determine the moment of inertia of the part about the reference axis.

Summation.   The moment of inertia of the entire area about the reference axis is determined by summing the results of its composite parts. In particular, if a composite part has a ''hole,'' its moment of inertia is found by ''subtracting'' the moment of inertia for the hole from the moment of inertia of the entire part including the hole.

Example 10–5

Compute the moment of inertia of the composite area shown in Fig. 10–9a about the x axis.

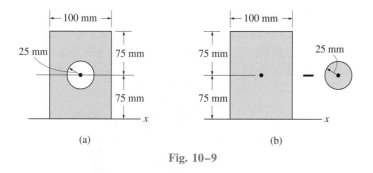

(a)   (b)

Fig. 10–9

SOLUTION

Composite Parts.   The composite area is obtained by *subtracting* the circle from the rectangle as shown in Fig. 10–9b. The centroid of each area is located in the figure.

Parallel-Axis Theorem.   The moments of inertia about the x axis are determined using the parallel-axis theorem and the data in the table on the inside back cover.

   Circle

$$I_x = \bar{I}_{x'} + Ad_y^2$$

$$= \frac{1}{4}\pi(25)^4 + \pi(25)^2(75)^2 = 11.4(10^6) \text{ mm}^4$$

   Rectangle

$$I_x = \bar{I}_{x'} + Ad_y^2$$

$$= \frac{1}{12}(100)(150)^3 + (100)(150)(75)^2 = 112.5(10^6) \text{ mm}^4$$

Summation.   The moment of inertia for the composite area is thus

$$I_x = -11.4(10^6) + 112.5(10^6)$$

$$= 101(10^6) \text{ mm}^4 \qquad\qquad Ans.$$

Example 10–6

Determine the moments of inertia of the beam's cross-sectional area shown in Fig. 10–10a about the x and y centroidal axes.

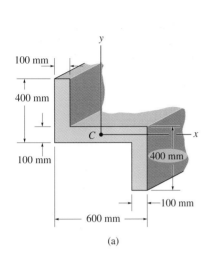

(a)

(b)

Fig. 10–10

SOLUTION

Composite Parts.   The cross section can be considered as three composite rectangular areas A, B, and D shown in Fig. 10–10b. For the calculation, the centroid of each of these rectangles is located in the figure.

Parallel-Axis Theorem.   From the table on the inside back cover, or Example 10–1, the moment of inertia of a rectangle about its centroidal axis is $\bar{I} = \frac{1}{12}bh^3$. Hence, using the parallel-axis theorem for rectangles A and D, the calculations are as follows:

Rectangle A

$$I_x = \bar{I}_{x'} + Ad_y^2 = \frac{1}{12}(100)(300)^3 + (100)(300)(200)^2$$
$$= 1.425(10^9) \text{ mm}^4$$

$$I_y = \bar{I}_{y'} + Ad_x^2 = \frac{1}{12}(300)(100)^3 + (100)(300)(250)^2$$
$$= 1.90(10^9) \text{ mm}^4$$

Rectangle B

$$I_x = \frac{1}{12}(600)(100)^3 = 0.05(10^9) \text{ mm}^4$$

$$I_y = \frac{1}{12}(100)(600)^3 = 1.80(10^9) \text{ mm}^4$$

Rectangle D

$$I_x = \bar{I}_{x'} + Ad_y^2 = \frac{1}{12}(100)(300)^3 + (100)(300)(200)^2$$
$$= 1.425(10^9) \text{ mm}^4$$

$$I_y = \bar{I}_{y'} + Ad_x^2 = \frac{1}{12}(300)(100)^3 + (100)(300)(250)^2$$
$$= 1.90(10^9) \text{ mm}^4$$

Summation.   The moments of inertia for the entire cross section are thus

$$I_x = 1.425(10^9) + 0.05(10^9) + 1.425(10^9)$$
$$= 2.90(10^9) \text{ mm}^4 \qquad \qquad \text{Ans.}$$

$$I_y = 1.90(10^9) + 1.80(10^9) + 1.90(10^9)$$
$$= 5.60(10^9) \text{ mm}^4 \qquad \qquad \text{Ans.}$$

PROBLEMS

10–26. Determine the moment of inertia of the beam's cross-sectional area about the x axis.

10–27. Determine the moment of inertia of the beam's cross-sectional area about the y axis.

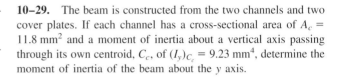

Probs. 10–26/10–27

*10–28. The beam is constructed from the two channels and two cover plates. If each channel has a cross-sectional area of $A_c = 11.8$ mm² and a moment of inertia about a horizontal axis passing through its own centroid, C_c, of $(\bar{I}_{\bar{x}})_{C_c} = 349$ mm⁴, determine the moment of inertia of the beam about the x axis.

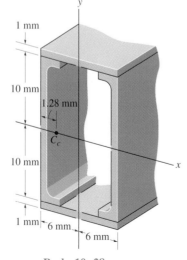

Prob. 10–28

10–29. The beam is constructed from the two channels and two cover plates. If each channel has a cross-sectional area of $A_c = 11.8$ mm² and a moment of inertia about a vertical axis passing through its own centroid, C_c, of $(I_y)_{C_c} = 9.23$ mm⁴, determine the moment of inertia of the beam about the y axis.

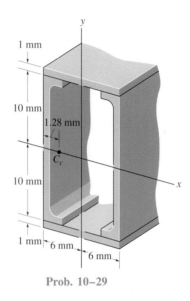

Prob. 10–29

10–30. Determine the distance $\bar{x}$ to the centroid for the beam's cross-sectional area, then find $\bar{I}_{y'}$.

10–31. Determine the moment of inertia of the beam's cross-sectional area about the x axis.

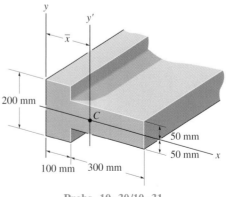

Probs. 10–30/10–31

*10–32. Determine the distance $\bar{x}$ to the centroid for the beam's cross-sectional area, then find $\bar{I}_{y'}$.

10–33. Determine the moment of inertia of the beams cross-sectional area about the x axis.

*10–36. Determine the distance $\bar{y}$ to the centroid for the beam's cross-sectional area; then find $\bar{I}_{x'}$.

10–37. Determine the moment of inertia of the beam's cross-sectional area about the y axis.

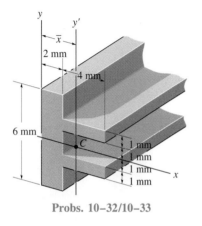

Probs. 10–32/10–33

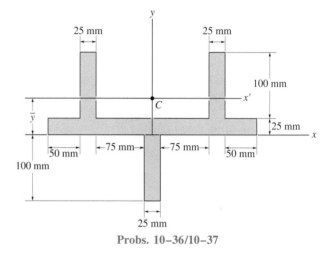

Probs. 10–36/10–37

10–34. Determine $\bar{y}$, which locates the centroid C of the wing channel, and then determine the moment of inertia $\bar{I}_{x'}$ about the centroidal x' axis. Neglect the effect of rounded corners. The material has a uniform thickness of 0.5 mm.

10–35. Determine the moment of inertia of the wing channel about the y axis.

10–38. Determine the moment of inertia of the parallelogram about the x axis.

10–39. Determine the moment of inertia of the parallelogram about the y axis.

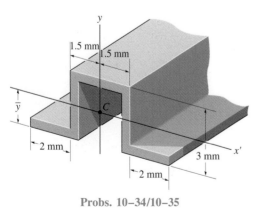

Probs. 10–34/10–35

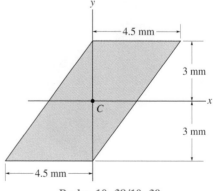

Probs. 10–38/10–39

*10–40. Locate the centroid $\bar{y}$ of the channel's cross-sectional area, and then determine the moment of inertia with respect to the x' axis passing through the centroid.

10–42. Determine the moment of inertia I_x of the shaded area about the x axis.

10–43. Determine the moment of inertia I_y of the shaded area about the y axis.

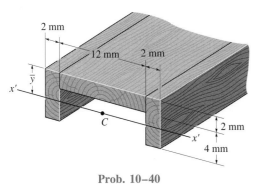

Prob. 10–40

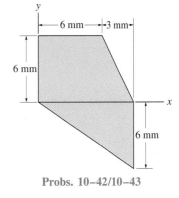

Probs. 10–42/10–43

*10–44. Compute the moments of inertia I_x and I_y for the shaded area about the x and y axes.

10–41. Determine $\bar{y}$, which locates the centroidal axis x' for the cross-sectional area of the T-beam, and then find the moments of inertia $\bar{I}_{x'}$ and $\bar{I}_{y'}$.

10–45. Determine the distance $\bar{y}$ to the centroid C of the beam's cross-sectional area and then compute the moment of inertia $\bar{I}_{x'}$ about the x' axis.

10–46. Determine the distance $\bar{x}$ to the centroid C of the beam's cross-sectional area and then compute the moment of inertia $\bar{I}_{y'}$ about the y' axis.

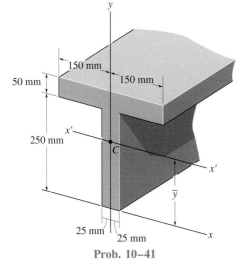

Prob. 10–41

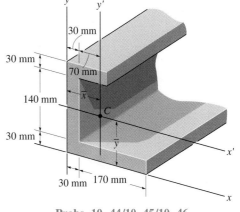

Probs. 10–44/10–45/10–46

10–47. Determine the moments of inertia I_x and I_y of the shaded area.

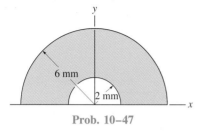

Prob. 10–47

*10–48. Determine the moment of inertia of the beam's cross-sectional area about the x axis, which passes through the centroid C.

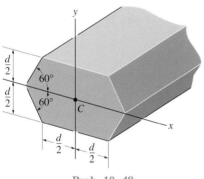

Prob. 10–48

10–49. Determine the moment of inertia of the beam's cross-sectional area about the y axis, which passes through the centroid C.

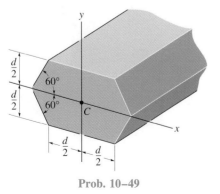

Prob. 10–49

10–50. Compute the polar moments of inertia J_O for the cross-sectional area of the solid shaft and tube. What percentage of J_O is contributed by the tube to that of the solid shaft?

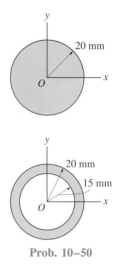

Prob. 10–50

10–51. Determine the moment of inertia of the rectangle about its diagonal axis xx.

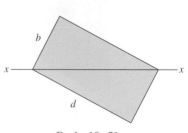

Prob. 10–51

*10–52. Determine the polar moment of inertia of the shaft's cross-sectional area about the center O.

10–54. Determine the moment of inertia of the parallelogram about the x' axis, which passes through the centroid C of the area.

10–55. Determine the moment of inertia of the parallelogram about the y' axis, which passes through the centroid C of the area.

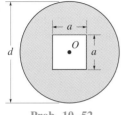

Prob. 10–52

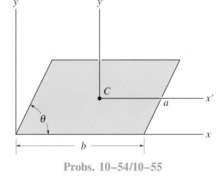

Probs. 10–54/10–55

10–53. Determine the radius of gyration k_x for the column's cross-sectional area.

*10–56. Determine the moments of inertia of the triangular area about the x' and y' axes, which pass through the centroid C of the area.

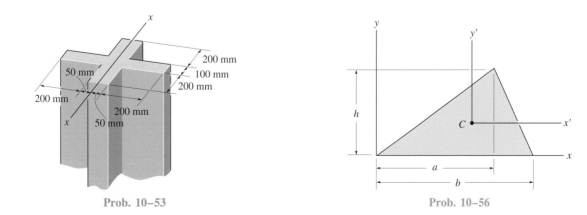

Prob. 10–53

Prob. 10–56

*10.6 Product of Inertia for an Area

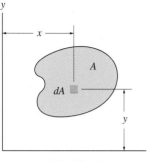

Fig. 10–11

In general, the moment of inertia for an area is different for every axis about which it is computed. In some applications of structural or mechanical design it is necessary to know the orientation of those axes which give, respectively, the maximum and minimum moments of inertia for the area. The method for determining this is discussed in Sec. 10.7. To use this method, however, one must first compute the product of inertia for the area as well as its moments of inertia for given x, y axes.

The product of inertia for an element of area located at point (x, y), Fig. 10–11, is defined as $dI_{xy} = xy\, dA$. Thus, for the entire area A, the *product of inertia* is

$$I_{xy} = \int_A xy\, dA \qquad (10\text{–}7)$$

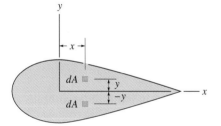

Fig. 10–12

If the element of area chosen has a differential size in two directions, as shown in Fig. 10–11, a double integration must be performed to evaluate I_{xy}. Most often, however, it is easier to choose an element having a differential size or thickness in only one direction, in which case the evaluation requires only a single integration (see Example 10–7).

Like the moment of inertia, the product of inertia has units of length raised to the fourth power, e.g., m^4, mm^4. However, since x or y may be a negative quantity, while the element of area is always positive, the product of inertia may be positive, negative, or zero, depending on the location and orientation of the coordinate axes. For example, the product of inertia I_{xy} for an area will be *zero* if either the x or y axis is an axis of *symmetry* for the area. To show this, consider the shaded area in Fig. 10–12, where for every element dA located at point (x, y) there is a corresponding element dA located at $(x, -y)$. Since the products of inertia for these elements are, respectively, $xy\, dA$ and $-xy\, dA$, the algebraic sum or integration of all the elements that are chosen in this way will cancel each other. Consequently, the product of inertia for the total area becomes zero. It also follows from the definition of I_{xy} that the "sign" of this quantity depends on the quadrant where the area is located. As shown in Fig. 10–13, if the area is rotated from one quadrant to another one, the sign of I_{xy} will change.

Parallel-Axis Theorem. Consider the shaded area shown in Fig. 10–14, where x' and y' represent a set of axes passing through the *centroid* of the area, and x and y represent a corresponding set of parallel axes. Since the product of inertia of dA with respect to the x and y axes is $dI_{xy} = (x' + d_x)$

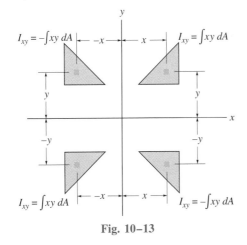

Fig. 10–13

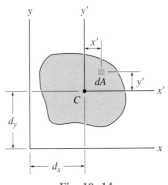

Fig. 10–14

$(y' + d_y)\ dA$, then for the entire area,

$$I_{xy} = \int_A (x' + d_x)(y' + d_y)\ dA$$

$$= \int_A x'y'\ dA + d_x \int_A y'\ dA + d_y \int_A x'\ dA + d_x d_y \int_A dA$$

The first term on the right represents the product of inertia of the area with respect to the centroidal axis, $\bar{I}_{x'y'}$. The integrals in the second and third terms are zero since the moments of the area are taken about the centroidal axis. Realizing that the fourth integral represents the total area A, the final result is therefore

$$I_{xy} = \bar{I}_{x'y'} + A d_x\ d_y \qquad (10\text{–}8)$$

The similarity between this equation and the parallel-axis theorem for moments of inertia should be noted. In particular, it is important that the *algebraic signs* for d_x and d_y be maintained when applying Eq. 10–8. As illustrated in Example 10–8, the parallel-axis theorem finds important application in determining the product of inertia of a *composite area* with respect to a set of x, y axes.

Example 10–7

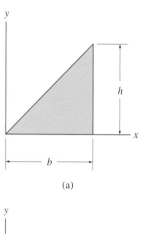

(a)

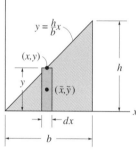

(b)

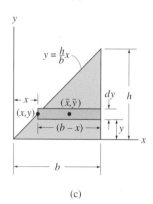

(c)

Fig. 10–15

Determine the product of inertia I_{xy} of the triangle shown in Fig. 10–15a.

SOLUTION I

A differential element that has a thickness dx, Fig. 10–15b, has an area $dA = y\,dx$. The product of inertia of the element about the x, y axes is determined using the parallel-axis theorem.

$$dI_{xy} = d\bar{I}_{x'y'} + dA\,\widetilde{x}\,\widetilde{y}$$

where $(\widetilde{x}, \widetilde{y})$ locates the *centroid* of the element or the origin of the x', y' axes. Since $d\bar{I}_{x'y'} = 0$, due to symmetry, and $\widetilde{x} = x$, $\widetilde{y} = y/2$, then

$$dI_{xy} = 0 + (y\,dx)x\left(\frac{y}{2}\right) = \left(\frac{h}{b}x\,dx\right)x\left(\frac{h}{2b}x\right)$$

$$= \frac{h^2}{2b^2}x^3\,dx$$

Integrating with respect to x from $x = 0$ to $x = b$ yields

$$I_{xy} = \frac{h^2}{2b^2}\int_0^b x^3\,dx = \frac{b^2h^2}{8} \qquad \textit{Ans.}$$

SOLUTION II

The differential element that has a thickness dy, Fig. 10–15c, and area $dA = (b - x)\,dy$ can also be used. The *centroid* is located at point $\widetilde{x} = x + (b - x)/2 = (b + x)/2$, $\widetilde{y} = y$, so the product of inertia of the element becomes

$$dI_{xy} = d\bar{I}_{x'y'} + dA\,\widetilde{x}\,\widetilde{y}$$

$$= 0 + (b - x)\,dy\left(\frac{b + x}{2}\right)y$$

$$= \left(b - \frac{b}{h}y\right)dy\left[\frac{b + (b/h)\,y}{2}\right]y = \frac{1}{2}y\left(b^2 - \frac{b^2}{h^2}y^2\right)dy$$

Integrating with respect to y from $y = 0$ to $y = h$ yields

$$I_{xy} = \frac{1}{2}\int_0^h y\left(b^2 - \frac{b^2}{h^2}y^2\right)dy = \frac{b^2h^2}{8} \qquad \textit{Ans.}$$

Example 10–8

Compute the product of inertia of the beam's cross-sectional area, shown in Fig. 10–16a, about the x and y centroidal axes.

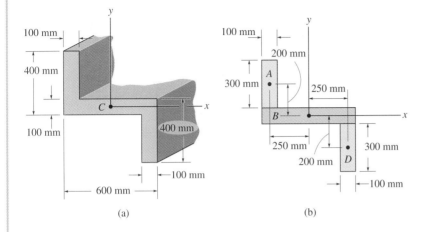

Fig. 10–16

(a)   (b)

SOLUTION

As in Example 10–6, the cross section can be considered as three composite rectangular areas A, B, and D, Fig. 10–16b. The coordinates for the centroid of each of these rectangles are shown in the figure. Due to symmetry, the product of inertia of *each rectangle* is *zero* about a set of x', y' axes that pass through the rectangle's centroid. Hence, application of the parallel-axis theorem to each of the rectangles yields

Rectangle A

$$I_{xy} = \bar{I}_{x'y'} + Ad_x d_y$$
$$= 0 + (300)(100)(-250)(200)$$
$$= -1.50(10^9) \text{ mm}^4$$

Rectangle B

$$I_{xy} = \bar{I}_{x'y'} + Ad_x d_y$$
$$= 0 + 0$$
$$= 0$$

Rectangle D

$$I_{xy} = \bar{I}_{x'y'} + Ad_x d_y$$
$$= 0 + (300)(100)(250)(-200)$$
$$= -1.50(10^9) \text{ mm}^4$$

The product of inertia for the entire cross section is therefore

$$I_{xy} = -1.50(10^9) + 0 - 1.50(10^9) = -3.00(10^9) \text{ mm}^4 \qquad \textit{Ans.}$$

*10.7 Moments of Inertia for an Area About Inclined Axes

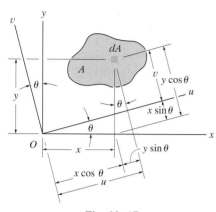

Fig. 10–17

In structural and mechanical design, it is sometimes necessary to calculate the moments and product of inertia I_u, I_v, and I_{uv} for an area with respect to a set of inclined u and v axes when the values for θ, I_x, I_y, and I_{xy} are *known*. To do this we will use *transformation equations* which relate the x, y and u, v coordinates. From Fig. 10–17, these equations are

$$u = x \cos \theta + y \sin \theta$$
$$v = y \cos \theta - x \sin \theta$$

Using these equations, the moments and product of inertia of dA about the u and v axes become

$$dI_u = v^2 \, dA = (y \cos \theta - x \sin \theta)^2 \, dA$$
$$dI_v = u^2 \, dA = (x \cos \theta + y \sin \theta)^2 \, dA$$
$$dI_{uv} = uv \, dA = (x \cos \theta + y \sin \theta)(y \cos \theta - x \sin \theta) \, dA$$

Expanding each expression and integrating, realizing that $I_x = \int y^2 \, dA$, $I_y = \int x^2 \, dA$, and $I_{xy} = \int xy \, dA$, we obtain

$$I_u = I_x \cos^2 \theta + I_y \sin^2 \theta - 2I_{xy} \sin \theta \cos \theta$$
$$I_v = I_x \sin^2 \theta + I_y \cos^2 \theta + 2I_{xy} \sin \theta \cos \theta$$
$$I_{uv} = I_x \sin \theta \cos \theta - I_y \sin \theta \cos \theta + I_{xy}(\cos^2 \theta - \sin^2 \theta)$$

These equations may be simplified by using the trigonometric identities $\sin 2\theta = 2 \sin \theta \cos \theta$ and $\cos 2\theta = \cos^2\theta - \sin^2\theta$, in which case

$$I_u = \frac{I_x + I_y}{2} + \frac{I_x - I_y}{2} \cos 2\theta - I_{xy} \sin 2\theta$$

$$I_v = \frac{I_x + I_y}{2} - \frac{I_x - I_y}{2} \cos 2\theta + I_{xy} \sin 2\theta \qquad (10\text{–}9)$$

$$I_{uv} = \frac{I_x - I_y}{2} \sin 2\theta + I_{xy} \cos 2\theta$$

If the first and second equations are added together, we can show that the polar moment of inertia about the z axis passing through point O is *independent* of the orientation of the u and v axes; i.e.,

$$J_O = I_u + I_v = I_x + I_y$$

Principal Moments of Inertia. From Eqs. 10–9, it may be seen that I_u, I_v, and I_{uv} depend on the angle of inclination, θ, of the u, v axes. We will now determine the orientation of the u, v axes about which the moments of inertia for the area, I_u and I_v, are maximum and minimum. This particular set of axes is called the *principal axes* of the area, and the corresponding moments of inertia with respect to these axes are called the *principal moments of inertia*.

In general, there is a set of principal axes for every chosen origin O, although in structural and mechanical design, the area's centroid is an important location for O.

The angle $\theta = \theta_p$, which defines the orientation of the principal axes for the area, may be found by differentiating the first of Eqs. 10–9 with respect to θ and setting the result equal to zero. Thus,

$$\frac{dI_u}{d\theta} = -2\left(\frac{I_x - I_y}{2}\right)\sin 2\theta - 2I_{xy}\cos 2\theta = 0$$

Therefore, at $\theta = \theta_p$,

$$\tan 2\theta_p = \frac{-I_{xy}}{(I_x - I_y)/2} \qquad (10\text{–}10)$$

This equation has two roots, θ_{p_1} and θ_{p_2}, which are 90° apart and so specify the inclination of the principal axes. In order to substitute them into Eq. 10–9, we must first find the sine and cosine of $2\theta_{p_1}$ and $2\theta_{p_2}$. This can be done using the triangles shown in Fig. 10–18, which are based on Eq. 10–10.

For θ_{p_1},

$$\sin 2\theta_{p_1} = -I_{xy}\bigg/\sqrt{\left(\frac{I_x - I_y}{2}\right)^2 + I_{xy}^2}$$

$$\cos 2\theta_{p_1} = \left(\frac{I_x - I_y}{2}\right)\bigg/\sqrt{\left(\frac{I_x - I_y}{2}\right)^2 + I_{xy}^2}$$

For θ_{p_2},

$$\sin 2\theta_{p_2} = I_{xy}\bigg/\sqrt{\left(\frac{I_x - I_y}{2}\right)^2 + I_{xy}^2}$$

$$\cos 2\theta_{p_2} = -\left(\frac{I_x - I_y}{2}\right)\bigg/\sqrt{\left(\frac{I_x - I_y}{2}\right)^2 + I_{xy}^2}$$

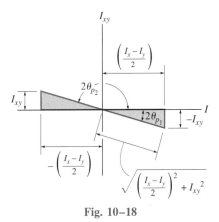

Fig. 10–18

Substituting these two sets of trigonometric relations into the first or second of Eqs. 10–9 and simplifying, we obtain

$$I_{\substack{max \\ min}} = \frac{I_x + I_y}{2} \pm \sqrt{\left(\frac{I_x - I_y}{2}\right)^2 + I_{xy}^2} \qquad (10\text{–}11)$$

Depending on the sign chosen, this result gives the maximum or minimum moment of inertia for the area. Furthermore, if the above trigonometric relations for θ_{p_1} and θ_{p_2} are substituted into the third of Eqs. 10–9, it can be shown that $I_{uv} = 0$; that is, the *product of inertia with respect to the principal axes is zero*. Since it was indicated in Sec. 10.6 that the product of inertia is zero with respect to any symmetrical axis, it therefore follows that *any symmetrical axis represents a principal axis of inertia for the area*.

Example 10–9

Determine the principal moments of inertia for the beam's cross-sectional area shown in Fig. 10–19a with respect to an axis passing through the centroid.

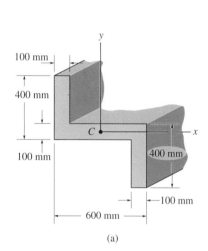

(a)

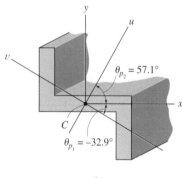

(b)

Fig. 10–19

SOLUTION

The moments and product of inertia of the cross section with respect to the x, y axes have been computed in Examples 10–6 and 10–8. The results are

$$I_x = 2.90(10^9) \text{ mm}^4 \qquad I_y = 5.60(10^9) \text{ mm}^4 \qquad I_{xy} = -3.00(10^9) \text{ mm}^4$$

Using Eq. 10–10, the angles of inclination of the principal axes u and v are

$$\tan 2\theta_p = \frac{-I_{xy}}{(I_x - I_y)/2} = \frac{3.00(10^9)}{[2.90(10^9) - 5.60(10^9)]/2} = -2.22$$

$$2\theta_{p_1} = -65.8° \qquad \text{and} \qquad 2\theta_{p_2} = 114.2°$$

Thus, as shown in Fig. 10–19b,

$$\theta_{p_1} = -32.9° \quad \text{and} \quad \theta_{p_2} = 57.1°$$

The principal moments of inertia with respect to the u and v axes are determined from Eq. 10–11. Hence,

$$I_{\substack{\max \\ \min}} = \frac{I_x + I_y}{2} \pm \sqrt{\left(\frac{I_x - I_y}{2}\right)^2 + I_{xy}^2}$$

$$= \frac{2.90(10^9) + 5.60(10^9)}{2}$$

$$\pm \sqrt{\left[\frac{2.90(10^9) - 5.60(10^9)}{2}\right]^2 + [-3.00(10^9)]^2}$$

$$I_{\substack{\max \\ \min}} = 4.25(10^9) \pm 3.29(10^9)$$

or

$$I_{\max} = 7.54(10^9) \text{ mm}^4 \qquad I_{\min} = 0.960(10^9) \text{ mm}^4 \qquad \textbf{\textit{Ans.}}$$

Specifically, the maximum moment of inertia, $I_{\max} = 7.54(10^9) \text{ mm}^4$, occurs with respect to the selected u axis, since *by inspection* most of the cross-sectional area is farthest away from this axis. Or, stated in another manner, $I_{\max}$ occurs about the u axis since it is located within $\pm 45°$ of the y axis, which has the largest value of I ($I_y > I_x$). Also, this may be concluded mathematically by substituting the data with $\theta = 57.1°$ into the first of Eqs. 10–9.

*10.8 Mohr's Circle for Moments of Inertia

Equations 10–9 to 10–11 have a graphical solution that is convenient to use and generally easy to remember. Squaring the first and third of Eqs. 10–9 and adding, it is found that

$$\left(I_u - \frac{I_x + I_y}{2}\right)^2 + I_{uv}^2 = \left(\frac{I_x - I_y}{2}\right)^2 + I_{xy}^2 \qquad (10\text{–}12)$$

In a given problem, I_u and I_{uv} are *variables,* and I_x, I_y, and I_{xy} are *known constants.* Thus, Eq. 10–12 may be written in compact form as

$$(I_u - a)^2 + I_{uv}^2 = R^2$$

When this equation is plotted on a set of axes that represent the respective moment of inertia and the product of inertia, Fig. 10–20, the resulting graph represents a *circle* of radius

$$R = \sqrt{\left(\frac{I_x - I_y}{2}\right)^2 + I_{xy}^2}$$

having its center located at point $(a, 0)$, where $a = (I_x + I_y)/2$. The circle so constructed is called *Mohr's circle,* named after the German engineer Otto Mohr (1835–1918).

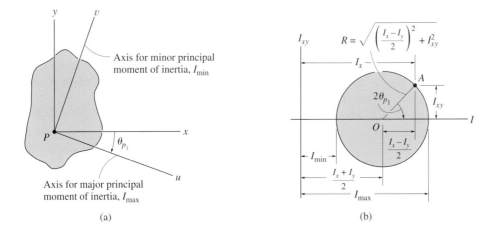

(a)

(b)

Fig. 10–20

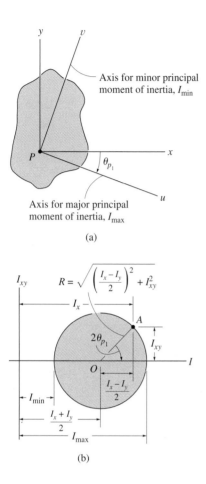

Axis for minor principal moment of inertia, I_{min}

θ_{p_1}

Axis for major principal moment of inertia, I_{max}

(a)

I_{xy} $R = \sqrt{\left(\dfrac{I_x - I_y}{2}\right)^2 + I_{xy}^2}$

I_x

$2\theta_{p_1}$ I_{xy}

O $\dfrac{I_x - I_y}{2}$ I

I_{min}

 $\dfrac{I_x + I_y}{2}$

I_{max}

(b)

Fig. 10–20

PROCEDURE FOR ANALYSIS

The main purpose in using Mohr's circle here is to have a convenient means for transforming I_x, I_y, and I_{xy} into the principal moments of inertia. The following procedure provides a method for doing this.

Determine I_x, I_y, I_{xy}.   Establish the x, y axes for the area, with the origin located at the point P of interest, and determine I_x, I_y, and I_{xy}, Fig. 10–20a.

Construct the Circle.   Construct a rectangular coordinate system such that the abscissa represents the moment of inertia I, and the ordinate represents the product of inertia I_{xy}, Fig. 10–20b. Determine the center of the circle, O, which is located at a distance $(I_x + I_y)/2$ from the origin, and plot the reference point A having coordinates (I_x, I_{xy}). By definition, I_x is always positive, whereas I_{xy} will be either positive or negative. Connect the reference point A with the center of the circle, and determine the distance OA by trigonometry. This distance represents the radius of the circle, Fig. 10–20b. Finally, draw the circle.

Principal Moments of Inertia.   The points where the circle intersects the abscissa give the values of the principal moments of inertia I_{min} and I_{max}. Notice that the *product of inertia will be zero at these points*, Fig. 10–20b.

Principal Axes.   To find the direction of the major principal axis, determine by trigonometry the angle $2\theta_{p_1}$, *measured from the radius OA to the positive I axis,* Fig. 10–20b. This angle represents *twice* the angle from the x axis of the area in question to the axis of maximum moment of inertia I_{max}, Fig. 10–20a. Both the angle on the circle, $2\theta_{p_1}$, and the angle to the axis on the area, θ_{p_1}, *must be measured in the same sense,* as shown in Fig. 10–20. The axis for minimum moment of inertia I_{min} is perpendicular to the axis for I_{max}.

Using trigonometry, the above procedure may be verified to be in accordance with the equations developed in Sec. 10.7.

Example 10–10

Using Mohr's circle, determine the principal moments of inertia for the beam's cross-sectional area, shown in Fig. 10–21a, with respect to an axis passing through the centroid.

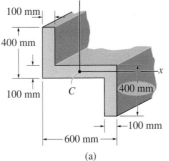

(a)

SOLUTION

Determine I_x, I_y, I_{xy}.   The moments of inertia and the product of inertia have been determined in Examples 10–6 and 10–8 with respect to the x, y axes shown in Fig. 10–21a. The results are $I_x = 2.90(10^9)$ mm^4, $I_y = 5.60(10^9)$ mm^4, and $I_{xy} = -3.00(10^9)$ mm^4.

Construct the Circle.   The I and I_{xy} axes are shown in Fig. 10–21b. The center of the circle, O, lies at a distance $(I_x + I_y)/2 = (2.90 + 5.60)/2 = 4.25$ from the origin. When the reference point $A(2.90, -3.00)$ is connected to point O, the radius OA is determined from the triangle OBA using the Pythagorean theorem.

$$OA = \sqrt{(1.35)^2 + (-3.00)^2} = 3.29$$

The circle is constructed in Fig. 10–21c.

Principal Moments of Inertia.   The circle intersects the I axis at points (7.54, 0) and (0.960, 0). Hence,

$$I_{\max} = 7.54(10^9) \text{ mm}^4 \qquad \textit{Ans.}$$
$$I_{\min} = 0.960(10^9) \text{ mm}^4 \qquad \textit{Ans.}$$

Principal Axes.   As shown in Fig. 10–21c, the angle $2\theta_{p_1}$ is determined from the circle by measuring counterclockwise from OA to the direction of the *positive I* axis. Hence,

$$2\theta_{p_1} = 180° - \sin^{-1}\left(\frac{|BA|}{|OA|}\right) = 180° - \sin^{-1}\left(\frac{3.00}{3.29}\right) = 114.2°$$

The principal axis for $I_{\max} = 7.54(10^9)$ mm^4 is therefore oriented at an angle $\theta_{p_1} = 57.1°$, measured *counterclockwise*, from the *positive x* axis to the *positive u* axis. The v axis is perpendicular to this axis. The results are shown in Fig. 10–21d.

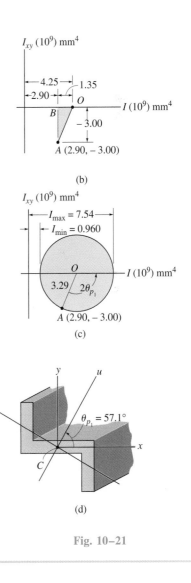

I_{xy} (10^9) mm^4

$-4.25-$ 1.35
$-2.90-$ O — I (10^9) mm^4
B
-3.00

A $(2.90, -3.00)$

(b)

I_{xy} (10^9) mm^4

$-I_{\max} = 7.54-$
$-I_{\min} = 0.960$

O — I (10^9) mm^4
3.29 $2\theta_{p_1}$

A $(2.90, -3.00)$

(c)

y u
v
$\theta_{p_1} = 57.1°$
x
C

(d)

Fig. 10–21

PROBLEMS

10–57.   Determine the product of inertia of the shaded area with respect to the x and y axes.

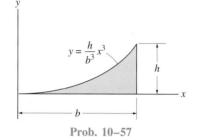

$$y = \frac{h}{b^3}x^3$$

Prob. 10–57

10–58.   Determine the product of inertia of the shaded area of the ellipse with respect to the x and y axes.

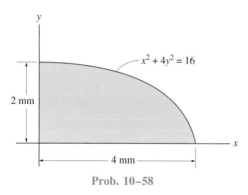

$$x^2 + 4y^2 = 16$$

2 mm

4 mm

Prob. 10–58

10–59.   Determine the product of inertia of the shaded area with respect to the x and y axes.

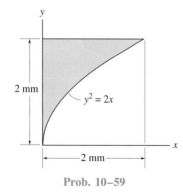

2 mm

$$y^2 = 2x$$

2 mm

Prob. 10–59

***10–60.**   Determine the product of inertia of the parabolic area with respect to the x and y axes.

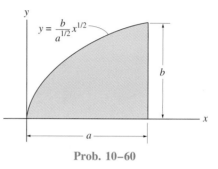

$$y = \frac{b}{a^{1/2}}x^{1/2}$$

b

a

Prob. 10–60

10–61.   Determine the product of inertia of the shaded area with respect to the x and y axes.

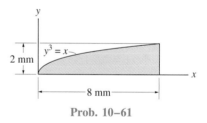

$$y^3 = x$$

2 mm

8 mm

Prob. 10–61

10–62.   Determine the product of inertia of the shaded area with respect to the x and y axes.

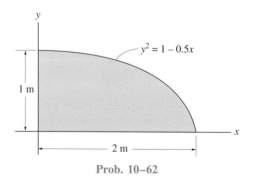

$$y^2 = 1 - 0.5x$$

1 m

2 m

Prob. 10–62

10–63.   Determine the product of inertia of the shaded area with respect to the x and y axes.

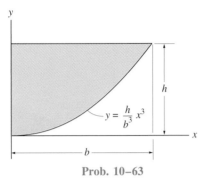

Prob. 10–63

***10–64.**   Determine the product of inertia of the shaded area with respect to the x and y axes.

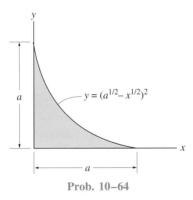

Prob. 10–64

■**10–65.**   Determine the product of inertia of the shaded area with respect to the x and y axes. Use Simpson's rule to evaluate the integral.

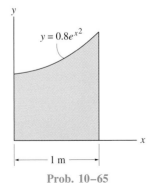

Prob. 10–65

10–66.   Determine the product of inertia of the beam's cross-sectional area with respect to the x and y axes.

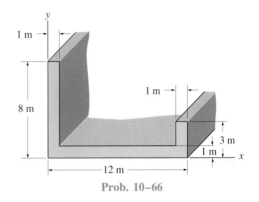

Prob. 10–66

10–67.   Locate the position $\bar{x}$, $\bar{y}$ for the centroid C of the beam's cross-sectional area, and then compute the product of inertia with respect to the x' and y' axes.

***10–68.**   Determine the product of inertia of the beam's cross-sectional area with respect to the x and y axes.

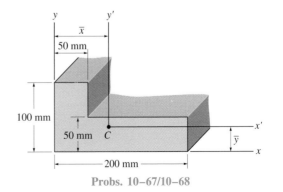

Probs. 10–67/10–68

10–69. Determine the product of inertia of the beam's cross-sectional area with respect to the x and y axes that have their origin located at the centroid C.

10–71. Determine the product of inertia for the angle with respect to the x and y axes passing through the centroid C. Assume all corners to be square.

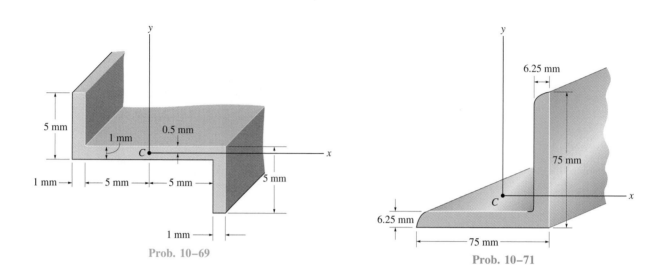

Prob. 10–69

Prob. 10–71

10–70. Determine the product of inertia of the beam's cross-sectional area with respect to the x and y axes that have their origin located at the centroid C.

*10–72. Determine the product of inertia of the beam's cross-sectional area with respect to the x and y axes.

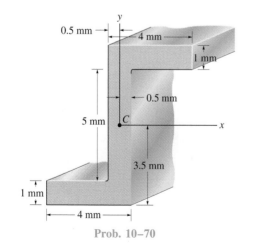

Prob. 10–70

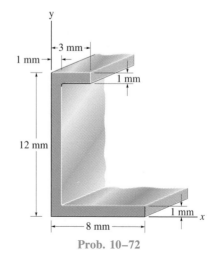

Prob. 10–72

10–73. Determine the product of inertia for the beam's cross-sectional area with respect to the u and v axes.

*10–76. Determine the distance $\bar{y}$ to the centroid of the area and then calculate the moments of inertia I_u and I_v of the channel's cross-sectional area. The u and v axes have their origin at the centroid C. For the calculation, assume all corners to be square.

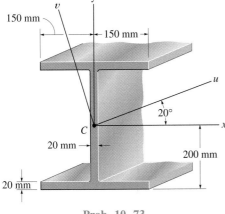

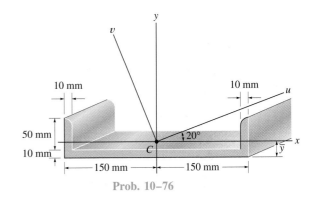

Prob. 10–76

Prob. 10–73

10–74. Determine the moments of inertia I_u and I_v and the product of inertia I_{uv} for the rectangular area. The u and v axes pass through the centroid C. Take $\theta = 30°$.

10–75. Determine the moments of inertia I_u and I_v and the product of inertia I_{uv} for the rectangular area. The u and v axes pass through the centroid C. Take $\theta = 20°$.

10–77. The area of the cross section of an airplane wing has the following properties about the x and y axes passing through the centroid C: $\bar{I}_x = 187.3 \times 10^{-6}\,\text{m}^4$, $\bar{I}_y = 720.0 \times 10^{-6}\,\text{m}^4$, $\bar{I}_{xy} = 57.4 \times 10^{-6}\,\text{m}^4$. Determine the orientation of the principal axes and the principal moments of inertia.

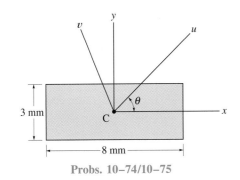

Probs. 10–74/10–75

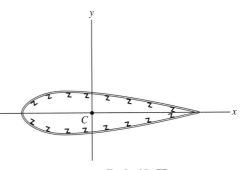

Prob. 10–77

10–78. Determine the moments of inertia I_u, I_v and the product of inertia I_{uv} of the beam's cross-sectional area. Take $\theta = 45°$.

10–79. Determine the moments of inertia I_u, I_v and the product of inertia I_{uv} of the beam's cross-sectional area. Take $\theta = 60°$.

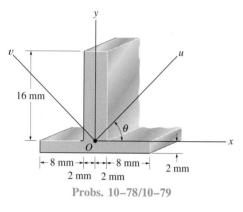

Probs. 10–78/10–79

10–81. Determine the principal moments of inertia of the beam's cross-sectional area about the principal axes that have their origin located at the centroid C. Use the equations developed in Section 10–7. For the calculation, assume all corners to be square.

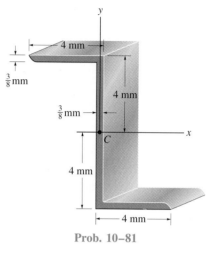

Prob. 10–81

*10–80. Determine the moments of inertia I_u and I_v of the beam's cross-sectional area.

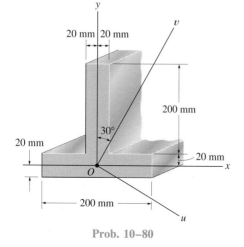

Prob. 10–80

10–82. Determine the principal moments of inertia for the angle's cross-sectional area with respect to a set of principal axes that have their origin located at the centroid C. Use the equation developed in Section 10–7. For the calculation, assume all corners to be square.

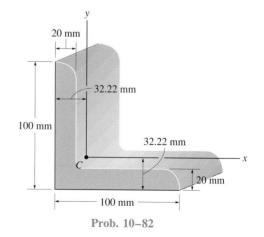

Prob. 10–82

10–83. Locate the centroid, $\bar{y}$, and determine the orientation of the principal centroidal axes for the composite area. What are the moments of inertia with respect to these axes?

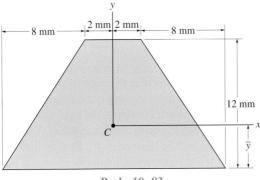

Prob. 10–83

10–85. Determine the directions of the principal axes with origin located at point O, and the principal moments of inertia for the rectangular area about these axes.

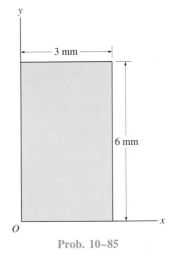

Prob. 10–85

*10–84. Determine the directions of the principal axes with origin located at point O, and the principal moments of inertia for the quarter-circular area about these axes.

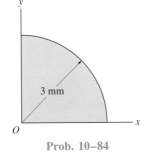

Prob. 10–84

10–86. Solve Prob. 10–77 using Mohr's circle.

10–87. Solve Prob. 10–81 using Mohr's circle.

*10–88. Solve Prob. 10–84 using Mohr's circle.

10–89. Solve Prob. 10–82 using Mohr's circle.

10–90. Solve Prob. 10–85 using Mohr's circle.

10.9 Mass Moment of Inertia

The mass moment of inertia of a body is a property that measures the resistance of the body to angular acceleration. Since it is used in dynamics to study rotational motion, methods for its calculation will now be discussed.

We define the *mass moment of inertia* as the integral of the "second moment" about an axis of all the elements of mass *dm* which compose the body.* For example, consider the rigid body shown in Fig. 10–22. The body's moment of inertia about the *z* axis is

$$I = \int_m r^2 \, dm \tag{10–13}$$

Here the "moment arm" *r* is the perpendicular distance from the axis to the arbitrary element *dm*. Since the formulation involves *r*, the value of *I* is *unique* for each axis *z* about which it is computed. However, the axis which is generally chosen for analysis passes through the body's mass center *G*. The moment of inertia computed about this axis will be defined as I_G. Realize that because *r* is squared in Eq. 10–13, the mass moment of inertia is always a *positive quantity*. Common units used for its measurement are kg · m².

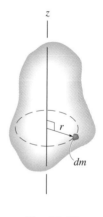

Fig. 10–22

---

*Another property of the body which measures the symmetry of the body's mass with respect to a coordinate system is the mass product of inertia. This property most often applies to the three-dimensional motion of a body and is discussed in *Engineering Mechanics: Dynamics* (Chapter 21).

PROCEDURE FOR ANALYSIS

For integration, we will consider only symmetric bodies having surfaces which are generated by revolving a curve about an axis. An example of such a body which is generated about the z axis is shown in Fig. 10–23.

If the body consists of material having a variable density, $\rho = \rho(x, y, z)$, the elemental mass dm of the body may be expressed in terms of its density and volume as $dm = \rho\, dV$. Substituting dm into Eq. 10–13, the body's moment of inertia is then computed using *volume elements* for integration; i.e.,

$$I = \int_V r^2 \rho\, dV \qquad (10\text{--}14)$$

In the special case of ρ being a *constant,* this term may be factored out of the integral and the integration is then purely a function of geometry:

$$I = \rho \int_V r^2\, dV \qquad (10\text{--}15)$$

When the elemental volume chosen for integration has differential sizes in all three directions, e.g., $dV = dx\,dy\,dz$, Fig. 10–23a, the moment of inertia of the body must be determined using "triple integration." The integration process can, however, be simplified to a *single integration* provided the chosen elemental volume has a differential size or thickness in only *one direction.* Shell or disk elements are often used for this purpose.

Shell Element.   If a *shell element* having a height z, radius y, and thickness dy is chosen for integration, Fig. 10–23b, then the volume $dV = (2\pi y)(z)\, dy$. This element may be used in Eq. 10–14 or 10–15 for determining the moment of inertia I_z of the body about the z axis, since the *entire element,* due to its "thinness," lies at the *same* perpendicular distance $r = y$ from the z axis (see Example 10–11).

Disk Element.   If a disk element having a radius y and a thickness dz is chosen for integration, Fig. 10–23c, then the volume $dV = (\pi y^2)\, dz$. In this case, however, the element is *finite* in the radial direction, and consequently its parts *do not* all lie at the *same radial distance* r from the z axis. As a result, Eq. 10–14 or 10–15 *cannot* be used to determine I_z. Instead, to perform the integration using this element, it is first necessary to determine the moment of inertia *of the element* about the z axis and then integrate this result (see Example 10–12).

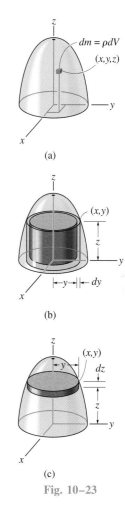

Fig. 10–23

Example 10–11

Determine the moment of inertia of the cylinder shown in Fig. 10–24a about the z axis. The density ρ of the material is constant.

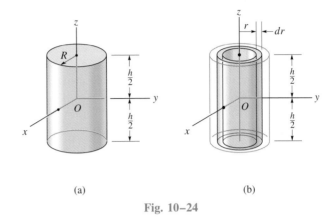

Fig. 10–24

SOLUTION

Shell Element. This problem may be solved using the *shell element* in Fig. 10–24b and single integration. The volume of the element is $dV = (2\pi r)(h)\, dr$, so that its mass is $dm = \rho\, dV = \rho(2\pi hr\, dr)$. Since the *entire element* lies at the same distance r from the z axis, the moment of inertia *of the element* is

$$dI_z = r^2\, dm = \rho 2\pi h r^3\, dr$$

Integrating over the entire region of the cylinder yields

$$I_z = \int_m r^2\, dm = \rho 2\pi h \int_0^R r^3\, dr = \frac{\rho\pi}{2} R^4 h$$

The mass of the cylinder is

$$m = \int_m dm = \rho 2\pi h \int_0^R r\, dr = \rho\pi h R^2$$

so that

$$I_z = \frac{1}{2} mR^2 \qquad\qquad\qquad Ans.$$

Example 10–12

A solid is formed by revolving the shaded area shown in Fig. 10–25a about the y axis. If the density of the material is 5 Mg/m^3, determine the moment of inertia about the y axis.

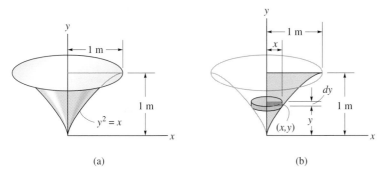

Fig. 10–25

SOLUTION

Disk Element.   The moment of inertia will be determined using a *disk element*, as shown in Fig. 10–25b. Here the element intersects the curve at the arbitrary point (x, y) and has a mass

$$dm = \rho \, dV = \rho(\pi x^2) \, dy$$

Although all portions of the element are *not* located at the same distance from the y axis, it is still possible to determine the moment of inertia dI_y *of the element* about the y axis. In Example 10–11 it was shown that the moment of inertia of a cylinder about its longitudinal axis is $I = \frac{1}{2}mR^2$, where m and R are the mass and radius of the cylinder. Since the height of the cylinder is not involved in this formula, we can also use it for a disk. Thus, for the disk element in Fig. 10–25b, we have

$$dI_y = \frac{1}{2}(dm)x^2 = \frac{1}{2}[\rho(\pi x^2) \, dy]x^2$$

Substituting $x = y^2$, $\rho = 5$ Mg/m^3, and integrating with respect to y, from $y = 0$ to $y = 1$ m, yields the moment of inertia for the entire solid:

$$I_y = \frac{5\pi}{2} \int_0^1 x^4 \, dy = \frac{5\pi}{2} \int_0^1 y^8 \, dy = 0.873 \text{ Mg} \cdot \text{m}^2 \qquad \textit{Ans.}$$

Parallel-Axis Theorem. If the moment of inertia of the body about an axis passing through the body's mass center is known, then the moment of inertia about any other *parallel axis* may be determined by using the *parallel-axis theorem*. This theorem can be derived by considering the body shown in Fig. 10–26. The z' axis passes through the mass center G, whereas the corresponding *parallel z axis* lies at a constant distance d away. Selecting the differential element of mass dm which is located at point (x', y') and using the Pythagorean theorem, $r^2 = (d + x')^2 + y'^2$, we can express the moment of inertia of the body about the z axis as

$$I = \int_m r^2 \, dm = \int_m [(d + x')^2 + y'^2] \, dm$$

$$= \int_m (x'^2 + y'^2) \, dm + 2d \int_m x' \, dm + d^2 \int_m dm$$

Since $r'^2 = x'^2 + y'^2$, the first integral represents I_G. The second integral equals *zero*, since the z' axis passes through the body's mass center, i.e., $\int x' \, dm = \bar{x} \int dm = 0$ since $\bar{x} = 0$. Finally, the third integral represents the total mass m of the body. Hence, the moment of inertia about the z axis can be written as

$$I = I_G + md^2 \tag{10–16}$$

where I_G = moment of inertia about the z' axis passing through the mass center G

m = mass of the body

d = perpendicular distance between the parallel axes

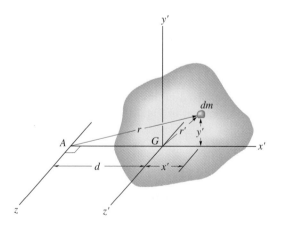

Fig. 10–26

Radius of Gyration. Occasionally, the moment of inertia of a body about a specified axis is reported in handbooks using the *radius of gyration, k*. This value has units of length, and when it and the body's mass m are known, the moment of inertia is determined from the equation

$$I = mk^2 \quad \text{or} \quad k = \sqrt{\frac{I}{m}} \qquad (10\text{--}17)$$

Note the *similarity* between the definition of k in this formula and r in the equation $dI = r^2\, dm$, which defines the moment of inertia of an elemental mass dm of the body about an axis.

Composite Bodies. If a body is constructed from a number of simple shapes such as disks, spheres, and rods, the moment of inertia of the body about any axis z can be determined by adding algebraically the moments of inertia of all the composite shapes computed about the z axis. Algebraic addition is necessary since a composite part must be considered as a negative quantity if it has already been included within another part—for example, a "hole" subtracted from a solid plate. The parallel-axis theorem is needed for the calculations if the center of mass of each composite part does not lie on the z axis. For the calculation, then, $I = \Sigma(I_G + md^2)$, where I_G for each of the composite parts is computed by integration or can be determined from a table, such as the one given on the inside back cover.

Example 10–13

If the plate shown in Fig. 10–27a has a density of 8000 kg/m³ and a thickness of 10 mm, compute its moment of inertia about an axis directed perpendicular to the page and passing through point O.

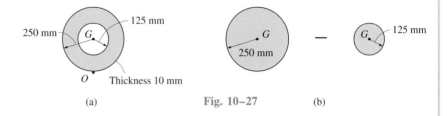

Fig. 10–27

(a)   (b)

SOLUTION

The plate consists of two composite parts, the 250-mm-radius disk *minus* a 125-mm-radius disk, Fig. 10–27b. The moment of inertia about O can be determined by computing the moment of inertia of each of these parts about O and then *algebraically* adding the results. The computations are performed by using the parallel-axis theorem in conjunction with the data listed in the table on the inside back cover.

Disk.   The moment of inertia of a disk about an axis perpendicular to the plane of the disk is $I_G = \frac{1}{2}mr^2$. The mass center of the disk is located at a distance of 0.25 m from point O. Thus,

$$m_d = \rho_d V_d = 8000 \text{ kg/m}^3[\pi(0.25 \text{ m})^2(0.01 \text{ m})] = 15.71 \text{ kg}$$

$$(I_O)_d = \tfrac{1}{2}m_d r_d^2 + m_d d^2$$

$$= \frac{1}{2}(15.71 \text{ kg})(0.25 \text{ m})^2 + (15.71 \text{ kg})(0.25 \text{ m})^2$$

$$= 1.473 \text{ kg} \cdot \text{m}^2$$

Hole.   For the 125-mm-radius disk (hole), we have

$$m_h = \rho_h V_h = 8000 \text{ kg/m}^3[\pi(0.125 \text{ m})^2(0.01 \text{ m})] = 3.93 \text{ kg}$$

$$(I_O)_h = \tfrac{1}{2}m_h r_h^2 + m_h d^2$$

$$= \frac{1}{2}(3.93 \text{ kg})(0.125 \text{ m})^2 + (3.93 \text{ kg})(0.25 \text{ m})^2$$

$$= 0.276 \text{ kg} \cdot \text{m}^2$$

The moment of inertia of the plate about point O is therefore

$$I_O = (I_O)_d - (I_O)_h$$

$$= 1.473 \text{ kg} \cdot \text{m}^2 - 0.276 \text{ kg} \cdot \text{m}^2$$

$$= 1.20 \text{ kg} \cdot \text{m}^2 \qquad\qquad\qquad \textit{Ans.}$$

Example 10–14

The pendulum consists of two thin rods each having a weight of 10 N and suspended from point O as shown in Fig. 10–28. Determine the pendulum's moment of inertia about an axis passing through (a) the pin at O, and (b) the mass center G of the pendulum.

SOLUTION

Part (a).   Using the table on the inside back cover, the moment of inertia of rod OA about an axis perpendicular to the page and passing through the end point O of the rod is $I_O = \frac{1}{3}ml^2$. Hence,

$$(I_{OA})_O = \frac{1}{3}ml^2 = \frac{1}{3}\left(\frac{10}{9.81}\right)(2)^2 = 1.359 \text{ kg} \cdot \text{m}^2$$

The same value may be computed using $I_G = \frac{1}{12}ml^2$ and the parallel-axis theorem; i.e.,

$$(I_{OA})_O = \frac{1}{12}ml^2 + md^2 = \frac{1}{12}\left(\frac{10}{9.81}\right)(2)^2 + \frac{10}{9.81}(1)^2$$
$$= 1.359 \text{ kg} \cdot \text{m}^2$$

For rod BC we have

$$(I_{BC})_O = \frac{1}{12}ml^2 + md^2 = \frac{1}{12}\left(\frac{10}{9.81}\right)(2)^2 + \frac{10}{9.81}(2)^2$$
$$= 4.417 \text{ kg} \cdot \text{m}^2$$

The moment of inertia of the pendulum about O is therefore

$$I_O = 1.359 + 4.417 = 5.776 \text{ kg} \cdot \text{m}^2 \qquad \textit{Ans.}$$

Part (b).   The mass center G will be located relative to the pin at O. Assuming this distance to be $\bar{y}$, Fig. 10–28, and using the formula for determining the mass center, we have

$$\bar{y} = \frac{\Sigma \tilde{y}m}{\Sigma m} = \frac{1(10/9.81) + 2(10/9.81)}{(10/9.81) + (10/9.81)} = 1.50 \text{ m}$$

The moment of inertia I_G may be computed in the same manner as I_O, which requires successive applications of the parallel-axis theorem in order to transfer the moments of inertia of rods OA and BC to G. A more direct solution, however, involves applying the parallel-axis theorem using the result for I_O determined above; i.e.,

$$I_O = I_G + md^2; \qquad 5.776 = I_G + \left(\frac{20}{9.81}\right)(1.50)^2$$
$$I_G = 1.189 \text{ kg} \cdot \text{m}^2 \qquad \textit{Ans.}$$

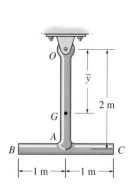

Fig. 10–28

PROBLEMS

10–91.   Determine the moment of inertia I_y for the slender rod. The rod's density ρ and cross-sectional area A are constant. Express the result in terms of the rod's total mass m.

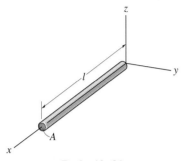

Prob. 10–91

*10–92.   Determine the moment of inertia of the thin ring about the z axis. The ring has a mass m.

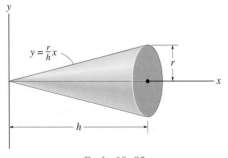

Prob. 10–92

10–93.   The right circular cone is formed by revolving the shaded area around the x axis. Determine the moment of inertia I_x and express the result in terms of the total mass m of the cone. The cone has a constant density ρ.

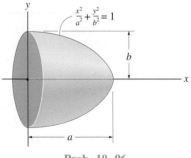

Prob. 10–93

10–94.   The sphere is formed by revolving the shaded area around the x axis. Determine the moment of inertia I_x and express the result in terms of the total mass m of the sphere. The sphere has a constant density ρ.

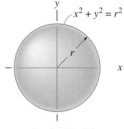

Prob. 10–94

10–95.   The paraboloid is formed by revolving the shaded area around the x axis. Determine the radius of gyration k_x. The material has a constant density ρ.

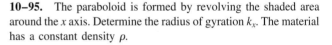

Prob. 10–95

*10–96.   A semiellipsoid is formed by rotating the shaded area about the x axis. Determine the moment of inertia with respect to the x axis and express the result in terms of the mass m of the semiellipsoid. The material has constant density ρ.

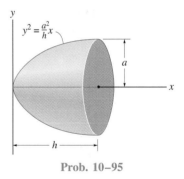

Prob. 10–96

10–97. An ellipsoid is formed by rotating the shaded area about the x axis. Determine the moment of inertia with respect to the x axis and express the result in terms of the mass m of the ellipsoid. The material has a constant density ρ.

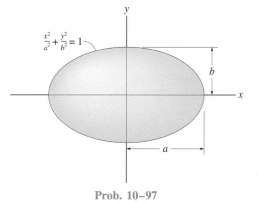

$$\frac{x^2}{a^2} + \frac{y^2}{b^2} = 1$$

Prob. 10–97

10–98. The concrete shape is formed by rotating the shaded area about the y axis. Determine the moment of inertia I_y. The specific weight of concrete is $\gamma = 25$ kN/m³.

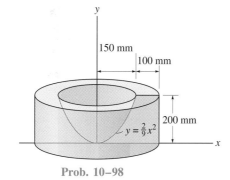

150 mm
100 mm
200 mm
$y = \frac{2}{9}x^2$

Prob. 10–98

10–99. The frustum is formed by rotating the shaded area around the x axis. Determine the moment of inertia I_x and express the result in terms of the total mass m of the frustum. The frustum has a constant density.

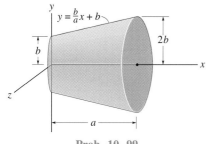

$y = \frac{b}{a}x + b$
$2b$
b
a

Prob. 10–99

*10–100.** The body is formed by revolving the shaded area around the x axis. Determine the radius of gyration k_x. The specific weight of the material is $\gamma = 60$ kN/m³.

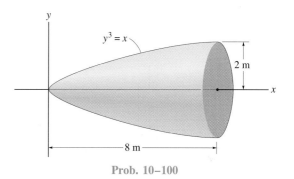

$y^3 = x$
2 m
8 m

Prob. 10–100

10–101. Determine the moment of inertia of the homogeneous triangular prism with respect to the y axis. Express the result in terms of the mass m of the prism. *Hint:* For integration, use thin plate elements parallel to the x-y plane and having a thickness dz.

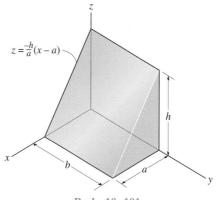

$z = \frac{-h}{a}(x - a)$
h
b
a

Prob. 10–101

10–102. The slender rods have a weight of 15 N/m. Determine the moment of inertia of the assembly about an axis perpendicular to the page and passing through point A.

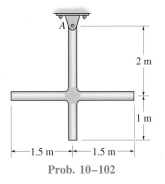

Prob. 10–102

10–103. Determine the moment of inertia I_z of the frustum of the cone which has a conical depression. The material has a density of 200 kg/m³.

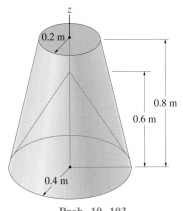

Prob. 10–103

*10–104.** Determine the moment of inertia of the wheel about an axis which is perpendicular to the page and passes through the center of mass G. The material has a specific weight of $\gamma = 15$ kN/m³.

10–105. Determine the moment of inertia of the wheel about an axis which is perpendicular to the page and passes through point O. The material has a specific weight of $\gamma = 15$ kN/m³.

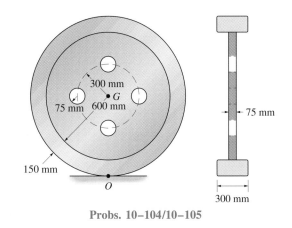

Probs. 10–104/10–105

10–106. The pendulum consists of two slender rods AB and OC which have a mass of 3 kg/m. The thin plate has a mass of 12 kg/m². Determine the location $\bar{y}$ of the center of mass G of the pendulum, then calculate the moment of inertia of the pendulum about an axis perpendicular to the page and passing through G.

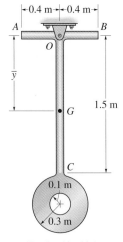

Prob. 10–106

10–107.   The pendulum consists of a disk having a mass of 6 kg and slender rods *AB* and *DC* which have a mass of 2 kg/m. Determine the length *L* of *DC* so that the center of mass is at the bearing *O*. What is the moment of inertia of the assembly about an axis perpendicular to the page and passing through point *O*?

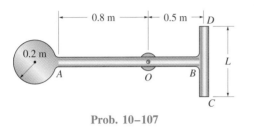

Prob. 10–107

10–109.   Determine the location $\bar{y}$ of the center of mass *G* of the assembly and then calculate the moment of inertia about an axis perpendicular to the page and passing through *G*. The block has a mass of 3 kg and the mass of the semicylinder is 5 kg.

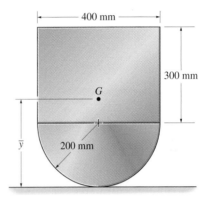

Prob. 10–109

*10–108.   Determine the moment of inertia of the wire triangle about an axis perpendicular to the page and passing through point *O*. Also, locate the mass center *G* and determine the moment of inertia about an axis perpendicular to the page and passing through point *G*. The wire has a mass of 0.3 kg/m. Neglect the size of the ring at *O*.

10–110.   Determine the moment of inertia I_z of the frustum of the cone which has a conical depression. The material has a density of 200 kg/m^3.

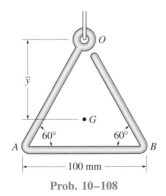

Prob. 10–108

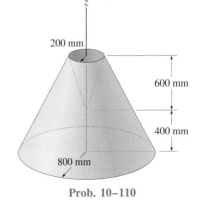

Prob. 10–110

REVIEW PROBLEMS

10–111. Determine the moments of inertia I_u and I_v and the product of inertia I_{uv} for the semicircular area.

10–113. Determine the moments of inertia I_x and I_y of the shaded area.

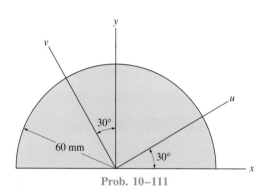

Prob. 10–111

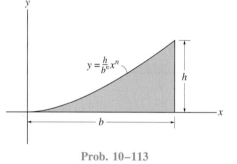

Prob. 10–113

10–114. Determine the moment of inertia of the shaded area about the x axis.

*10–112. The paraboloid is formed by revolving the shaded area around the x axis. Determine the radius of gyration k_x. The density of the material is $\rho = 5 \text{ Mg/m}^3$.

10–115. Determine the moment of inertia of the shaded area about the y axis.

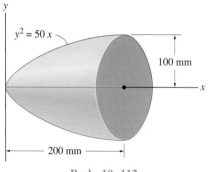

Prob. 10–112

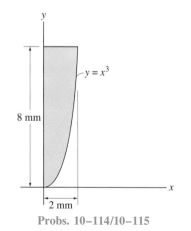

Probs. 10–114/10–115

*10–116. Determine the moment of inertia of the shaded area about the x axis.

10–117. Determine the moment of inertia of the shaded area about the y axis.

*10–120. Determine the moment of inertia of the triangular area about (a) the x axis, and (b) the centroidal x' axis.

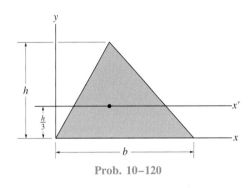

Prob. 10–120

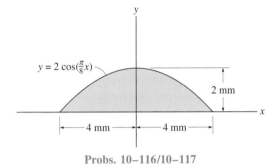

$y = 2\cos(\frac{\pi}{8}x)$

2 mm

4 mm   4 mm

Probs. 10–116/10–117

10–121. Determine the moments of inertia I_x and I_y of the shaded area.

10–122. Determine the product of inertia I_{xy} of the shaded area with respect to the x and y axes.

10–123. Determine the directions of the principal axes with origin located at point O, and the principal moments of inertia for the shaded area about these axes. Here $I_x = 0.1667\ \text{m}^4$, $I_y = 0.0333\ \text{m}^4$, $I_{xy} = 0.0625\ \text{m}^4$.

10–118. Determine the moment of inertia of the shaded area about the x axis.

10–119. Determine the moment of inertia of the shaded area about the y axis.

*10–124. Determine the principal moments of inertia for the shaded area using Mohr's circle. Here $I_x = 0.1667\ \text{m}^4$, $I_y = 0.0333\ \text{m}^4$, $I_{xy} = 0.0625\ \text{m}^4$.

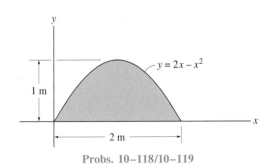

$y = 2x - x^2$

1 m

2 m

Probs. 10–118/10–119

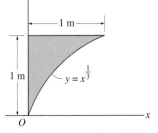

1 m

1 m

$y = x^{\frac{1}{3}}$

O

Probs. 10–121/10–122/10–123/10–124

Equilibrium and stability of this articulated crane boom as a function of its position can be analyzed using methods based on work and energy, which are explained in this chapter.

11

Virtual Work

In this chapter we will use the principle of virtual work and the potential-energy method to determine the equilibrium position of a series of connected rigid bodies. Although application requires more mathematical sophistication than using the equations of equilibrium, it will be shown that, once the equation of virtual work or the potential-energy function is established, the solution may be obtained *directly*, without having to dismember the system in order to obtain relationships between forces occurring at the connections. Furthermore, by using the potential-energy method, we will be able to investigate the "type" of equilibrium or the stability of the configuration.

11.1 Definition of Work and Virtual Work

Work of a Force. In mechanics a force **F** does work only when it undergoes a displacement in the direction of the force. For example, consider the force **F** in Fig. 11–1, which is located on the path s specified by the position vector **r**. If the force moves along the path to a new position $\mathbf{r}' = \mathbf{r} + d\mathbf{r}$, the displacement is $d\mathbf{r}$ and therefore the work dU is a *scalar quantity*, defined by the dot product

$$dU = \mathbf{F} \cdot d\mathbf{r}$$

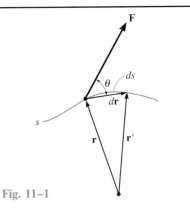

Fig. 11–1

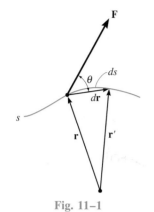

Fig. 11–1

Because $d\mathbf{r}$ is infinitesimal, the magnitude of $d\mathbf{r}$ can be represented by ds, the differential arc segment along the path. If the angle between the tails of $d\mathbf{r}$ and $\mathbf{F}$ is θ, Fig. 11–1, then by definition of the dot product, the above equation may also be written as

$$dU = F\,ds\,\cos\theta$$

Work expressed by this equation may be interpreted in one of two ways: either as the product of $\mathbf{F}$ and the component of displacement in the direction of the force, i.e., $ds\cos\theta$; or as the product of ds and the component of force in the direction of displacement, i.e., $F\cos\theta$. Note that if $0° \le \theta < 90°$, then the force component and the displacement have the *same sense,* so that the work is *positive;* whereas if $90° < \theta \le 180°$, these vectors have an *opposite sense,* and therefore the work is *negative.* Also, $dU = 0$ if the force is *perpendicular* to displacement, since $\cos 90° = 0$, or if the force is applied at a *fixed point,* in which case the displacement $ds = 0$.

The basic unit for work combines the units of force and displacement. In the SI system a *joule* (J) is equivalent to the work done by a force of 1 newton which moves 1 meter in the direction of the force (1 J = 1 N · m). In the FPS system, work is defined in units of ft · lb. The moment of a force has the same combination of units; however, the concepts of moment and work are in no way related. A moment is a vector quantity, whereas work is a scalar.

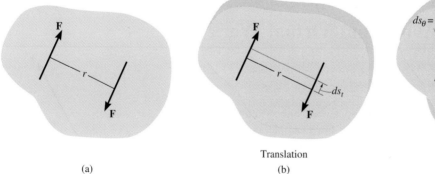
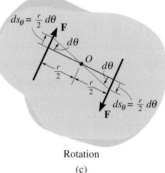

(a)

Translation

(b)

Rotation

(c)

Fig. 11–2

Work of a Couple. The two forces of a couple do work when the couple *rotates* about an axis perpendicular to the plane of the couple. To show this, consider the body in Fig. 11–2a, which is subjected to a couple whose moment has a magnitude $M = Fr$. Any general differential displacement of the body can be considered as a combination of a translation and rotation. When the body *translates* such that the *component of displacement* along the line of action of each force is ds_t, clearly the "positive" work of one force ($F\,ds_t$) *cancels* the "negative" work of the other ($-F\,ds_t$), Fig. 11–2b. Consider now a differential *rotation* $d\theta$ of the body about an axis perpendicular to the plane of the couple, which intersects the plane at point O, Fig. 11–2c. (For the derivation, any other point in the plane may also be considered.) As shown, each force undergoes a displacement $ds_\theta = (r/2)\,d\theta$ in the direction of the force; hence, the work of both forces is

$$dU = F\left(\frac{r}{2}\,d\theta\right) + F\left(\frac{r}{2}\,d\theta\right) = (Fr)\,d\theta$$

or

$$dU = M\,d\theta$$

The resultant work is *positive* when the sense of $\mathbf{M}$ is the *same* as that of $d\boldsymbol{\theta}$, and negative when they have an opposite sense. As in the case of the moment vector, the *direction and sense* of $d\boldsymbol{\theta}$ are defined by the right-hand rule, where the fingers of the right hand follow the rotation or "curl" and the thumb indicates the direction of $d\boldsymbol{\theta}$. Hence, the line of action of $d\boldsymbol{\theta}$ will be *parallel* to the line of action of $\mathbf{M}$ if movement of the body occurs in the *same plane*. If the body rotates in space, however, the *component* of $d\boldsymbol{\theta}$ in the direction of $\mathbf{M}$ is required. Thus, in general, the work done by a couple is defined by the dot product, $dU = \mathbf{M} \cdot d\boldsymbol{\theta}$.

Virtual Work. The definitions of the work of a force and a couple have been presented in terms of *actual movements* expressed by differential displacements having magnitudes of ds and $d\theta$. Consider now an *imaginary* or *virtual movement*, which indicates a displacement or rotation that is *assumed* and *does not actually exist*. These movements are first-order differential quantities and will be denoted by the symbols δs and $\delta\theta$ (delta s and delta θ), respectively. The *virtual work* done by a force undergoing a virtual displacement δs is

$$\delta U = F\cos\theta\;\delta s \tag{11–1}$$

Similarly, when a couple undergoes a virtual rotation $\delta\theta$ in the plane of the couple forces, the *virtual work* is

$$\delta U = M\;\delta\theta \tag{11–2}$$

11.2 Principle of Virtual Work for a Particle and a Rigid Body

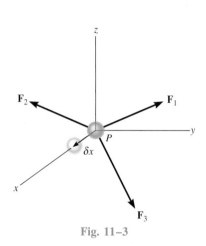

Fig. 11–3

If a particle is in equilibrium, the resultant of the force system acting on it must be equal to zero. Hence, if the particle undergoes an imaginary or virtual displacement in the x, y, or z direction, the virtual work (δU) done by the force system must be equal to zero since the components $\Sigma F_x = 0, \Sigma F_y = 0, \Sigma F_z = 0$. Alternatively, this may be expressed as

$$\delta U = 0$$

For example, if the particle in Fig. 11–3 is given a virtual displacement $\delta \mathbf{x}$, only the x components of the forces acting on the particle do work. (No work is done by the y and z components, since they are perpendicular to the displacement.) The virtual-work equation is therefore

$$\delta U = 0; \qquad F_{1x}\,\delta x + F_{2x}\,\delta x + F_{3x}\,\delta x = 0$$

Factoring out δx, which is common to every term, yields

$$(F_{1x} + F_{2x} + F_{3x})\,\delta x = 0$$

Since $\delta x \neq 0$, this equation is satisfied only if the sum of the force components in the x direction is equal to zero, i.e., $\Sigma F_x = 0$. Two other virtual work equations can be written by assuming virtual displacements $\delta \mathbf{y}$ and $\delta \mathbf{z}$ in the y and z directions, respectively. Doing this, however, amounts to satisfying the equilibrium equations $\Sigma F_y = 0$ and $\Sigma F_z = 0$ for the particle.

In a similar manner, a rigid body that is subjected to a coplanar force system will be in equilibrium provided $\Sigma F_x = 0, \Sigma F_y = 0,$ and $\Sigma M_O = 0$. We can also write a set of three virtual work equations for the body, each of which requires $\delta U = 0$. If these equations involve separate virtual translations in the x and y directions and a virtual rotation about an axis perpendicular to the x–y plane and passing through point O, then it can be shown that they will correspond to the above-mentioned three equilibrium equations. When writing these equations, it is *not necessary* to include the work done by the *internal forces* acting within the body, since a rigid body *does not deform* when subjected to an external loading, and furthermore, when the body moves through a virtual displacement, the internal forces occur in equal but opposite collinear pairs, so that the corresponding work done by each pair of forces *cancels*.

As in the case of a particle, however, no added advantage would be gained by solving rigid-body equilibrium problems using the principle of virtual work. This is because for each application of the virtual-work equation the virtual displacement, common to every term, factors out, leaving an equation that could have been obtained in a more *direct manner* by applying the equations of equilibrium.

11.3 Principle of Virtual Work for a System of Connected Rigid Bodies

The method of virtual work is most suitable for solving equilibrium problems that involve a system of several *connected* rigid bodies such as the ones shown in Fig. 11–4. Before we can apply the principle of virtual work to these systems, however, we must first specify the number of degrees of freedom for a system and establish coordinates that define the position of the system.

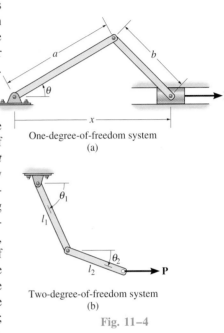

One-degree-of-freedom system
(a)

Two-degree-of-freedom system
(b)

Fig. 11–4

Degrees of Freedom. A system of connected bodies takes on a unique shape that can be specified provided we know the position of a number of specific points on the system. These positions are defined using *independent coordinates q,* which are measured from fixed reference points. For every coordinate established, the system will have a *degree of freedom* for displacement along the coordinate axis such that it is consistent with the constraining action of the supports. Thus, an *n*-degree-of-freedom system requires *n* independent coordinates q_n to specify the location of all its members. For example, the link and sliding-block arrangement shown in Fig. 11–4a is an example of a one-degree-of-freedom system. The independent coordinate $q = \theta$ may be used to specify the location of the two connecting links and the block. The coordinate *x* could also be used as the independent coordinate. However, since the block is constrained to move within the slot, *x* is not independent of θ; rather, it can be related to θ using the cosine law, $b^2 = a^2 + x^2 - 2ax \cos \theta$. The double-link arrangement, shown in Fig. 11–4b, is an example of a two-degree-of-freedom system. To specify the location of each link, the coordinate angles θ_1 and θ_2 must be known, since a rotation of one link is independent of a rotation of the other.

Principle of Virtual Work. The principle of virtual work for a system of rigid bodies whose connections are *frictionless* may be stated as follows: *A system of connected rigid bodies is in equilibrium provided the virtual work done by all the external forces and couples acting on the system is zero for each independent virtual displacement of the system.* Mathematically, this may be expressed as

$$\delta U = 0 \qquad\qquad (11\text{–}3)$$

where δU represents the virtual work of all the external forces (and couples) acting on the system during any independent virtual displacement.

As stated above, if a system has *n* degrees of freedom it takes *n* independent coordinates q_n to completely specify the location of the system. Hence, for the system it is possible to write n independent virtual-work equations, one for every virtual displacement taken along each of the independent coordinate axes, while the remaining $n - 1$ independent coordinates are held *fixed*.*

PROCEDURE FOR ANALYSIS

The following procedure provides a method for applying the equation of virtual work to solve problems involving a system of frictionless connected rigid bodies having a single degree of freedom.

Free-Body Diagram.   Draw the free-body diagram of the entire system of connected bodies and define the *independent coordinate q*. Sketch the "deflected position" of the system on the free-body diagram when the system undergoes a *positive* virtual displacement δq. From this, specify the "active" forces and couples, that is, those that do work.

Virtual Displacements.   Indicate *position coordinates s_i,* measured from a *fixed point* on the free-body diagram to each of the i number of "active" forces and couples. Each coordinate axis should be parallel to the line of action of the "active" force to which it is directed, so that the virtual work along the coordinate axis can be calculated.

Relate each of the position coordinates s_i to the independent coordinate q; then *differentiate* these expressions in order to express the virtual displacements δs_i in terms of δq.

Virtual-Work Equation.   Write the *virtual-work equation* for the system assuming that, whether possible or not, all the position coordinates s_i undergo *positive* virtual displacements δs_i. Using the relations for δs_i, express the work of each "active" force and couple in the equation in terms of the single independent virtual displacement δq. By factoring out this common displacement, one is left with an equation that generally can be solved for an unknown force, couple, or equilibrium position.

If the system contains n degrees of freedom, n independent coordinates q_n must be specified. In this case, follow the above procedure and let *only one* of the independent coordinates undergo a virtual displacement, while the remaining $n - 1$ coordinates are held fixed. In this way, n virtual-work equations can be written, one for each independent coordinate.

The following examples should help to clarify application of this procedure.

---

*This method of applying the principle of virtual work is sometimes called the *method of virtual displacements,* since a virtual displacement is applied, resulting in the calculation of a real force. Although it is not to be used here, realize that we can also apply the principle of virtual work as a method of virtual forces. This method is often used to determine the displacements of points on deformable bodies. See R. C. Hibbeler, *Mechanics of Materials,* 2nd edition, Macmillan Publishing Company, New York, 1994.

Example 11–1

Determine the angle θ for equilibrium of the two-member linkage shown in Fig. 11–5a. Each member has a mass of 10 kg.

SOLUTION

Free-Body Diagram. The system has only one degree of freedom, since the location of both links may be specified by the single independent coordinate $(q =) \theta$. As shown on the free-body diagram in Fig. 11–5b, when θ undergoes a *positive* (clockwise) virtual rotation $\delta\theta$, only the active forces, **F** and the two 98.1-N weights, do work. (The reactive forces $\mathbf{D}_x$ and $\mathbf{D}_y$ are fixed, and $\mathbf{B}_y$ does not move along its line of action.)

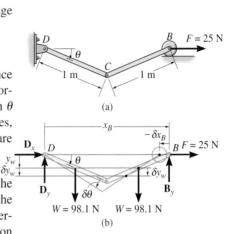

(a)

(b)

Fig. 11–5

Virtual Displacements. If the origin of coordinates is established at the *fixed* pin support D, the location of **F** and **W** may be specified by the position coordinates x_B and y_w, as shown in the figure. In order to determine the work, note that these coordinates are parallel to the lines of action of their associated forces.

Expressing the position coordinates in terms of the independent coordinate θ and taking the derivatives yields

$$x_B = 2(1 \cos \theta) \text{ m} \qquad \delta x_B = -2 \sin \theta \, \delta\theta \text{ m} \qquad (1)$$

$$y_w = \tfrac{1}{2}(1 \sin \theta) \text{ m} \qquad \delta y_w = 0.5 \cos \theta \, \delta\theta \text{ m} \qquad (2)$$

It is seen by the *signs* of these equations, and indicated in Fig. 11–5b, that an *increase* in θ (i.e., $\delta\theta$) causes a *decrease* in x_B and an *increase* in y_w.

Virtual-Work Equation. If the virtual displacements δx_B and δy_w were *both positive*, then the forces **W** and **F** would do positive work since the forces and their corresponding displacements would have the same sense. Hence, the virtual-work equation for the displacement $\delta\theta$ is

$$\delta U = 0; \qquad W \, \delta y_w + W \, \delta y_w + F \, \delta x_B = 0 \qquad (3)$$

Substituting Eqs. 1 and 2 into Eq. 3 in order to relate the virtual displacements to the common virtual displacement $\delta\theta$ yields

$$98.1(0.5 \cos \theta \, \delta\theta) + 98.1(0.5 \cos \theta \, \delta\theta) + 25(-2 \sin \theta \, \delta\theta) = 0$$

Notice that the "negative work" done by **F** (force in the opposite sense to displacement) has been *accounted for* in the above equation by the "negative sign" of Eq. 1. Factoring out the *common displacement* $\delta\theta$ and solving for θ, noting that $\delta\theta \neq 0$, yields

$$(98.1 \cos \theta - 50 \sin \theta) \, \delta\theta = 0$$

$$\theta = \tan^{-1} \frac{98.1}{50} = 63.0° \qquad \qquad Ans.$$

If this problem had been solved using the equations of equilibrium, it would have been necessary to dismember the links and apply three scalar equations to *each* link. The principle of virtual work, by means of calculus, has eliminated this task so that the answer is obtained directly.

Example 11–2

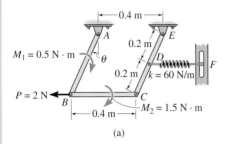

(a)

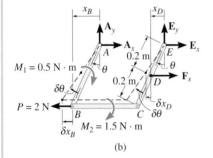

(b)

Fig. 11–6

Using the principle of virtual work, determine the angle θ required to maintain equilibrium of the mechanism shown in Fig. 11–6a. Neglect the weight of the links. The spring is unstretched when $\theta = 0°$ and it maintains a horizontal position due to the roller.

SOLUTION

Free-Body Diagram. The mechanism has one degree of freedom, and therefore the location of each member may be specified using the independent coordinate θ. When θ undergoes a *positive* virtual displacement $\delta\theta$, as shown on the free-body diagram in Fig. 11–6b, links *AB* and *EC* rotate by the same amount since they have the same length, and link *BC* only translates. Since a couple moment does work *only* when it rotates, the work done by $\mathbf{M}_2$ is zero. The reactive forces at *A* and *E* do no work. Why?

Virtual Displacements. The position coordinates x_B and x_D are *parallel* to the lines of action of **P** and $\mathbf{F}_s$, and these coordinates locate these forces with respect to the *fixed points* *A* and *E*. From Fig. 11–6b,

$$x_B = 0.4 \sin\theta \text{ m}$$
$$x_D = 0.2 \sin\theta \text{ m}$$

Thus,

$$\delta x_B = 0.4 \cos\theta \, \delta\theta \text{ m}$$
$$\delta x_D = 0.2 \cos\theta \, \delta\theta \text{ m}$$

Virtual-Work Equation. Applying the equation of virtual work, noting that, for positive virtual displacements, $\mathbf{F}_s$ is opposite to $\delta\mathbf{x}_D$ and hence does negative work, we obtain

$$\delta U = 0; \qquad M_1 \, \delta\theta + P \, \delta x_B - F_s \, \delta x_D = 0$$

Relating each of the virtual displacements to the *common* virtual displacement $\delta\theta$ yields

$$0.5 \, \delta\theta + 2(0.4 \cos\theta \, \delta\theta) - F_s(0.2 \cos\theta \, \delta\theta) = 0$$
$$(0.5 + 0.8 \cos\theta - 0.2 F_s \cos\theta) \, \delta\theta = 0 \qquad (1)$$

For the arbitrary angle θ, the spring is stretched a distance of $x_D = (0.2 \sin\theta)$ m; and therefore, $F_s = 60 \text{ N/m}(0.2 \sin\theta)$ m $= (12 \sin\theta)$ N. Substituting into Eq. 1 and noting that $\delta\theta \neq 0$, we have

$$0.5 + 0.8 \cos\theta - 0.2(12 \sin\theta) \cos\theta = 0$$

Since $\sin 2\theta = 2 \sin\theta \cos\theta$, then

$$1 = 2.4 \sin 2\theta - 1.6 \cos\theta$$

Solving for θ by trial and error yields

$$\theta = 36.3° \qquad \qquad \textit{Ans.}$$

Example 11–3

Using the principle of virtual work, determine the horizontal force that the pin at C must exert in order to hold the mechanism shown in Fig. 11–7a in equilibrium when $\theta = 45°$. Neglect the weight of the members.

SOLUTION

Free-Body Diagram. The reaction $\mathbf{C}_x$ can be obtained by *releasing* the pin constraint at C in the x direction and allowing the frame to be displaced in this direction. The system then has only one degree of freedom, defined by the independent coordinate θ, Fig. 11–7b. When θ undergoes a *positive* virtual displacement $\delta\theta$, only $\mathbf{C}_x$ and the 200-N force do work.

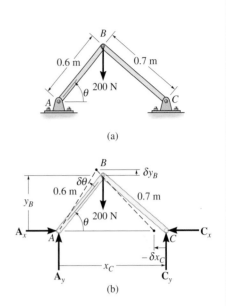

(a)

Virtual Displacements. Forces $\mathbf{C}_x$ and 200 N are located from the fixed origin A using position coordinates y_B and x_C. From Fig. 11–7b, x_C can be related to θ by the "law of cosines." Hence,

$$(0.7)^2 = (0.6)^2 + x_C^2 - 2(0.6)x_C \cos\theta \tag{1}$$

$$0 = 0 + 2x_C\,\delta x_C - 1.2\,\delta x_C \cos\theta + 1.2 x_C \sin\theta\,\delta\theta$$

$$\delta x_C = \frac{1.2 x_C \sin\theta}{1.2 \cos\theta - 2x_C}\,\delta\theta \tag{2}$$

Also,

$$y_B = 0.6 \sin\theta$$

$$\delta y_B = 0.6 \cos\theta\,\delta\theta \tag{3}$$

(b)

Fig. 11–7

Virtual-Work Equation. When y_B and x_C undergo *positive* virtual displacements δy_B and δx_C, $\mathbf{C}_x$ and 200 N do *negative work,* since they both act in the opposite sense to δy_B and δx_C. Hence,

$$\delta U = 0; \qquad -200\,\delta y_B - C_x\,\delta x_C = 0$$

Substituting Eqs. 2 and 3 into this equation, factoring out $\delta\theta$, and solving for C_x yields

$$-200(0.6 \cos\theta\,\delta\theta) - C_x \frac{1.2 x_C \sin\theta}{1.2 \cos\theta - 2x_C}\,\delta\theta = 0$$

$$C_x = \frac{-120 \cos\theta(1.2 \cos\theta - 2x_C)}{1.2 x_C \sin\theta} \tag{4}$$

At the required equilibrium position $\theta = 45°$, the corresponding value of x_C can be found by using Eq. 1, in which case

$$x_C^2 - 1.2 \cos 45° x_C - 0.13 = 0$$

Solving for the positive root yields

$$x_C = 0.981 \text{ m}$$

Thus, from Eq. 4,

$$C_x = 114 \text{ N} \qquad\qquad Ans.$$

Example 11–4

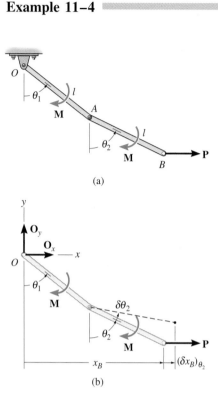

(a)

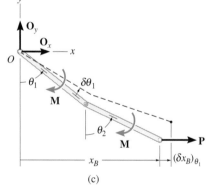

(b)

(c)

Fig. 11–8

Using the principle of virtual work, determine the equilibrium position of the two-bar linkage shown in Fig. 11–8a. Neglect the weight of the links.

SOLUTION

The system has two degrees of freedom, since the *independent coordinates* θ_1 and θ_2 must be known to locate the position of both links. The position coordinate x_B, measured from the fixed point O, is used to specify the location of **P,** Fig. 11–8b and c.

If θ_1 is held *fixed* and θ_2 varies by an amount $\delta\theta_2$, as shown in Fig. 11–8b, the virtual-work equation becomes

$$[\delta U = 0]_{\theta_2}; \qquad P(\delta x_B)_{\theta_2} - M\,\delta\theta_2 = 0 \qquad (1)$$

Here P and M represent the magnitudes of the applied force and couple moment acting on link AB.

When θ_2 is held *fixed* and θ_1 varies by an amount $\delta\theta_1$, as shown in Fig. 11–8c, the virtual-work equation becomes

$$[\delta U = 0]_{\theta_1}; \qquad P(\delta x_B)_{\theta_1} - M\,\delta\theta_1 - M\,\delta\theta_1 = 0 \qquad (2)$$

The *position coordinate* x_B may be related to the independent coordinates θ_1 and θ_2 by the equation

$$x_B = l\sin\theta_1 + l\sin\theta_2 \qquad (3)$$

To obtain the variation of δx_B in terms of $\delta\theta_2$, it is necessary to take the *partial derivative* of x_B with respect to θ_2 since x_B is a function of both θ_1 and θ_2. Hence,

$$\frac{\partial x_B}{\partial\theta_2} = l\cos\theta_2 \qquad (\delta x_B)_{\theta_2} = l\cos\theta_2\,\delta\theta_2$$

Substituting into Eq. 1, we have

$$(Pl\cos\theta_2 - M)\,\delta\theta_2 = 0$$

Since $\delta\theta_2 \neq 0$, then

$$\theta_2 = \cos^{-1}\left(\frac{2M}{Pl}\right) \qquad \text{Ans.}$$

Using Eq. 3 to obtain the variation of x_B with θ_1 yields

$$\frac{\partial x_B}{\partial\theta_1} = l\cos\theta_1 \qquad (\delta x_B)_{\theta_1} = l\cos\theta_1\,\delta\theta_1$$

Substituting into Eq. 2, we have

$$(Pl\cos\theta_1 - M)\,\delta\theta_1 = 0$$

Since $\delta\theta_1 \neq 0$, then

$$\theta_1 = \cos^{-1}\left(\frac{2M}{Pl}\right) \qquad \text{Ans.}$$

PROBLEMS

11–1. The crankshaft is subjected to a torque of $M = 50 \text{ N} \cdot \text{m}$. Determine the horizontal compressive force F applied to the piston for equilibrium when $\theta = 60°$.

11–2. The crankshaft is subjected to a torque of $M = 50 \text{ N} \cdot \text{m}$. Determine the horizontal compressive force F and plot the result of F (ordinate) versus θ (abscissa) for $0° \le \theta \le 90°$.

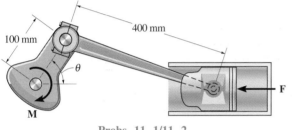

Probs. 11–1/11–2

11–3. The toggle joint is subjected to the load **P**. Determine the compressive force F it creates on the cylinder at A as a function of θ.

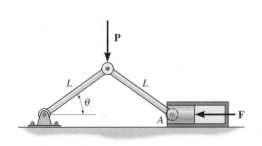

Prob. 11–3

*11–4. The pin-connected mechanism is constrained at A by a pin and at B by a roller. If $P = 200 \text{ N}$, determine the angle θ for equilibrium.

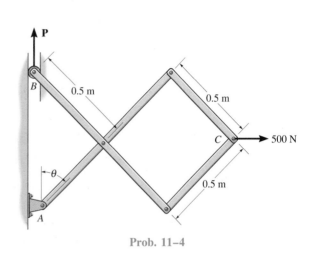

Prob. 11–4

■11–5. Each member of the pin-connected mechanism has a mass of 8 kg. If the spring is unstretched when $\theta = 0°$, determine the angle θ for equilibrium. Set $k = 2500 \text{ N/m}$ and $M = 50 \text{ N} \cdot \text{m}$.

11–6. Each member of the pin-connected mechanism has a mass of 8 kg. If the spring is unstretched when $\theta = 0°$, determine the required stiffness k so that the mechanism is in equilibrium when $\theta = 30°$. Set $M = 0$.

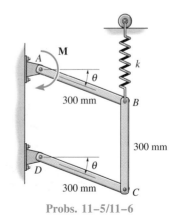

Probs. 11–5/11–6

11–7. The thin rod of weight W rests against the smooth wall and floor. Determine the magnitude of the couple moment **M** needed to hold it in equilibrium for a given angle θ.

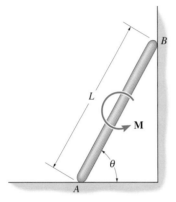

Prob. 11–7

■11–10. The 4-m members of the mechanism are pin-connected at their centers. If vertical forces act at C and E as shown, determine the angle θ for equilibrium. The spring is unstretched when $\theta = 45°$. Neglect the weight of the members. Take $P_1 = 40$ N, $P_2 = 20$ N.

■11–11. The 4-m members of the mechanism are pin-connected at their centers. If vertical forces $P_1 = P_2 = 30$ N act at C and E as shown, determine the angle θ for equilibrium. The spring is unstretched when $\theta = 45°$. Neglect the weight of the members.

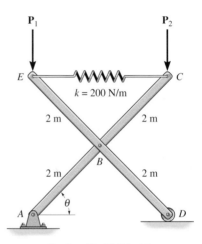

Probs. 11–10/11–11

***■11–8.** Determine the angle θ for equilibrium of the uniform 7-kg bar AC. Due to the roller guide at B, the spring remains vertical and is unstretched when $\theta = 0°$. Set $k = 200$ N/m.

11–9. Determine the required stiffness k so that the uniform 7-kg bar AC is in equilibrium when $\theta = 30°$. Due to the roller guide at B, the spring remains vertical and is unstretched when $\theta = 0°$.

***11–12.** If each of the three links of the mechanism has a weight of 20 N, determine the angle θ for equilibrium of the spring, which, due to the roller guide, always remains horizontal and is unstretched when $\theta = 0°$.

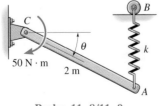

Probs. 11–8/11–9

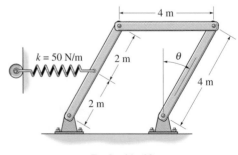

Prob. 11–12

11–13. If each of the three links of the mechanism has a weight of 20 N, determine the angle θ for equilibrium. The spring, which always remains vertical due to the roller guide, is unstretched when $\theta = 0°$. Set **P = 0**.

11–15. The two-bar linkage is subjected to a couple moment $M = 100$ N $\cdot$ m and vertical force $P = 500$ N. Determine the angle θ for equilibrium. The spring is unstretched when $\theta = 45°$. Neglect the mass of each bar.

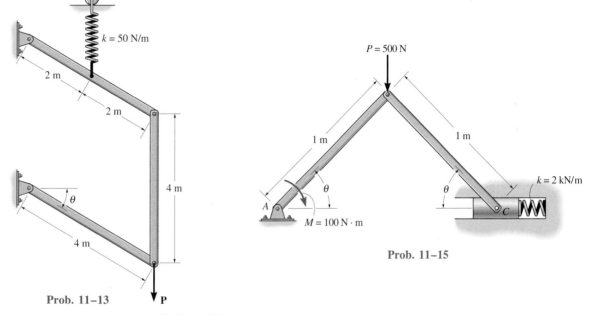

Prob. 11–15

11–14. The spring is unstretched when $\theta = 0°$. If $P = 8$ N, determine the angle θ for equilibrium. Due to the roller guide, the spring always remains vertical. Neglect the weight of the links.

***11–16.** The two-bar linkage is subjected to a couple moment $M = 100$ N $\cdot$ m and vertical force $P = 500$ N. Also, each bar is uniform and has a mass of 10 kg. Determine the angle θ for equilibrium. The spring is unstretched when $\theta = 45°$.

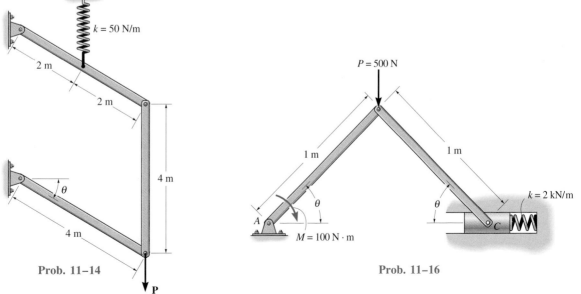

Prob. 11–13

Prob. 11–14

Prob. 11–16

11–17. Determine the force F needed to lift the bucket having a weight of 100 N. *Hint:* Note that the coordinates s_A and s_B can be related to the *constant* vertical length l of the cord.

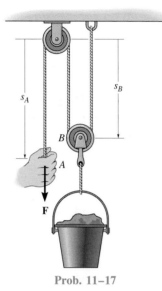

Prob. 11–17

11–18. Determine the force F needed to lift the block having a weight of 100 N. *Hint:* Note that the coordinates s_A and s_B can be related to the *constant* vertical length l of the cord.

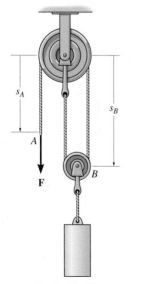

Prob. 11–18

11–19. The assembly is used for exercise. It consists of four pin-connected bars, each of length L, and a spring of stiffness k and unstretched length a ($<2L$). If horizontal forces **P** and $-$**P** are applied to the handles so that θ is slowly decreased, determine the angle θ at which the magnitude of **P** becomes a maximum.

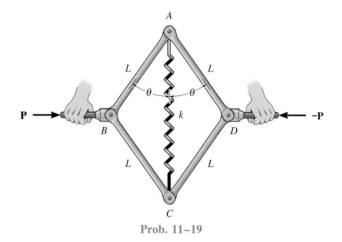

Prob. 11–19

*11–20.** The scissors jack supports a load **P**. Determine the axial force in the screw necessary for equilibrium when the jack is in the position θ. Each of the four links has a length L and is pin-connected at its center. Points B and D can move horizontally.

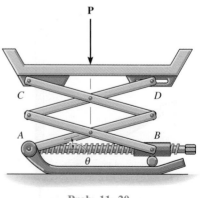

Prob. 11–20

11–21. Determine the horizontal force F required to maintain equilibrium of the slider mechanism when $\theta = 60°$. Set $M = 6$ N · m.

11–22. Determine the moment M that must be applied to the slider mechanism in order to maintain the equilibrium position $\theta = 60°$ when the horizontal force $\mathbf{F} = 100$ N is applied at D.

11–23. Solve Prob. 11-22 if the force $\mathbf{F}$ acts vertically upward.

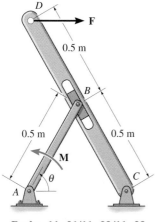

Probs. 11–21/11–22/11–23

***11–24.** Rods AB and BC have a center of mass located at their midpoints. If all contacting surfaces are smooth and BC has a mass of 100 kg, determine the appropriate mass of AB required for equilibrium.

11–25. Determine the mass of A and B required to hold the 400-g desk lamp in balance for any angles θ and ϕ. Neglect the weight of the mechanism and the size of the lamp.

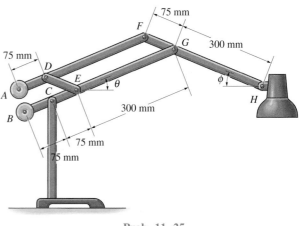

Prob. 11–25

11–26. If a force $P = 40$ N is applied perpendicular to the handle of the toggle press, determine the vertical compressive force developed at C; $\theta = 30°$.

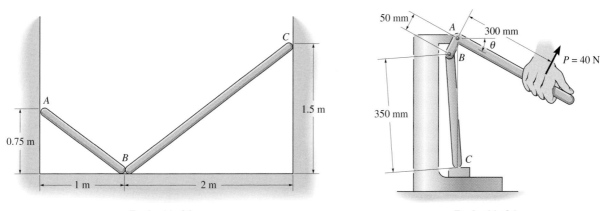

Prob. 11–24

Prob. 11–26

*11.4 Conservative Forces

The work done by a force when it undergoes a *differential displacement* has been defined as $dU = F \cos \theta \, ds$. If the force is displaced over a path that has a *finite length s,* the work is determined by integrating over the path; i.e.,

$$U = \int_s F \cos \theta \, ds$$

To evaluate the integral, it is necessary to obtain a relationship between F and the component of displacement $ds \cos \theta$. In some instances, however, the work done by a force will be *independent* of its path and, instead, will depend only on the initial and final locations of the force along the path. A force that has this property is called a *conservative force.*

Weight. Consider the body in Fig. 11–9, which is initially at P'. If the body is moved *down* along the *arbitrary path A* to the dashed position, then, for a given displacement ds along the path, the displacement component in the direction of **W** has a magnitude of $dy = ds \cos \theta$, as shown. Since both the force and displacement are in the same direction, the work is positive; hence,

$$U = \int_s W \cos \theta \, ds = \int_0^y W \, dy$$

or

$$U = Wy$$

In a similar manner, the work done by the weight when the body moves up a distance y back to P', along the arbitrary path A', is

$$U = -Wy$$

Why is the work negative?

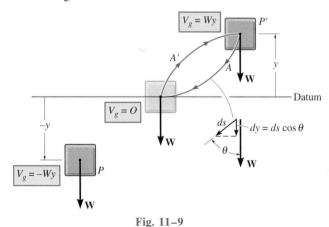

Fig. 11–9

The weight of a body is therefore a conservative force, since the work done by the weight depends *only* on the body's *vertical displacement* and is independent of the path along which the body moves.

Elastic Spring. The force developed by an elastic spring ($F_s = ks$) is also a conservative force. If the spring is attached to a body and the body is displaced along *any path,* such that it causes the spring to elongate or compress from a position s_1 to a further position s_2, the work will be negative, since the spring exerts a force $\mathbf{F}_s$ *on the body* that is opposite to the body's displacement ds, Fig. 11–10. For either extension or compression, the work is independent of the path and is simply

$$U = \int_{s_1}^{s_2} F_s \, ds = \int_{s_1}^{s_2} (-ks) \, ds$$
$$= -(\tfrac{1}{2}ks_2^2 - \tfrac{1}{2}ks_1^2)$$

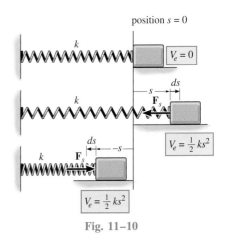

Fig. 11–10

Friction. In contrast to a conservative force, consider the force of *friction* exerted on a moving body by a fixed surface. The work done by the frictional force depends on the path; the longer the path, the greater the work. Consequently, frictional forces are *nonconservative,* and the work done is dissipated from the body in the form of heat.

*11.5 Potential Energy

When a conservative force acts on a body, it gives the body the capacity to do work. This capacity, measured as *potential energy,* depends on the location of the body.

Gravitational Potential Energy. If a body is located a distance y *above* a fixed horizontal reference or datum, Fig. 11–9, the weight of the body has *positive* gravitational potential energy V_g since $\mathbf{W}$ has the capacity of doing positive work when the body is moved back down to the datum. Likewise, if the body is located a distance y *below* the datum, V_g is *negative* since the weight does negative work when the body is moved back up to the datum. At the datum, $V_g = 0$.

Measuring y as *positive upward,* the gravitational potential energy of the body's weight $\mathbf{W}$ is thus

$$V_g = Wy \tag{11–4}$$

Elastic Potential Energy. The elastic potential energy V_e that a spring produces on an attached body, when the spring is elongated or compressed from an undeformed position ($s = 0$) to a final position s, is

$$V_e = \tfrac{1}{2}ks^2 \tag{11–5}$$

Here V_e is *always positive*, since in the deformed position the spring has the capacity of doing *positive work* in *returning* the body back to the spring's undeformed position, Fig. 11–10.

Potential Function. In the general case, if a body is subjected to *both* gravitational and elastic forces, the *potential energy or potential function V* of the body can be expressed as the algebraic sum

$$V = V_g + V_e \tag{11–6}$$

where measurement of V depends on the location of the body with respect to a selected datum in accordance with Eqs. 11–4 and 11–5.

In general, if a system of frictionless connected rigid bodies has a *single degree of freedom* such that its position from the datum is defined by the independent coordinate q, then the potential function for the system can be expressed as $V = V(q)$. The work done by all the conservative forces acting on the system in moving it from q_1 to q_2 is measured by the *difference* in V; i.e.,

$$U_{1-2} = V(q_1) - V(q_2) \tag{11–7}$$

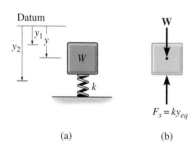

Datum

Fig. 11–11

For example, the potential function for a system consisting of a block of weight **W** supported by a spring, Fig. 11–11a, can be expressed in terms of its independent coordinate ($q =$) y, measured from a fixed datum located at the unstretched length of the spring; we have

$$\begin{aligned} V &= V_g + V_e \\ &= -Wy + \tfrac{1}{2}ky^2 \end{aligned} \tag{11–8}$$

If the block moves from y_1 to a farther downward position y_2, then the work of **W** and $\mathbf{F}_s$ is

$$U_{1-2} = V(y_1) - V(y_2) = -W[y_1 - y_2] + \tfrac{1}{2}ky_1^2 - \tfrac{1}{2}ky_2^2$$

*11.6 Potential-Energy Criterion for Equilibrium

System Having One Degree of Freedom. When the displacement of a frictionless connected system is *infinitesimal*, i.e., from q to $q + dq$, Eq. 11–7 becomes

$$dU = V(q) - V(q + dq)$$

or

$$dU = -dV$$

Furthermore, if the system undergoes a *virtual displacement* δq, rather than an actual displacement dq, then $\delta U = -\delta V$. For equilibrium, the principle of virtual work requires that $\delta U = 0$ and therefore, provided the potential function for the system is known, this also requires that $\delta V = 0$. We can also express this requirement as

$$\frac{dV}{dq} = 0 \qquad (11\text{–}9)$$

Hence, *when a frictionless connected system of rigid bodies is in equilibrium, the first variation or change in V is zero.* This change is determined by taking the *first derivative* of the potential function and setting it equal to zero. For example, using Eq. 11–8 to determine the equilibrium position for the spring and block in Fig. 11–11a, we have

$$\frac{dV}{dy} = W - ky = 0$$

Hence, the equilibrium position $y = y_{eq}$ is

$$y_{eq} = \frac{W}{k}$$

Of course, the *same result* is obtained by applying $\Sigma F_y = 0$ to the forces acting on the free-body diagram of the block, Fig. 11–11b.

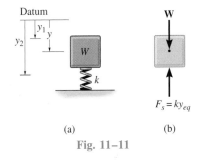

(a)

(b)

Fig. 11–11

System Having *n* Degrees of Freedom.

When the system of connected bodies has *n* degrees of freedom, the total potential energy stored in the system will be a function of *n* independent coordinates q_n; i.e., $V = V(q_1, q_2, \ldots, q_n)$. In order to apply the equilibrium criterion $\delta V = 0$, it is necessary to determine the change in potential energy δV by using the "chain rule" of differential calculus; i.e.,

$$\delta V = \frac{\partial V}{\partial q_1}\,\delta q_1 + \frac{\partial V}{\partial q_2}\,\delta q_2 + \cdots + \frac{\partial V}{\partial q_n}\,\delta q_n = 0$$

Since the virtual displacements $\delta q_1, \delta q_2, \ldots, \delta q_n$ are independent of one another, the equation is satisfied provided

$$\frac{\partial V}{\partial q_1} = 0, \quad \frac{\partial V}{\partial q_2} = 0, \quad \ldots, \quad \frac{\partial V}{\partial q_n} = 0$$

Hence *it is possible to write n independent equations for a system having n degrees of freedom.*

*11.7   Stability of Equilibrium

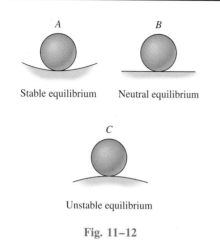

Stable equilibrium   Neutral equilibrium

Unstable equilibrium

Fig. 11–12

Once the equilibrium configuration for a body or system of connected bodies is defined, it is sometimes important to investigate the "type" of equilibrium or the stability of the configuration. For example, consider the position of a ball resting at a point on each of the three paths shown in Fig. 11–12. Each situation represents an equilibrium state for the ball. When the ball is at A, it is said to be in *stable equilibrium* because if it is given a small displacement up the hill, it will always *return* to its original, lowest, position. At A, its total potential energy is a *minimum*. When the ball is at B, it is in *neutral equilibrium*. A small displacement to either the left or right of B will not alter this condition. The ball *remains* in equilibrium in the displaced position, and therefore its potential energy is *constant*. When the ball is at C, it is in *unstable equilibrium*. Here a small displacement will cause the ball's potential energy to be *decreased*, and so it will roll farther *away* from its original, highest position. At C, the potential energy of the ball is a *maximum*.

Type of Equilibrium. The example just presented illustrates that one of three types of equilibrium positions can be specified for a body or system of connected bodies.

1. *Stable equilibrium* occurs when a small displacement of the system causes the system to return to its original position. In this case the original potential energy of the system is a minimum.
2. *Neutral equilibrium* occurs when a small displacement of the system causes the system to remain in its displaced state. In this case the potential energy of the system remains constant.
3. *Unstable equilibrium* occurs when a small displacement of the system causes the system to move farther away from its original position. In this case the original potential energy of the system is a maximum.

System Having One Degree of Freedom. For *equilibrium* of a system having a single degree of freedom, defined by the independent coordinate q, it has been shown that the first derivative of the potential function for the system must be equal to zero; i.e., $dV/dq = 0$. If the potential function $V = V(q)$ is plotted, Fig. 11–13, the first derivative (equilibrium position) is represented as the slope dV/dq, which is zero when the function is maximum, minimum, or an inflection point.

If the *stability* of a body at the equilibrium position is to be investigated, it is necessary to determine the *second derivative* of V and evaluate it at the equilibrium position $q = q_{eq}$. As shown in Fig. 11–13a, if $V = V(q)$ is a *minimum*, then

$$\frac{dV}{dq} = 0, \quad \frac{d^2V}{dq^2} > 0 \quad \text{stable equilibrium} \qquad (11\text{–}10)$$

If $V = V(q)$ is a *maximum,* Fig. 11–13b, then

$$\frac{dV}{dq} = 0, \quad \frac{d^2V}{dq^2} < 0 \quad \text{unstable equilibrium} \qquad (11\text{–}11)$$

If the second derivative is zero, it will be necessary to investigate *higher-order* derivatives to determine the stability. In particular, stable equilibrium will occur if the order of the lowest remaining nonzero derivative is *even* and the sign of this nonzero derivative is positive when it is evaluated at $q = q_{eq}$; otherwise, it is unstable.

If the system is in neutral equilibrium, Fig. 11–13c, it is required that

$$\frac{dV}{dq} = \frac{d^2V}{dq^2} = \frac{d^3V}{dq^3} = \cdots = 0 \quad \text{neutral equilibrium} \qquad (11\text{–}12)$$

since then V must be constant at and around the "neighborhood" of q_{eq}.

System Having Two Degrees of Freedom.

A criterion for investigating stability becomes increasingly complex as the number of degrees of freedom for the system increases. For a system having two degrees of freedom, defined by independent coordinates (q_1, q_2), it may be verified (using the calculus of functions of two variables) that equilibrium and stability occur at a point (q_{1eq}, q_{2eq}) when

$$\frac{\partial V}{\partial q_1} = \frac{\partial V}{\partial q_2} = 0$$

$$\left[\left(\frac{\partial^2 V}{\partial q_1\, \partial q_2} \right)^2 - \left(\frac{\partial^2 V}{\partial q_1^2} \right) \left(\frac{\partial^2 V}{\partial q_2^2} \right) \right] < 0$$

$$\left(\frac{\partial^2 V}{\partial q_1^2} + \frac{\partial^2 V}{\partial q_2^2} \right) > 0$$

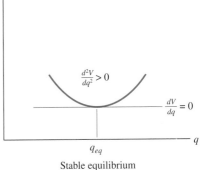

$$\frac{d^2V}{dq^2} > 0$$
$$\frac{dV}{dq} = 0$$

q_{eq}

Stable equilibrium

(a)

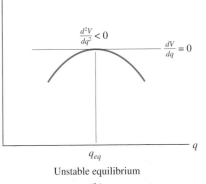

$$\frac{d^2V}{dq^2} < 0$$
$$\frac{dV}{dq} = 0$$

q_{eq}

Unstable equilibrium

(b)

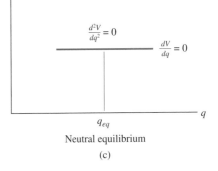

$$\frac{d^2V}{dq^2} = 0$$
$$\frac{dV}{dq} = 0$$

q_{eq}

Neutral equilibrium

(c)

Fig. 11–13

Both equilibrium and instability occur when

$$\frac{\partial V}{\partial q_1} = \frac{\partial V}{\partial q_2} = 0$$

$$\left[\left(\frac{\partial^2 V}{\partial q_1 \, \partial q_2}\right)^2 - \left(\frac{\partial^2 V}{\partial q_1^2}\right)\left(\frac{\partial^2 V}{\partial q_2^2}\right)\right] < 0$$

$$\left(\frac{\partial^2 V}{\partial q_1^2} + \frac{\partial^2 V}{\partial q_2^2}\right) < 0$$

PROCEDURE FOR ANALYSIS

Using potential-energy methods, the equilibrium positions and the stability of a body or a system of connected bodies having a single degree of freedom can be obtained by applying the following procedure.

Potential Function.   Formulate the potential function $V = V_g + V_e$ for the system. To do this, sketch the system so that it is located at some *arbitrary position* specified by the independent coordinate q. A horizontal *datum* is established through a *fixed point,** and the *gravitational potential energy* V_g is expressed in terms of the weight W of each member and its vertical distance y from the datum, $V_g = Wy$, Eq. 11–4. The elastic potential energy V_e of the system is expressed in terms of the stretch or compression, s, of any connecting spring and the spring's stiffness k, $V_e = \frac{1}{2}ks^2$, Eq. 11–5. Once V has been established, express the *position coordinates* y and s in terms of the independent coordinate q.

Equilibrium Position.   The equilibrium position is determined by taking the first derivative of V and setting it equal to zero, $\delta V = 0$, Eq. 11–9.

Stability.   Stability at the equilibrium position is determined by evaluating the second or higher-order derivatives of V as indicated by Eqs. 11–10 to 11–12.

   *The location of the datum is *arbitrary* since only the *changes* or differentials of V are required for investigation of the equilibrium position and its stability.

The following examples illustrate this procedure numerically.

Example 11–5

The uniform link shown in Fig. 11–14a has a mass of 10 kg. The spring is unstretched when $\theta = 0°$. Determine the angle θ for equilibrium and investigate the stability at the equilibrium position.

SOLUTION

Potential Function.   The datum is established at the top of the link when the *spring is unstretched*, Fig. 11–14b. When the link is located at the arbitrary position θ, the spring increases its potential energy by stretching and the weight decreases its potential energy. Hence,

$$V = V_e + V_g = \frac{1}{2}ks^2 - W\left(s + \frac{l}{2}\cos\theta - \frac{l}{2}\right)$$

Since $l = s + l\cos\theta$ or $s = l(1 - \cos\theta)$, then

$$V = \frac{1}{2}kl^2(1 - \cos\theta)^2 - \frac{Wl}{2}(1 - \cos\theta)$$

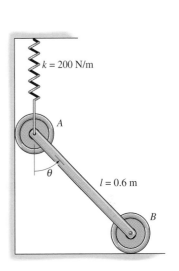

(a)

Equilibrium Position.   The first derivative of V gives

$$\frac{dV}{d\theta} = kl^2(1 - \cos\theta)\sin\theta - \frac{Wl}{2}\sin\theta = 0$$

or

$$l\left[kl(1 - \cos\theta) - \frac{W}{2}\right]\sin\theta = 0$$

This equation is satisfied provided

$$\sin\theta = 0 \qquad \theta = 0° \qquad\qquad Ans.$$

$$\theta = \cos^{-1}\left(1 - \frac{W}{2kl}\right) = \cos^{-1}\left[1 - \frac{10(9.81)}{2(200)(0.6)}\right] = 53.8° \quad Ans.$$

Stability.   Determining the second derivative of V gives

$$\frac{d^2V}{d\theta^2} = kl^2(1 - \cos\theta)\cos\theta + kl^2\sin\theta\sin\theta - \frac{Wl}{2}\cos\theta$$

$$= kl^2(\cos\theta - \cos 2\theta) - \frac{Wl}{2}\cos\theta$$

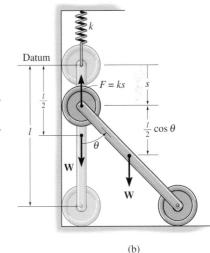

(b)

Fig. 11–14

Substituting values for the constants, with $\theta = 0°$ and $\theta = 53.8°$, yields

$$\frac{d^2V}{d\theta^2}\bigg|_{\theta=0°} = 200(0.6)^2(\cos 0° - \cos 0°) - \frac{10(9.81)(0.6)}{2}\cos 0°$$

$$= -29.4 < 0 \qquad \text{(unstable equilibrium at } \theta = 0°) \qquad Ans.$$

$$\frac{d^2V}{d\theta^2}\bigg|_{\theta=53.8°} = 200(0.6)^2(\cos 53.8° - \cos 107.6°) - \frac{10(9.81)(0.6)}{2}\cos 53.8°$$

$$= 46.9 > 0 \qquad \text{(stable equilibrium at } \theta = 53.8°) \qquad Ans.$$

Example 11–6

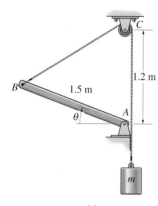

(a)

Fig. 11–15

Determine the mass m of the block required for equilibrium of the uniform 10-kg rod shown in Fig. 11–15a when $\theta = 20°$. Investigate the stability at the equilibrium position.

SOLUTION

Potential Function.   The datum is established through point A, Fig. 11–15b. When $\theta = 0°$, the block is assumed to be suspended $(y_W)_1$ below the datum. Hence, in the position θ,

$$V = V_e + V_g = 98.1\left(\frac{1.5 \sin \theta}{2}\right) - m(9.81)(\Delta y) \qquad (1)$$

The distance $\Delta y = (y_W)_2 - (y_W)_1$ may be related to the independent coordinate θ by measuring the difference in cord lengths $B'C$ and BC. Since

$$B'C = \sqrt{(1.5)^2 + (1.2)^2} = 1.92$$
$$BC = \sqrt{(1.5 \cos \theta)^2 + (1.2 - 1.5 \sin \theta)^2} = \sqrt{3.69 - 3.60 \sin \theta}$$

then

$$\Delta y = B'C - BC = 1.92 - \sqrt{3.69 - 3.60 \sin \theta}$$

Substituting the above result into Eq. 1 yields

$$V = 98.1\left(\frac{1.5 \sin \theta}{2}\right) - m(98.1)(1.92 - \sqrt{3.69 - 3.60 \sin \theta}) \qquad (2)$$

Equilibrium Position

$$\frac{dV}{d\theta} = 73.6 \cos \theta - \left[\frac{m(98.1)}{2}\right]\left(\frac{3.60 \cos \theta}{\sqrt{3.69 - 3.60 \sin \theta}}\right) = 0$$

$$\frac{dV}{d\theta}\bigg|_{\theta=20°} = 69.16 - 10.58m = 0$$

$$m = \frac{69.16}{10.58} = 6.54 \text{ kg} \qquad \textit{Ans.}$$

Stability.   Taking the second derivative of Eq. 2, we obtain

$$\frac{d^2V}{d\theta^2} = -73.6 \sin \theta - \left[\frac{m(9.81)}{2}\right]\left(\frac{-1}{2}\right)\frac{(3.60 \cos \theta)^2}{(3.69 - 3.60 \sin \theta)^{3/2}}$$

$$- \frac{m(9.81)}{2}\left(\frac{-3.60 \sin \theta}{\sqrt{3.69 - 3.60 \sin \theta}}\right)$$

For the equilibrium position $\theta = 20°$, also $m = 6.54$ kg, so

$$\frac{d^2V}{d\theta^2} = 47.6 > 0 \qquad \text{(stable equilibrium at } \theta = 20°) \qquad \textit{Ans.}$$

Example 11–7

The homogeneous block having a mass m rests on the top surface of the cylinder, Fig. 11–16a. Show that this is a condition of unstable equilibrium if $h > 2R$.

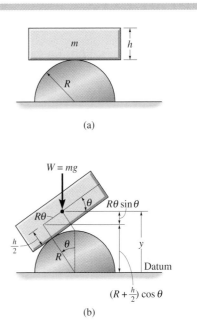

(a)

SOLUTION

Potential Function.   The datum is established at the base of the cylinder, Fig. 11–16b. If the block is displaced by an amount θ from the equilibrium position, the potential function may be written in the form

$$V = V_e + V_g$$
$$= 0 + mgy$$

From Fig. 11–16b,

$$y = \left(R + \frac{h}{2}\right)\cos\theta + R\theta\sin\theta$$

Thus,

$$V = mg\left[\left(R + \frac{h}{2}\right)\cos\theta + R\theta\sin\theta\right]$$

(b)

Fig. 11–16

Equilibrium Position

$$\frac{dV}{d\theta} = mg\left[-\left(R + \frac{h}{2}\right)\sin\theta + R\sin\theta + R\theta\cos\theta\right] = 0$$

$$= mg\left(-\frac{h}{2}\sin\theta + R\theta\cos\theta\right) = 0$$

Obviously, $\theta = 0°$ is the equilibrium position that satisfies this equation.

Stability.   Taking the second derivative of V yields

$$\frac{d^2V}{d\theta^2} = mg\left(-\frac{h}{2}\cos\theta + R\cos\theta - R\theta\sin\theta\right)$$

At $\theta = 0°$,

$$\left.\frac{d^2V}{d\theta^2}\right|_{\theta=0°} = -mg\left(\frac{h}{2} - R\right)$$

Since all the constants are positive, the block is in unstable equilibrium if $h > 2R$, for then $d^2V/d\theta^2 < 0$.

PROBLEMS

11–27. If the potential function for a conservative one-degree-of-freedom system is $V = (8x^3 - 2x^2 - 10)$ J, where x is given in meters, determine the positions for equilibrium and investigate the stability at each of these positions.

*11–28. If the potential function for a conservative one-degree-of-freedom system is $V = (12 \sin 2\theta + 15 \cos \theta)$ J, where $0° < \theta < 180°$, determine the positions for equilibrium and investigate the stability at each of these positions.

11–29. If the potential function for a conservative one-degree-of-freedom system is $V = (10 \cos 2\theta + 25 \sin \theta)$ J, where $0° < \theta < 180°$, determine the positions for equilibrium and investigate the stability at each of these positions.

11–30. If the potential function for a conservative two-degree-of-freedom system is $V = (9y^2 + 18x^2)$ J, where x and y are given in meters, determine the equilibrium position and investigate the stability at this position.

11–31. Solve Prob. 11–6 using the principle of potential energy, and investigate the stability at the equilibrium position.

*11–32. Solve Prob. 11–12 using the principle of potential energy, and investigate the stability at the equilibrium position.

11–33. Solve Prob. 11–13 using the principle of potential energy, and investigate the stability at the equilibrium position.

11–34. Solve Prob. 11–24 using the principle of potential energy.

■11–35. The two bars each have a weight of 8 N. Determine the angle θ for equilibrium and investigate the stability at the equilibrium position. The spring has an unstretched length of 1 m. Take $k = 30$ N/m.

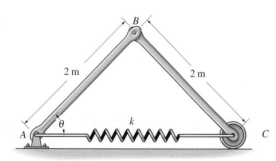

Prob. 11–35

*11–36. The two bars each have a weight of 8 N. Determine the required stiffness k of the spring so that the two bars are in neutral equilibrium when $\theta = 30°$. The spring has an unstretched length of 1 m.

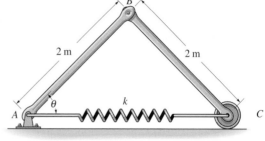

Prob. 11–36

11–37. The uniform beam has a weight W. If the contacting surfaces are smooth, determine the angle θ for equilibrium. The spring is uncompressed when $\theta = 90°$.

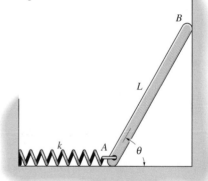

Prob. 11–37

11–38. The uniform beam has a weight W. If the contacting surfaces are smooth, determine the angle θ for equilibrium. The spring is unstretched when $\theta = 0°$.

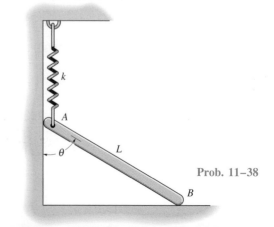

Prob. 11–38

11–39.   The bar supports a weight of $W = 500$ N at its end. If the springs are originally unstretched when the bar is vertical, and $k_1 = 300$ N/m, $k_2 = 500$ N/m, investigate the stability of the bar when it is in the vertical position.

*11–40.   The bar supports a weight of $W = 500$ N at its end. If the springs are originally unstretched when the bar is vertical, determine the required stiffness $k_1 = k_2 = k$ of the springs so that the bar is in neutral equilibrium when it is vertical.

■11–42.   The spring has a stiffness $k = 400$ N/m and an unstretched length of 0.3 m. Determine the angle θ for equilibrium if the uniform links each have a mass of 5 kg.

11–43.   The spring has an unstretched length of 0.3 m. Determine its stiffness k so that it is in equilibrium when $\theta = 20°$. Each uniform link has a mass of 5 kg. Investigate the stability at this position.

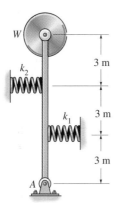

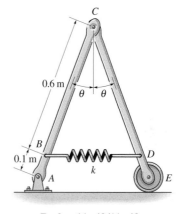

Probs. 11–42/11–43

Probs. 11–39/11–40

11–41.   The rod supports a disk B having a weight of 800 N. It is held in equilibrium using four springs for which $k = 400$ N/m. Determine the minimum distance d between the springs so that the rod remains in stable equilibrium when it is vertical.

*11–44.   Determine the angle θ for equilibrium and investigate the stability at this position. The bars each have a mass of 3 kg and the suspended block D has a mass of 7 kg. Cord DC has a total length of 1 m.

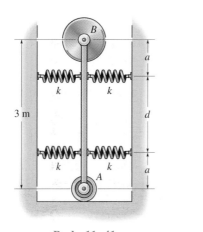

Prob. 11–41

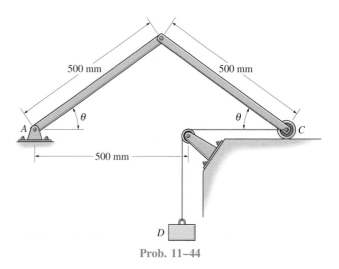

Prob. 11–44

11–45. The door has a uniform weight of 50 N. It is hinged at A and is held open by the 30-N weight and the pulley. Determine the angle θ for equilibrium.

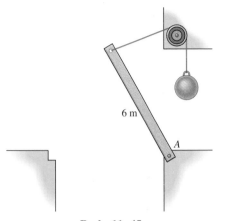

Prob. 11–45

11–46. The spring of the scale has an unstretched length a. Determine the angle θ for equilibrium when a weight W is supported on the platform. Neglect the weights of the members. What value of W would be required to keep the scale in neutral equilibrium when $\theta = 0°$?

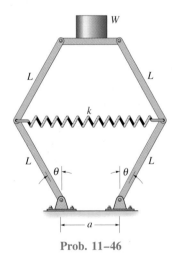

Prob. 11–46

11–47. The block weighs W and is supported by links AB and BC. Determine the necessary spring stiffness k required to hold the system in neutral equilibrium. The springs are subjected to an initial compression F_0 when the links are vertical as shown.

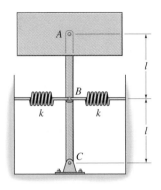

Prob. 11–47

∗11–48. The Roberval balance is in equilibrium when no weights are placed on the pans A and B. If two masses m_A and m_B are placed at *any* location a and b on the pans, show that neutral equilibrium is maintained if $m_A d_A = m_B d_B$.

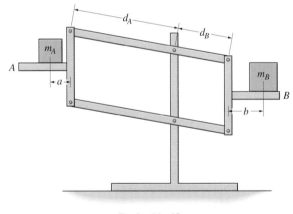

Prob. 11–48

11–49. The cup has a hemispherical bottom and a mass m. Determine the position h of the center of mass G so that the cup is in neutral equilibrium.

Prob. 11–49

11–50. The homogeneous cylinder has a conical cavity cut into its base as shown. Determine the depth d of the cavity so that the cylinder balances on the pivot and remains in neutral equilibrium.

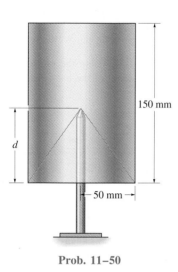

Prob. 11–50

11–51. The uniform right circular cone having a mass m is suspended from the cord as shown. Determine the angle θ at which it hangs from the wall for equilibrium. Is the cone in stable equilibrium?

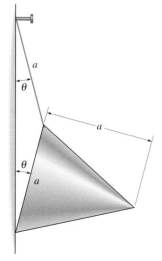

Prob. 11–51

*11–52.** The triangular block of weight W rests on the smooth corners which are a distance a apart. If the block has three equal sides of length d, determine the angle θ for equilibrium.

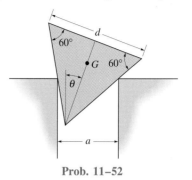

Prob. 11–52

11–53. Two uniform bars, each having a weight W, are pin-connected at their ends. If they are placed over a smooth cylindrical surface, show that the angle θ for equilibrium must satisfy the equation $\cos \theta / \sin^3 \theta = a/2r$.

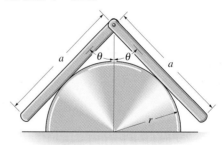

Prob. 11–53

11–54. The homogeneous block has a mass of 10 kg and rests on the smooth corners of two ledges. Determine the angle θ for placement that will cause the block to be stable.

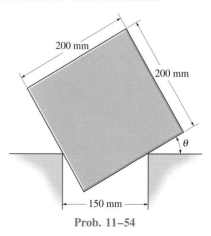

Prob. 11–54

REVIEW PROBLEMS

11–55. The three-bar mechanism of negligible weight is subjected to a couple moment $M_A = 8 \text{ N} \cdot \text{m}$. Determine the magnitude of the couple moment $\mathbf{M}_D$ needed to maintain the equilibrium position $\theta = 30°$, $\phi = 90°$.

11–57. The uniform bar AB weighs 10 N. If the attached spring is unstretched when $\theta = 90°$, use the method of virtual work and determine the angle θ for equilibrium. Note that the spring always remains in the vertical position due to the roller guide.

11–58. Solve Prob. 11–57 using the principle of potential energy. Investigate the stability of the bar when it is in the equilibrium position.

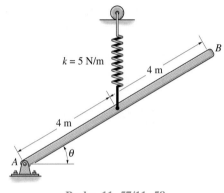

Probs. 11–57/11–58

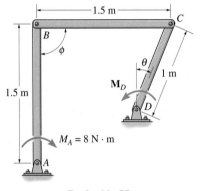

Prob. 11–55

11–59. The uniform rod AB has a weight of 10 N. If the spring DC is unstretched when $\theta = 90°$, determine the angle θ for equilibrium using the principle of virtual work. The spring is always in the horizontal position because of the roller guide at D.

***11–56.** Compute the force developed in the spring required to keep the 6-kg rod in equilibrium when $\theta = 30°$. The spring remains horizontal due to the roller guide.

***11–60.** Solve Prob. 11–59 using the principle of potential energy. Investigate the stability of the rod when it is in the equilibrium position.

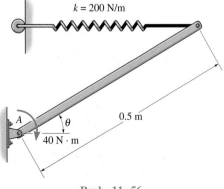

Prob. 11–56

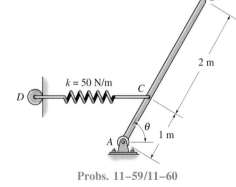

Probs. 11–59/11–60

11–61.   The chain puller is used to draw two ends of a chain together in order to attach the "master link." The device is operated by turning the screw S, which pushes the bar AB downward, thereby drawing the tips C and D towards one another. If the sliding contacts at A and B are smooth, determine the force F maintained by the screw at E, which for the position shown is required to develop a drawing tension of 1.2 kN in the chain.

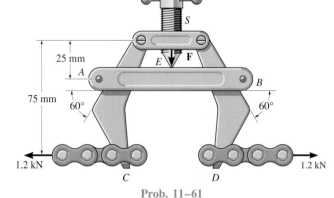

Prob. 11–61

11–62.   A disk of weight W is attached to the end of rod ABC. If the rod is supported by a smooth slider block at C and rod BD, determine the angle θ for equilibrium. Neglect the weight of the rods and the slider.

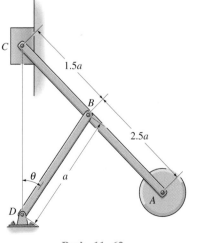

Prob. 11–62

11–63.   If $P = 45$ N, determine the angle θ for equilibrium of the pin-connected mechanism. The spring is unstretched when $\theta = 45°$. Neglect the weight of the members.

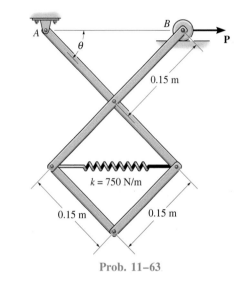

Prob. 11–63

*11–64.   Each member of the pin-connected mechanism has a mass of 8 kg. If the spring is unstretched when $\theta = 0°$, determine the angle θ for equilibrium.

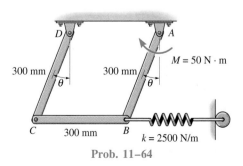

Prob. 11–64

A

Mathematical Expressions

Quadratic Formula

If $ax^2 + bx + c = 0$, then $x = \dfrac{-b \pm \sqrt{b^2 - 4ac}}{2a}$

Hyperbolic Functions

$$\sinh x = \frac{e^x - e^{-x}}{2}, \quad \cosh x = \frac{e^x + e^{-x}}{2}, \quad \tanh x = \frac{\sinh x}{\cosh x}$$

Trigonometric Identities

$$\sin \theta = \frac{A}{C}, \quad \csc \theta = \frac{C}{A}$$

$$\cos \theta = \frac{B}{C}, \quad \sec \theta = \frac{C}{B}$$

$$\tan \theta = \frac{A}{B}, \quad \cot \theta = \frac{B}{A}$$

$$\sin^2 \theta + \cos^2 \theta = 1$$

$$\sin (\theta \pm \phi) = \sin \theta \cos \phi \pm \cos \theta \sin \phi$$

$$\sin 2\theta = 2 \sin \theta \cos \theta$$

$$\cos (\theta \pm \phi) = \cos \theta \cos \phi \mp \sin \theta \sin \phi$$

$$\cos 2\theta = \cos^2 \theta - \sin^2 \theta$$

$$\cos \theta = \pm\sqrt{\frac{1 + \cos 2\theta}{2}}, \quad \sin \theta = \pm\sqrt{\frac{1 - \cos 2\theta}{2}}$$

$$\tan \theta = \frac{\sin \theta}{\cos \theta}$$

$$1 + \tan^2 \theta = \sec^2 \theta \qquad\qquad 1 + \cot^2 \theta = \csc^2 \theta$$

Power-Series Expansions

$$\sin x = x - \frac{x^3}{3!} + \cdots, \quad \cos x = 1 - \frac{x^2}{2!} + \cdots$$

$$\sinh x = x + \frac{x^3}{3!} + \cdots, \quad \cosh x = 1 + \frac{x^2}{2!} + \cdots$$

Derivatives

$$\frac{d}{dx}(u^n) = nu^{n-1}\frac{du}{dx} \qquad \frac{d}{dx}(\sin u) = \cos u \frac{du}{dx}$$

$$\frac{d}{dx}(uv) = u\frac{dv}{dx} + v\frac{du}{dx} \qquad \frac{d}{dx}(\cos u) = -\sin u \frac{du}{dx}$$

$$\frac{d}{dx}\left(\frac{u}{v}\right) = \frac{v\dfrac{du}{dx} - u\dfrac{dv}{dx}}{v^2} \qquad \frac{d}{dx}(\tan u) = \sec^2 u \frac{du}{dx}$$

$$\frac{d}{dx}(\cot u) = -\csc^2 u \frac{du}{dx} \qquad \frac{d}{dx}(\sinh u) = \cosh u \frac{du}{dx}$$

$$\frac{d}{dx}(\sec u) = \tan u \sec u \frac{du}{dx} \qquad \frac{d}{dx}(\cosh u) = \sinh u \frac{du}{dx}$$

$$\frac{d}{dx}(\csc u) = -\csc u \cot u \frac{du}{dx}$$

Integrals

$$\int x^n \, dx = \frac{x^{n+1}}{n+1} + C, \ n \neq -1$$

$$\int \frac{dx}{a+bx} = \frac{1}{b} \ln{(a+bx)} + C$$

$$\int \frac{dx}{a+bx^2} = \frac{1}{2\sqrt{-ba}} \ln{\left[\frac{\sqrt{a} + 2\sqrt{-b}}{\sqrt{a} - x\sqrt{-b}} \right]} + C,$$
$$a > 0, \ b < 0$$

$$\int \frac{x \, dx}{a+bx^2} = \frac{1}{2b} \ln{(bx^2 + a)} + C$$

$$\int \frac{x^2 \, dx}{a+bx^2} = \frac{x}{b} - \frac{a}{b\sqrt{ab}} \tan^{-1}{\frac{x\sqrt{ab}}{a}} + C$$

$$\int \frac{dx}{a^2 - x^2} = \frac{1}{2a} \ln{\left[\frac{a+x}{a-x} \right]} + C, \ a^2 > x^2$$

$$\int \sqrt{a+bx} \, dx = \frac{2}{3b} \sqrt{(a+bx)^3} + C$$

$$\int x\sqrt{a+bx} \, dx = \frac{-2(2a - 3bx)\sqrt{(a+bx)^3}}{15b^2} + C$$

$$\int x^2 \sqrt{a+bx} \, dx = $$
$$\frac{2(8a^2 - 12abx + 15b^2 x^2)\sqrt{(a+bx)^3}}{105b^3} + C$$

$$\int \sqrt{a^2 - x^2} \, dx = \frac{1}{2} \left[x\sqrt{a^2 - x^2} + a^2 \sin^{-1}{\frac{x}{a}} \right] + C,$$
$$a > 0$$

$$\int x\sqrt{a^2 - x^2} \, dx = -\frac{1}{3} \sqrt{(a^2 - x^2)^3} + C$$

$$\int x^2 \sqrt{a^2 - x^2} \, dx = -\frac{x}{4} \sqrt{(a^2 - x^2)^3}$$
$$+ \frac{a^2}{8} \left(x\sqrt{a^2 - x^2} + a^2 \sin^{-1}{\frac{x}{a}} \right) + C, \ a > 0$$

$$\int \sqrt{x^2 \pm a^2} \, dx = $$
$$\frac{1}{2} \left[x\sqrt{x^2 \pm a^2} \pm a^2 \ln{(x + \sqrt{x^2 \pm a^2})} \right] + C$$

$$\int x\sqrt{x^2 \pm a^2} \, dx = \frac{1}{3} \sqrt{(x^2 \pm a^2)^3} + C$$

$$\int x^2 \sqrt{x^2 \pm a^2} \, dx = \frac{x}{4} \sqrt{(x^2 \pm a^2)^3}$$
$$\pm \frac{a^2}{8} x\sqrt{x^2 \pm a^2} - \frac{a^4}{8} \ln{(x + \sqrt{x^2 \pm a^2})} + C$$

$$\int \frac{dx}{\sqrt{a+bx}} = \frac{2\sqrt{a+bx}}{b} + C$$

$$\int \frac{x \, dx}{\sqrt{x^2 \pm a^2}} = \sqrt{x^2 \pm a^2} + C$$

$$\int \frac{dx}{\sqrt{a+bx+cx^2}} = \frac{1}{\sqrt{c}} \ln{\left[\sqrt{a+bx+cx^2} + \right.}$$
$$\left. x\sqrt{c} + \frac{b}{2\sqrt{c}} \right] + C, \ c > 0$$
$$= \frac{1}{\sqrt{-c}} \sin^{-1}{\left(\frac{-2cx - b}{\sqrt{b^2 - 4ac}} \right)} + C, \ c < 0$$

$$\int \sin x \, dx = -\cos x + C$$

$$\int \cos x \, dx = \sin x + C$$

$$\int x \cos{(ax)} \, dx = \frac{1}{a^2} \cos{(ax)} + \frac{x}{a} \sin{(ax)} + C$$

$$\int x^2 \cos{(ax)} \, dx = \frac{2x}{a^2} \cos{(ax)} + \frac{a^2 x^2 - 2}{a^3} \sin{(ax)} + C$$

$$\int e^{ax} \, dx = \frac{1}{a} e^{ax} + C$$

$$\int x e^{ax} \, dx = \frac{e^{ax}}{a^2} (ax - 1) + C$$

$$\int \sinh x \, dx = \cosh x + C$$

$$\int \cosh x \, dx = \sinh x + C$$

B

Numerical and Computer Analysis

Occasionally the application of the laws of mechanics will lead to a system of equations for which a closed-form solution is difficult or impossible to obtain. When confronted with this situation, engineers will often use a numerical method which in most cases can be programmed on a microcomputer or "programmable" pocket calculator. Here we will briefly present a computer program for solving a set of linear algebraic equations and three numerical methods which can be used to solve an algebraic or transcendental equation, evaluate a definite integral, and solve an ordinary differential equation. Application of each method will be explained by example, and an associated computer program written in Microsoft BASIC, which is designed to run on most personal computers, is provided.* A text on numerical analysis should be consulted for further discussion regarding a check of the accuracy of each method and the inherent errors that can develop from the methods.

B.1 Linear Algebraic Equations

Application of the equations of static equilibrium or the equations of motion sometimes requires solving a set of linear algebraic equations. The computer program listed in Fig. B–1 can be used for this purpose. It is based on the method of a Gaussian elimination and can solve at most 10 equations with 10

*Similar types of programs can be written or purchased for programmable pocket calculators.

552

```
1 PRINT"Linear system of equations":PRINT    20 PRINT"Unknowns"                          39 NEXT I
2 DIM A(10,11)                                21 FOR I = 1 TO N                           40 FOR I = M+1 TO N
3 INPUT"Input number of equations : ",N       22 PRINT "X(";I;")=";A(I,N+1)               41 FC=A(I,M)/A(M,M)
4 PRINT                                        23 NEXT I                                   42 FOR J = M+1 TO N+1
5 PRINT"A  coefficients"                       24 END                                      43 A(I,J)=A(I,J)-FC*A(M,J)
6 FOR I = 1 TO N                               25 REM Subroutine Guassian                  44 NEXT J
7 FOR J = 1 TO N                               26 FOR M=1 TO N                             45 NEXT I
8 PRINT "A(";I;",";J;                          27 NP=M                                     46 NEXT M
9 INPUT")=",A(I,J)                             28 BG=ABS(A(M,M))                           47 A(N,N+1)=A(N,N+1)/A(N,N)
10 NEXT J                                      29 FOR I = M TO N                           48 FOR I = N-1 TO 1 STEP -1
11 NEXT I                                      30 IF ABS(A(I,M))<=BG THEN 33               49 SM=0
12 PRINT                                       31 BG=ABS(A(I,M))                           50 FOR J=I+1 TO N
13 PRINT"B  coefficients"                      32 NP=I                                     51 SM=SM+A(I,J)*A(J,N+1)
14 FOR I = 1 TO N                              33 NEXT I                                   52 NEXT J
15 PRINT "B(";I;                               34 IF NP=M THEN 40                          53 A(I,N+1)=(A(I,N+1)-SM)/A(I,I)
16 INPUT")=",A(I,N+1)                          35 FOR I = M TO N+1                         54 NEXT I
17 NEXT I                                      36 TE=A(M,I)                                55 RETURN
18 GOSUB 25                                    37 A(M,I)=A(NP,I)
19 PRINT                                       38 A(NP,I)=TE
```

Fig. B–1

unknowns. To do so, the equations should first be written in the following general format:

$$A_{11}x_1 + A_{12}x_2 + \cdots + A_{1n}x_n = B_1$$
$$A_{21}x_1 + A_{22}x_2 + \cdots + A_{2n}x_n = B_2$$
$$\cdot$$
$$\cdot$$
$$\cdot$$
$$A_{n1}x_1 + A_{n2}x_2 + \cdots + A_{nn}x_n = B_n$$

The "A" and "B" coefficients are "called" for when running the program. The output presents the unknowns $x_1, \ldots, x_n$.

Example B–1

Solve the two equations

$$3x_1 + x_2 = 4$$
$$2x_1 - x_2 = 10$$

SOLUTION

When the program begins to run, it first calls for the number of equations (2); then the A coefficients in the sequence $A_{11} = 3$, $A_{12} = 1$, $A_{21} = 2$, $A_{22} = -1$; and finally the B coefficients $B_1 = 4$, $B_2 = 10$. The output appears as

Unknowns

$X(1) = 2.8$ *Ans.*

$X(2) = -4.4$ *Ans.*

B.2 Simpson's Rule

Simpson's rule is a numerical method that can be used to determine the area under a curve given as a graph or as an explicit function $y = f(x)$. Likewise, it can be used to compute the value of a definite integral which involves the function $y = f(x)$. To do so, the area must be subdivided into an *even number* of strips or intervals having a width h. The curve between three consecutive ordinates is approximated by a parabola, and the entire area or definite integral is then determined from the formula

$$\int_{x_0}^{x_n} f(x)\, dx \simeq \frac{h}{3}[y_0 + 4(y_1 + y_3 + \cdots + y_{n-1})$$
$$+ 2(y_2 + y_4 + \cdots + y_{n-2}) + y_n] \quad \text{(B-1)}$$

The computer program for this equation is given in Fig. B–2. For its use, we must first specify the function (on line 6 of the program). The upper and lower limits of the integral and the number of intervals are called for when the program is executed. The value of the integral is then given as the output.

```
1 PRINT"Simpson's rule":PRINT
2 PRINT" To execute this program :":PRINT
3 PRINT"   1- Modify right-hand side of the equation given below,
4 PRINT"      then press RETURN key"
5 PRINT"   2- Type  RUN 6":PRINT:EDIT 6
6 DEF FNF(X)=LOG(X)
7 PRINT:INPUT" Enter Lower Limit = ",A
8 INPUT" Enter Upper Limit = ",B
9 INPUT" Enter Number (even) of Intervals = ",N%
10 H=(B-A)/N%:AR=FNF(A):X=A+H
11 FOR J%=2 TO N%
12 K=2*(2-J%+2*INT(J%/2))
13 AR=AR+K*FNF(X)
14 X=X+H:NEXT J%
15 AR=H*(AR+FNF(B))/3
16 PRINT" Integral = ",AR
17 END
```

Fig. B–2

Example B–2

Evaluate the definite integral

$$\int_2^5 \ln x \, dx$$

SOLUTION

The interval $x_0 = 2$ to $x_6 = 5$ will be divided into six equal parts ($n = 6$), each having a width $h = (5 - 2)/6 = 0.5$. We then compute $y = f(x) = \ln x$ at each point of subdivision.

n	x_n	y_n
0	2	0.693
1	2.5	0.916
2	3	1.099
3	3.5	1.253
4	4	1.386
5	4.5	1.504
6	5	1.609

Thus, Eq. B–1 becomes

$$\int_2^5 \ln x \, dx \approx \frac{0.5}{3} [0.693 + 4(0.916 + 1.253 + 1.504)$$

$$+ 2(1.099 + 1.386) + 1.609]$$

$$\approx 3.66 \qquad\qquad \textit{Ans.}$$

This answer is equivalent to the exact answer to three significant figures. Obviously, accuracy to a greater number of significant figures can be improved by selecting a smaller interval h (or larger n).

Using the computer program, we first specify the function $\ln x$, line 6 in Fig. B–2. During execution, the program input requires the upper and lower limits 2 and 5, and the number of intervals $n = 6$. The output appears as

$$\text{Integral} = 3.66082 \qquad\qquad \textit{Ans.}$$

B.3 The Secant Method

The secant method is used to find the real roots of an algebraic or transcendental equation $f(x) = 0$. The method derives its name from the fact that the formula used is established from the slope of the secant line to the graph $y = f(x)$. This slope is $[f(x_n) - f(x_{n-1})]/(x_n - x_{n-1})$, and the secant formula is

$$x_{n+1} = x_n - f(x_n)\left[\frac{x_n - x_{n-1}}{f(x_n) - f(x_{n-1})}\right] \qquad \text{(B-2)}$$

For application it is necessary to provide two initial guesses, x_0 and x_1, and thereby evaluate x_2 from Eq. B–2 ($n = 1$). One then proceeds to reapply Eq. B–2 with x_1 and the calculated value of x_2 and obtain x_3 ($n = 2$), etc., until the value $x_{n+1} \simeq x_n$. One can see this will occur if x_n is approaching the root of the function $f(x) = 0$, since the correction term on the right of Eq. B–2 will tend toward zero. In particular, the larger the slope, the smaller the correction to x_n, and the faster the root will be found. On the other hand, if the slope is very small in the neighborhood of the root, the method leads to large corrections for x_n, and convergence to the root is slow and may even lead to a failure to find it. In such cases other numerical techniques must be used for solution.

A computer program based on Eq. B–2 is listed in Fig. B–3. We must first specify the function on line 7 of the program. When the program is executed, two initial guesses, x_0 and x_1, must be entered in order to approximate the solution. The output specifies the value of the root. If it cannot be determined, this is so stated.

```
 1 PRINT"Secant method":PRINT
 2 PRINT" To execute this program :":PRINT
 3 PRINT"    1) Modify right hand side of the equation given below,"
 4 PRINT"       then press RETURN key."
 5 PRINT"    2) Type  RUN 7"
 6 PRINT:EDIT 7
 7 DEF FNF(X)=.5*SIN(X)-2*COS(X)+1.3
 8 INPUT"Enter point #1 =",X
 9 INPUT"Enter point #2 =",X1
10 IF X=X1 THEN 14
11 EP=.00001:TL=2E-20
12 FP=(FNF(X1)-FNF(X))/(X1-X)
13 IF ABS(FP)>TL THEN 15
14 PRINT"Root can not be found.":END
15 DX=FNF(X1)/FP
16 IF ABS(DX)>EP THEN 19
17 PRINT "Root = ";X1;"       Function evaluated at this root = ";FNF(X1)
18 END
19 X=X1:X1=X1-DX
20 GOTO 12
```

Fig. B–3

Example B–3

Determine the root of the equation

$$f(x) = 0.5 \sin x - 2 \cos x + 1.30 = 0$$

SOLUTION

Guesses of the initial roots will be $x_0 = 45°$ and $x_1 = 30°$. Applying Eq. B–2,

$$x_2 = 30° - (-0.1821)\frac{(30° - 45°)}{(-0.1821 - 0.2393)} = 36.48°$$

Using this value in Eq. B–2, along with $x_1 = 30°$, we have

$$x_3 = 36.48° - (-0.0108)\frac{36.48° - 30°}{(-0.0108 + 0.1821)} = 36.89°$$

Repeating the process with this value and $x_2 = 36.48°$ yields

$$x_4 = 36.89° - (0.0005)\left[\frac{36.89° - 36.48°}{(0.0005 + 0.0108)}\right] = 36.87°$$

Thus $x = 36.9°$ is appropriate to three significant figures.

If the problem is solved using the computer program, first we specify the function, line 7 in Fig. B–3. During execution, the first and second guesses must be entered in radians. Choosing these to be 0.8 rad and 0.5 rad, the result appears as

Root = 0.6435022.

Function evaluated at this root = 1.66893E–06.

This result converted from radians to degrees is therefore

$$x = 36.9° \qquad\qquad Ans.$$

Answers

1–1. (*a*) 78.5 N,   (*b*) 0.392 mN,   (*c*) 7.46 MN
1–4. (*a*) 45.3 MN,   (*b*) 56.8 km,   (*c*) 5.63 μg
1–5. (*a*) 0.185 Mg^2,   (*b*) 4 μg^2,   (*c*) 0.0122 km^3
1–6. (*a*) kN · m,   (*b*) Gg/m,   (*c*) μN/s^2,   (*d*) GN/s
1–8. 4.96 μN
1–10. 44.9(10)$^{-3}$ N^2, 2.79(10^3) s^2, 23.4 s
1–11. (*a*) 15.9 mm/s,   (*b*) 3.69 Mm · s/kg,   (*c*) 1.14 km · kg

Chapter 2

2–1. 867 N, 108°
2–2. 308 N, 91.9°
2–3. 393 N, 353°
2–5. 45.8 N, 162°
2–6. 605 N, 85.4°
2–7. F_{1u} = 205 N, F_{1v} = 160 N
2–9. 115 N
2–10. F_u = 91.9 N, F_v = 80.8 N
2–11. $F_{\parallel}$ = 46.5 N, $F_{\perp}$ = 99.7 N
2–13. F_{AC} = 43.9 N, F_{AB} = 53.8 N
2–14. 79.5°, 90.8 N
2–15. 10.8 kN, 3.16°

2–17. F_a = 30.6 N, F_b = 26.9 N
2–18. F = 19.6 N, F_b = 26.4 N
2–19. 60°
2–21. T = 877 N, F_R = 1.34 kN
2–22. 744 N, 23.8°
2–23. (*a*) F_n = −14.1 N, F_t = 14.1 N,   (*b*) F_x = 19.3 N, F_y = 5.18 N
2–25. F_2 = 8.35 N, F_1 = 14.0 N
2–26. 70.5°
2–27. 19.2 N, 2.37° ◥
2–29. 53.5°, 621 N
2–30. 38.3°
2–31. F_A = 3.66 kN, F_B = 7.07 kN
2–33. F_x = 514 N, F_y = −613 N
2–34. 747 N, 85.5°
2–35. 97.8 N, 46.5°
2–37. 546 N, 253°
2–38. $\mathbf{F}_1$ = {−15.0**i** − 26.0**j**} kN, $\mathbf{F}_2$ = {−10.0**i** + 24.0**j**} kN
2–39. 25.1 kN, 185°
2–41. 217 N, 87.0°
2–42. 867 N, 108°
2–43. 308 N, 91.9°
2–45. 10.8 kN, 3.16°
2–46. 19.2 N, 2.37° ◣

2–47. $F_{1x} = -200$ N, $F_{1y} = 0$, $F_{2x} = 320$ N, $F_{2y} = -240$ N, $F_{3x} = 180$ N, $F_{3y} = 240$ N, $F_{4x} = -300$ N, $F_{4y} = 0$

2–49. 68.3 kN, 1.44° ⬈

2–50. $\mathbf{F}_1 = \{90\mathbf{i} - 120\mathbf{j}\}$ N, $\mathbf{F}_2 = \{-275\mathbf{j}\}$ N, $\mathbf{F}_3 = \{-37.5\mathbf{i} - 65.0\mathbf{j}\}$ N, 463 N

2–51. 37.0°, 889 N

2–53. $0 \le P \le 1.62$ kN

2–54. 54.3°, 686 N

2–55. 1.23 kN, 6.08°

2–57. 389 N, 42.7°

2–58. $\mathbf{F}_1 = \{F_1 \cos \theta\mathbf{i} + F_1 \sin \theta\mathbf{j}\}$ N, $\mathbf{F}_2 = \{350\mathbf{i}\}$ N, $\mathbf{F}_3 = \{-100\mathbf{j}\}$ N, 67.0°, $F_1 = 434$ N

2–59. 117°, $1.12 F_1$

2–61. 1.03 kN, 87.9°

2–62. 5.96 kN

2–63. $\mathbf{F} = \{217\mathbf{i} + 85.5\mathbf{j} - 91.2\mathbf{k}\}$ N

2–65. 50 N, $\alpha = 74.1°$, $\beta = 41.3°$, $\gamma = 53.1°$

2–66. 90°, $\{-30\mathbf{i} - 52.0\mathbf{k}\}$ N

2–67. $\mathbf{F}_1 = \{4.80\mathbf{i} + 6.40\mathbf{j}\}$ kN, $\mathbf{F}_2 = \{3.00\mathbf{i} - 3.00\mathbf{j} + 4.24\mathbf{k}\}$ kN, $\mathbf{F}_R = \{7.80\mathbf{i} + 3.40\mathbf{j} + 4.24\mathbf{k}\}$ kN, $F_R = 9.51$ kN, $\alpha = 34.9°$, $\beta = 69.0°$, $\gamma = 63.5°$

2–69. $\mathbf{F}_1 = \{53.1\mathbf{i} - 44.5\mathbf{j} + 40\mathbf{k}\}$ N, $\alpha_1 = 48.4°$, $\beta_1 = 124°$, $\gamma_1 = 60°$, $\mathbf{F}_2 = \{-130\mathbf{k}\}$ N, $\alpha_2 = 90°$, $\beta_2 = 90°$, $\gamma_2 = 180°$

2–70. $\mathbf{F}_1 = \{175\mathbf{i} + 175\mathbf{j} - 247\mathbf{k}\}$ N, $\mathbf{F}_2 = \{173\mathbf{i} - 100\mathbf{j} + 150\mathbf{k}\}$ N

2–71. 369 N, $\alpha = 19.5°$, $\beta = 78.3°$, $\gamma = 105°$

2–73. $\mathbf{F}_1 = \{176\mathbf{j} - 605\mathbf{k}\}$ N, $\mathbf{F}_2 = \{125\mathbf{i} - 177\mathbf{j} + 125\mathbf{k}\}$ N, $\mathbf{F}_R = \{125\mathbf{i} - 0.377\mathbf{j} - 480\mathbf{k}\}$ N, $F_R = 496$ N, $\alpha = 75.4°$, $\beta = 90.0°$, $\gamma = 165°$

2–74. $\alpha_1 = 45.6°$, $\beta_1 = 53.1°$, $\gamma_1 = 66.4°$

2–75. $\alpha_1 = 90°$, $\beta_1 = 53.1°$, $\gamma_1 = 66.4°$

2–77. $\alpha = 124°$, $\beta = 71.3°$, $\gamma = 140°$

2–78. $F_x = 1.28$ kN, $F_y = 2.60$ kN, $F_z = 0.776$ kN

2–79. $F = 2.02$ kN, $F_y = 0.523$ kN

2–81. $F_x = 40$ N, $F_y = 40$ N, $F_z = 56.6$ N

2–82. $F = 32.7$ N, $F_y = 16.3$ N

2–83. 32.4 N, $\alpha_2 = 122°$, $\beta_2 = 74.5°$, $\gamma_2 = 144°$

2–85. 7 m, $\alpha = 31.0°$, $\beta = 107°$, $\gamma = 115°$

2–86. $\{-2.35\mathbf{i} + 3.93\mathbf{j} + 3.71\mathbf{k}\}$ m, 5.89 m, $\alpha = 113°$, $\beta = 48.2°$, $\gamma = 51.0°$

2–87. 2.11 m

2–89. 4.42 m

2–90. 467 mm

2–91. 732 mm

2–93. 12.3 km

2–94. $r_{AD} = 1.50$ m, $r_{BD} = 1.50$ m, $r_{CD} = 1.73$ m

2–95. $\{-37.7\mathbf{i} - 15.6\mathbf{j} + 44.0\mathbf{k}\}$ N, $\alpha = 129°$, $\beta = 105°$, $\gamma = 42.8°$

2–97. $\mathbf{F}_1 = \{38.0\mathbf{i} + 104\mathbf{j} - 101\mathbf{k}\}$ N, $\mathbf{F}_2 = \{119\mathbf{i} - 19.9\mathbf{j} - 159\mathbf{k}\}$ N

2–98. 316 N, $\alpha = 60.1°$, $\beta = 74.6°$, $\gamma = 146°$

2–99. $\mathbf{F}_1 = \{-3.79\mathbf{i} + 11.4\mathbf{k}\}$ N, $\mathbf{F}_2 = \{-6.65\mathbf{i} - 11.8\mathbf{j} + 11.8\mathbf{k}\}$ N

2–101. $x = 7.65$ m, $y = 4.24$ m, $z = 3.76$ m

2–102. $x = 8.67$ m, $y = 1.89$ m

2–103. $\{-34.3\mathbf{i} + 22.9\mathbf{j} - 68.6\mathbf{k}\}$ N

2–105. $\{98.1\mathbf{i} + 269\mathbf{j} - 201\mathbf{k}\}$ N

2–106. $\{59.4\mathbf{i} - 88.2\mathbf{j} - 83.2\mathbf{k}\}$ N, $\alpha = 63.9°$, $\beta = 131°$, $\gamma = 128°$

2–107. $\mathbf{F}_1 = \{-26.2\mathbf{i} - 41.9\mathbf{j} + 62.9\mathbf{k}\}$ N, $\mathbf{F}_2 = \{13.4\mathbf{i} - 26.7\mathbf{j} - 40.1\mathbf{k}\}$ N, $F_R = 73.5$ N, $\alpha = 100°$, $\beta = 159°$, $\gamma = 71.9°$

2–109. 757 N, $\alpha = 149°$, $\beta = 90.0°$, $\gamma = 59.0°$

2–110. 492 mm, $\{-13.2\mathbf{i} - 18.3\mathbf{j} + 19.8\mathbf{k}\}$ N

2–111. $\mathbf{F}_{EA} = \{12\mathbf{i} - 8\mathbf{j} - 24\mathbf{k}\}$ kN, $\mathbf{F}_{EB} = \{12\mathbf{i} + 8\mathbf{j} - 24\mathbf{k}\}$ kN, $\mathbf{F}_{EC} = \{-12\mathbf{i} + 8\mathbf{j} - 24\mathbf{k}\}$ kN, $\mathbf{F}_{ED} = \{-12\mathbf{i} - 8\mathbf{j} - 24\mathbf{k}\}$ kN, $\mathbf{F}_R = \{-96\mathbf{k}\}$ kN

2–113. $\{476\mathbf{i} + 329\mathbf{j} - 159\mathbf{k}\}$ N

2–114. $x = -6.41$ m, $y = 13.4$ m

2–115. 1.50 kN, $\alpha = 77.6°$, $\beta = 90.6°$, $\gamma = 168°$

2–117. 121°

2–118. $\mathbf{r}_1 = |2.57$ m$|$, $\mathbf{r}_2 = |3.60$ m$|$

2–119. 109°

2–121. $F_B = 566$ N, $F_A = 293$ N, $F_{oa} = 693$ N, $F_{ob} = 773$ N

2–122. 82.0°

2–123. 2.67 m

2–125. 115°

2–126. $F_1 = 20$ N, $F_2 = 31.6$ N

2–127. $F_1 = 333$ N, $F_2 = 373$ N

2–129. $|50.6$ N$|$

2–130. 97.3°

2–131. $\mathbf{F}_\parallel = 99.1$ N, $F_\perp = 592$ N

2–133. $\theta = 74.4°$, $\phi = 55.4°$

2–134. 22.7 N, 14.8 N

2–135. 70.5°

2–137. 10.5 N

2–138. 143°

2–139. $T = 54.7$ N, $P = 42.6$ N

2–141. $F_u = 320$ N, $F_v = 332$ N

2–142. 428 N, $\alpha = 88.3°$, $\beta = 20.6°$, $\gamma = 69.5°$

2–143. 250 N, $\alpha = 87.0°$, $\beta = 143°$, $\gamma = 53.1°$

2–145. $\mathbf{F}_1 = \{43.3\mathbf{i} + 25\mathbf{j}\}$ N, $\mathbf{F}_2 = \{-14.8\mathbf{i} - 31.7\mathbf{j}\}$ N

2–146. 29.3 N, 347°

2–147. $\theta = 74.0°$, $\phi = 33.9°$

2–149. $F_x = -13.2$ kN, $F_y = 18.8$ kN

Chapter 3

3–1. $F_1 = 439$ N, $F_2 = 233$ N

3–2. 16.4°, 137 N

3–3. 31.8°, 4.94 kN

3–5. $F_1 = 1.83$ kN, $F_2 = 9.60$ kN

3–6. 4.69°, 4.31 kN

3–7. 1.32 kN

3–9. 1.13 mN

3–10. 158 N

3–11. 1.56 m

3–13. $m = 12.8$ kg

3–14. $\dfrac{1}{k_1} + \dfrac{1}{k_2}$

3–15. 78.7°, 127 N

3–17. 43.1°, 20.5 kg

3–18. 11.5°

3–19. $F_{BC} = 70.7$ N, $F_{AB} = 50$ N, $F_{AD} = 70.7$ N, 7.67 m

3–21. 60°, 34.6 N

3–22. 60°, 46.2 N

3–23. $F_A = 34.6$ N, $F_B = 57.3$ N

3–25. 6 N

3–26. 35.0°

3–27. 2.66 m

3–29. $\dfrac{50}{\cos \theta}$

3–30. 40.8 N

3–31. 2.46 m

3–33. 43.0°

3–34. 88.8 N

3–35. $T_{AB} = 340$ N, $T_{AE} = 170$ N, $T_{BD} = 490$ N, $T_{BC} = 562$ N

3–37. 40.2°

3–38. 4.98 m

3–39. 0 and 6.59 m

3–41. $F_1 = 5.10$ kN, $F_2 = 11.8$ kN, $F_3 = 3.92$ kN

3–42. $F_1 = 800$ N, $F_2 = 147$ N, $F_3 = 564$ N

3–43. $F_1 = 5.60$ kN, $F_2 = 8.55$ kN, $F_3 = 9.44$ kN

3–45. 55.8 N

3–46. $F_{AB} = 441$ N, $F_{AC} = 515$ N, $F_{AD} = 221$ N

3–47. $F_{AB} = 348$ N, $F_{AC} = 413$ N, $F_{AD} = 174$ N

3–49. 771 N

3–50. $F_{AD} = 1.20$ kN, $F_{AC} = 0.40$ kN, $F_{AB} = 0.80$ kN

3–51. $F_{AB} = 0.980$ kN, $F_{AC} = 0.463$ kN, $F_{AD} = 1.55$ kN

3–53. 138 N

3–54. $F_{AC} = 92.9$ N, $F_{AD} = 364$ N, $F_{AO} = 757$ N

3–55. $F_{AC} = 574$ N, $F_{AB} = 500$ N, $F_{CD} = 1.22$ kN, $F_{CE} = F_{CF} = 416$ N

3–57. $F_{AD} = 1.70$ kN, $F_{AC} = 0.744$ kN, $F_{AB} = 1.37$ kN

3–58. $F_{AD} = 1.42$ kN, $F_{AC} = 0.914$ kN, $F_{AB} = 1.47$ kN

3–59. $F_{AB} = 469$ N, $F_{AC} = F_{AD} = 331$ N

3–61. 267 N

3–62. 1.64 m

3–63. 120 N

3–65. $T_B = 109$ N, $T_C = 47.4$ N, $T_D = 87.9$ N

3–66. yes, yes

3–67. $l'_{AB} = 0.452$ m, $l'_{AC} = 0.658$ m

3–69. 4.11 m

3–70. $T = 25.3$ N, $F = 22.3$ N

3–71. 20°, 305 N

3–73. 240 N

Chapter 4

4–5. 8.00 kN · m $\downarrow$

4–6. 18.5 kN · m $\curvearrowright$

4–7. 1.68 kN · m $\curvearrowright$

4–9. 3.57 kN · m $\curvearrowright$

4–10. 3.15 kN · m $\curvearrowright$

4–11. 2.42 kN · m $\downarrow$

4–13. (a) 73.9 N · m $\downarrow$, (b) 82.2 N $\leftarrow$

4–14. 37.9°, 79.8 N · m, 128°, 0

4–15. $M_B = 90.6$ N · m $\curvearrowright$, $M_C = 141$ N · m $\curvearrowright$

4–17. $\left(\dfrac{l}{d + l}\right) M$

4–18. 2.53 kN · m

4–19. 1.59 kN

4–21. 1.41 kN

4–22. (a) 330 N · m, 76.0°,   (b) 0, 166°

4–23. 80 kN · m, 24.0 m

4–25. (a) 13.0 N · m,   (b) 35.2 N

4–26. (a) 3.51 kN · m ↓,   (b) 1.18 kN · m ↓

4–27. 414 N · m, 2.93 kN

4–29. 104 N

4–30. M_A = 6.88 N · m, M_B = 9.03 N · m ↓

4–31. {−84**i** − 8**j** − 39**k**} kN · m

4–33. {260**i** + 180**j** + 510**k**} N · m

4–34. {440**i** + 220**j** + 990**k**} N · m

4–35. {−31.6**i** + 18.3**j**} N · m

4–37. M_A = {−1.90**i** + 6.00**j**} kN · m

4–38. {61.2**i** + 81.6**j**} N · m

4–39. 176 N · m

4–41. {−37.6**i** + 90.7**j** − 155**k**} N · m

4–42. {−840**i** + 360**j** − 660**k**} N · m

4–43. {−720**i** + 120**j** − 660**k**} N · m

4–45. 18.6 N

4–46. y = 2 m, z = 1 m

4–47. y = 1 m, z = 3 m, d = 1.15 m

4–49. {218**j** + 163**k**} N · m

4–50. {−43.0**i** − 39.5**j** + 14.3**k**} kN · m

4–51. {26.1**i** − 15.1**j**} N · m

4–53. 0.277 N · m

4–54. {−78.4**j**} N · m

4–55. 17.5 N

4–57. 165 N · m

4–58. 226 N · m

4–59. 3.84 N · m

4–61. 4.43 N

4–62. 14.8 N · m

4–63. 20.2 N

4–65. 5.66 N

4–66. 18.3 kN · m ↾

4–67. 17.6 kN · m ↾

4–69. 720 N · m ↾

4–70. 108 N

4–71. 39.7 N · m

4–73. 167 kN, resultant couple can act anywhere

4–74. 348 kN · m, resultant couple can act anywhere

4–75. 2.03 m

4–77. (a) {126**k**} N · m,   (b) 126 N · m

4–78. 0.909 kN

4–79. 139 kN

4–81. {557**i** − 7.14**j** + 857**k**} N · m

4–82. {37.5**i** − 25**j**} N · m, 45.1 N · m

4–83. 832 N

4–85. (a) {−5**i** + 8.75**j**} N · m,   (b) {−5**i** + 8.75**j**} N · m

4–86. 992 N

4–87. 59.9 N · m, α = 99.0°, β = 106°, γ = 18.3°

4–89. {−50**i** + 60**j**} N · m, 78.1 N · m

4–90. {7.01**i** + 42.1**j**} N · m

4–91. 35.1 N

4–93. 375 N, 737 N · m ↾

4–94. 80 N, 399 N · m ↓

4–95. 80 N, 696 N · m ↓

4–97. 274 N, 5.24° ◿, 5.48 kN · m ↾

4–98. 2.10 kN, 81.6° ◿, 10.6 kN · m ↓

4–99. 2.10 kN, 81.6° ◿, 16.8 kN · m ↓

4–101. 375 N ↑, 2.47 m (to the left)

4–102. {−1**i** − 2**j** − 5**k**} kN, {0.650**i** + 19.75**j** − 9.05**k**} kN · m

4–103. 798 N, 67.9° ◺, 7.43 m

4–105. 1302 N, 84.5° ◺, 7.36 m

4–106. 1302 N, 84.5°, 1.36 m (to the right)

4–107. 922 N, 77.5° ◺, 3.56 m

4–109. 991 N, 1.78 m

4–110. 991 N, 2.64 m

4–111. F_R = {−3.11**j** − 9.79**k**} kN,
        M_{RA} = {26.6**i** − 29.3**j** + 11.4**k**} kN · m

4–113. {−26**i** + 357**j** + 127**k**} N · m

4–114. {−40**j** − 40**k**} N, {−12**j** + 12**k**} N · m

4–115. {−28.3**j** − 68.3**k**} N, {−20.5**j** + 8.49**k**} N · m

4–117. F_B = 163 N, F_C = 223 N

4–118. {270**k**} N, {−2.22**i**} N · m

4–119. {270**k**} N, y = −8.22 mm, x = 0

4–121. 140 kN, x = 6.43 m, y = 7.29 m

4–122. {−180**k**} N, 1.06 m

4–123. {−210**k**} N, {−15**i** + 225**j**} N · m

4–125. 0.20 m, {−40**i**} N, {−30**i**} N · m

4–126. 53.3 N, {−40**i**} N, {−30**i**} N · m

4–127. 990 N, 3.07 kN · m, x = 1.16 m, y = 2.06 m

4–129. 13.2 kN↓, 3.09 m

4–130. F_R = 0, M_{R_O} = 1.35 kN · m

4–131. 18.0 kN↓, 11.7 m

4–133. 90 kN↓, 338 kN · m ↙

4–134. 0.525 kN, 0.171 m

4–135. 10.6 kN↓, 0.479 m

4–137. 3.60 kN↓, 16.2 kN · m ↙

4–138. 1.35 kN, 42.0° ↗, 0.1 m

4–139. 1.35 kN, 42.0° ↗, 0.556 m

4–141. 17.35 kN↑, 3.43 m

4–142. 1.5 m, 175 N/m

4–143. 22.4 kN, 7.50 m

4–145. 107 kN, 2.40 m

4–146. 14.9 kN, 2.27 m

4–147. 43.6 N, 3.27 m

4–149. 73.3 N, 2.22 m

4–150. $F_R = \dfrac{w_0}{a}(e^{aL} - 1), \; x = \dfrac{(e^{aL}aL - e^{aL} + 1)}{a(e^{aL} - 1)}$

4–151. $\{-5.39\mathbf{i} + 13.1\mathbf{j} + 11.4\mathbf{k}\}$ N · m

4–153. 13.8 kN · m ↙

4–154. 10.1 kN · m ↖

4–155. $\{15\mathbf{i} + 35.4\mathbf{j} + 25\mathbf{k}\}$ N · m, 45.8 N · m, $\alpha = 70.9°$, $\beta = 39.5°$, $\gamma = 56.9°$

4–157. −809 N · m

4–158. 19.8 N

4–159. $\mathbf{M}_x = \{15\mathbf{i}\}$ N · m, $\mathbf{M}_y = \{4\mathbf{j}\}$ N · m, $\mathbf{M}_z = \{36\mathbf{k}\}$ N · m

Chapter 5

5–11. 10 N

5–13. $B_y = 642$ N, $A_x = 192$ N, $A_y = 180$ N

5–14. $F_B = 3.76$ kN, $A_x = 3.26$ kN, $A_y = 3.12$ kN

5–15. $N_B = 2.14$ kN, $A_y = 1.49$ kN, $A_x = 1.29$ kN

5–17. $N_C = 57.7$ N, $N_A = 237$ N, $N_B = 122$ N

5–18. $C_y = 586$ N, $F_A = 413$ N

5–19. $F_D = 480$ kN, $F_L = 750$ kN, 3.53 m

5–21. $R_A = 667$ N, $R_B = 220$ N, $R_C = 440$ N

5–22. $F_{CD} = 1950$ N, $A_x = 974$ N, $A_y = 312$ N

5–23. $F_H = 257$ N, $T_B = 292$ N

5–25. $F = 22$ kN, $A_x = 30$ kN, $A_y = 16$ kN

5–26. 78.6 N

5–27. $F_B = 6.38$ N, $A_x = 3.19$ N, $A_y = 2.48$ N

5–29. $B_x = 6.67$ kN, $A_x = 6.67$ kN, $A_y = 8.00$ kN

5–30. 14.4 kN

5–31. $A_x = 1462$ N, $F_B = 1.66$ kN

5–33. $D_x = 0$, $D_y = 1.65$ kN, $M_D = 1.40$ kN · m, $(M_D)_{max} = 3.00$ kN · m

5–34. 6 m, 267 N/m

5–35. 105 N

5–37. $T_{CD} = 1.87$ kN, $F_A = 2.07$ kN, 81.4° ↘

5–38. 4.73 m

5–41. $F_{CB} = 782$ N, $A_x = 625$ N, $A_y = 681$ N

5–42. $F_2 = 724$ N, $F_1 = 1.45$ kN, $F_A = 1.75$ kN

5–43. $R_A = 1.06$ kN, $R_B = 1.42$ kN, $R_C = 0.501$ kN

5–45. $T_{BC} = 16.4$ kN, $T = 5$ kN, $F_A = 20.6$ kN

5–46. 41.4°

5–47. $R_A = 151$ kN, $R_B = 679$ kN

5–49. $F_A = 404$ N, $R = 808$ N

5–50. $F_A = 1.85$ kN, $F_B = 2.02$ kN, $F_C = 391$ N

5–51. $F_B = 0.3P$, $F_C = 0.6P$, $x_C = \dfrac{0.6P}{k}$

5–53. $R_C = 284$ N, $R_B = 53.1$ N, $R_A = 115$ N

5–54. $\dfrac{a}{\cos^3 \theta}$

5–55. $F_C = 8.50$ N, $F_B = 16.6$ N, $F_A = 1.90$ N

5–57. $\tan^{-1} \dfrac{b}{a}$

5–59. 13.0 m

5–61. 568 mm

5–62. 47.5°

5–63. $A_x = 0$, $A_y = -200$ N, $A_z = 150$ N, $(M_A)_x = 100$ N · m, $(M_A)_y = 0$, $(M_A)_z = 500$ N · m

5–65. $T_B = 2.75$ kN, $T_C = 1.38$ kN, $T_A = 1.38$ kN

5–66. $T_A = 1.05$ kN, $T_B = 1.29$ kN, $T_C = 1.66$ kN

5–67. 3.00 kN, $x = 1.58$ m, $y = 1.60$ m

5–69. $B_z = 373$ N, $A_x = 0$, $A_y = 0$, $A_z = 333$ N, $T_{CD} = 43.5$ N

5–70. $O_x = 0$, $O_y = -84.9$ N, $O_z = 80$ N, $M_{O_x} = 948$ N · m, $M_{O_y} = 0$, $M_{O_z} = 0$

5–71. $P = 75$ N, $A_y = 0$, $B_z = 75$ N, $A_z = 75$ N, $B_x = 112$ N, $A_x = 37.5$ N

5–73. $T_{BC} = 205$ N, $T_{ED} = 629$ N, $A_x = 32.4$ N, $A_y = 107$ N, $A_z = 1.28$ kN

5–74. $A_x = 0$, $A_y = 1.50$ kN, $A_z = 750$ N, $T = 919$ N

5–75. $F = 1.31$ kN, $A_x = 0$, $A_y = 1.31$ kN, $A_z = 653$ N

5–77. $A_z = 5$ kN, $A_x = 0$, $A_y = 16.7$ kN

5–78. $T_{BC} = T_{BD} = 17$ kN, $A_y = 11.3$ kN, $A_x = 0$, $A_z = -15.7$ kN

5–79. $F_{DE} = 0.721$ kN, $F_{BC} = 2.16$ kN, $A_x = -0.309$ kN, $A_y = 1.55$ kN, $A_z = -1.21$ kN

5–81. $T = 58$ N, $C_z = 87$ N, $C_y = -28.8$ N, $D_y = -79.2$ N, $D_z = 58$ N, $D_x = 0$

5–82. $T = 58$ N, $C_z = 77.6$ N, $C_y = -24.9$ N, $D_x = 0$, $D_y = -68.5$ N, $D_z = 32.1$ N

5–83. $F_{AC} = F_{BC} = 6.13$ kN, $F_{DE} = 19.6$ kN

5–85. $T_{BC} = 524$ N, $T_{BD} = 2.04$ kN, $A_x = 0$, $A_y = 0$, $A_z = 2.35$ kN

5–86. $B_z = 11\,670$ N, $C_z = 7340$ N, $A_z = 16\,000$ N

5–87. $F_{BC} = 0$, $A_y = 0$, $A_z = 800$ N, $(M_A)_x = 4.80$ kN · m, $(M_A)_y = 0$, $(M_A)_z) = 0$

5–89. $F_{BDC} = 620$ N, $F_{CE} = 1100$ N, $A_x = 194$ N, $A_y = 1920$ N, $A_z = -258$ N

5–90. $A_x = 633$ N, $A_y = -141$ N, $B_x = -721$ N, $B_z = 895$ N, $C_y = 200$ N, $C_z = -506$ N

5–91. $F_2 = 674$ N

5–93. 43.75 N · m

5–94. $A_z = 106$ N, $D_y = -2.30$ N, $A_y = 2.30$ N, $D_x = 51.7$ N, $A_x = 54.4$ N, $M = 4.59$ N · m

5–95. $A_x = 0$, $F_{BD} = 208$ N, $F_{BC} = 792$ N, $A_z = 0$, $M_{Ax} = 0$, $M_{Az} = 700$ N · m

5–97. $R_A = 105$ N, $B_x = 97.4$ N, $B_y = 269$ N

5–98. $kR(\cot \theta - \cos \theta)$

5–99. $A_y = 390$ N, $B_x = 0$, $B_y = 60$ N

5–101. $F_{BC} = 175$ N, $A_x = 130$ N, $A_y = -10$ N, $M_{Ax} = -30$ N · m, $M_{Ay} = 0$, $M_{Az} = -72$ N · m

5–102. 105 N

5–103. $w_2 = 137$ N, $w_1 = 549$ N

5–105. $A_z = 49.0$ N, $B_y = 24.5$ N, $A_y = -24.5$ N, $T_{BC} = 24.5$ N, $A_x = 24.5$ N

5–106. $A_y = 8$ kN, $B_y = 5$ kN, $B_x = 5.20$ kN

Chapter 6

6–1. $F_{AD} = 849$ N (C), $F_{AB} = 600$ N (T), $F_{BD} = 600$ N (C), $F_{BC} = 600$ N (T), $F_{DC} = 1.41$ kN (T), $F_{DE} = 1.60$ kN (C)

6–2. $F_{AD} = 1.13$ kN (C), $F_{AB} = 800$ N (T), $F_{BD} = 0$, $F_{BC} = 800$ N (T), $F_{DC} = 1.13$ kN (C), $F_{DE} = 1.60$ kN (C)

6–3. $F_{AD} = 9.90$ kN (C), $F_{AB} = 7$ kN (T), $F_{DB} = 4.95$ kN (T), $F_{DC} = 14.8$ kN (C), $F_{CB} = 10.5$ kN (T)

6–5. $F_{AG} = 2.10$ kN (C), $F_{AB} = 1.48$ kN (T), $F_{BG} = 0$, $F_{BC} = 1.48$ kN (T), $F_{DE} = 4.20$ kN (C), $F_{DC} = 2.97$ kN (T), $F_{EC} = 2.97$ kN (T), $F_{EG} = 2.97$ kN (T), $F_{CG} = 2.10$ kN (T)

6–6. $F_{AG} = 5.23$ kN (C), $F_{AB} = 3.70$ kN (T), $F_{BC} = 3.70$ kN (T), $F_{BG} = 2.22$ kN (T), $F_{DE} = 7.33$ kN (C), $F_{DC} = 5.18$ kN (T), $F_{EC} = 5.18$ kN (T), $F_{EG} = 5.18$ kN (C),

$F_{CG} = 2.11$ kN (T)

6–7. $F_{GB} = 27.5$ kN (T), $F_{AF} = 15.0$ kN (C), $F_{AB} = 28.0$ kN (C), $F_{BF} = 25.0$ kN (T), $F_{BC} = 15.0$ kN (T), $F_{FC} = 21.2$ kN (C), $F_{FE} = 0$, $F_{ED} = 0$, $F_{EC} = 15.0$ kN (T), $F_{DC} = 0$

6–9. $F_{BC} = 3$ kN (C), $F_{BA} = 8$ kN (C), $F_{AC} = 1.46$ kN (C), $F_{AF} = 4.17$ kN (T), $F_{CD} = 4.17$ kN (C), $F_{CF} = 3.12$ kN (C), $F_{EF} = 0$, $F_{ED} = 13.1$ kN (C), $F_{DF} = 5.21$ kN (T)

6–10. $F_{AB} = 33.0$ kN (C), $F_{AF} = 7.94$ kN (T), $F_{BF} = 23.3$ kN (T), $F_{BC} = 23.3$ kN (C), $F_{FC} = 4.71$ kN (C), $F_{FE} = 11.3$ kN (T), $F_{EC} = 30.0$ kN (T), $F_{ED} = 11.3$ kN (T), $F_{CD} = 37.7$ kN (C)

6–11. $F_{AB} = 37.7$ kN (C), $F_{AF} = 19.0$ kN (T), $F_{BF} = 26.7$ kN (T), $F_{BC} = 26.7$ kN (C), $F_{FC} = 18.9$ kN (T), $F_{FE} = 5.67$ kN (T), $F_{EC} = 0$, $F_{ED} = 5.67$ kN (T), $F_{CD} = 18.9$ kN (C)

6–13. 849 N

6–14. 849 N

6–15. $F_{AE} = 8.94$ kN (C), $F_{AB} = 8$ kN (T), $F_{BC} = 8$ kN (T), $F_{BE} = 8$ kN (C), $F_{EC} = 8.94$ kN (T), $F_{ED} = 17.9$ kN (C), $F_{DC} = 8$ kN (T)

6–17. $F_{AB} = 7.5$ kN (T), $F_{AE} = 4.5$ kN (C), $F_{ED} = 4.5$ kN (C), $F_{EB} = 8$ kN (T), $F_{BD} = 19.8$ kN (C), $F_{BC} = 18.5$ kN (T)

6–18. $F_{AB} = 196$ N (T), $F_{AE} = 118$ N (C), $F_{ED} = 118$ N (C), $F_{EB} = 216$ N (T), $F_{BD} = 1.04$ kN (C), $F_{BC} = 857$ N (T)

6–19. $F_{CD} = 3.61$ kN (C), $F_{CB} = 3$ kN (T), $F_{BA} = 3$ kN (T), $F_{BD} = 3$ kN (C), $F_{DA} = 2.70$ kN (T), $F_{DE} = 6.31$ kN (C)

6–21. $F_{CB} = 400$ N (C), $F_{CD} = 693$ N (C), $F_{BD} = 667$ N (T), $F_{BA} = 1.13$ kN (C)

6–22. $F_{EF} = 0.667P$ (T), $F_{FD} = 1.67P$ (T), $F_{AB} = 0.471P$ (C), $F_{AE} = 1.67P$ (T), $F_{AC} = 1.49P$ (C), $F_{BF} = 1.41P$ (T), $F_{BD} = 1.49P$ (C), $F_{EC} = 1.41P$ (T), $F_{CD} = 0.471$ P (C)

6–23. $F_{BC} = 1.89P$ (C), $F_{CD} = 1.37P$ (T), $F_{AB} = 0.471P$ (C), $F_{BD} = 1.67P$ (T), $F_{DA} = 1.37P$ (T)

6–25. $F_{BA} = P \csc 2\theta$ (C), $F_{BC} = P \cot 2\theta$ (C), $F_{CA} = (\cot \theta \cos \theta - \sin \theta + 2 \cos \theta)P$ (T), $F_{CD} = (\cot 2\theta + 1)P$ (C), $F_{DA} = (\cot 2\theta + 1)(\cos 2\theta)(P)$ (C)

6–26. 732 N

6–27. $F_{CB} = 0$, $F_{CD} = 0$, $F_{AB} = 2.40P$ (C), $F_{AF} = 2.00P$ (T), $F_{EF} = 1.86P$ (T), $F_{ED} = 0.373P$ (C), $F_{FB} = 1.86P$ (T), $F_{FD} = 0.333P$ (T), $F_{DB} = 0.373P$ (C)

6–29. $F_{DE} = 16.3$ kN (C), $F_{DC} = 8.40$ kN (T), $F_{EA} = 8.85$ kN (C), $F_{EC} = 6.20$ kN (C), $F_{AB} = 3.11$ kN (T), $F_{AF} = 6.20$ kN (T), $F_{BC} = 2.20$ kN (T), $F_{BF} = 6.20$ kN (C), $F_{CF} = 8.77$ kN (T)

6–30. $F_{DE} = 18.7$ kN (C), $F_{DC} = 9.60$ kN (T), $F_{EA} = 10.1$ kN (C), $F_{EC} = 4.80$ kN (C),

$F_{AB} = 6.79$ kN (T), $F_{AF} = 4.80$ kN (T),
$F_{BC} = 4.80$ kN (T), $F_{BF} = 4.80$ kN (C),
$F_{CF} = 6.79$ kN (T)

6–31. $F_{HG} = 29$ kN (C), $F_{BC} = 20.5$ kN (T),
$F_{HC} = 12.0$ kN (T)

6–33. $F_{CD} = 50$ kN (T), $F_{HD} = 7.07$ kN (C), $F_{GD} = 5$ kN (T)

6–34. $F_{HI} = 35$ kN (C), $F_{BC} = 50$ kN (T), $F_{HB} = 21.2$ kN (C)

6–35. $F_{KJ} = 13.3$ kN (T), $F_{BC} = 14.9$ kN (C), $F_{CK} = 0$

6–37. $F_{KJ} = 113$ kN (T), $F_{CD} = 93.8$ kN (C),
$F_{CJ} = 31.3$ kN (C), $F_{DJ} = 0$

6–38. $F_{JI} = 75$ kN (T), $F_{EI} = 25$ kN (C)

6–39. $F_{CD} = 25$ kN (T), $F_{FE} = 35$ kN (T),
$F_{CE} = 25$ kN (C)

6–41. $F_{GF} = 6.71$ kN (C), $F_{GB} = 6.71$ kN (T)

6–42. $F_{BC} = 7.0$ kN (T), $F_{FC} = 0.769$ kN (T)

6–43. $F_{BC} = 10.4$ kN (C), $F_{HG} = 9.16$ kN (T),
$F_{HC} = 2.24$ kN (T)

6–45. 11.6 kN (T)

6–46. 8.90 kN (C)

6–47. 4.45 kN (T)

6–49. $F_{AB} = F_{BC} = F_{CD} = F_{DE} = F_{HI} = F_{GI} = 0$,
$F_{IC} = 5.62$ kN (C), $F_{CG} = 9.00$ kN (T)

6–50. $F_{AB} = F_{BC} = F_{CD} = F_{DE} = F_{HI} = F_{GI} = 0$,
$F_{JE} = 9.38$ kN (C), $F_{GF} = 5.625$ kN (T)

6–51. $F_{EF} = P$ (C), $F_{CB} = 1.12P$ (T), $F_{BE} = 0.5P$ (T)

6–53. $F_{EH} = 3.33$ kN (T), $F_{IH} = 5.33$ kN (C),
$F_{EF} = 1.33$ kN (T)

6–54. $F_{KD} = 0$, $F_{CD} = 4.00$ kN (T), $F_{KJ} = 5.33$ kN (C)

6–55. $F_{HG} = 12.7$ kN (T), $F_{BC} = 15.1$ kN (C),
$F_{HC} = 1.50$ kN (T)

6–57. $F_{BC} = 9.55$ kN (C), $F_{GC} = 6.91$ kN (T),
$F_{GF} = 2.92$ kN (T)

6–58. $F_{DE} = 18.7$ kN (C), $F_{JI} = 18.7$ kN (T),
$F_{DO} = 11.7$ kN (C)

6–59. $F_{CD} = 25.3$ kN (C), $F_{KJ} = 25.3$ kN (T)

6–61. $F_{DC} = F_{DA} = 2.59$ kN (C), $F_{DB} = 3.85$ kN (C),
$F_{BC} = F_{BA} = 0.890$ kN (T), $F_{AC} = 0.616$ kN (T)

6–62. $F_{CA} = F_{CB} = 122$ N (C), $F_{CD} = 173$ N (T), $F_{BD} = 86.6$ N (T), $F_{BA} = 0$, $F_{DA} = 86.6$ N (T)

6–63. $F_{BC} = 0$, $F_{CD} = 0$, $F_{CF} = 8$ kN (C), $F_{BD} = 0$,
$F_{BA} = 6$ kN (C), $F_{AD} = 0$, $F_{DF} = 0$, $F_{DE} = 9$ kN (C),
$F_{EF} = 0$, $F_{EA} = 0$, $F_{AF} = 0$

6–65. $F_{BF} = 0$, $F_{BC} = 0$, $F_{BE} = 500$ N (T), $F_{AB} = 300$ N (C),
$F_{AC} = 972$ N (T), $F_{AD} = 0$, $F_{AE} = 367$ N (C), $F_{DE} = 0$,
$F_{EF} = 300$ N (C), $F_{CD} = 500$ N (C), $F_{CF} = 300$ N (C),
$F_{DF} = 424$ N (T)

6–66. $F_{BE} = 900$ N (T), $F_{BC} = 0$, $F_{AB} = 600$ N (C),
$F_{DE} = F_{CD} = F_{AD} = F_{CE} = F_{AC} = F_{BC} = 0$,
$F_{AE} = 671$ N (C)

6–67. $F_{AD} = 686$ N (T), $F_{BD} = 0$, $F_{CD} = 615$ N (C),
$F_{BC} = 229$ N (T), $F_{EC} = 457$ N (C), $F_{AC} = 343$ N (T)

6–69. $F_{BC} = F_{BD} = 1.34$ kN (C), $F_{AB} = 2.4$ kN (C),
$F_{AG} = F_{AE} = 1.01$ kN (T), $F_{BG} = 1.80$ kN (T),
$F_{BE} = 1.80$ kN (T)

6–70. $R_B = 267$ N, $A_x = 0$, $A_y = 347$ N

6–71. 1300 N

6–73. $R_E = 177$ N, $R_A = 128$ N

6–74. $R_B = 113$ N, $R_A = 144$ N, $R_D = 79.8$ N, $R_C = 79.8$ N

6–75. $A_x = C_x = 5$ kN, $A_y = C_y = 6.67$ kN, $B_x = 5$ kN,
$B_y = 1.33$ kN, $D_x = 0$, $D_y = 8$ kN, 10 kN · m

6–77. $F_B = 907$ N, $P = 156$ N

6–78. 464 N, 25.6 N · m

6–79. $A_x = 5.61$ kN, $A_y = 1.91$ kN, $C_x = 2.39$ kN,
$C_y = 1.91$ kN, $B_x = 2.39$ kN, $B_y = 6.91$ kN

6–81. $A_x = 6.43$ kN, $A_y = 2.62$ kN, $B_x = 1.57$ kN,
$B_y = 2.62$ kN, $C_x = 1.57$ kN, $C_y = 2.62$ kN

6–82. $A_x = 1500$ N, $A_y = 600$ N

6–83. 743 N

6–85. $P = 21.8$ N, $R_A = 43.6$ N, $R_B = 43.6$ N, $R_C = 131$ N

6–86. $P = 81.8$ N, $R_A = 183$ N, $R_B = 183$ N, $R_C = 441$ N

6–87. $P = 25$ N, $R_A = 25$ N, $R_B = 60$ N

6–89. 100 N, 14.6°

6–90. $C_x = 75$ N, $C_y = 100$ N

6–91. $B_x = 75$ N, $B_y = 300$ N, $E_x = 225$ N, $E_y = 600$ N,
$D_x = 300$ N, $D_y = 300$ N

6–93. $A_x = 240$ N, $A_y = 20$ N, $N_C = 50.0$ N

6–94. $A_x = 80$ N, $A_y = 80$ N, $B_y = 133$ N, $B_x = 333$ N,
$C_x = 413$ N, $C_y = 53.3$ N

6–95. $C_y = 34.4$ N, $C_x = 16.7$ N, $B_x = 66.7$ N, $B_y = 15.6$ N

6–97. 75 N

6–98. 56.3 N

6–99. 0.071 m

6–101. $F_B = 223$ N, $F_A = 386$ N

6–102. $m = 366$ kg, $F_A = 2.93$ kN

6–103. 1.11 Mg

6–105. $A_y = 657$ N, $C_y = 229$ N, $C_x = 0$, $B_x = 0$, $B_y = 429$ N

6–106. $E_x = 300$ N, $D_x = 300$ N, $D_y = E_y = 42.9$ N

6–107. 417 N, frictional force stops wheel

6–109. (a) $F = 911$ N, $N_C = 1.69$ kN, (b) $F = 456$ N,
$N_C = 323$ N

6–110. $A_x = 0$, $A_y = 34.0$ N, $C_y = 6.54$ N, $C_x = 0$, 292 mm, $B_x = 0$, $B_y = 1.06$ N

6–111. $N_C = 20$ N, $B_x = 34$ N, $B_y = 62$ N, $A_x = 34$ N, $A_y = 12$ N, 336 N · m

6–113. 22.1 kN

6–114. $T = W$

6–115. 109 mm

6–117. 660 N

6–118. 24.2 N

6–119. 28.8 kN, 39.9 kN

6–121. $F_{CA} = 129$ kN, $F_{AB} = 119$ kN, $F_{AD} = 23.9$ kN

6–122. $W_1 = 3$ N, $W_2 = 21$ N, $W_3 = 75$ N

6–123. 15.5°

6–125. 13.9 N

6–126. $\dfrac{kL}{2 \sin \theta \tan \theta}(2 - \csc \theta)$

6–127. $2.41P/L$

6–129. 3.86 kg

6–130. $F_{DE} = 525$ N, $B_z = 0$, $B_x = -214$ N, $B_y = 288$ N

6–131. $F_{DE} = 270$ N, $B_z = 0$, $B_x = -30$ N, $B_y = -13.3$ N

6–133. $P = 283$ N, $B_z = 283$ N, $D_z = 283$ N, $B_y = 283$ N, $D_y = 283$ N, $D_x = -70.7$ N, $B_x = -70.7$ N

6–134. $F_{AB} = 1.56$ kN, $M_{Ex} = 0.5$ kN · m, $M_{Ey} = 0$, $E_y = 0$, $E_x = 0$

6–135. $M_{Cx} = 0$, $C_x = 0$, $F_{BA} = 1.54$ kN, $C_z = -0.18$ kN, $C_y = -1.17$ kN, -4.14 kN · m, $A_x = 0$, $A_y = 1.44$ kN, $A_z = 0.540$ kN

6–137. $A_x = 1.40$ kN, $A_y = 250$ N, $C_x = 500$ N, $C_y = 1.70$ kN

6–138. $F_{CB} = 3$ kN (T), $F_{CD} = 2.60$ kN (C), $F_{DE} = 2.60$ kN (C), $F_{DB} = 2$ kN (T), $F_{BE} = 2$ kN (C), $F_{BA} = 5$ kN (T)

6–139. $C_y = 5$ kN, $B_y = 15$ kN, $A_x = 0$, $A_y = 5$ kN, 30 kN · m

6–141. $R_B = R_C = 49.5$ N, $F_C = 49.5$ N

6–142. $F_{BE} = 1.53$ kN, $F_{CD} = 350$ N

6–143 16.1°

6–145. $F_{DC} = 0.577P$ (C), $F_{DE} = 0.289P$ (T), $F_{CE} = 0.577P$ (T), $F_{CB} = 0.577P$ (C), $F_{BE} = F_{CE} = 0.577P$ (T), $F_{AB} = F_{DC} = 0.577P$ (C), $F_{AE} = F_{DE} = 0.289P$ (T)

6–146. $F_{DC} = 2.89W$ (C), $F_{DE} = 1.44W$ (T), $F_{CE} = 1.16W$ (T), $F_{CB} = 2.02W$ (C), $F_{BE} = F_{CE} = 1.16W$ (T), $F_{AB} = F_{DC} = 2.89W$ (C), $F_{AE} = F_{DE} = 1.44W$ (T)

6–147. 160 mm, 240 mm

6–149. $F_{AC} = 25.1$ kN (C), $F_{AB} = 30.8$ kN (C), $F_{AD} = 34.3$ kN (T)

Chapter 7

7–1. $T_A = 680$ N · m, $T_B = 350$ N · m, $T_C = 800$ N · m, $T_D = 0$

7–2. $T_C = 0$, $T_D = 400$ N · m, $T_E = 950$ N · m

7–3. $T_A = 550$ N · m, $T_B = 400$ N · m, $T_C = 600$ N · m, $T_D = 0$

7–5. $N_C = 0$, $V_C = 0.5$ kN, 3.6 kN · m

7–6. -292 kN · m, $N_C = -20$ kN, $V_C = 70.6$ kN

7–7. $N_D = -800$ N, $V_D = 0$, 1.20 kN · m

7–9. 48 kN · m, 6 kN

7–10. $N_A = 86.6$ N, $V_A = 150$ N, 1800 N · mm

7–11. 52.5 N · m

7–13. $N_A = 0$, $V_A = 2.7$ kN, $M_A = 2.03$ kN · m, $N_B = 0$, $V_B = 5.1$ kN, $M_B = 11.4$ kN · m, $V_C = 0$, $N_C = 7.2$ kN, $M_C = 14.6$ kN · m

7–14. $N_D = 1.92$ kN, $V_D = 100$ N, $M_D = 900$ N · m

7–15. $N_E = -1.92$ kN, $V_E = 800$ N, $M_E = 2.40$ kN · m

7–17. $P = 0.533$ kN, $N_C = -2$ kN, $V_C = 0.533$ kN, $M_C = 0.400$ kN · m

7–18. $V_D = 0$, $N_D = 0$, $M_D = 0$, $V_E = 2.37$ kN, $N_E = 5.34$ kN, $M_E = 2.17$ kN · m, $V_F = 0$, $N_F = 10.68$ kN, $M_F = 4.34$ kN · m

7–19. $N_C = -406$ N, $V_C = 903$ N, 1.35 kN · m

7–21. $N_D = 0$, $V_D = 0$, $M_D = 9$ kN · m, $N_E = 0$, $V_E = -7$ kN, $M_E = -12$ kN · m

7–22. 3.75 kN, 30 kN · m

7–23. $N_D = 0$, $V_D = 0.75$ kN, $M_D = 13.5$ kN · m, $N_E = 0$, $V_E = -9$ kN, $M_E = -24.0$ kN · m

7–25. $N_E = -250$ N, $V_E = 245$ N, $M_E = -490$ N · m, $N_F = 0$, $V_F = -308$ N, $M_F = -1.23$ kN · m

7–26. $N_B = 59.8$ N, $V_B = -496$ N, $M_B = -480$ N · m, $N_C = -495$ N, $V_C = 70.7$ N, $M_C = -1.59$ kN · m

7–27. $V = 0.293rw_0$, $N = -0.707rw_0$, $M = -0.0783r^2w_0$

7–29. 1.69 kN, 10.0 kN · m

7–30. $V_{Ax} = 0$, $N_{Ay} = 0$, $V_{Az} = 2.5$ kN, $M_{Ax} = 0.563$ kN · m, $M_{Ay} = 0.563$ kN · m, $M_{Az} = 0$, $V_{Bx} = 0$, $V_{By} = 0$, $N_{Bz} = 2.5$ kN, $M_{Bx} = 1.31$ kN · m, $M_{By} = 0.563$ kN · m, $M_{Bz} = 0$

7–31. $C_x = -150$ N, $C_y = -350$ N, $C_z = 700$ N, $M_{Cx} = 1.40$ kN · m, $M_{Cy} = -1.20$ kN · m, $M_{Cz} = -750$ N · m

7–33. $D_x = -116$ kN, $D_y = 65.6$ kN, $D_z = 0$, $M_{Dx} = -49.2$ kN · m, $M_{Dy} = -87.0$ kN · m, $M_{Dz} = -26.2$ kN · m

7–34. (a) $V = (1 - a/L)P$, $M = (1 - a/L)Px$, $V = -(a/L)P$, $M = P(a - a/L)x$

7–35. (a) $V = P$, $M = Px$, $V = 0$, $M = Pa$, $V = -P$,

$M = P(L - x)$,   (b) 800 N, $800x$ N · m, 0, 4000 N · m, -800 N, $(9600 - 800x)$ N · m

7–37. 2 kN · m

7–38. (a) $V = \dfrac{w}{2}(L - 2x)$, $M = \dfrac{w}{2}(Lx - x^2)$,

(b) $(2500 - 500x)$ N, $(2500x - 250x^2)$ N · m

7–39. 400 N/m

7–41. $V = 2.5 - 2x$, $M = 2.5x - x^2$, $V = -7.5$, $M = -7.5x + 25$

7–42. $V = 250(10 - x)$, $M = 25(100x - 5x^2 - 6)$

7–43. $V = 133.75 - 40x$, $M = 133.75x - 20x^2$, $V = 20$, $M = 20x - 370$

7–45. (a) $V = \dfrac{w}{8}(7L - 8x)$, $M = -\dfrac{w}{8}(4x^2 - 7Lx + 3L^2)$,

(b) $V = 5(10.5 - x)$, $M = -2.5(x^2 - 21x + 108)$

7–46. $V = 3 - \dfrac{x^2}{4}$, $M = 3x - \dfrac{x^3}{12}$

7–47. $0.366L$

7–49. 22.2 N/m

7–50. $x = \dfrac{L}{2}$, $P = \dfrac{4M_{max}}{L}$

7–51. $\dfrac{L}{3}$

7–53. $wr \sin \theta$, $wr(\cos \theta - 1)$, $M = wr^2(1 - \cos \theta)$

7–54. $N = \dfrac{P}{5}(4 \cos \theta + 3 \sin \theta)$, $V = \dfrac{P}{5}(4 \sin \theta - 3 \cos \theta)$,

$M = \dfrac{Pr}{5}(4 - 4 \cos \theta - 3 \sin \theta)$

7–55. See Prob. 7–35

7–57. See Prob. 7–38

7–58. See Prob. 7–42

7–59. See Prob. 7–46

7–61. $x = 0$, $V = 1200$, $M = 0$; $x = 24^-$, $V = 400$, $M = 19200$

7–62. $x = 0$, $V = 3.75$, $M = 0$; $x = 4^+$, $V = -3.25$, $M = 13$

7–63. $x = 0$, $V = 1.525$, $M = 0$; $x = 0.8^-$, $V = 0.025$, $M = 0.395$

7–65. $x = 0$, $V = -400$, $M = 0$; $x = 14.5$, $V = 0$, $M = 3912$

7–66. $x = 0$, $V = -200$, $M = 0$; $x = 2.75$, $V = 0$, $M = -46.9$

7–67. $x = 0$, $V = -14.3$, $M = 0$; $x = 3^+$, $V = -22.3$, $M = -22.9$

7–69. $x = 0$, $V = 0$, $M = 0$; $x = 6^+$, $V = 800$, $M = -1200$

7–70. 2 kN/m

7–71. $x = 0$, $V = 52.8$, $M = 0$, $x = 9^-$, $V = -172.2$, $M = -200$

7–73. $x = 0$, $V = wL/2$, $M = -5wL^2/24$; $x = L/2$, $V = wL/4$, $M = -wL^2/24$.

7–74. $x = 0$, $V = 0$, $M = 0$; $x = 18$, $V = 0$, $M = 0$

7–75. $x = 0$, $V = 0$, $M = 0$; $x = 6^+$, $V = 9$, $M = -36$

7–77. $x = 0$, $V = 1100$, $M = -10400$; $x = 8^+$, $V = 400$, $M = -1600$

7–78. $x = 0$, $V = 7wL/18$, $M = 0$; $x = L^+$, $V = wL/2$, $M = -72wL^2/648$

7–79. $x = 0$, $V = wL/3$, $M = 0$; $x = 0.528L$, $V = 0$, $M = 0.0940wL^2$

7–81. 2.43 m, 62.9 N

7–82. $y_B = 8.67$ m, $y_D = 7.04$ m

7–83. 6.44 m, 658 N

7–85. 71.4 N

7–86. $T_{AB} = 166$ N, $T_{CD} = 176$ N, $T_{BC} = 93.4$ N, 20.2 m

7–87. 3.98 m

7–89. $T_A = 61.7$ kN, $T_B = F_H = 36.5$ kN, $T_C = 50.7$ kN

7–90. 51.9 N/m

7–91. $y = \dfrac{x^2}{7813}\left(75 - \dfrac{x^2}{200}\right)$ m, 9.28 kN

7–93. $(38.5x^2 + 577x)(10^{-3})$ m, 5.20 kN

7–94. 0.141

7–97. 302 m

7–98. $L = 94.3$ m, $h = 25.3$ m

7–99. 292 N

7–101. 5.14 m

7–102. $(T_{max})_B = 2.73$ kN, $(T_{max})_C = 2.99$ kN

7–103. 5.18 m

7–105. 66.6 N

7–106. 121 N, 5.91 m

7–107. 170 N, 150 m

7–109. 35.8 m

7–110. 238 m, $h = 93.8$ m

7–111. $x = 0$, $V = 10$, $M = -30$, $x = 5$, $V = 0$, $M = -5$

7–113. $x = 0$, $V = 1022$, $M = 0$; $x = 0.6$, $V = 0$, $M = 407$

7–114. 11.1 N, 23.5 m

7–115. $x = 0$, $V = 9.11$, $M = 0$, $x = 9$, $V = -5.89$, $M = 0$

7–117. $x = 0$, $V = 2.5$, $M = 0$, $x = 1.25$, $V = 0$, $M = 1.56$

7–118. $V_D = M_D = 0$, $N_D = F_{CD} = 86.6$ N, $N_E = 0$, $V_E = 28.9$ N, $M_E = 86.6$ N · m

7–119. $\{16 - 4y\}$ N, $\{-2y^2 + 16y - 40\}$ N · m

Chapter 8

8–1. 224 N

8–2. 76.4 N $\leq P \leq$ 144 N

8–3. $F_C = 27.4$ N, $N_C = 309$ N

8–5. 140 N, 500 mm

8–6. $F_C = 30.5$ N, $N_C = 152$ N, 0.794 m

8–7. 0.4L

8–9. (a) $P = 30$ N < 39.8 N no,   (b) $P = 70$ N > 39.8 N yes

8–10. (a) $P = 30$ N < 34.26 N no,   (b) $P = 70$ N > 34.26 N yes

8–11. will *not slip*

8–13. 36.7 kN

8–14. 21.8°

8–15. 120 N

8–17. 0.268

8–18. hoop slips first

8–19. 15.3 kN

8–21. 227 N, 0.662

8–22. 54.9 kg

8–23. (a) 1.27 kN,   (b) 1.44 kN

8–25. $\dfrac{h}{\mu_s} + \dfrac{d}{2}$

8–26. 90 N, 0.15

8–27. 121 N, 0.195

8–29. 200 N

8–30. car A *will not* move

8–31. 2.70 m

8–33. 2.89 m

8–34. $\theta = 16.7°$, $\phi = 40.5°$

8–35. slipping at A

8–37. 1.02 m

8–38. 0.962 N

8–39. 0.962 N

8–41. 149 N

8–42. 400 N, 0.45 m

8–43. 63.4°

8–45. 33.4°

8–46. 0.176

8–47. $P = 355$ N

8–49. 77.3 N · m

8–50. 71.4 N

8–51. 13.3 N

8–53. 28.1 N

8–54. 375 N

8–55. 589 N

8–58. $\alpha = -\theta$, $\phi = \theta$, $P = W \sin (\alpha + \phi)$

8–59. $(l^2 - a^2 - h^2)^{1/2}(l^2 - a^2)^{1/2}/ha$

8–61. any number

8–62. 1.61 kN, 1.34 m

8–63. $P' = 4.96$ kN, $P = 8.16$ kN

8–65. 34.5 N

8–66. 5.53 kN, yes

8–67. 9.36 kN

8–69. 304 N

8–70. 32.9 mm

8–71. 25.3°

8–73. 56.9 N · mm

8–74. 7.19 kN

8–75. self locking

8–77. 1.98 kN

8–78. 0.202 N · m

8–79. 0.0637

8–81. 31.0 N · m

8–82. 72.7 N

8–83. $T_A = 1.87$ kN, $T_B = 1.37$ kN

8–85. (a) 1.31 kN,   (b) 372 N

8–86. (a) 4.60 kN,   (b) 16.2 kN

8–87. approx. 2 turns (695°)

8–89. 8.28 m

8–90. 73.3 N

8–91. 99.2°

8–93. 12.7 N

8–94. 15.4 N, 5.47 N

8–95. 42.3 N

8–97. 17.1 N

8–99. 25.6 kg

8–101. 7.6 N · m

8–102. 0.189

8–103. 8.87 N · m

8–105. 107 N

8–106. 118 N

8–107. $\frac{1}{2}\mu PR$

8–109. 0.521$P\mu R$

8–110. 2.91 kN/m², 1.98 kN

8–111. 2.00 N

8–113. 826 N

8–114. 814 N

8–115. 0.215, 0.211 (approx.), 60 N

8–117. 13.8 N

8–118. 29.0 N

8–119. 68.2°, 0.0455 N · m

8–121. 44.6 N

8–122. 394 N

8–123. 245 N

8–126. 2.25 kN

8–127. 1.94 kN

8–129. 0.0626

8–130. cam *cannot* support broom

8–131. 38.5 mm

8–133. 1.55 N

8–134. $F_{BC} = 1.38$ kN (T), $F_{BD} = 1.73$ kN (C),
$F_{AB} = 1.04$ kN (T), $F_{AC} = 1.73$ kN (C),
$F_{CD} = 1.04$ kN (T), $F_{AD} = 1.38$ kN (T)

8–135. 3.62 N · m

Chapter 9

9–1. $\dfrac{4}{\pi}$, $B_x = 1$ N, $A_x = 1$ N, $A_y = 3.14$ N

9–2. $0.75l$

9–3. $\dfrac{r \sin \alpha}{\alpha}$

9–5. 0.531 m, $O_x = 0$, $O_y = 0.574$ N, 0.305 N · m

9–6. 0.183 m

9–7. $\bar{x} = 0$, $\bar{y} = 1.60$ m

9–9. $\bar{x} = \frac{3}{8}b$, $\bar{y} = \frac{3}{5}h$

9–10. $\bar{x} = \frac{4}{7}a$, $\bar{y} = \dfrac{a}{5}$

9–11. $\bar{y} = \dfrac{a\pi}{8}$, $\bar{x} = \dfrac{L}{2}$

9–13. $\bar{x} = \dfrac{n+1}{2(n+2)}a$, $\bar{y} = \dfrac{n}{2n+1}h$

9–14. $\bar{x} = \frac{3}{8}a$, $\bar{y} = \frac{3}{5}a$

9–15. $\bar{y} = \dfrac{4b}{3\pi}$, $\bar{x} = \dfrac{4a}{3\pi}$

9–17. $\bar{x} = \dfrac{4r}{3\pi}$, $\bar{y} = \dfrac{4r}{3\pi}$

9–18. $\bar{y} = 0.223r$, $\bar{x} = 0.223r$

9–19. $\bar{x} = 0$, $\bar{y} = \dfrac{4r}{3\pi}$

9–21. 0.649 m

9–22. 2.04 m

9–23. $\bar{x} = 1.26$ m, $\bar{y} = 0.143$ m, $N_B = 47.9$ kN, $A_x = 33.9$ kN,
$A_y = 73.9$ kN

9–25. 0.980 m

9–26. 0.587 m

9–27. 0.404 m

9–29. $1.11a$

9–30. $\bar{y} = \frac{3}{8}b$, $\bar{x} = \bar{z} = 0$

9–31. $\bar{z} = \frac{2}{3}a$, $\bar{x} = \bar{y} = 0$

9–33. $\bar{x} = \dfrac{a}{2}$, $\bar{y} = \bar{z} = 0$

9–34. $\frac{2}{3}h$

9–37. $\bar{z} = \dfrac{h}{4}$, $\bar{x} = \bar{y} = \dfrac{a}{\pi}$

9–38. $0.4a$

9–39. $\dfrac{h}{4}\left(\dfrac{R^2 + 2Rr + 3r^2}{R^2 + Rr + r^2}\right)$

9–41. 2 kg, 5 m

9–42. $\bar{x} = 1.44$ m, $\bar{y} = 2.22$ m, $\bar{z} = 2.67$ m

9–43. 67.9 mm

9–45. 3 m

9–46. $\bar{x} = 1.60$ m, $\bar{y} = 7.04$ m, $A_x = 0$, $A_y = 149$ N,
$M_A = 502$ N · m

9–47. $\bar{x} = 60$ mm, $\bar{y} = 17.8$ mm

9–49. $\bar{x} = 3.52$ m, $\bar{y} = 4.09$ m, 6840 kN

9–50. 135 mm

9–51. $\dfrac{h}{3}\left(\dfrac{2b_1 + b_2}{b_1 + b_2}\right)$

9–53. $\bar{x} = 30$ mm, $\bar{y} = 20$ mm

9–54. 210 mm

9–55. 154 mm

9–57. $\bar{x} = 4.62$ mm, $\bar{y} = 1.00$ mm

9–58. 37.5 mm

9–59. $\bar{x} = 432$ mm, $\bar{y} = -339$ mm

9–61. 37.8°

9–62. 0.743 m, 38.9°

9–63. $\dfrac{\dfrac{2}{3}r \sin^3 \alpha}{\alpha - \dfrac{\sin 2\alpha}{2}}$

9–65. 1.625 m

9–66. $\bar{x} = 0.853$ m, $\bar{y} = 0.528$ m, $N_B = 359$ N, $N_A = 436$ N

9–67. $\bar{x} = 0.147$ m, $\bar{y} = 0.268$ m, $\bar{z} = 0.284$ m

9–69. 2.00 m

9–70. 0.165 m

9–71. 323 mm

9–73. 101 mm

9–74. 385 mm

9–75. 122 mm

9–77. 0.667 m^2, 0.375 m, 1.57 m^3

9–78. 1.39 m^2, 0.541 m, 4.71 m^3

9–79. 826 mm^2

9–81. 3.33 m^2, 1.2 m, 25.1 m^3

9–82. 3.14 m^3

9–83. 3.49 m^3

9–85. 2.25 m^2

9–86. 4.25(10^6) mm^3

9–87. 72.2 N

9–89. 20.0 MN

9–90. 1660 kN

9–91. 28.7 liters

9–93. 36.3 kN

9–94. 0.770 N

9–95. 143(10)$^{-6}$ m^3

9–97. 29.9 mm

9–98. 207 m^3, 188 m^2

9–99. 6.79(10^6) mm^3

9–101. 119(10^3) mm^2

9–102. 1.41 MN, 4 m

9–103. 157 kN, 235 kN, 4.22 m

9–105. 73.1 kN, 1.57 m, 79.8 kN

9–106. 391 kN/m

9–107. 45.9 kN, 1.41 m

9–109. $B_x = 2.20$ MN, $B_y = 0.859$ MN, $A_y = 2.51$ MN

9–110. 6.67 m

9–111. 7.73 m

9–113. $F_{R_1} = 2.45$ kN, 1.02 m, $F_{R_2} = 9.75$ kN, at center of plate B

9–114. 331 kN/m

9–115. 78.7 kN

9–117. 3.85 kN, 0.625 m

9–118. 120 kN, -1.77 m

9–119. 4.00 MN, -6.49 m

9–121. 1250 N, $\bar{x} = 2.33$ m, $\bar{y} = 4.33$ m

9–122. $\bar{x} = 0$, $\bar{y} = 2.40$ m, $F_R = 42.7$ kN, $B_y = C_y = 12.8$ kN, $A_y = 17.1$ kN

9–123. 24 kN, $\bar{x} = 2$ m, $\bar{y} = 1.33$ m

9–125. 1.25 m^2

9–126. 39.5 kg

9–127. 4.42 mm

9–129. $\bar{x} = 1.22$ m, $\bar{y} = 0.778$ m, $\bar{z} = 0.778$ m, $M_{Ax} = 16$ kN $\cdot$ m, $M_{Ay} = 57.1$ kN $\cdot$ m, $M_{Az} = 0$, $A_x = 0$, $A_y = 0$, $A_z = 20.6$ kN

9–130. 6.0 mm

9–131. 901(10^3) mm^3, 57.7(10^3) mm^2

9–133. $\bar{x} = 4.56$ m, $\bar{y} = 3.07$ m, $R_B = 4.66$ kN, $R_A = 5.99$ kN

9–134. $\bar{y} = 150$ mm, $\bar{x} = 33.9$ mm

Chapter 10

10–1. 973 mm^4

10–2. 15.4(10)^6 mm^4

10–3. 39.0 m^4

10–5. 10.7 mm^4

10–6. 307 mm^4

10–7. 3.20 m^4

10–9. $\frac{2}{15}bh^3$

10–10. $\frac{2}{7}b^3h$

10–11. (a) 23.8 m^4, (b) 23.8 m^4

10–13. 6.87 m^4

10–14. 5.81 m^4

10–15. 0.533 m^4

10–17. 3.20 m^4

10–18. 0.762 m^4

10–19. $\dfrac{\sqrt{3}a^4}{96}$

10–21. 0.935 m^4

10–22. 333 mm^4

10–23. 32.4 mm^4

10–25. 0.176 m^4

10–26. 2.26(10^3) mm^4

10–27. 341 mm^4

10–29. 832 mm^4

10–30. 170 mm, 722(10^6) mm^4

10–31. 91.7(10^6) mm^4

10–33. 54.7 mm^4

10–34. 1.29 mm, 6.74 mm^4

10–35. 18.3 mm^4

10–37. $122(10^6) \text{ mm}^4$

10–38. 81 mm^4

10–39. 91.1 mm^4

10–41. $207 \text{ mm}, \bar{I}_{x'} = 222(10^6) \text{ mm}^4, I_y = 115(10^6) \text{ mm}^4$

10–42. 648 mm^4

10–43. 1971 mm^4

10–45. $80.7 \text{ mm}, 67.6(10^6) \text{ mm}^4$

10–46. $61.6 \text{ mm}, 41.2(10^6) \text{ mm}^4$

10–47. 503 mm^4

10–49. $0.187d^4$

10–50. $251(10^3) \text{ mm}^4, 172(10^3) \text{ mm}^4, 68.4\%$

10–51. $\dfrac{b^3d^3}{6(b^2 + d^2)}$

10–53. 109 mm

10–54. $\frac{1}{12}a^3b \sin^3 \theta$

10–55. $\dfrac{ab \sin \theta}{12}(b^2 + a^2 \cos^2 \theta)$

10–57. $\dfrac{h^2b^2}{16}$

10–58. 8 mm^4

10–59. 1.33 mm^4

10–61. 48 mm^4

10–62. 0.333 m^4

10–63. $\frac{3}{16}b^2h^2$

10–65. 3.30 m^4

10–66. 97.8 mm^4

10–67. $\bar{x} = 85.0 \text{ mm}, \bar{y} = 35.0 \text{ mm}, -7.50(10^6) \text{ mm}^4$

10–69. -110 mm^4

10–70. 36 mm^4

10–71. $289(10^3) \text{ mm}^4$

10–73. $135(10)^6 \text{ mm}^4$

10–74. $I_u = 274 \text{ mm}^4, I_v = 571 \text{ mm}^4, I_{uv} = -258 \text{ mm}^4$

10–75. $I_u = 195 \text{ mm}^4, I_v = 650 \text{ mm}^4, I_{uv} = -191 \text{ mm}^4$

10–77. $6.08°, I_{max} = 0.726(10^{-3}) \text{ m}^4, I_{min} = 181(10^{-6}) \text{ m}^4$

10–78. $I_u = 3.47(10^3) \text{ mm}^4, I_v = 3.47(10^3) \text{ mm}^4,$
$I_{uv} = 2.05(10^3) \text{ mm}^4$

10–79. $I_u = 2.44(10^3) \text{ mm}^4, I_v = 4.49(10^3) \text{ mm}^4,$
$I_{uv} = 1.77(10^3) \text{ mm}^4$

10–81. $I_{max} = 64.1 \text{ mm}^4, I_{min} = 5.33 \text{ mm}^4$

10–82. $I_{max} = 4.92(10^6) \text{ mm}^4, I_{min} = 1.36(10^6) \text{ mm}^4$

10–83. x and y principal axes, $4.67 \text{ mm}, I_x = 1.47(10^3) \text{ mm}^4,$
$I_y = 2.50(10^3) \text{ mm}^4$

10–85. $-22.5°, I_{max} = 250 \text{ mm}^4, I_{min} = 20.4 \text{ mm}^4$

10–86. $I_{max} = 0.726(10^{-3}) \text{ m}^4, I_{min} = 181(10^{-6}) \text{ m}^4$

10–87. $I_{max} = 64.1 \text{ mm}^4, I_{min} = 5.33 \text{ mm}^4$

10–89. $I_{max} = 4.92(10^6) \text{ mm}^4, I_{min} = 1.36(10^6) \text{ mm}^4$

10–90. $I_{max} = 250 \text{ mm}^4, I_{min} = 20.4 \text{ mm}^4$

10–91. $\frac{1}{3}ml^2$

10–93. $\frac{3}{10}mr^2$

10–94. $\frac{2}{5}mr^2$

10–95. $\dfrac{a}{\sqrt{3}}$

10–97. $\frac{2}{5}mb^2$

10–98. $3.13 \text{ kg} \cdot \text{m}^2$

10–99. $\frac{93}{70}mb^2$

10–101. $\dfrac{m}{6}(a^2 + h^2)$

10–102. $35.6 \text{ kg} \cdot \text{m}^2$

10–103. $1.53 \text{ kg} \cdot \text{m}^2$

10–105. $390 \text{ kg} \cdot \text{m}^2$

10–106. $0.888 \text{ m}, 5.61 \text{ kg} \cdot \text{m}^2$

10–107. $6.39 \text{ m}, 53.2 \text{ kg} \cdot \text{m}^2$

10–109. $0.203 \text{ m}, 0.230 \text{ kg} \cdot \text{m}^2$

10–110. $34.2 \text{ kg} \cdot \text{m}^2$

10–111. $I_u = 5.09(10^6) \text{ mm}^4, I_v = 5.09(10^6) \text{ mm}^4, I_{uv} = 0$

10–113. $I_x = \dfrac{1}{3(3n + 1)} bh^3, I_y = \dfrac{1}{n + 3}b^3h$

10–114. 307 mm^4

10–115. 10.7 mm^4

10–117. 30.9 mm^4

10–118. 0.305 m^4

10–119. 1.60 m^4

10–121. $I_x = 0.167 \text{ m}^4, I_y = 0.0333 \text{ m}^4$

10–122. 0.0625 m^4

10–123. $\theta_{P_1} = -21.6°, \theta_{P_2} = 68.4°, I_{max} = 0.191 \text{ m}^4,$
$I_{min} = 0.00859 \text{ m}^4$

Chapter 11

11–1. 512 N

11–2. $\dfrac{500\sqrt{0.04 \cos^2 \theta + 0.6}}{(0.2 \cos \theta + \sqrt{0.04 \cos^2 \theta + 0.6}) \sin \theta}$

11–3. $\dfrac{P}{2 \tan \theta}$

11–5. $27.4°$

11–6. 1.05 kN/m

11–7. $\frac{1}{2}Wl\cos\theta$

11–9. 63.2 N/m

11–10. 16.6°

11–11. 16.6°

11–13. 53.1°

11–14. 9.21°

11–15. 35.2°

11–17. 50 N

11–18. 50 N

11–19. $\cos^{-1}\left(\dfrac{a}{2L}\right)^{1/3}$

11–21. 13.9 N

11–22. 43.3 N · m

11–23. 25.0 N · m

11–25. $m_B = 2$ kg, $m_A = 1.60$ kg

11–26. 427 N

11–27. $x = 0$ unstable, $x = 0.167$ m stable

11–29. 38.7° unstable, 90° stable

11–30. (0, 0), unstable

11–31. 1.05 kN/m, stable

11–33. 53.1° stable

11–34. 100 kg

11–35. 2.55°, unstable

11–37. 90°, $\sin^{-1}\left(\dfrac{W}{2kL}\right)$

11–38. 0°, $\cos^{-1}\left(1 - \dfrac{W}{2kL}\right)$

11–39. 0°, stable

11–41. 2.45 m

11–42. 15.5°

11–43. 94.3 N/m, stable

11–45. 30°

11–46. 0°, $\cos^{-1}\left(\dfrac{W}{2kL}\right)$, $2kL$

11–47. $\dfrac{W}{L}$

11–49. 0

11–50. 87.9 mm

11–51. 9.46°, stable

11–53. $\dfrac{a}{2r}$

11–54. 45°

11–55. 4.62 N · m

11–57. 90°, 30°

11–58. 90° unstable, 30° stable

11–59. 90°, 17.5°

11–61. 3.12 kN

11–62. 0°, 33.1°

11–63. 24.9°

Index

572

Geometric Properties of Line and Area Elements

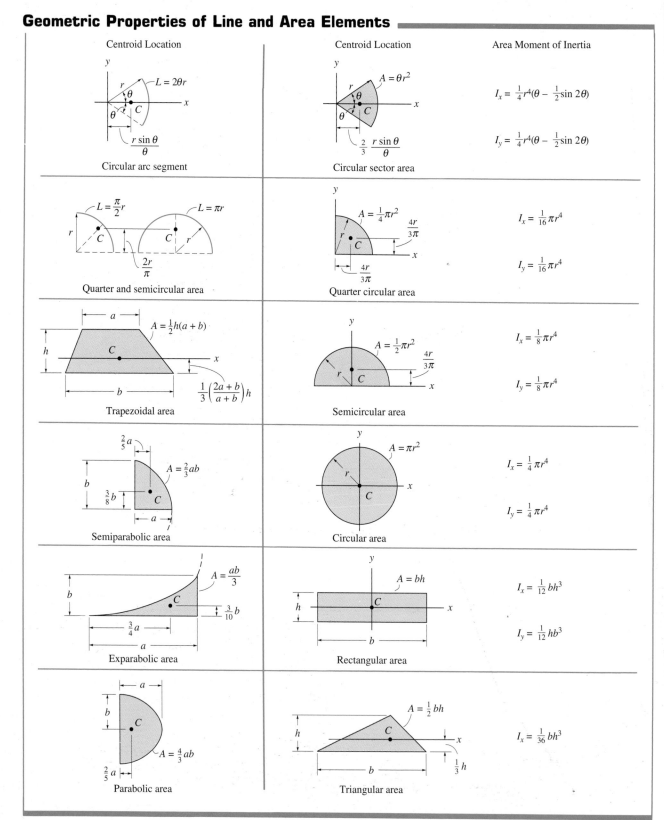

Centroid Location

$L = 2\theta r$

$$\frac{r\sin\theta}{\theta}$$

Circular arc segment

Centroid Location

$A = \theta r^2$

$$\frac{2}{3}\frac{r\sin\theta}{\theta}$$

Circular sector area

Area Moment of Inertia

$$I_x = \tfrac{1}{4}r^4(\theta - \tfrac{1}{2}\sin 2\theta)$$

$$I_y = \tfrac{1}{4}r^4(\theta - \tfrac{1}{2}\sin 2\theta)$$

$L = \tfrac{\pi}{2}r$ $\qquad$ $L = \pi r$

$$\frac{2r}{\pi}$$

Quarter and semicircular area

$A = \tfrac{1}{4}\pi r^2$

$$\frac{4r}{3\pi}$$

$$\frac{4r}{3\pi}$$

Quarter circular area

$$I_x = \tfrac{1}{16}\pi r^4$$

$$I_y = \tfrac{1}{16}\pi r^4$$

---

a

$A = \tfrac{1}{2}h(a + b)$

h

b

$$\frac{1}{3}\left(\frac{2a + b}{a + b}\right)h$$

Trapezoidal area

$A = \tfrac{1}{2}\pi r^2$

$$\frac{4r}{3\pi}$$

Semicircular area

$$I_x = \tfrac{1}{8}\pi r^4$$

$$I_y = \tfrac{1}{8}\pi r^4$$

$\tfrac{2}{5}a$

$A = \tfrac{2}{3}ab$

b

$\tfrac{3}{8}b$

a

Semiparabolic area

$A = \pi r^2$

r

Circular area

$$I_x = \tfrac{1}{4}\pi r^4$$

$$I_y = \tfrac{1}{4}\pi r^4$$

---

b

$A = \dfrac{ab}{3}$

$\tfrac{3}{10}b$

$\tfrac{3}{4}a$

a

Exparabolic area

$A = bh$

h

b

Rectangular area

$$I_x = \tfrac{1}{12}bh^3$$

$$I_y = \tfrac{1}{12}hb^3$$

---

a

b

$A = \tfrac{4}{3}ab$

$\tfrac{2}{5}a$

Parabolic area

$A = \tfrac{1}{2}bh$

h

$\tfrac{1}{3}h$

b

Triangular area

$$I_x = \tfrac{1}{36}bh^3$$